Tropentag 2019

International Conference on Research on Food Security, Natural Resource Management and Rural Development

Tropentag 2019

International Research on Food Security, Natural Resource Management and Rural Development

Filling gaps and removing traps for sustainable resources management

Book of abstracts

Editor: Eric Tielkes

Reviewers/scientific committee: Folkard Asch, Jan Banout, Andreas Buerkert, Marc Cotter, Uta Dickhöfer, Falko Feldmann, Yvonne Franke, Sven Goenster-Jordan, Ariane Götz, Renuka Suddapuli Hewage, Christian Herzig, Jürgen Hummel, Lisa Jäckering, Brigitte Kaufmann, Gudrun B. Keding, Katja Kehlenbeck, Thomas Kopp, Dagmar Mithöfer, Simone Pfeiffer, Eva Schlecht, Stefan Sieber, Andreas Thiel, Martin Wiehle, Stephan Winter

Editorial assistance: Joana Albrecht, Esther Asare

Impressum

Bibliografische Information der Deutschen Nationalbibliothek
Die Deutsche Nationalbibliothek verzeichnet diese Publikation in der Deutschen Nationalbibliografie; detailierte bibliografische Daten sind im Internet über http://dnb.ddb.de abrufbar.

Tropentag 2019: Filling gaps and removing traps for sustainable resources management, Tielkes, E. (ed.) Aufl. - Göttingen: Cuvillier, 2019

Nonnenstieg 8, 37075 Göttingen
Telefon: 0551-54724-0
Telefax: 0551-54724-21
http://www.cuvillier.de

Gedruckt auf umweltfreundlichem, säurefreiem Papier aus nachhaltiger Forstwirtschaft.

ISBN: 978-3-7369-7083-0
eISBN: 978-3-7369-6083-1

Online-Version: http://www.tropentag.de/

Preface

Tropentag is the largest interdisciplinary conference in Europe on research in sub-/tropical agriculture, food security, natural resource management and rural development. Taking place annually, *Tropentag* 2019 is jointly organised by the Centre for International Rural Development at the University of Kassel and the Centre of Biodiversity and Sustainable Land Use at the University of Göttingen, and takes place at the University of Kassel's main campus from 18 to 20 September 2019.

Since more than a decade *Tropentag* conferences are addressing the socio-political, economic and ecological consequences of global change phenomena for rural - and peri/urban - smallholder farmers and other groups whose livelihood depend on the use of natural resources, be it directly or indirectly. A lot of science-based yet practical solutions to site-specific problems of natural resources management and the sustainable provision of food and non-food commodities through agriculture and forestry have been provided, addressing ecological, technical, economic, social and political aspects. Yet undamped population growth, in particular in sub-Saharan Africa, continued rural-to-urban migration, such as in many Asian countries, and increasingly serious consequences of climate variability and change sometimes overrun our newly gained insights. In addition, obvious deficits or gaps, such as in quality education, equal share of responsibilities and incomes between women and men, in marketing opportunities between rural and urban farmers, between potential and actual crop yields and livestock performances, still need to be resolved in many regions of the world. Certainly, our research is limited in time and space, by financial means and available human power, and therefore it is all the more important to *a priori* identify - and avoid - common traps, such as neglect of relevant stakeholders, incomplete risk assessment, short-lived innovation testing, and fragmental systems assessment in our scientific endeavours.

"Filling gaps and removing traps for sustainable resources management" is therefore the theme of *Tropentag* 2019, and from the more than 930 contributions initially submitted 470 were found to provide very relevant aspects of this overarching theme - they are compiled in this book. A very prominent goal of *Tropentag* conference is the international - disciplinary and interdisciplinary - exchange and mutual learning. Prominent examples of successful research and development approaches towards sustainable resources management may provide us with new ideas for our own work.

The five plenary keynote contributions to this years' *Tropentag* showcase such successful strategies and concepts, be it the low-cost farmer managed natural land restoration technique for which Tony Rinaudo and Yacouba Sawadogo were awarded the Right Livelihood Award 2018, the use of voluntary sustainability standards in global food production on which Miet Maertens reflects, or the successful transfer of scientific knowledge into day-to-day policies on which Immaculate Njuthe Maina shares her insights - all of them are characterized by a holistic approach towards problem analysis and solution and provide inspiration. A special session chaired by the International Center for Agricultural Research in the Dry Areas (ICARDA), this year's CGIAR feature center, focusses on cereals and pulses for climate-smart agri-food systems, thereby also drawing attention to the aggravating problem of water scarcity in the dryer parts of the sub-/tropics.

Next to the five plenary keynote lectures the conference theme is addressed in 113 oral presentations and 362 poster contributions organised in 22 thematic sessions, and we wish to thank all participants for their scientific contributions and our colleagues of the scientific committee for reviewing all abstracts and acting as chairs for oral and poster sessions during *Tropentag* 2019. Special thanks go to Eric Tielkes for his very valuable support in compiling this book and organising the conference, and to all donors (listed on the back cover) for their financial and in kind contributions which allow us to keep conference fees at a modest level, especially for junior scientists.

We welcome you coming from the many different parts of the world to *Tropentag* 2019 in Kassel and wish you an inspiring and enriching conference with a lot of new insights, fruitful discussions and a stimulating exchange of knowledge and experiences.

On behalf of the local organising team of *Tropentag* 2019:

Eva Schlecht, Bernhard Brümmer, Simone Pfeiffer, Martin Wiehle

Kassel-Witzenhausen and Göttingen, September 2019

Contents

Plenary speeches

Getting Your Assumptions Right

TONY RINAUDO

World Vision, Australia

Our assumptions about a problem will influence our expenditure, choice of interventions and ultimately our success. It is very important to get our assumptions correct — and to be flexible enough to adjust them as required as new information comes to light. In the early 1980's, Niger Republic was in the grip of desertification which threatened the livelihoods of it's inhabitants. The default response of both state and non-state actors was to mount massive tree replanting schemes. The approach failed socially, technically and economically and by the late 1980's, appetites for land restoration through tree planting had largely disappeared. Identifying the technical gaps and addressing the real constraints - lack of tree ownership, theft of trees and false perceptions about trees - preventing people from restoring tree cover — opened the door for a spontaneous, wildly successful social reforestation movement.

Contact Address: Tony Rinaudo, World Vision, Victoria, Australia, e-mail: Tony.Rinaudo@worldvision.com.au

What Is the Flavour of Good Intentions?: On Interest Divergence and Responsibility in the Cacao-Chocolate Industry

CARLA D. MARTIN

Harvard University and Fine Cacao and Chocolate Institute, United States

In recent years, the cacao-chocolate industry has increasingly focused on taste - quality of flavour and the ability to discern quality - as a vehicle of market differentiation. At the same time, many of the marketing tools of specialty chocolate rely on linking good taste to social and economic goodness. The links among these elements remain blurry, are not yet clearly defined by specialty cacao or chocolate producers, and often rely on interpretations of work from related but different industries. A critical examination of the value chain indicates that the interests of those who produce chocolate frequently diverge from the interests of those who produce cacao, creating gaps and traps in work on social and environmental responsibility. The often contrary goals that impact the industry must thus be interrogated to determine the place of sustainability, and in whose best interests different players act. We must ask: what is the flavour of good intentions?

Contact Address: Carla D. Martin, Harvard University and Fine Cacao and Chocolate Institute, Cambridge, United States, e-mail: cdmartin@post.harvard.edu

Sustainable Food System and Job Creation under Water Scarcity

JACQUES WERY

International Center for Agricultural Research in the Dry Areas (ICARDA), Egypt

Agri-food systems in the DryArc region (between Southern Europe, North Africa, Sub-Saharan Africa and China) face a complex combination of challenges including water scarcity, rainfall variability, increased temperatures, land degradation, desertification, high population growth and migration, widespread poverty, malnutrition and unemployment. This region is expected to be among those worst affected by climate change, where reduced agricultural productivity, increased poverty, higher dependence on food imports, and increased competition for scarce natural resources will ultimately threaten the viability of agriculture and rural livelihoods. These constraints also present opportunities to transform agri-food systems across the DryArc and scaled in the frame of the diverse range of the region's agro-ecosystems: rainfed, irrigated, agro-pastoral and desert farming. Synergies across these agro-ecosystems can be leveraged among SDGs related to nutrition security, Natural Resource Management (including soils and water) and rural development, provided the potential of the agro-biodiversity is fully utilised and properly managed. This agro-ecological transformation of fields, farms, landscape, value chains and policies will be illustrated in the cereal-based agri-food systems of the DryArc region, showing the key role that food legumes, forages, livestock and trees can play in long-term sustainable use of water and soils. The implications for water, land and labour productivity under climate change and its effects on livelihoods of the people across this region is a critical part of this discussion, informing the future direction of research for development in the context of the CGIAR's DryArc initiative.

Contact Address: Jacques Wery, International Center for Agricultural Research in the Dry Areas (ICARDA), Cairo, Egypt, e-mail: j.wery@cgiar.org

Filling Gaps and Removing Traps in the Use of Voluntary Sustainability Standards

MIET MAERTENS

KU Leuven, Earth and Environmental Sciences, Belgium

Voluntary or private sustainability standards cover a large number of tropical products and are increasingly important in global markets. The rapid spread of voluntary sustainability standards is sometimes erroneously interpreted as a sign of increased sustainability in the food system. Yet, the actual impact of such standards matters. There is a potential trap that private standards satisfy consumers' demand for more ethically and sustainably produced food products and fulfil companies' sustainable sourcing strategies while not actually contributing to improved sustainability in global food systems — thereby merely easing consumers' conscience, diverting donor money to certification programs and extracting rents from food supply chains through expensive monitoring systems. Despite substantial research efforts on this topic, important knowledge gaps remain to be filled. This presentation highlights the results from recent research on the sustainability impact of voluntary sustainability standards, with evidence from different tropical countries and sectors. The presentation includes a focus on all three components of sustainability, including social, economic and environmental issues and sustainability trade-offs. The research findings, along with earlier findings described in the literature, entail important implications towards farmers, food companies, standard-setters, policy-makers, donors and consumers on how to avoid the above-mentioned trap of widespread use of private standards without actual improvement in food system sustainability. Finally, the presentation points to some remaining research gaps that need to be addressed in order to develop a more effective system of private standards.

Contact Address: Miet Maertens, KU Leuven, Earth and Environmental Sciences, Celestijnenlaan 200e - box 2411, 3001 Leuven, Belgium, e-mail: Miet.Maertens@kuleuven.be

Scientific Knowledge Transfer and the Science-Policy Interface: Bridging the Gaps and Overcoming the Traps

IMMACULATE NJUTHE MAINA
Nakuru County, Kenya, County Minister of Agriculture, Livestock and Fisheries, Kenya

Scientific knowledge transfer and the science-policy interface is complex and challenging. It includes contestations about the credibility and legitimacy of scientific knowledge and the strategic use of such knowledge. The cognitive distance between the scientists and science experts on one hand and the policy making institutions on the other, enhances the complexities. In addition, the willingness of scientists to engage in knowledge transfer and exchange with policy is often related to individual capacities, level of training and career trajectories, and motivations. These realities often hamper the successful knowledge transfer between the spheres of science and policy-making leading to under-exploitation of the potential for science to support decision-making. The policy-science interface is further compounded by the fact that policies often address complex, far reaching and large-scale, multi-faceted problems that present social, institutional as well as natural resources contexts. To be more effective many policy makers would need to develop scientific competences to interact more effectively with scientific experts. The policy-making process is politically-driven and involves various arms of government. Interest groups and lobbyists with different points of view are often involved. Scientists need to enhance their level of knowledge of the internal workings of policy-making processes. Once policies are enacted, there often is a lack of coordination among agencies responsible for implementing policy and this contributes to fragmentation. This presentation presents insights on the possible strategies that can fill the gaps and remove some traps in the science-policy-practice continuum for sustainable resources management. They include processes of public participation; a definition of intended outcomes; the theory of change; determination of policy anchors and enablers; as well as provisions on measurement of the transformational impact of a policy. The insights presented are drawn from recent publications on this topic as well as from practice. Inference is drawn from the Kenyan Agriculture Sector Transformation and Growth Strategy (ASTGS). This 10-year strategy supports policies that address food and nutrition security while embracing sustainable exploitation, utilisation, management and conservation of the environment and natural resources.

Contact Address: Immaculate Njuthe Maina, Nakuru County, Kenya, County Minister of Agriculture, Livestock and Fisheries, Nakuru, Kenya, e-mail: immaculate.n.maina@gmail.com

Plants and crops

Crop and cropping systems

Oral Presentations

Posters

Difficult Time for Maize Cropping: How Can we Sustain it under Climate Extremes?

NUTTAPON KHONGDEE[1], THOMAS HILGER[1], WANWISA PANSAK[2], GEORG CADISCH[1]

[1] *University of Hohenheim, Inst. of Agric. Sci. in the Tropics (Hans-Ruthenberg-Institute), Germany*

[2] *Naresuan University, Dept. of Agricultural Science, Thailand*

Climate significantly affects maize (*Zea mays*) productivity. Therefore, the objective of present study was to assess growth, physiology and yield components of maize under different cropping patterns in areas prone to weather variability. The experiment was carried out in an upland maize production area of Thailand, using a split plot design with three replicates. The treatments were (i) June planted maize monocrop, (ii) July planted maize monocrop (farmers' practice) and (iii) July planted maize relay cropped with mung beans (*Vigna radiata*). Maize growth and physiological parameters (grain yield, yield components and δ^{13}C of maize grains) and soil moisture were assessed. Weather data showed that rainfall was mostly falling from June to mid-August and mean maximum temperature was 35°C. During drought periods, the maximum temperature exceeded 40°C. As a consequence, June planted maize was less affected by extreme weather conditions during sensitive periods of maize growing, while July planted maize regardless of mono or relay cropping were negatively affected. The results showed that June planted maize had a significantly better growth performance and finally a higher yield than the other two treatments tested. N uptake of June planted maize was also significantly higher compared to the other two treatments. Light transmission ratio of June planted maize was higher, reaching up to 20 %, while July planted maize treatments were above 40 %. As July planted maize treatments were highly affected by extreme climate, maize-mung bean relay cropping was higher in demanding water than maize monocrop as indicated by soil moisture depletion. Stomatal conductance of these two treatments were not significantly different during normal condition, but during extreme climate, relay cropping was still able to keep stomata open ($P \leq 0.05$). Therefore, relay copping performed better than sole cropping when maize was planted in July. Moreover, δ^{13}C of maize grains confirmed that June planted maize had a much better water access than July planted maize. Nevertheless, July planted maize relay crop had also less water stress than the monocrop. This study indicated that maize-mung bean relay cropping can mitigate extreme weather while using a proper planting period enhances productivity of maize mono cropping.

Keywords: Climate change mitigation, mung bean, relay cropping

Contact Address: Thomas Hilger, University of Hohenheim, Inst. of Agric. Sci. in the Tropics (Hans-Ruthenberg-Institute), Garbenstr. 13, 70593 Stuttgart, Germany, e-mail: thomas.hilger@uni-hohenheim.de

Land Rental Markets in Rural Vietnam – Determinants and Welfare Effects of Renting in Crop Plots

OLIVER SCHULTE

Leibniz University Hannover, Institute for Environmental Economics and World Trade, Germany

It is often suggested that land rental markets contribute to efficiency and equity, especially in developing countries that only recently introduced privately managed agriculture and liberalised land markets. However, our understanding of the determinants of participation in those markets remains limited, affecting the estimation of welfare effects. This paper aims at analysing the determinants and welfare effects of renting in additional plots, thereby identifying possible ways to tap into rental markets' potential. A descriptive analysis is followed by a probit regression to identify the most important determinants of the decision to rent in, while an endogenous switching regression is proposed to analyse welfare effects. The analysis is based on data from 937 households owning rice plots in rural Vietnam from 2016 and 2017, collected in the context of the DFG-funded project 283672937 (TVSEP). Households are split between those renting in additional plots and those only cultivating the plots they own (autarkic). The descriptive analysis shows that households additionally renting in achieved higher yields and net income per hectare from crop production compared to autarkic households. Moreover, their per capita home consumption of their produce was of higher value, while the ratio of home consumption to sales was smaller. These observations point to possibly significant effects of renting in additional land on productivity and food security. The results from the probit regression indicate that households with more members, smaller land endowment, and higher farming ability were more likely to rent in, while minorities were less likely to do so. The results of the selection equation in the endogenous switching regression support these findings. The results from the endogenous switching regression furthermore suggest that autarkic households achieve a lower net income per hectare from crop production than a randomly selected household, while households renting in are able to increase their profit per hectare primarily through the expansion of planted area. The results highlight the importance of households' self-selection into rental markets based on their anticipated gains. Policies should facilitate this process and increase the credibility of rental agreements by providing credit and safeguarding rental payments, which could be targeted specifically at minorities.

Keywords: Cropland rental market, determinants, endogenous switching regression, Vietnam, welfare effects

---

Contact Address: Oliver Schulte, Leibniz University Hannover, Institute for Environmental Economics and World Trade, Königsworther Platz 1, 30167 Hannover, Germany, e-mail: schulte@iuw.uni-hannover.de

Impact of Urbanisation on Water Bodies in Agro-Ecosystems of Bengaluru-Metropolitan City of India

DHANUSH REDDY, DEVAKUMAR AUSTIN

University of Agricultural Sciences, Forestry & Environmental Sciences, India

Water contamination is one of the many consequential concerns of urbanisation that need immediate attention in most cosmopolitan cities. In Bengaluru, one of the fast growing metropolitan cities in India, most of the water bodies are contaminated. However, under unabated, rapid land use changes and severe water scarcity, it is inevitable to utilise the available water for all possible uses depending on the quality of water. Thus it is critical to ascertain the extent of as well the source of contamination. Among the various causes of contamination agriculture is one of the major factors. Due to urbanisation agriculture has undergone changes in and around Bengaluru in terms of crops grown as well as cultivation practices. This study is an attempt to assess the changes in the agro-biodiversity related water quality in the agro-ecosystems along rural urban transition zones (RUT) of Bengaluru. In this study suitability of water for major utilities such as drinking, irrigation and for livestock use is assessed. Water quality is determined based on specific physical, chemical and biological indicators in ground water (n=30) as well as surface water (n=30) bodies located within one kilometer radius of agriculture lands. The mean water quality index (developed based on 22 water quality parameters) of urban surface and ground water suggest that it is not suitable for drinking but, fairly suitable for irrigation and industrial use (C3 and C4 category), while in the transition and rural area it is again not suitable for drinking purpose but suitable for irrigation and industrial use (C2 category). Piper trilinear diagram indicate that majority of the ground and surface water samples belong to mixed Ca^{2+}–Mg^{2+}– Cl^{-}–$SO_2{}^{4-}$ type, and continuous use of this water in future may lead to soil degradation and crop damage. Results indicate that water quality across the RUT of Bengaluru is deteriorating and hence there is an immediate need for improving the water resource management.

Keywords: Agro-ecosystems, drinking standards, irrigation quality, livestock quality, urbanisation, water quality index

Contact Address: Devakumar Austin, University of Agricultural Sciences, Forestry & Environmental Sciences, 560065 Bengaluru, India, e-mail: asdevakumar@gmail.com

Grain and Tuber Yield of the African Yam Bean Intercropped with Cassava

STEFAN HAUSER[1], DENIS BUNGU[2], MICHAEL ABBERTON[3]

[1]*International Institute of Tropical Agriculture (IITA), Root & Tuber Cropping Systems Agronomy, Nigeria*

[2]*University of Kinshasa, Crop Science, DR Congo*

[3]*International Institute of Tropical Agriculture (IITA), Nigeria*

The African yam bean (*Sphenostylis stenocarpa* (Hoechst ex. A. Rich.) Harms.) can be considered an orphan crop, yet interest in the crop has increased recently. In DR Congo, farmers still grow the crop and keep a few varieties for grain and tuber production. The Genetic Resource Unit at the International Institute of Tropical Agriculture, Ibadan, Nigeria is keeping African yam bean (AYB) germplasm and exchanged some with DR Congo. Knowledge on AYB varietal traits and agronomic performance are lacking. A trial was conducted in 2017–2018 season on the Plateau de Bateke to assess the suitability of 4 AYB varieties for grain and tuber production and intercropping with cassava, the most common crop on the plateau. Varieties Feshi, Ngidinga, TSS10 and '209013' were seeded as sole crop and with cassava variety TME 419. Plots were $5\times5\,m^{-2}$, plant density was 2 AYB m^{-2} and 1 cassava m^{-2}. Soils are poor, coarse textured sand, receiving over 1200 mm rainfall, yet vegetation is grassland. Crop establishment and survival to harvest of the AYB was poor in TSS10 (0.5 plants m^{-2}) and moderate for the other varieties (0.9 plants m^{-2}). Grain yield was higher in AYB sole crops and highest in variety '209013' (268 kg ha^{-1}) followed by Feshi (191 kg ha^{-1}), TSS10 and Ngidinga produced 102 and 89 kg ha^{-1}, respectively. When intercropped grain yields were 24–60 % lower with highest losses in '209013'. Tuber yields were highest in Feshi sole crop (13.02 t ha^{-1}) followed by Feshi intercropped (10.16 t ha^{-1}). All other varieties produced significantly less (in t/ha): Ngidinga 2.85 (sole), 2.26 (intercropped); TSS10 0.09 t ha^{-1} (sole), 0.31 (intercropped) and '209013' 0.32 (sole) and 0.07 (intercropped). Feshi is the only variety with a reasonable grain yield to keep seed and a sufficient tuber yield that matches that of cassava. The cassava root yield did not respond to the presence of AYB varieties '209013', Ngidinga and TSS10 but produced significantly less (7.67 t ha^{-1} fresh roots) when intercropped with Feshi. Cassava sole crop attained 11.2 t ha^{-1} fresh roots. AYB tubers fetch a higher price than cassava roots thus AYB Feshi is an alternative to cassava.

Keywords: DR Congo, intercropping, orphan crop, *Sphenostylis stenocarpa*

Contact Address: Stefan Hauser, International Institute of Tropical Agriculture (IITA), Root & Tuber Cropping Systems Agronomy, Oyo Road, 200001 Ibadan, Nigeria, e-mail: s.hauser@cgiar.org

Impact of Climate on the Cassava Yield and Biomass Gap Variability in Sub-Saharan Africa - A Case Study in Nigeria

AMIT KUMAR SRIVASTAVA, THOMAS GAISER, FRANK EWERT
University of Bonn, Inst. Crop Sci. and Res. Conserv. (INRES), Germany

Cassava (*Manihot esculenta* L.) production is vital to the economy of Nigeria as the country is the world's largest producer of the commodity, contributes almost 19 % of the total world production. We investigated the impact of climatic variables on yield gap variability across the three states in Nigeria using the crop model LINTUL5 embedded into a modelling framework, SIMPLACE (Scientific Impact Assessment and Modelling Platform for Advanced Crop and Ecosystem Management). The simulations were run using a cassava variety (TME 419) and historical weather data (1995–2010). Yield gap was estimated as a difference between simulated water-limited yield and farmer's yield (i.e., observed yield), whereas, biomass gaps were estimated as a difference between a simulated water-limited condition and the simulated actual biomass (i.e., under water and nutrient-limited condition). To examine whether variations in crop yield and biomass were related to variations in a specific climate variable, a multiple linear regression was performed for each district-crop combination with yield as the dependent variable and radiation, mean temperature and precipitation as independent variables. The estimated DM yield gaps were 8.1 Mg ha^{-1}, 6.4 Mg ha^{-1}, and 4.0 Mg ha^{-1} in Edo, Ogun and Kwara states respectively, whereas, DM biomass gap was 10.8 Mg ha^{-1}, 9.7 Mg ha^{-1}, and 3.6 Mg ha^{-1} in the respective states. Average farmer's yield could be increased by 176.4 %, 104.7 % and 80 % respectively in the abovementioned states under water-limited conditions. The spatial and temporal variability in cassava yield gap and biomass gap was not correlated with the climate variables (i.e., precipitation, radiation, minimum and maximum temperature) during the crop growing period. Closing the yield gaps will require in the first place adequate supply of nutrients, and reliable seasonal weather forecasts would be required to allow farmers to manage each seasonal potential, i.e., overcoming season-specific yield limitations.

Keywords: Cassava, LINTUL5, Nigeria, variability, yield gap

Contact Address: Amit Kumar Srivastava, University of Bonn, Institute of Crop Science and Resource Conservation (INRES), Katzenburgweg 5, 53115 Bonn, Germany, e-mail: amit@uni-bonn.de

Potential for Productivity Improvements of Wheat Crop by Means of Irrigation Optimisation in Lower Kabul River Basin, Afghanistan

ATIQURRAHMAN JALIL[1], NAEEMULLAH SALARZAI[2]
[1]*Min. of Agriculture, Irrigation and Livestock, On Farm Water Management, Afghanistan*
[2]*Ministry of Energy and Water, Smart Waters Project, Afghanistan*

Afghanistan lies in the arid to semi-arid climatic zone, where efficient irrigated agriculture is the only hope to meet the locally growing food demand. Around 90 percent of irrigated land across the country is being managed conventionally which is associated with unsustainable practices resulting in to the inefficiencies and poor productivity in the system. Therefore, the purpose of this study was to assess the current water productivity of wheat crop under the conventional irrigation system to simulate potential options for improvement in water productivity. The current study was carried out in the Attawor irrigation scheme located in the lower reaches of the Kabul River Basin (KRB). The scheme has a command area of 260 ha, around 23 % of which remained uncultivated due to insufficient use of available water for irrigation. The field work for this study is split into two distinct categories: (i) the conventional irrigation system was monitored from Nov-2016 to May-2017 by measuring field-inflow to the research plots. (ii) AquaCrop model was used to simulate the possible irrigation optimisation options for the scheme level operation. For conventional irrigation, the average water productivity was 2 $kg\,m^{-3}$ and the water use efficiency was in the range of 0.58–0.66 $kg\,m^{-3}$ GI (gross inflow). While looking at the farm application and water productivity, there is great potential for expansion of irrigated area with the existing water, provided some farm level interventions through technological incentives to the farmers as well as infrastructure development are made. With the available water at scheme level, the AquaCrop simulations show that by adopting deficit irrigation and irrigating the crop when the soil moisture content depletes to 130 % of the available water, the dry yield can be increased up to 3.35 $kg\,m^{-3}$ from the existing 1.2 $kg\,m^{-3}$ The available water in Attawor irrigation scheme has a potential to irrigate about 474 ha of land at field capacity level and at unstressed crop conditions. The findings of this study will be helpful for on-farm policy development and raising crop-water productivity and eventually attain the "more crop per drop" slogan to meet the local food demand.

Keywords: Afghanistan, AquaCrop, conventional irrigation, irrigation optimisation, on-farm water management, water productivity, water use efficiency

---

Contact Address: Atiqurrahman Jalil, Min. of Agriculture, Irrigation and Livestock, On Farm Water Management, Pamir Building 7th Floor Jamal Mina, Kabul, Afghanistan, e-mail: atiq001@gmail.com

Farmer's Perception Regarding Effectiveness of Drip Irrigation System in Attock, Pakistan

MUHAMMAD ALI MIAN[1], DANIELA LUKEŠOVÁ[1], VLADIMIR KREPL[2]

[1]*Czech University of Life Sciences Prague, Fac. Tropical Agrisciences, Dept. of Animal Science and Food Processing, Czech Republic*

[2]*Czech University of Life Sciences Prague, Fac. Tropical AgriSciences, Dept. of Sustainable Technologies, Czech Republic*

Agriculture is consigned as the most important element of Pakistan. It contributes 20.9 % to GDP and almost 43.5 % of the entire labour of the country is involved in this sector. It supplies total 60 % of its contribution in the economy by exports and provides raw materials for different industries. Pakistan has abundant water resources, but the misery is that we are not getting proper benefit from these resources. The only way to overcome this situation is to increase water productivity by adopting modern and efficient technologies for sustainable agriculture, which can lead to poverty reduction, profitability and improved food safety with job opportunities. Drip irrigation, water is supplied to the soil with the help of mechanical devices called as the emitters (located on water pipes along selected points). The two types of methods used in drip irrigation technique are surface drip irrigation system (water is delivered at or near the root zone of plants, drop by drop) and subsurface drip irrigation system. We aimed at obtaining information about the awareness of farmers regarding new techniques of irrigation system use in Pakistan. In the year 2015 the questionnaire data was gathered from 120 farmers who came mainly from the Attock district (89.2 %), region Punjab. Their analysis was carried out using SPSS Software. About 45.0 % of farmers were old aged (>50 years old). About one-third of respondents (34.2 %) had a bachelor's degree. More than one quarter (28.3 %) had five years of farming experience. One-third (33.3 %) reported sand and clay type of soil in their field. The majority (69.2 %) were small farmers (< 12.5 acres) and only 3.3 % had large farm (>25 acres) size. The vast majority of respondents (84.2 %) believed that the drip irrigation system should be accepted. One third of respondents (38.3 %) said their work ratio was reduced to 1: 3 while less than one-third (32.5 %) of respondents said their work ratio had dropped to one quarter. In Pakistan agriculture this would be advisable because it minimises evaporation and water drainage. Through this system fertilisers can be delivered which helps in improving yields.

Keywords: Irrigation, Pakistan, questionnaire, sustainable development, water use

Contact Address: Muhammad Ali Mian, Czech University of Life Sciences Prague, Fac. Tropical Agrisciences, Dept. of Animal Science and Food Processing, F 101 Koleje Fgh. 16521 Prague 6-Suchdol., 160 00-169 99 Prague, Czech Republic, e-mail: mohammadalimian18@gmail.com

Analysis of Drought Impact on Sugarcane Bagasse Based Electricity Generation under Climate Change Scenarios

Jazmin Campos Zeballos[1], Vinicius Bof Bufon[2], Zita Sebesvari[1]

[1]*United Nations University (UNU), Institute for Environment and Human Security (EHS), Germany*

[2]*Brazilian Agricultural Research Corporation (Embrapa), Research and Development, Brazil*

The Rio dos Patos Basin in the Cerrado Biome in Brazil is experiencing land use change from (degraded) pasture to sugarcane fields. While the Cerrado has a low share of land under sugarcane cultivation (2 %), it contains half of Brazil's sugarcane area and presents the highest sugarcane expansion rate due to land availability, flat topography and climatological characteristics. Sugarcane in Brazil is a relevant crop for sugar and biofuel production and, nowadays increasingly, for electricity production. Electricity demand in Brazil is covered mostly by hydropower plants (65 %). However, during dry and low river discharge periods, thermal power plants (oil and natural gas) cover most of the gap (14 %). Some sugarcane mills had experience selling electricity surplus prior 2005, when the first auction to 'new energies' was held. Thereafter, electricity generation based on sugarcane bagasse covers 8 % of annual demand, mostly delivered during dry season. Recent unprecedented drought events in Brazil, showed the electricity production vulnerability to droughts. Drought risk assessments including weather trend analysis and climate change scenarios can help to identify regions at risk and help to point opportunities to better cope and adapt to future drought events. In this context, the potential of sugarcane bagasse based electricity generation was evaluated in the Rio dos Patos basin. Weather data from 36 stations within the area and downscaled climate change projections under RCP 4.5 showed a decrease on precipitation jeopardising water availability. Using the Soil & Water Assessment Tool (SWAT) two scenarios were modelled: 1) without sugarcane expansion and 2) with 45 % expansion to sugarcane suitable land. Drought impacts on sugarcane and electricity production will be discussed under the mentioned scenarios. Considering the biome susceptibility to similar drought events, it is important to consider different adaptation strategies for both, sugarcane crops and energy generation, towards energy and water security.

Keywords: Cerrados, climate change, drought risk, energy, sugarcane

---

Contact Address: Jazmin Campos Zeballos, United Nations University (UNU), Institute for Environment and Human Security (EHS), Platz der Vereinten Nationen 1, 53113 Bonn, Germany, e-mail: campos@ehs.unu.edu

Pearl Millet Yield Stability and Susceptibility to Abiotic Stresses in Semiarid; A Modelling Perspective

FESTO RICHARD SILUNGWE[1], FRIEDER GRAEF[1], MARCOS LANA[2], SONOKO DOROTHEA BELLINGRATH-KIMURA[1]

[1]*Leibniz Centre for Agric. Landscape Res. (ZALF), Germany*

[2]*Swedish University of Agricultural Sciences, Crop Production Ecology, Sweden*

Over 500 million people depend on pearl millet for their lives worldwide. The crop is reported to be resilient to climate change due to its inherent adaptability to drought and high temperatures. These traits make pearl millet a very important crop in fighting hunger. However, the performance and resilience of pearl millet to different scenarios of future climates have been rarely explored in sub-Saharan Africa. Data from experiments conducted in two consecutive seasons (2015/2016 and 2016/2017) to determine the yield responses to different fertiliser application levels for Okoa pearl millet variety in Dodoma a semi-arid region in Tanzania were used to calibrate and validate the DSSAT model (CERES-Millet). A validated model was evaluated for 49 synthetic scenarios of climate change constructed by incremental method on historical series of observations, with temperature increments ranging from +0.5 up to +3.0°C and precipitation changes from -30 up to +30 %. Nine planting dates from very early planting 5th December to late planting 25th February after every 10 days were used for simulations of pearl millet yields response to the synthetic scenarios. Results show that the model reproduced the phenology and yield of Okoa pearl millet cultivar with relative root mean square error (rrmse) values for calibration (anthesis days (0.0 %), maturity days (0.8 %), tops weight (7.3 %) and grain yield (6.4 %)) and validation (anthesis days (2 %), maturity days (1.2 %), tops weight (3.5 %) and grain yield (11.8 %)). From evaluation, the yield response surfaces indicated that both very early and late planting dates produced lower yields with a higher risk of crop failure. The best planting window with stable yield was between 25th December and 15th January. Our results demonstrate that Okoa pearl millet variety planted before or after best planting window is susceptible to abiotic stresses under the constructed scenarios.

Keywords: Climate change, DSSAT CERES-Millet, pearl millet, planting dates

Contact Address: Festo Richard Silungwe, Leibniz Centre for Agric. Landscape Res. (ZALF), Mücheberg, Germany, e-mail: festo.richard@zalf.de

Organic Crop Productivity Demonstration Using Tillage and Biofertiliser Management Approaches at Kulumsa/Arisi, Southeastern Ethiopia

Wogayehu Worku[1], Mekuria Tadesse[1], D. Abebe[1], A. Almaz[1], Daniel Neuhoff[2], Gerold Rahmann[3], Magnus Schmid[4]

[1]*Ethiopian Institute of Agricultural Research (EIAR), KARC/Arsi, Ethiopia*

[2]*University of Bonn, Inst. Crop Sci. and Res. Conserv. (INRES), Germany*

[3]*Thuenen Institute of Organic Farming, Germany*

[4]*Deutsche Gesellschaft für Internationale Zusammenarbeit (GIZ) GmbH, Green Innovation Center, Ethiopia*

Modern crop production consider systemic knowledge based farming system which focused on nutrient balances, soil fertility and crop health effects. Such integrated crop management approaches are seldom practised in Ethiopia. Particularly, in the Arsi region farmers mainly grow wheat (75–90 %) and to a smaller extent of some other crops. Most farmers owing average farm-size of 1–2 ha have no understanding of modern organic farming strategies and also rarely use external farm inputs. Most of the training and support is on increasing external farm input. Hence, Green innovation Center/ GIZ funded EIAR initiated Crop Productivity Demonstration in Organic farming systems approach in Kulumusa/Arisi Region in 2016.

A total of 20 ha were allocated for the project by KARC. Interventions were deep plowing of the site and soil harmonisation using planting of alfalfa and Lathyrus seed collected from local vendors. Then all the planted biomass were converted to the soil using second heavy plowing before planting of organic crop productivity demonstration trials.

Non-replicated different separate sets of large block field experiments were undertaken in 2017 main cropping season. Tillage type experiments using mechanised and oxen plowing treatments were compared in faba bean and tef demonstration fields of 2.28 ha in each crop. With respect to tillage, faba bean productivity was higher in oxen plow treatments as compared to tractor. Contrarily, despite the insignificant differences, the two bio fertiliser strains application gave better faba bean yield with tractor than oxen plowed field. Tef agronomic yield was better in oxen than tractor plowed field, that may be associated to the crop required compact seed bed which failed to have in heavy plowing using tractor. Similarly, five crops were evaluated for their productivity in organic model farm experiment under green manure and with additional decomposed animal manure each in 4.5 ha (0.75 ha for each crop). Organic wheat productivity (3300 kg ha^{-1}) showed above national average (2.4 t ha^{-1}) with great optimistic future. Similarly, productivity of potato was 24,444 kg ha^{-1} and that of fresh biomass of alfalfa was 11,640 kg ha^{-1} and other tested crops were above the national average productivity levels.

Keywords: Alfalfa, *Eragrostis tef*, faba bean, organic farming, tillage

Contact Address: Daniel Neuhoff, University of Bonn, Inst. Crop Sci. and Res. Conserv. (INRES), Katzenburgweg 3, Bonn, Germany, e-mail: d.neuhoff@uni-bonn.de

Effects of P Nutrition and VPD on Rice Leaf Morphology and Photosynthesis

OLIVER KNOPF, SABINE STÜRZ, MARC SCHMIERER, FOLKARD ASCH

University of Hohenheim, Inst. of Agric. Sci. in the Tropics (Hans-Ruthenberg-Institute), Germany

The first time in a decade, global hunger is on the rise again driven by conflicts and climate change. The rice cultivation system as such is a significant contributor to greenhouse gas emissions and a major consumer of phosphate (PO_4^{3-}) a non- renewable resource and the most limiting nutrient for plant growth after nitrogen. In order to meet the sustainable development goals, rice production has to be increased by 50–110 % and at the same time, the impact on environment and water usage has to be reduced. To address this issue, a greenhouse experiment was conducted comprising 32 rice plants of two varieties (IR64, Chomrong). Plants were grown in a hydroponic system and manipulated using a nutrient solution with two different phosphate levels (low and optimal P) and two levels of vapour pressure deficit (low and high VPD). Destructive samplings, leaf phosphate analysis, stomatal imprints were performed during the vegetative phase of the plants. Additionally, gas exchange measurements were conducted on young and on older leaves.

Plants subjected to the low phosphate treatment had a significantly smaller leaf area, a lower leaf phosphate concentration, higher root to shoot ratio and generally a higher stomatal density.

While in young leaves P nutrition did not significantly affect assimilation rates, in old leaves reduced P nutrition led to higher assimilation rates in IR64 at high VPD due to higher stomatal conductance. Since under P deficiency, inorganic phosphate (Pi) is translocated from old to young leaves, stomatal control is probably lost in highly P deficient old leaves of IR64. In contrast, Chomrong was able to maintain its stomatal control even in its old leaves, which could be an interesting trait for plant breeding for P limited environments.

Keywords: Climate change, phosphate, phosphor, photosynthesis, resource management, rice, stomata, sustainability, vapor pressure deficit

---

Contact Address: Oliver Knopf, University of Hohenheim, Inst. of Agric. Sci. in the Tropics (Hans-Ruthenberg-Institute), Stuttgart, Germany, e-mail: oliver.knopf@uni-hohenheim.de

Mechanical Weed Control in Cassava: Effects on Weed Biomass, Labour Requirements and Root Yields

Stefan Hauser[1], Friday Ekeleme[1], Adeyemi Olojede[2], Patience Olorunmaiye[3], Mary Agada[4], Alfred Gilbert Dixon[1]

[1]*International Institute of Tropical Agriculture (IITA), Weed Science Center, Nigeria*

[2]*National Root Crops Research Institute (NRCRI), Nigeria*

[3]*Federal University of Agriculture Abeokuta, Crop Science, Nigeria*

[4]*University of Agriculture Makurdi, Crop science, Nigeria*

Weed control is a major labour input in cassava production and commonly burdens women and children with drudgery of bending down and weeding with short handled hoes. Chemical weed control has not been investigated as well as in other crops and mechanical weeding is uncommon. Multi-location trials in major Nigerian cassava growing areas investigated if manually operated or engine driven weeders can attain weeding quality and root yields as manual weeding. Over two years 7 treatments were tested: short handled hoe (SHH), long handled hoe (LHH), rotary weeder (RW), Spike weeder (SW), small motorized tiller (SMT), large motorized tiller (LMT) and motorized brushcutter (BC). Weeding was conducted at 4, 8, 12 and 24 weeks after planting, cassava density was 12,500 plants $^{-1}$. Net-plots measured 4×8 m. Each implement was tested on ridged versus flat soil and by female and male operators. The rotary and spike weeder and brushcutter were eliminated due to poor quality and high time requirements (60–90 minutes plot^{-1}). The LMT was too difficult to handle. SHH, LHH and SMT required 30–40 minutes plot^{-1}. Ridging the soil was an advantage reducing weed biomass. Small plot data appeared to underestimate the time requirement and did not tally with data reported at field level. In 2017 four large plot trials (1250 m^2 per plot) were established comparing the SHH and the SMT on ridged soil. At first and second weeding the SMT required significantly less time to weed than the SHH. Cassava root yield was not significantly affected by the weeding implement or the operators' sex. However, fields weeded by male operators attained higher yields (19.7t ha^{-1} fresh roots) than female operators (15.9t ha^{-1} fresh roots). The difference between the implements was marginal (17.9t ha^{-1} SHH versus 17.7t ha^{-1} SMT). The larger time requirement rendered the SHH the more cost intensive weeding method despite the fuel requirement of the SMT, yet not regarding the purchase cost (600 US$) and depreciation of the SMT. Labour requirements in the large plots were higher than those measured in small plots. Delayed weeding caused weeding time to increase up to 4 fold.

Keywords: Labour time, motorized weeder, weed control

Contact Address: Stefan Hauser, International Institute of Tropical Agriculture (IITA), Root & Tuber Cropping Systems Agronomy, Oyo Road, 200001 Ibadan, Nigeria, e-mail: s.hauser@cgiar.org

Nitrogen Fixation in White Yam (*Dioscorea rotundata*) Using Naturally Abundant ^{15}N

KANAKO TAKADA[1], RYO MATSUMOTO[2], HARUKI ISHIKAWA[3], ASRAT ASFAW[2], PACHAKKIL BABIL[1], HIDEHIKO KIKUNO[1], HIRONOBU SHIWACHI[1]

[1]*Tokyo University of Agriculture, International Agricultural Development, Japan*

[2]*International Institute of Tropical Agriculture (IITA), Yam Breeding, Nigeria*

[3]*International Institute of Tropical Agriculture (IITA), Grain Legume, Nigeria*

White yam (*Dioscorea rotundata*), a tuber crop cultivated and utilised as a staple food in West Africa, plays an important role in the daily diet of millions of people in the region. However, in spite of the growing local demand for white yam, little progress has been made in increasing supply by optimising fertilisation methods to improve yield. Several experiments conducted throughout the region to determine the optimal quantity of nitrogen fertiliser needed for the crop arrived at contradictory conclusions, suggesting the need to take a different approach to elucidate the nitrogen absorption mechanism and to apply it to cultivation. Here we report on the nitrogen absorption mechanism in white yam with a special focus on its nitrogen fixation ability using naturally abundant ^{15}N ($\delta^{15}N$).

A field experiment was conducted at the International Institute of Tropical Agriculture in Ibadan, Nigeria using six white yam accessions with squash (*Cucurbita moschat*) plants as a reference. Experimental plots with three replicates for each accession were prepared without the application of fertiliser in a randomised block design. The percentage of plant N derived from atmospheric N_2 (%Ndfa) was calculated based on $\delta^{15}N$ values.

A wide range of variation was observed in %Ndfa values among the six white yam accessions. The highest value was accession DrDRS074 (45.5 %) and the lowest was DrDRS042 (-4.7 %). Moreover, the %Ndfa of accession DrDRS074 (45.5 %) was statistically significant than that of DrDRS058 (7.2 %) and DrDRS042 (-4.7 %). The results obtained in this study suggest the ability of white yam to fix atmospheric nitrogen; however, intra-specific variation must be also considered. To ascertain the nitrogen fixation ability of different white yam varieties, further research on the presence of symbiotic bacteria as well as the effect of nitrogen fixation on plant growth should be conducted.

Keywords: Low soil fertility, nitrogen fixation, white yam

Contact Address: Kanako Takada, Tokyo University of Agriculture, International Agricultural Development, 1-1-1 Sakuragaoka, 156-8502 Setagaya-ku, Japan, e-mail: 46317402@nodai.ac.jp

Experimental Substantiation of the Use of Agricultural Waste in Obtaining Biodiesel

ASKHAT MARATOV
Kazakh-German University, Fac. of Physics and Economics, Kazakhstan

The territory of the Turkestan region of the Republic of Kazakhstan is the most promising for growing pumpkins. The climate of the region is characterised by high spring-summer-autumn temperatures, and the well-developed root system of this melon crop is perfectly adapted for cultivation on irrigated lands. The peel formed during pumpkin processing can be considered as a promising specific raw material for the production of products with higher value added, for example, as an energy source. This study substantiate the use of the rind and pumpkin seeds of *Cucurbita pepo* L. as a potential source of biofuels.
To test the effectiveness of the technique, an experiment was conducted in the laboratory. The rind of *Cucurbita pepo* L. (Cucurbitacea) was washed with detergent, rinsed with clean water, reduced in size $0.5 \times 0.5\,cm^{-2}$ and dried in a Nabertherm muffle furnace at 110textdegreeC for 72 hours. After drying, the peel was crushed using a hand blender (Philips HR 2102 White).
The sample was extracted with methanol : chloroform (1:2) according to the Folch method. The solid and non-lipid material was removed, the solvent was dried. As a result of the experiment, it was found hat per 100 grams of pumpkin peel there are 3 grams of the total lipid fraction, and per 100 grams of seeds 8 grams. Based on the experimental results obtained in this work, it was concluded that the potential use of pumpkin peel waste as a biofuel is possible on an industrial scale in the Republic of Kazakhstan.

Keywords: Biodiesel, biofuel, Folch method, pumpkin

Contact Address: Askhat Maratov, Kazakh-German University, Fac. of Physics and Economics, Saina/Tole-Bi House 4a Flat 40, A10A9C8 Almaty, Kazakhstan, e-mail: askhatn1@gmail.com

Architecture of Potato Roots (*Solanum tuberosum*, group Phureja) in Early Stage

PEDRO ARTURO WAGNER RODRÍGUEZ[1], TERESA DE JESUS MOSQUERA VASQUEZ[2]

[1]*National University of Colombia, Fac. of Engineering, Colombia*

[2]*National University of Colombia, Fac. of Agricultural Sciences, Colombia*

We evaluated the morphological response of 5 varieties of diploid potato (*Solanum tuberosum* L., Phureja group) subjected to *in vitro* conditions. The hypothesis proposed was based on the root architecture of each of the varieties, presenting a differential response in the angle of insertion as a diagnostic character associated with tolerance to water stress.

The objective of this study was to determine differences in the endogenous early response of the tolerant and non-tolerant materials to water stress previously studied. For this a morphological characterisation was carried out.

The experiment was carried out in the Faculty of Agrarian Sciences of Universidad Nacional de Colombia, Bogotá headquarters with geographic coordinates 4° 35'56"57 N and 74° 04'51"30 W and altitude of 2,600 meters above sea level. Treatments were arranged in a completely random design with three repetitions. Three varieties of *S. tuberosum* group Phureja susceptible to water deficit were evaluated: Dorada, Ocarina and Colombia, and two tolerant to water stress: Violet and Milagros.

They were evaluated during 4 months with an interval of 3 days to measure the variables of angle of insertion, average length of primary roots and length of secondary roots, maximum length, number of roots and number of secondary roots and perimeter area. As a result, the description of the potato root architecture of the five contrasting varieties was obtained with respect to the response to water stress in the field, whereby no differences were found in the variables evaluated.

Keywords: Root architecture, root morphology, *Solanum tuberosum*, drought tolerance, *in vitro* propagation

Contact Address: Pedro Arturo Wagner Rodríguez, National University of Colombia, Fac. of Engineering, Carrera 20 No. 9-73 Cuarto Piso, 250251 Zipaquirá, Colombia, e-mail: parodriguezwa@unal.edu.co

Agricultural Extension for the Production of Quinoa in the Andean Highlands of Perú

RAINER ZACHMANN[1], ELMER VALDEIGLESIAS[2], IGNACIO BENAVENT[3]
[1]*Ex-CGIAR (CIP, IITA), Peru*
[2]*APRODES, Peru*
[3]*Universidad de Piura, Peru*

In the southern Andes of Perú, particularly around Lake Titicaca up to over 4000 m altitude, the Inca domesticated many plant species for their daily diet. These plants include quinoa (*Chenopodium quinoa*), known since 7000 years. Quinoa causes increasing worldwide demand due to its nutritive value. Quinoa has been considered by the FAO as one of the crops for food security because of both the nutritional quality, and its ability to grow in difficult environments, showing tolerance to frost, hail, snow, heat, salinity and drought. Quinoa seeds are rich in protein, dietary fiber, B vitamins, and dietary minerals in amounts greater than in many grains. It is gluten-free. With hundreds - may be more than three thousand - native varieties, Perú and Bolivia present the highest genetic diversity of quinoa, and in 2008 contributed to 92 % of world production, followed by USA, Ecuador, Argentina and Canada. In 2012, Perú produced 44 000 tons on 38 500 hectares. Since 1996, the NGO Asociación Peruana para la Promoción de Desarrollo Sostenible (APRODES; www.aprodes.org), with support from national and international institutions, assists farm families with social, economic and environmental development. APRODES is oriented towards the entire country. In the south, APRODES began in 2007 with production and commercialisation of Andean crops (kiwicha, quinoa, tara, chia, aguaymanto), in addition to food security, livestock production and basic sanitation. There, APRODES has been working with over 800 beneficiary families, distributed over more than 25 Andean communities. APRODES contributes to personal, family and social well-being, promotes sustainable economic development, and protects the environment and sustainable management of natural resources. APRODES improves the nutrition and health of vulnerable families with emphasis on child population from a focus on disease prevention.

Recently, in contribution to agricultural extension, APRODES organised a workshop for extension personnel on the design of a field book (Cuaderno de Campo) for organic quinoa production at Izcuchaca, Provincia de Anta, Departamento de Cusco.

Keywords: Agricultural relevance, agricultural extension, Andean highlands, biodiversity, cultural relevance, NGO APRODES, quinoa (*Chenopodium quinoa*), sustainable development

Contact Address: Rainer Zachmann, Ex-CGIAR (CIP, IITA), Malecón Cisneros 450, Lima, Peru, e-mail: rainerzachmann@hotmail.com

Socioeconomic and Technological Profile of Rice Growers in Sao Mateus do Maranhao Municipality, Maranhao State, Brazil

UELSON SERRA GARCIA[1], CARLOS MAGRI FERREIRA[2], LUCIANO CAVALCANTE MUNIZ[3], CLEYZER ADRIAN CUNHA[4], ALCIDO ELENOR WANDER[2]

[1] *Federal University of Goias (UFG), Postgraduate Program in Agribusiness, Brazil*
[2] *Brazilian Agricultural Research Organization (EMBRAPA), National Rice and Beans Research Center (CNPAF), Brazil*
[3] *State University of Maranhao (UEMA), Brazil*
[4] *Federal University of Goias (UFG), School of Economics, Brazil*

This study aimed to study the socioeconomic and technological profile of rice growers in São Mateus do Maranhão, Maranhão state, Brazil. Therefore, structured questionnaires were applied, as well as the use of secondary data with bibliographical and documentary research. In order to determine the number of producers that would be interviewed, a sample was defined based on IBGE's agricultural census data (2017), which pointed out that the municipality of São Mateus do Maranhão has 269 rice growing farms. From this population, a sample of 75 farmers were interviewed (95 % of confidence level, with 10 % error). With the data collection, it was possible to observe that the activity of rice production in São Mateus do Maranhão municipality is developed by small, medium and large producers. From interviewed rice growers, 73 % are organised in associations, 24 % in societies of limited responsibility, and 3 % in isolated family business. From all interviewed farmers, 31 % grow rice on leased land. Also, 12 % of rice growers lease part of their land to other farmers to grow rice. Rice growers use two cropping systems: (a) the rainfed paddy cultivation in areas with favourable rainfall distribution (96 %); and (b) the irrigated paddy rice (4 %). The results show that the marketing of production and acquisition of new technologies represent the main obstacles for producers, especially the small ones, who have difficulties in obtaining technologies to improve the activity and little access to marketing channels available in the region. This shows to the need for public policies related to technical assistance directed at marketing channels and, above all, effective technical monitoring.

Keywords: Associations, marketing, technological innovation, rice cultivation

Contact Address: Alcido Elenor Wander, Brazilian Agricultural Research Corporation (EMBRAPA), Rodovia GO-462, km 12, 75375-000 Santo Antonio de Goias, Brazil, e-mail: alcido.wander@embrapa.br

Effect of Farming on Watershed Area of Inle Lake, Myanmar

Tin Amy Chit

World Wide Fund for Nature (WWF-Myanmar), Enterprise and Livelihood Development, Myanmar

The present research was carried to unfold the awareness of Inle Lake dwellers on the adverse effects of farming activities on the watersheds. Study areas comprised seven villages in four village tracts around the lake. A total of randomly selected 81 farmers were interviewed during March 2016. Information from experts and printed materials from the department of agriculture were also collected. The dependent variable, farmers' awareness of certain environmental effects of farming was further split into six dimensions to get a clear picture of the awareness. Probit regression analysis was performed to determine the determinants of farmers' awareness of certain environmental effects of farming on the watersheds. Findings show that 72, 100, 23, 81, 17 and 59 % of the sampled farmers are aware of sedimentation increase, water quality declining, biodiversity loss, fisheries resources decline, drought increase and air pollution increase, respectively, that are caused by farming activities in the lake areas. They have the highest awareness of water quality declining and the lowest awareness of drought as a consequence of farming activities. However, awareness on different aspects of adverse farming effects on the watersheds is influenced by different factors. In general, education, family size, farm size, frequency of manure application and frequency of pesticide application contribute to improve the farmers' awareness of adverse farming effects. On the other hand, cropping intensity (mono-cropping) and duration of high yielding variety (HYV) cultivation negatively impact on this awareness. However, age does not have any significant influence to build awareness of adverse farming effects on the watersheds. This study emphasises the increased awareness as a prerequisite to discourage farmers to practice such farm activities that could potentially harm the watersheds. In order to increase the farmers' awareness of adverse farming effects, government should invest to improve the educational qualifications of the farmers regarding adverse environmental effects of farming activities. Agricultural extension services should be focused on delivering supports to the farmers on increasing the manure (organic fertilisers) application and cropping intensity (indirect way of increasing farm size in terms of total cropped area).

Keywords: Environmental awareness, farmers' attitude, farming, watershed area

Contact Address: Tin Amy Chit, World Wide Fund for Nature (WWF-Myanmar), Enterprise and Livelihood Development, House no (128), Myawinn Street, Dawei, Myanmar, e-mail: tinamy.chit@wwf.org.mm

Contribution of Resilient Agronomic Techniques to Increased Maize (*Zea mays* L.) Productivity on Degraded Dry Lands in Karamoja, Uganda

Samuel N. Kayongo[1], Agnes Amongin[2], Christine Ilemut[1,2], Stephen Walyaula [1], Paul Okullo[1]

[1]*National Agricultural Research Organization (NARO), Uganda*

[2]*Nabuin Zonal Agricultural Research and Development Inst. (Nabuin ZARDI), Uganda*

Application of selected resilient agronomic techniques on degraded lands enhanced increased productivity of maize for smallholder farmers in Karamoja. The land is structurally sandy, poor in soil fertility with high moisture deficit due to drought stress. Technologies for improving crop productivity were adopted, and focused on maize productivity due to food and income importance to smallholder farmers. Thus enhancing improved productivity for smallholder farmers, explored interventions aimed at (i) assessing agronomic techniques enhancing biomass development and crop growth on degraded land (ii) Identify drought stress tolerance mechanisms ideal for plant survival (iii) Assess plant nutrition accumulation from soil fertility replenishments and (iv) quantify maize yield. Applied agronomic techniques comprised of *Zai* pits, furrows and integration of velvet bean legume (*Mucuna pruriens*) for revitalisation through biodegradation. Results showed improvement in biomass development with 42 % increase in plant vigour, 17–50 % (leaf area), 12–19 % (stem), 14–30 % (root exploration) for *Zai* pits than furrows and legume integration techniques. Leaf folding ranged significantly from 115°–73° in *mucuna* intercrop indicating mechanisms for maintaining leaf turgidity, and 42 % stay green from leaf senescence for plant survival during drought stress. Maize leaf chlorophyll was significantly quantified at 39.3 $mg\,g^{-1}$ (furrows) and 45.2 $mg\,g^{-1}$ (*Zai* pits) indicating proper nutrient uptake correlating positively with plant vigour and leaf area. Maize output across agronomic techniques varied significantly with *Zai* pit techniques enhancing the highest yield at 3.1 MT ha^{-1}, furrow (2.6 MT ha^{-1}) compared with sole maize (0.74 MT ha^{-1}), indicating an increase of 40–65 % in maize productivity. Therefore growing maize in *Zai* pits and furrows on degraded lands enhanced increased productivity hence food security.

Keywords: Degraded land, furrows, maize, resilience, *Zai* pits

Contact Address: Samuel N. Kayongo, National Agricultural Research Organization (NARO), Box 132, Moroto, Uganda, e-mail: jskayongo@yahoo.com

Plant genetic resources - a tool to fill gaps and remove traps

Oral Presentations

Posters

Defining Thresholds for Identifying Genetically Redundant Accessions within CIAT's Global Cassava Collection

MONICA CARVAJAL, GUSTAVO CARDONA, PETER WENZL

International Center for Tropical Agriculture (CIAT), Genetic Resource Progr., Colombia

Crop diversity is critical for food security, climate adaptation, dietary quality and poverty reduction. Cassava is a critical source of carbohydrates for 500 million people in Africa and elsewhere. The International Center for Tropical Agriculture conserves the globally largest collection of cassava (*Manihot esculenta*) and its wild relatives (6,155 accessions). The collection, currently conserved in-vitro under slow-growth conditions, will be cryopreserved once genetically redundant accessions have been removed. We conducted an experiment to determine thresholds for identifying genetically redundant accessions by genotyping technical and biological replicates from a group of accessions. Twenty-five cassava accessions were selected for DNA extraction. Three types of replicates were analysed to quantify genetic distances (Gd): (i) different individuals from the same accession "Individual-Reps", (ii) different DNA samples extracted from the same individual "Extract-Reps", and (iii) the same DNA sample analysed twice "DNA-Reps". In addition, 17 accessions were randomly selected from a core collection to compare Gd between replicate pairs using a more diverse group of accessions. DNA samples were analysed using the DArT-seq genotyping platform (DArT-PL) to generate dense profiles of SNP and 'in-silico-DArT' markers (the latter being the presence/absence of sequence-tags in the genome representation). Marker identification and allele-calling were performed using DS14 software (KDCompute plug-in system). A total of 188 samples and 114,545 markers were obtained (59,519 SNP/55,026 'in-silicoDArT). SNP markers were used to compute distance matrices after applying additional marker-specific quality filters, such as sequencing depth, call rate and polymorphism content. The variation in the average Gd between replicate pairs (Individual-Reps, Extract-Reps and DNA-Reps) was used as a criterion to adjust marker-quality thresholds to retain a maximum number of markers while producing close-to-zero genetic-distance estimates for replicate pairs. The thresholds values determined in this study will be used to identify a genetically non-redundant subset of DArTseq-characterised accessions for cryopreservation to reduce long-term conservation costs.

Keywords: Cassava, genetic diversity, genotyping

Contact Address: Monica Carvajal, Intern. Center for Tropical Agriculture (CIAT), Genetic Resource Program, Km 17 recta Cali Palmira, Cali, Colombia, e-mail: m.carvajal@cgiar.org

Seed Distribution in Rural Communities of Central America

RICARDO ROMERO-PEREZGROVAS[1], IRENE ALVARADO VAN DER LAAT[2], JENNIFER TORRES VEGA[2], GLORIA UWIZEYIMANA[2], DAVID DE LA ROSA[3], JUDITH HECK[4]

[1] *Sustainable Harvest International, Field Programs, Mexico*

[2] *EARTH University, Costa Rica*

[3] *CIMAT, Applied Statistics, Mexico*

[4] *Deutsche Gesellschaft für Internationale Zusammenarbeit, PROCAMBIO, Honduras*

The importance of *in-situ* agricultural biodiversity has increased over the last two decades. Local crops and varieties have been lost at an alarming rate since the 1970s. Efforts to preserve them ex-situ in germplasm banks are just a partial solution: without a continuous cycle of planting and selection, agricultural biodiversity stops its dynamic cycle. Since the beginning of agriculture, seed exchange between farmers has been one of the main mechanisms of distribution and conservation of agricultural biodiversity.

In this study, we tracked and mapped how rural communities are preserving and exchanging seeds. We analysed and compared how seed networks operate for: 1) grains and legumes; 2) horticulture and vegetables; 3) tubers and root crops, and 4) forest and fruit trees. Using a structured survey, the research took place in a total of 18 rural communities in Belize, Honduras and Panama. Using a "following the seed" method, the total number of farmers surveyed was 427. Some of the results indicate that the grains and legumes and the tubers and roots networks were more dynamic and decentralised in the three countries when compared to the horticulture and vegetables networks now controlled by few actors.

Maize and bean varieties were selected, preserved and distributed between farmers with high intensity and without the need for external actors. In parallel, horticulture and vegetable networks are heavily concentrated, and their seed distribution is dependent on external actors such as NGOs, government agencies or private enterprises. The difficulties to select and preserve seeds of different crops is one of the main causes of a concentrated network dependent on external actors. If in-situ agricultural biodiversity is to be preserved in these countries, there is a need for greater understanding of how and why farmers are preserving and exchanging seeds.

Keywords: Central America, crop biodiversity, seed networks

Contact Address: Irene Alvarado Van der Laat, EARTH University, Environmental Economics, Guacimo, Costa Rica, e-mail: ialvarad@earth.ac.cr

The Contribution of CIAT Genebank in the Development of High-Iron Bean Varieties and Farm-Level Impacts in Rwanda

Stefania Sellitti[1,4], Ricardo Labarta[1], Kate Vaiknoras[2], Melinda Smale[3], Nelissa Jamora[4], Robert Andrade[1], Peter Wenzl[1]

[1] *International Center for Tropical Agriculture (CIAT), Colombia*

[2] *Virginia Tech, Dept. of Agricultural and Applied Economics, USA*

[3] *Michigan State University, Agricultural, Food and Resources Economics, USA*

[4] *Crop Trust, Germany*

Genebanks play an essential role in a world where a substantial part of agricultural biodiversity has been lost from farming habitats, malnutrition persists as the global population continues to rise, and farm productivity is vulnerable to climate change. We demonstrate the importance of the genebank of the International Center for Tropical Agriculture (CIAT) in the development of seven iron-biofortified climbing bean varieties (CAB2, RWV3316, RWV3317, RWV3006, RWV2887, MAC44, MAC42) and the impact of their adoption on farm households in Rwanda. First, we link iron-biofortified climbing varieties directly to the genebank through pedigree analysis and key informant interviews with the breeders who developed them. Second, we apply an econometric model to test the impact of their adoption on the yield, consumption and purchase of beans by farming households in Rwanda. Analysis is based on a dataset collected from nearly 1400 households in 2015 by Harvest Plus. We find that the scope of the genetic diversity housed in the bean collection at CIAT was fundamental to developing successful iron-biofortified beans. In contrast to the findings of Vaiknoras (2019), which focused on a most extensively adopted variety of bush bean, we found no statistically significant effects. There could be various explanation for these results, such as the comparatively lower rate of adoption of climbing beans or their different characteristics and the difference in the way they are harvested. Our results suggest that it is possible to track the journey of an accession from its introduction in the genebank to its final use by farmers. Further research is needed to understand why these climbing varieties had less impact than the bush variety previously studied.

Keywords: Breeding, genebank, high-iron beans, malnutrition, Rwanda

Contact Address: Stefania Sellitti, International Center for Tropical Agriculture (CIAT), DAPA / Genetic Resources Program, Carrera 6 # 4-36 San Antonio, Cali, Colombia, e-mail: stefaniasellitti@outlook.com

Screening Sweetpotato Crop Wild Relatives for Sweetpotato Virus Disease Resistance

BETTINA HEIDER[1], SEGUNDO FUENTES[2], JAN KREUZE[2]
[1]*International Potato Centre (CIP), Genetics, Genomics and Crop Improvement, Peru*
[2]*International Potato Centre (CIP), Crop and Systems Science Division, Peru*

Sweetpotato Virus Disease (SPVD) is a major constraint to sweetpotato production worldwide and particularly in sub-Saharan Africa. SPVD is caused by the synergistic interaction of the whitefly-borne Sweetpotato Chlorotic Stunt Virus (SPCSV) and the aphid-borne Sweetpotato Feathery Mottle Virus (SPFMV). Approaches to control SPVD are limited to phytosanitary measures and the use of virus-free planting material. Due to the lack of extreme natural resistance to SPVD in the sweetpotato genepool, crop wild relatives are a potential alternative source of resistance genes. However, no resistance could be found in species of the *Ipomoea* series *Batatas* (Choisy) D.F. Austin complex, the group of crop wild relatives most closely related to sweetpotato. To evaluate sweetpotato crop wild relatives for resistance to SPCSV and SPFMV 53 accessions of nine species belonging to the *I. batatas* complex were subjected to repetitive cycles of grafting with infected stem cuttings and subsequent enzyme-linked immunosorbent assay (ELISA) screenings. Resistance of accessions that remained negative to SPFMV or SPCSV was finally validated by grafting plant material on indicator plants (*I. setosa*) followed by ELISA analysis. Two *Ipomoea cordatotriloba* accessions (CIP 460296 and CIP 460164) were identified showing stabile resistance to SPCSV and one *Ipomoea tiliacea* accession with resistance to SPFMV (CIP 460531) was found. No resistance to both viruses was detected. These findings demonstrate that sweetpotato crop wild relatives are a potential sources of true and durable resistance to SPVD useful for sweetpotato pre-breeding initiatives. Since crossing barriers of species of the Batatas complex with sweetpotato exist further research is required to explore pre-breeding approaches such as the use of bridge species.

Keywords: *Ipomoea* series *Batatas*, pre-breeding, sweetpotato chlorotic stunt virus, sweetpotato feathery mottle virus, sweetpotato virus disease

Contact Address: Bettina Heider, International Potato Centre (CIP), Genetics, Genomics and Crop Improvement, Av. La Molina 1895, Lima, Peru, e-mail: b.heider@cgiar.org

Potential for Domestication of *Borassus aethiopum* Mart.: A Wild Multipurpose Palm Species in Sub-Saharan Africa

VALÈRE SALAKO[1], AHUÉFA MAURICEL KÉGBÉ[1], FLORA JOSIANE CHADARE[2], KONOUTAN KAFOUTCHONI[1], AUBIN AMANGNIDE[1], CASTRO RODRIGUE GBÈDOMON[1], ACHILLE ASSOGBADJO[3], CLÉMENT AGBANGLA[4], ROMAIN GLÈLÈ KAKAÏ[5]

[1]*University of Abomey-Calavi, Lab. of Biomathematics and Forest Estimations, Benin*

[2]*University of Abomey-Calavi, Lab. of Food Science, Benin*

[3]*University of Abomey-Calavi, Lab. of Applied Ecology, Benin*

[4]*University of Abomey-Calavi, Genetics and Biotechnology , Benin*

[5]*University of Abomey-Calavi, Dept. of Natural Resources Management, Benin*

Borassus aethiopum Mart. is a dioecious palm species native to mainland Africa. It is a multipurpose tree daily used by local communities but is unfortunately threatened by anthropogenic pressures in many parts of Africa. Fruits and young shoots represent the most exploited parts putting the species at risk. Since the domestication of species offers a good alternative for its long-term benefit, the present study aimed at assessing the environmental-induced diversity in morphological traits of fruits and evaluated differences in growth and weight of hypocotyls from one-seeded, two-seeded as well as three-seeded fruits from different provenances in the three climatic regions of Benin. A total of 5,400 fruits were collected from 180 trees in six populations and fruit and tree morphological traits were measured. A randomized complete block design with three replications was used for the experimentation in each climatic region. The results showed that variation in fruit morphological traits was not influenced by climatic regions. Furthermore, the greatest variation (65 - 94 %) in fruit morphological traits was located at tree level, highlighting that selection of many fruits and individual trees within a few populations would capture a large variation of fruit traits. Tree diameter at breast height (18.5 - 52 cm), total height (6.4 - 19.6 m) and bole height (4.8 - 17.6 m), fruit length (7.00 - 20.50 cm), dry weight (98 - 2552 g), shape index (0.59 - 2.80), and number of seeds per fruit (1 - 3) were the most discriminative traits of the studied populations. Clustering of the trees resulted in five different morphotypes based on discriminating traits. Morphotypes 1 and 2 showed high performance for fruit and seed production and are consequently recommended for selection and breeding programs. Irrespective of the provenances, the best performance of hypocotyls were observed in the humid region. The study provides important baseline information for domestication and sustainable conservation of *B. aethiopum* in Benin.

Keywords: Climatic regions, domestication, hypocotyl, morphological traits

---

Contact Address: Ahuéfa Mauricel Kégbé, University of Abomey-Calavi, Lab. of Biomathematics and Forest Estimations, Abomey-Calavi, Benin, e-mail: kegbem@gmail.com

Observations on Seed Embryo and Germination, Seedling Morphology and Development of *Vitellaria paradoxa* (C.F.Gaertn.)

ABDULAI IDDRISU, WILFRED ELEGBA, GEORGE KLU, KENNETH DANSO
Ghana Atomic Energy Commission, Biotechnology and Nuclear Agriculture Research Institute, Ghana

The pyrophytic features and high longevity of shea tree (*Vitellaria paradoxa*) make it an excellent species for combating desertification and mitigating climate change in sub-Saharan Africa. In this study, our aim was to describe precisely the cotyledon morphology, seedling development process and seedling morphology that account for cryptohypogeal germination of the species. To identify the embryonic axis, fresh and partially dry seeds were deshelled and their distal ends transversely cut off. Cut seeds were immersed in 1.0 % tetrazolium chloride (TTC) solution for 24 hours and kept in the dark for staining. To study cotyledon morphology, 50 seeds were deshelled, air-dried for 3-5 days and split open from the distal end. Another batch of 50 seeds were sown directly. The sprouted seeds or seedlings were sampled at 5-day intervals and their cotyledonary tubes sectioned to observe morphological and anatomical features. To study seedling development, 180 seeds were classified into three groups based on their sizes and then sown on seedbeds in randomized complete block design with 60 in each of the 3 replicates. The TTC solution stained the proximal ends of only the fresh seeds. Morphologically the seed bears a cotyledonary raphe, which lies vertically with distally schizocotylous and proximally syncotylous cotyledons. Seed size significantly ($P < 0.05$) affected the various stages of seedling development. Large seeds sprouted earlier (7 days) followed by seedling emergence within 61 days, while small seeds sprouted almost a week later (12 days) with seedling emergence within 75 days. The location of the embryos at the proximal ends cause germinating seeds to produce long cotytedonary tubes which bury the plumules deep into the soil as a protection against bushfires. This germination study will enhance nursery establishment for plantation development of *V. paradoxa*.

Keywords: Cotyledon, crytohypogeal, establishment, germination, pyrophytic, seedling emergence, shea, *Vitellaria paradoxa*

Contact Address: Wilfred Elegba, Ghana Atomic Energy Commission, Biotechnology and Nuclear Agriculture Research Institute, Box LG 80, Kwabenya, Ghana, e-mail: welegba@gmail.com

Constraints and Opportunities of Cassava Seed Systems in South India: An Exploratory Case Study

SEBASTIAN MENGEL, KARIN ZBINDEN GYSIN, ALESSANDRA GIULIANI
Bern University of Applied Sciences (BFH), School of Agricultural, Forest and Food Sciences (HAFL), Switzerland

In India, most cassava is produced in the federal states of Kerala and Tamil Nadu. While the crop is generally valued for being well adapted to marginal cropping conditions, the cassava mosaic disease (CMD) is widespread in India, resulting from the indiscriminate use of infected planting material, a lack of commercial interest, and the long-prevailing non-availability of improved cassava varieties. Under the Indo-Swiss Collaboration in Biotechnology (ISCB), Indian and Swiss research institutes have increasingly worked together on the development of improved CMD-resistant cassava varieties. Yet, as for many other tuber crops, cassava propagation generally follows a vegetative approach, resulting in a relatively slow seed dissemination. Hence, the pressing question is how new varieties can be effectively disseminated within reasonably short time, so that farmers, the ultimate beneficiaries, can access them.

In 2018, qualitative in-depth interviews and group discussions were conducted with cassava farmers, breeders, extension and agricultural scientists, agricultural economists, and cassava processing industry representatives in Kerala and Tamil Nadu to first, identify the main constraints and bottlenecks for the dissemination of new cassava varieties to farmers and second, develop recommendations to sustainably improve the dissemination process.

The research shows that cassava is generally of low and highly fluctuating economic value, which constitutes a key challenge for the development of a seed system driven by a larger demand of farmers. This is interrelated with a low willingness to pay (WTP) of farmers for new varieties, a perceived high opportunity cost (risk) of acquiring new planting material, inadequate rules and regulations for the re-use of infected seed and the legal status and support of cassava, as well as the absence of a clear mandate and adequate capacities for seed multiplication and dissemination. The study recommends a multi-perspective and multi-stakeholder approach to (i) formally clarify the mandate for planting material multiplication and dissemination, (ii) strengthen extension capacities for agricultural technology transfer, (iii) explore the potential of farmers' associations engaging as actors in an improved, functional cassava seed system, and (iv) consider changes at policy level for the re-use and replacement of planting material and the allocation of specific subsidies.

Keywords: Cassava, cassava mosaic disease (CMD), dissemination, India, manioc, multiplication, planting material, seed system

Contact Address: Sebastian Mengel, Bern University of Applied Sciences (BFH), School of Agricultural, Forest and Food Sciences (HAFL), Länggasse 85, 3052 Zollikofen, Switzerland, e-mail: sebastian.mengel@bfh.ch

Response of Drought-Inducible Proline Accumulation in Barley Genotypes to Seed Set and Filling

FELIX FRIMPONG[1], ALI NAZ[2], CAREL WINDT[1], MICHAEL FREI[2], DAGMAR VAN DUSSCHOTEN[1], FABIO FIORANI[1]

[1]*Forschungszentrum Juelich, Inst. of Bio- and Geosciences, IBG-2: Plant Sciences, Germany*

[2]*University of Bonn, Inst. Crop Sci. and Res. Conserv. (INRES), Germany*

Drought as a consequence of climate change is projected to become more prevalent in the future to cause yield decline of about 33 % in barley. In this study we focus on drought response of five different spring barley genotypes; four 2-row types with *Barke* and *Scarlett* as elite German cultivars, two introgression lines *IL143* & *IL141* which bears the ancestral allele *Pyrroline-5-carboxylate synthase1- P5cs1* and one 6-row type *HOR10151*. Plants were grown until booting stage in the greenhouse in pots with peat soil and then subjected to 18 days of well-watered (100 % field capacity) and water-stress (30 % field capacity) treatments. For all genotypes significant reductions due to the treatment were observed for phenotypic, biochemical and physiological traits. Averaged over all genotypes, a significant reduction in plant height (17 %), tiller number (20 %), spike number (56.3 %), grain number (28.8 %), spike length (19 %), grain weight (76.5 %), relative leaf water content (4.4 %), CO_2 assimilation (55.6 %), stomatal conductance (73 %), transpiration rate (63.6 %) and electron transport rate (49.3 %) was recorded due to drought. A positive correlation was found between the treatment effect on spike length and the reduction in grain number ($r=0.73$, $P\leq0.01$) and grain weight ($r=0.37$, $P\leq0.05$), respectively. Reduced green leaves area was observed for all genotypes, with the two introgression lines showing milder drought symptoms. The reduction in relative leaf water content was negatively correlated ($r=-0.61$, $P\leq0.001$) with the reduction in shoot dry weight. Again, the reduction in relative leaf water content positively correlated ($r=0.65$, 0.53, 0.51, $P\leq0.01$) with reductions in transpiration rate, stomatal conductance and CO_2 assimilation respectively. Across genotypes, the grain filling duration was not affected. However, preliminary MRI scans indicated that there might be differences in floret abortion among drought treated elite cultivars compared to introgression lines. The 6-row barley *HOR10151* was severely affected while *IL143* showed better drought tolerance by staying green, maintaining relative leaf water content and remaining photosynthetically active. These results support and extend recent findings that barley genotypes harboring wild variant of P5cs1 might have an advantage in tolerating low water availability.

Keywords: Barley, drought-inducible proline, filling, seed set

Contact Address: Felix Frimpong, Forschungszentrum Juelich, Inst. of Bio- and Geosciences, IBG-2: Plant Sciences, Wilhem Johnen Str., Juelich, Germany, e-mail: f.frimpong@fz-juelich.de

Exploring Farmers' Preference for Varietal Traits in Ethiopia: Stated Choice Analysis

SHIMELIS ARAYA GEDA, RAINER KÜHL

Justus Liebig University Giessen, Agribusiness Management and Food Econ., Germany

Plant breeding is one of the key tools to reduce vulnerability associated with abiotic and biotic stress challenges and improve food security. In Ethiopia, public breeding programs are releasing large number of varieties mainly based on yield superiority as the major criterion. The existing strategy follow 'one size fits for all' approach. There is a question of; do breeders really provide what farmers want? Breeding impacts depend on farmers' adoption decision. However, drought-prone sites are benefited little from breeding and new varieties are not adequately adopted.

The literature has emphasized the lack of access to improved varieties and credit constraints as the major bottlenecks. Accordingly, there exists little research on whether the variety traits are compatible with farmers' preferences. Against this background, the objective of this study is to examine whether the traits fit enough with farmer preferences located in drought-prone areas.

A choice experiment with attributes of productivity, high adaptation to stress conditions, bean color, and seed cost was used to conduct survey with randomly selected 167 beans growers in southern Ethiopia. The choice survey was supplemented with focus group studies and a separate household survey questionnaire to identify farmers' socioeconomic characteristics. Random parameter logit model was used to estimate farmers' trait preferences for variety selection.

Results from the focus group studies show that farmers have a higher propensity towards drought-tolerant varieties. The model analysis also shows that farmers have highly opted for a trait associated with adaptation to stress conditions. Furthermore, the analysis provides evidence for significant preference heterogeneity across farmers. Overall, higher preference towards yield stabilizing attributes has strong welfare enhancing arguments to set breeding priorities. This implies the need for re-orientation of the existing breeding approach to establish demand-driven and context-specific breeding strategies to enhance adoption of new varieties.

Keywords: Breeding strategy, choice experiment, common bean, drought-tolerant, random parameter logit, traits preference heterogeneity

Contact Address: Shimelis Araya Geda, Justus Liebig University Giessen, Agribusiness Management and Food Economics, Senckenbergstr. 3, 35390 Gießen, Germany, e-mail: araya.gedam@gmail.com

Triticum araraticum: A Wild Tetraploid Wheat Species with Potential Implications in Crop Breeding Programs

ESRAA GABAL[1], BENJAMIN KILIAN[2], MICHAEL FREI[1], HAKAN ÖZKAN [3]

[1] *University of Bonn, Inst. Crop Sci. and Res. Conserv. (INRES), Germany*

[2] *Global Crop Diversity Trust, Germany*

[3] *Cukurova University , Field Crop Dept., Turkey*

Wheat is one of the major crops, globally known since its early domestication in the Fertile Crescent 12000 years ago. Its evolutionary process and species diversification are of interest to many breeders. With the help of molecular markers, it became feasible to reveal the information about crop genomes and detect their genomic diversification. Our research aimed at characterising one of the least explored wheat species. This wild wheat species was first identified in the 1930s and given its botanical name *Triticum araraticum*, Jakubz in 1947 by Jakubziner, with a tetraploid genome (GGAuAu). Wheat Ararat is the wild progenitor of domesticated tetraploid *T. timopheevii* and it is included among the group of hulled wheat with tenacious glumes and disarticulated spikes. Our research was conducted on 79 different genotypes of *T. araraticum*, originated from Turkey, Iraq, Iran, Israeli, Syria, Azerbaijan, and Armenia. The genotypes were sampled from their natural habitat by H. Özkan and yet kept in the genebanks. With the implementation of a retrotransposon-based iPBS marker system, we could identify the genomic diversification among the 79 genotypes arranging them into two main clusters. Moreover, we estimated a number of phenotypic measurements, revealing that *T. araraticum* has an average plant height of 95 cm, spike length of 11.5 cm, peduncle length of 40 cm, spike weight of 1.25 gm, seed weight per spike of 0.4 gm, and a number of 30 seed grains per spike. Additionally, most of the plants required up to 115 days until starting the heading (spikes production). Our research revealed substantial phenotypic diversity in *T. araraticum*. Moreover, the nutrient analysis revealed wild wheat accessions had an average zinc content of 96 mg kg^{-1}, phosphate concentration of 5.5 g kg^{-1}, and phytate content of 15 g kg^{-1}. Hence, we concluded that *T. araraticum* accessions involved in our study showed diversification at the genomic, phenotypic, and nutritional levels ensuring the diversified genomic pool of Ararat that can help in enriching the genomic diversity of wheat cultivars if being implemented in breeding programs.

Keywords: Genomic diversification, iPBS retrotransposons-based markers, wheat evolution, wild relatives

---

Contact Address: Esraa Gabal, University of Bonn, Inst. Crop Sci. and Res. Conserv. (INRES), Karlrobert-Kreiten-Strasse 13, 53119 Bonn, Germany, e-mail: esraa.gabal93@yahoo.com

Large Scale Screening for Potential Abiotic Stresses Tolerant Rice Germplasm in Vietnam

TRUNG DUC TRAN, LINH HONG TA, DANG QUANG BUI

Vietnam Academy of Agricultural Sciences, Science and International Cooperation, Vietnam

Global climate change has potentially grave affects for rice production and, consequently, global food security. Its phenomena, such as sea-level rise leading to salinity intrusion, temperature increment and unpredicted weather causing flooding and drought, occur more frequently in the irrigated intensively managed rice areas and is accounted for a 15 % decrease in rice yields worldwide. Rice production of Vietnam recently has been under the influence of the same circumstance. In order to mitigate and adapt to increasing severe unfavoured conditions, breeding for new rice varieties which effectively tolerant to abiotic stresses such as submergence, salinity and drought is a priority in the restructuring strategy of Vietnam agriculture sector.
Vietnam is one of the world's genetic diversity centres of rice. Sufficient rice germplasm evaluation is considered as a crucial premise which, therefrom, facilitates effective exploitation of such gene pool for rice breeding programme in the themes of food security guarantee and climate change mitigation.
In this presented study, a panel of 146 Vietnam rice accessions that included landraces, elite and popular cultivars, and breeding lines developed from various rice breeding programme across Vietnam were screened both phenotypically and genetically. Phenotypic evaluation under unfavourable abiotic stresses identified five accessions highly tolerant to submergence, one accession moderately tolerant to 9‰ salinity and 16 accessions exhibiting potential yield performance under drought stress. In parallel, genotyping experiment using the 7K SNP Chip revealed a high diversity among screened rice accessions. Genotypes which associate with characterised genes/QTLs conferring targeted abiotic stresses tolerance were observed for most of potential rice accessions except that showing tolerance to salinity. It implies that, there are other genetic elements may involving in salinity tolerance in rice. Notably, clustering analysis based on the selected 52 SNP markers representing submergence, salinity and drought tolerance genes/QTLs also discriminated groups of landraces and improved accessions. These results provided a useful insight into the genetic diversity of Vietnam rice population.

Keywords: Drought, rice, salinity, screening, SNP, submergence

---

Contact Address: Trung Duc Tran, Vietnam Academy of Agricultural Sciences, Science and International Cooperation, Quynh Do Road Thanh Tri District, Hanoi, Vietnam, e-mail: ductrung83@gmail.com

Genetic Diversity in Napier Grass (*Cenchrus purpureus*) Collections as Revealed by Genotyping-by-Sequencing Method of the DArTseq Platform

MEKI SHEHABU MUKTAR[1], ERMIAS HABTE[1], CHRIS S. JONES[2]
[1] *International Livestock Research Institute (ILRI), Ethiopia*
[2] *International Livestock Research Institute (ILRI), Kenya*

Napier grass is one of the most important fodder crops, particularly in Eastern and Central African countries and used as a cut-and-carry feed with high potential as a bio-fuel crop. The ILRI (International Livestock Research Institute) genebank holds a diverse set of Napier grass accessions and also has a collection contributed by the Brazilian Agricultural Research Corporation (EMBRAPA). One hundred and five accessions were subjected to genotyping by sequencing using the DArTseq platform, which generated 116,190 SilicoDArT and 85,452 SNP high-density and polymorphic markers together with short sequence reads. The short sequence reads, with an average of 54 nucleotides, were mapped to the pearl millet reference genome, which is the closest related species to Napier grass. Around 17 % of the SNP and 33 % of the SilicoDArT markers were mapped and, based on the map position, the closest genes aligned with the markers were identified and the corresponding annotation information extracted. In turn, these data were used to select candidate genes for important forage traits based on functional annotations and sequence similarity. A total of 980 highly polymorphic SNP markers distributed across the genome and mostly independent were used to assess population structure and diversity. Up to seven subgroups were identified using phylogenetic analysis and the major ones were supported by the admixture model in STRUCTURE and principal component analysis (PCA). A few representative Napier grass accessions were subsetted from the diversity with the objective to distribute a representative subset of a manageable size for adaptation/evaluation in different production systems and agroecological conditions. Genome-wide linkage disequilibrium (LD) analyses revealed a fast LD-decay, on average at about 2.54 kbp, in the overall population with the LD-decay slower in the ILRI material compared to the EMBRAPA collection. This genotyping initiative generated high-density markers with a reasonable distribution across the genome. The diversity analysis revealed the existence of a substantial amount of variation, particularly in the ILRI collection and identified some unique materials from the EMBRAPA collection, demonstrating the suitability of the overall population for further genetic and marker-trait association studies.

Keywords: DArTseq, diversity analysis, elephant grass, forage, genebank, linkage disequilibrium

Contact Address: Meki Shehabu Muktar, International Livestock Research Institute (ILRI), Feed and Forage Development, Addis Ababa, Ethiopia, e-mail: mshehabu@gmail.com

Insights into the *Batatas* Complex: A Crossing Study and its Significance for Breeding

BETTINA HEIDER[1], RICHARD ELTON MILLER[2], CRAIG G. YENCHO[3]

[1] *International Potato Centre (CIP), Genetics, Genomics and Crop Improvement, Peru*

[2] *Independent Expert in Evolutionary Biology, United States of America*

[3] *North Carolina State Univ., Dept. of Horticultural Science, United States of America*

Sweetpotato crop wild relatives (*Ipomoea* series *Batatas* (Choisy) D.F. Austin) are an untapped and potentially valuable source of genetic diversity for crop improvement. However, the use of wild relatives in sweetpotato breeding remains largely unexplored as obstacles specific to this group must be overcome to fully unleash the resources provided by the sweetpotato crop wild relatives (SP CWR). These difficulties include, for example, unclear species delimitation, lack of diagnostic characters enabling differentiation among taxa, different ploidy levels and unknown levels of cross compatibility. Thus, the objective of this study was to test the compatibility of SWP CWR through a systematic interspecific and intraspecific crossing study to provide information critical for the design of a pre-breeding approach. The crossing study included 46 accessions from eight out of 14 CWR species of the series Batatas: *I. cordatotriloba, I. cynanchifolia, I. grandifolia, I. leucantha, I. ramosissima, I. splendor-sylvae, I. trifida*, and *I. triloba*. The crossing design was a full diallel with 2,070 possible cross combinations and a target of 20 crosses completed per cross combination. Three crossing groups with high levels of interfertility (>40 %) within the diallel were identified that correspond to three distinct syngameons. Crossing group 1 (CG1, 24 accessions) contained *I. cordatotriloba, I. cynanchifolia, I. grandifolia, I. leucantha* and *I. triloba* while CG2 (12 accessions) included *I. trifida* and one *I. ramosissima* accession. CG3 encompassed four *I. ramosissima* accessions. Crosses between accessions of CG1 and CG3 exhibited zero to low levels (<27 %) of interfertility but no absolute barriers to reproduction. In contrast, accessions from CG2 did not form fertile offsprings with CG3 indicating reproductive isolation. Accessions from CG1 (self-compatible species) as females crossed at zero to low levels with accessions from CG2 (self-incompatible species) as males, but the reverse shows no interfertility. This distinction provides us with a roadmap for future pre-breeding approaches. *I. trifida* accessions from CG2 (as males) may serve as bridge species to introgress desired traits from CG1 (as females) into the sweetpotato genepool.

Keywords: Cross compatibility, crossing barriers, *Ipomoea*, pre-breeding, sweetpotato crop wild relatives

Contact Address: Bettina Heider, International Potato Centre (CIP), Genetics, Genomics and Crop Improvement, Av. La Molina 1895, Lima, Peru, e-mail: b.heider@cgiar.org

Evaluation of Agronomic Traits Variation in White Yam (*Dioscorea rotundata*)

RYO MATSUMOTO[1], HARUKI ISHIKAWA[2], ASRAT ASFAW[1]

[1]*International Institute of Tropical Agriculture (IITA), Yam Breeding, Nigeria*

[2]*International Institute of Tropical Agriculture (IITA), Grain Legume, Nigeria*

Yams (*Dioscorea* spp.) are grown globally in Africa, Asia, Latin America, Caribbean and Oceania. Especially in the West African, there is a great demand for yam as staple food as well as a cash crop with increasing population. Among the 20 or more species of yam, white yam (*D. rotundata*) is the most important species as food material in West Africa. The growth characteristic of white yam such as large plant size and the long growing period makes it difficult to conduct field experiments and evaluations of the agronomic traits. Moreover, systematic breeding has hardly been carried out so far due to lack of plant physiological information. Although the genetic resource material held by IITA may include individuals with high breeding value, the diversity of their agronomic traits has not been clarified. Therefore, to obtain basic information for promoting white yam breeding, the evaluation of agricultural traits in the genetic resource material was conducted. The experiment was conducted at 2017 cropping season (May to Dec) in IITA Ibadan, Nigeria. Thirty genotypes of white yam were selected from IITA genetic resource material and used in this experiment. Destructive sampling was carried out to understand the tuber index (tuber weight / total plant weight) at early September. Growth period was calculated from date of sprouting to date of aerial part 100 % senescence. Tubers were harvested in December. The varietal difference between genotypes was clearly observed in agronomic traits. The genotypes with high tuber index in September tended to have a short growth period. The tuber yield was correlated with the aerial part biomass in September, suggesting that the biomass of the aerial part up to September are key traits to obtain a high tuber yield. From these results, it is expected to contribute to white yam breeding and development of improved varieties suitable for the region.

Keywords: Agronomic traits, varietal difference, yam

Contact Address: Ryo Matsumoto, International Institute of Tropical Agriculture (IITA), Yam Breeding, Ibadan, Nigeria, e-mail: r.matsumoto@cgiar.org

Cassava Varietal Performance and Productivity Analysis under Subjective and Objective Measurements - A Case of Malawi

Muzee L. Kazamwali[1], John Ilukor[2]

[1] *Evangelical University in Africa, Department of Rural Economy, Uganda*

[2] *The World Bank, Development Data Group - Survey Unit, Uganda*

Cassava is the second most important crop to maize in Malawi. Since 1995 per hectare yield of cassava has more than quadrupled and production has increased more than nine times between 1997 and 2007. This is mainly attributed to breeding and dissemination of improved varieties through collaborative programs between the government of Malawi and the International Institute of Tropical Agriculture (IITA). However, despite all the efforts put into breeding and dissemination of improved cassava varieties, several weaknesses in evaluating the performance of these crop technologies at plot level still persist. These include for instance high variability in on-farm yield, piecemeal harvesting, and varietal misidentification, implying that both adoption and impact estimates of improved cassava varieties cannot be trusted. Although studies have investigated the effect of systematic measurement errors in agricultural surveys, only few have assessed their effect on varietal performance and productivity analysis. The current study therefore investigates the effect of measurement errors on varietal performance and productivity analysis. The study used data from methodological experiment on cassava varietal identification and productivity measurement (CVIP 2015) collected over 1129 households in five districts in Malawi. Using DNA fingerprinting and crop-cutting as benchmarks, descriptive statistics showed that farmer estimates were more volatile in estimating yield on selected varieties compared to crop-cutting. The latter, on the other side, over-estimated yield by 30 percentage point on average. Econometric estimations showed under crop-cutting that though cassava has always been advertised as a woman's crop, female-headed households were still less efficient as compared to those of their male counterparts. Furthermore, while adoption appeared to have significant effect on technical efficiency under farmer estimates, crop-cutting showed on the other side that this variable was not of crucial importance on farmers' technical efficiency. Study findings support therefore investment in advanced quality data collection through more reliable methods such that yield can be tracked to specific variety; but also the role played by socioeconomic factors, especially gender in adoption and impact of food security crops. Hence, periodic dissemination of planting materials for resource-constrained groups such as women as well as training of extension workers for better varietal identification are mainly recommended.

Keywords: Measurements, productivity analysis, varietal performance

---

Contact Address: Muzee L. Kazamwali, Evangelical University in Africa, Department of Rural Economy, Kampala, Uganda, e-mail: kazamwali@gmail.com

Phenotypic Diversity of Rice Landraces Collected from Ayeyarwady Region, Myanmar Using Agro-Morphological Characterisation

Aye Aye Thant[1], Hein Zaw[2], Htay Htay Aung[2], Marie Kalousová[1], Rakesh Kumar Singh[3], Bohdan Lojka[1]

[1] *Czech University of Life Sciences Prague, Fac. Tropical AgriSciences, Czech Republic*

[2] *Plant Biotechnology Center, Czech Republic*

[3] *The International Centre for BioSaline Agriculture, Crop Diversity and Genetics Section, United Arab Emirates*

An investigation was conducted to determine the agro-morphological characterisation and relationship between rice accessions collected from the Ayeyarwady region for their utilisation in breeding programs. The use of agro-morphological traits, both qualitative and quantitative, is the most common and traditional approach utilised to estimate relationships between genotypes. In this study, 117 accessions were characterised for 13 qualitative and 15 quantitative traits as per IRRI rice descriptors. Qualitative traits were analysed using Shannon diversity and cluster analysis by NTSYS, whereas quantitative characters were subjected to principal component analysis (PCA) and coefficient matrix. The standardised Shannon-Weaver diversity index ranged from 0.41 (least polymorphic) for sterile lemma colour to 0.84 (highly polymorphic) for amylose content with an average of 0.7. Cluster analysis separated the different varieties into various groups. Truncating the tree at the Euclidean distance of 1.18 resulted in 15 clusters. In the truncated tree, 3 clusters had single accession, 5 clusters had two to three accessions, 6 clusters had five to eighteen accessions and the largest cluster had 44 accessions. Four accessions (TLNKYAUK, MSEIK, MSWE and KYTUN) clustered together in the genetic distances (0.00) due to their similar qualitative traits. For analysis of 15 quantitative traits, the coefficient of variation was more than 10 % for most of the characters, the highest one being the number of panicles per plant (24.54 %). Correlation analyses among quantitative traits showed a strong positive correlation in some traits such as long and short sterile lemmas, grain length and length-width ratio, harvest index and yield. The principal coordinate analysis showed similar groupings as in the cluster analysis. PCA has revealed six major components (Eigenvalue > 1), which altogether explained 80.4 % of the total variation. Component loadings for each principal component showed quantitative traits, such as grain width, yield and harvest index that were among the phenotypic traits contributing positive projections in three principal components that explained 54 % of total variation in the characterised rice accessions. The present study indicated that diversity in agro-morphological traits were useful for preliminary evaluation for crop improvement programme and can be used for assessing genetic distance among morphologically distinguishable rice landraces.

Keywords: Agro-morphological characterisation, landraces, Myanmar, *Oryza sativa*

Contact Address: Aye Aye Thant, Czech University of Life Sciences Prague, Fac. Tropical AgriSciences, Prague, Czech Republic, e-mail: aye18988@gmail.com

Genetic Diversity of Native Guatemalan Avocado Using SSR Molecular Markers

José Alejandro Ruiz-Chután[1,2], Marie Kalousová[1], Julio Ernesto Berdúo-Sandoval[2], Amílcar Sánchez-Pérez [2], Bohdan Lojka[1]

[1] *Czech University of Life Sciences, Fac. of Tropical AgriSciences, Dept. of Crop Sciences and Agroforestry, Guatemala*

[2] *Universidad de San Carlos de Guatemala, Facultad de Agronomía, Guatemala*

Avocado is one of the most widely consumed crops worldwide and originated in the Sierra Nevada region in California eight to five million years ago. In Guatemala, a high genetic diversity has been reported by morphological characterisations, but due the cross-pollination and interracial hybridisations, these characterisations present complications and it is not possible to detail the total genetic richness. Molecular markers are a useful tool for this type of studies because they are free of environmental effects and focus on differences at the DNA level. The main objective of this study was to analyse the genetic diversity of 189 materials of native Guatemalan avocado collected in seven populations located in eight departments of the country, using 11 SSR molecular marker. A total number of 262 polymorphic alleles were detected with a mean number of 10.701 per population. Among population, the expected heterozygosity varied from 0.738 to 0.837. The analysis of molecular variance (AMOVA) showed that only 2 % genetic variation existed among population, while 65 % existed within individuals and 33 % existed among individuals . Moderate differentiations among the analysed populations were indicated by Fst index 0.018. Structure analysis suggested two and four cluster for the whole avocado materials. The lack of population structure is attributed to the process of selection, the system of cross-pollination, the exchange of seeds and the process of domestication to which the avocado is still subject. We therefore suggest the protection of the phytogenetic resource, as well as a complementary agromorphological characterisation, to establish an adequate strategy of exploitation through plant breeding programs.

Keywords: Avocado, genetic diversity, Guatemala, SSR

Contact Address: José Alejandro Ruiz-Chután, Czech University of Life Sciences, Fac. of Tropical AgriSciences, Dept. of Crop Sciences and Agroforestry, 15 Calle B 16-61 Zona 6, 01006 Guatemala, Guatemala, e-mail: josealejandro.ruiz@icloud.com

In vitro Conservation and Recovery of *Ullucus tuberosus* (Loz.) after Reduced Growth of Microshoots

Stacy Denise Hammond Hammond[1], Iva Viehmannová[1], Jiri Zamecnik[2], Bart Panis[3], Petra Hlasna Cepkova [4]

[1]*Czech University of Life Sciences Prague, Fac. of Tropical AgriSciences, Dept. of Crop Sciences and Agroforestry, Czech Republic*

[2]*Crop Research Institute, Plant Physiology and Cryobiology Laboratory, Czech Republic*

[3]*Bioversity International, c/o KU Leuven, Belgium*

[4]*Crop Research Institute, Gene Bank, Czech Republic*

In vitro conservation by the use of reduced growth is considered to be a reliable biotechnological tool for medium-term conservation of plant germplasm while ensuring its immediate availability. In the present study, we assessed reduced growth condition using five culture media supplements, a cultivation temperature of 4°C and a 24 h *in vitro* growth in complete darkness, to establish a simple and reliable *in vitro* conservation protocol for ulluco (*Ullucus tuberosus*). Ulluco is an Andean tuberous crop rich in carbohydrates and vitamin C and it represents a staple crop for local people. For this experiment, individual nodal segments of ulluco were precultured for 28 days on half-strength Murashige and Skoog (MS) medium maintained in a culture room under a 16/8 h light/dark regime at 17°C, and at a photosynthetic photon flux density of 35 μmol m^{-2} s^{-1} provided by cool-white fluorescent tubes. They were then transferred to half-strength MS medium supplemented with mannitol (10–30 g l^{-1}), sorbitol (10–30 g l^{-1}), sucrose (10–120 g l^{-1}), chlorcholinchlorid (CCC; 300–700 mg l^{-1}) or abscisic acid (ABA; 1–3 mg l^{-1}) and were placed in a cultivation temperature of 4 °C and a 24 h dark conditions for 24 months. Based on survival percentage and number and size of MTs, three superior treatments were selected for further experiment on microtuber germination: mannitol (20 g l^{-1}), sorbitol (30 g l^{-1}) and sucrose (90 g l^{-1}). Three regrowth media were tested: MS, half-strength MS and MS supplemented with 0.5 mg $^{-1}$ GA3. After 3-months survival, MT germination and morphological characteristics were evaluated. Results showed that MS cultivation medium supplemented with GA3 and MTs originated from conservation medium supplemented with 90 g l^{-1} sucrose showed the fastest regrowth and provided overall superior characteristics over plants from other conservation treatments and tested regrowth media. The protocol optimised in this study provides minimal labour and efficient method of ulluco conservation for 24 months.

This research was financially supported by the Internal Grant Agency of Faculty of Tropical AgriSciences, Czech University of Life Sciences Prague IGA (Project No. 20195002) and the MZe RO0418 project. BP gratefully acknowledges the Gene Bank CGIAR Research Programme and the CGIAR Research Programme on Roots, Tubers and Bananas (RTB) and the Directorate-General for Development, Belgium (DGD) for financial support of the project 'Safeguarding vegetatively-propagated crop diversity to nourish people now and in the future'.

Keywords: Gibberellic acid, *in vitro* conservation, microtubers, sucrose, ulluco

Contact Address: Stacy Denise Hammond Hammond, Czech University of Life Sciences Prague, Fac. of Tropical AgriSciences, Dept. of Crop Sciences and Agroforestry, Kamycka 1281, Kolej F204, 16521 Prague, Czech Republic, e-mail: hammondstacy9@gmail.com

Survival Is Insufficient: A Cryopreservation Case Study on Cassava

Gwendal Plumier[1], Bart Panis[2], Hannes Wilms[1]

[1]*KU Leuven, Belgium*
[2]*Bioversity International, c/o KU Leuven, Belgium*

Food security is one of the sustainable development goals of the United Nations. Most of the regions that are affected by this insecurity are areas with infertile or dry soils, often located in the tropics. An essential crop in such environments that provides food security is cassava, an important source of carbohydrates and the third most important crop in the tropics. Even though there is only one cassava species, there are plenty of varieties, each with its own characteristics, allowing the species to survive in many different areas. Having access to such a diversity within the species is important for farmers since it allows them to adapt to different situations.
The cassava diversity can be kept in field and *in vitro* genebanks but the ultimate way of preserving cassava for the long term is using cryopreservation. In these conditions plant tissues are cooled to the temperature of liquid nitrogen (-196°C). At this temperature all biological and chemical processes come to a halt. However for cassava this cryopreservation protocol is not yet optimised. In this study, different parameters during the cryopreservation protocol were varied, such as a the application of a sucrose preculture, different loading solutions and the use of apical versus axillar meristems and this on three different cassava varieties. One month after recovering from the liquid nitrogen exposure, this resulted in survival rates of up to 91 % for some cultivars. Under survival we understand that new green leaves were formed. However after this initial growth, the shoots completely stopped growing and this for both control and cryopreserved meristems, suggesting that the storage in liquid nitrogen is not the main bottleneck. Regeneration towards normal rooted plants is thus the next hurdle to overcome, before we can speak about a cryopreservation protocol that is applicable to an *in vitro* collection.

Keywords: Cassava, cryopreservation, genebanks, *in vitro* culture

Contact Address: Hannes Wilms, KU Leuven, Biosystems, willem decroylaan 42, 3001 Leuven, Belgium, e-mail: hannes.wilms@kuleuven.be

Plant protection and plant health (DPG session)

Agrobiodiversity of Homegardens in Pyay District, Myanmar

Su Myat Thwe[1], Lukas Pawera[1], Cory Whitney[2], Zbynek Polesny[1]

[1] *Czech University of Life Sciences Prague, Fac. of Tropical AgriSciences, Dept. of Crop Sciences and Agroforestry, Czech Republic*

[2] *University of Bonn, Inst. Crop Sci. and Res. Conserv. (INRES), Germany*

This survey was conducted in 96 randomly sampled home gardens across eight villages of the Thegon Township situated in Bago region. The research aimed to assess the species diversity of cultivated and wild plants to uncover typology of local home gardens and finally to document traditional botanical knowledge on useful plants and plant management practices. Data were collected during August and September 2018. The survey included a household survey, vegetation survey and plant artefact interviews. A total number of 246 species belonging to 70 botanical families were identified and the value of Shannon-Wiener index is ranging from 0.1 to 2 with an average value of 1.03. The best-represented families with the highest number of species were *Fabaceae*, *Euphorbiaceae*, *Rutaceae*, *Zingiberaceae*, *Apocynaceae* and *Asteraceae*. We identified 63 species as an ecologically important species. In our study area, type and structure of local home gardens for all households was a combination of subsistence and market-oriented gardens with different levels of commercialisation. There was a positive correlation between the number of plant species and home garden size, natural vegetation, species diversity, and a number of species commercialised. On the other hand, the home garden age was not correlated with species diversity. The results showed that there were no significant environmental and management effects on species diversity. In most of the home gardens, plants were managed by both chemical and biological control methods. While biological methods were applied for most of the crops, agrochemicals were applied only in insect pests control of marketed crops to improve the cash crop yields.

Keywords: Biodiversity, ethnobotany, traditional knowledge, useful plants

Contact Address: Zbynek Polesny, Czech University of Life Sciences Prague, Fac. of Tropical AgriSciences, Kamýcká 129, 16500 Praha - Suchdol, Czech Republic, e-mail: polesny@ftz.czu.cz

Conventional and Organic Farmers in Iran: A Comparison Study on Perceptions, Socio-Economic and Demographic Characteristics

AREZOU BABAJANI[1], ZEYNAB JOUZI[2]

[1] *Tehran University, Department of Agricultural Development and Management, Germany*
[2] *North Carolina State University, College of Natural Resources, United States of America*

Organic farming in Iran has great potential for contribution in sustainable agriculture. In order to support the implementation of organic farming, policy makers need to have better understanding of farmers. The purpose of this study is to analyse the socio-economic characteristics of organic versus conventional farmers. In addition, we aim to investigate farmer's perception and attitude towards organic farming. The sample included 141 pioneer farmers (organic and non-organic farmers) in five provinces in Iran (Kerman, Golestan, Khorasan Razavi, Kermanshah and Fars). These regions have the highest number of organic farmers in the country. For conducting the analysis, we used Mann-Whitney Test, Chi-Square Test and Discriminant Analysis.

The results of Mann-Whitney Test show that perceptions of farmers about environmental risks and human health effects of chemical fertilisers and pesticides were significantly different among organic and conventional farmers. In addition, our results indicate that more than 60 % of farmers in the sample had very limited or no knowledge about organic farming. With respect to educational program, 77.3 % of respondents reported that they never enrolled in organic educational courses. Concerning the organic certification, our results show that 77.3 % of farmers were unfamiliar with the process of certification. Regarding organic market, almost all farmers (93.6 %) reported very limited or no knowledge about the available markets. Respecting demographic factors, our findings revealed that, organic farmers were more socially active and relatively younger than the conventional farmers. However, regarding the educational level, we did not observe significant difference between two groups. Using discriminant analysis in SPSS software, we analysed the differences between organic and conventional farmers. Four discriminating factors include having experience of organic agriculture, engagement in social activities, the experience of IPM methods on their own farm, and attending the sustainable agriculture courses were entered in discriminant function of organic farmers' characteristics. The results of our study suggest that in order to support the development of organic farming in Iran, policy makers, experts and extension agents' efforts should focus on extensional, informational and market-oriented plans.

Keywords: Conventional farmers, organic agriculture development, organic farmers, policy makers, sustainable agriculture

Contact Address: Arezou Babajani, Tehran University, Department of Agricultural Development and Management, Wengertsteige 1,, 71088 Holzgerlingen, Germany, e-mail: babajani_a@yahoo.com

Impact of Organic and Conventional Farming Systems on Termite Presence, Diversity and Maize Crop Damage

JOHN J ANYANGO[1], DAVID BAUTZE[2], NOAH ADAMTEY[2]

[1]*Kenyan Agricultural and Livestock Research Organization (KALRO), Kenya*

[2]*Research Institute of Organic Agriculture (FiBL), Dept. of International Cooperation, Switzerland*

Termites are major soil macrofauna and within the literature, they are either depict as 'pests' or an important indicator for environmental sustainability. It is worthwhile to understand the extent to which termites can be managed to avoid crop damage and to improve the sustainability of farming systems. Therefore, the objectives of the study were to assess the effect of organic and conventional farming systems on termite presence, diversity, activity and crop damage. To achieve these objectives, we conducted a study in the maize crops on the on-going long-term systems comparisons trials (SysCom) at two sites in the Central Highlands of Kenya. The trial is comparing organic and conventional farming systems at two input levels: low input representing smallholder farmer practice and high input representing commercial scale practice.
The results showed higher termite abundance, incidence, activity, and diversity in the organic high input farming system compared to the other farming systems. However, the overall the damage patterns due to termites appeared to be a function of farming systems, plant growth stages, trial site, type and amount of fertiliser and/or organic material applied. During the study period, we identified nine different termite genera that belong to three subfamilies: (i) Macrotermitinae (genera: Allodontotermes, Ancistrotermes, Macrotermes, Microtermes, Odontotermes, and Pseudocanthotermes), (ii) Termitinae (Amitermes and Cubitermes) and (iii) Nasutitiermitinae (Trinervitermes). Our findings demonstrate that certain farming systems attract termites, which are an important, and often beneficial, component of soil fauna. Nonetheless, damage patterns were not necessarily higher in these farming systems and thus not generally lead to higher yield loss.

Keywords: Farming systems, organic agriculture, termites

Contact Address: David Bautze, Research Institute of Organic Agriculture, International Cooperation, Ackerstr. 113, 5070 Frick, Switzerland, e-mail: david.bautze@fibl.org

Rice Farmers' Perception of the Three-Controls Technology (3CT) in Guangdong Province, China

Helena Wehmeyer[1,2], Annalyn H. de Guia[2], Xuhua Zhong[3], Melanie Connor[1]

[1]*University of Basel, Dept. of Environmental Sciences, Philippines*

[2]*International Rice Research Institute, Sustainable Impact Platform, Philippines*

[3]*Guangdong Academy of Agricultural Sciences, The Rice Research Institute, China*

Over the past decades, the overuse of fertiliser in China has led to decreased fertiliser use efficiency, stagnating yields, and heavy environmental pollution causing disastrous effects on the climate, soils, water, and human health. In order to combat this problem in rice production, the Three-Controls Technology (3CT) was developed and released in Guangdong Province in 2007 as a best management practice to reduce rice farmers' fertiliser use, while decreasing the number of unproductive rice tillers and controlling for diseases and insects. By use of the new technology, nitrogen fertiliser input is typically reduced by 20 % while rice yields increased by 10 %. From an agronomic perspective, the technology has been successful and thus has also been introduced to other Chinese provinces. However, the aspects regarding farmers' satisfaction with the 3CT and their long-term adoption behaviour have not yet been studied. Hence, the main objective of this study is to investigate farmers' perception of sustainable development through 3CT. In particular, the aim is to examine how the three pillars of sustainability – economic, social and environmental – have changed since 3CT introduction. A quantitative survey questionnaire has been developed and data will be collected via tablets with the CommCare application. In total, 150 farmers will be surveyed in Huicheng county, Guangdong Province, in April 2019. Farmers' perceived personal benefits or drawbacks regarding their financial situation, changes in biodiversity, knowledge, and capacity building will be evaluated and presented. The results will be analysed using uni- and multivariate statistics. Additionally, an econometric and psychometric statistical analysis will be performed to determine the significance of farmers' perceptions with regard to adoption satisfaction. Data will be examined taking into account three key elements of development: (1) economic level, which is output oriented, (2) social level, which is behaviourally oriented, and (3) intervention level, which is modification oriented. The implications will be discussed with regard to the long-term adoption of 3CT, which may lead to a better understanding of effective sustainable best management practices that benefit the environment as well as the farmers' businesses and livelihoods.

Keywords: Best management practices, China, fertiliser, input use, perception

Contact Address: Helena Wehmeyer, University of Basel, Dept. of Environmental Sciences current address: International Rice Research Institute, 4031 Los Baños, Philippines, e-mail: h.wehmeyer@irri.org

Farmers' Knowledge and Practices of Potato Bacterial Wilt Management in Ethiopia

Shiferaw Tafesse[1], Elias Damtew[2], Barbara Van Mierlo[2], Rico Lie [2], Berga Lemaga[3], Kalpana Sharma[4], Cees Leeuwis[2], Paul C. Struik[1]

[1] *Wageningen University and Research, Centre for Crop Systems Analysis, The Netherlands*

[2] *Wageningen University and Research, Knowledge, Technology and Innovation Group, The Netherlands*

[3] *International Potato Center, CGIAR Research Program on Roots, Tubers and Bananas (RTB), Ethiopia*

[4] *CGIAR Research Program on Roots, Tubers and Bananas (RTB), International Potato Center, Kenya*

Potato (*Solanum tuberosum* L.) is an increasingly important crop for food and nutrition security in Ethiopia. It is also a vital source of income and more than 3.7 million smallholder farmers are involved in potato production in the country. However, bacterial wilt is currently causing an overwhelming impact on the country's potato production systems, threatening food and nutrition security initiatives.

A survey of 261 randomly selected smallholder farmers was carried out in three major potato growing districts in the central highlands of Ethiopia to examine farmers' knowledge and management practices of bacterial wilt, and to analyse the role of relevant knowledge in their practices. Considering their different characteristics, three groups of farmers were distinguished: producers of quality declared seed, producers of normal seed and producers of ware potatoes. The results of the study indicated that most farmers (72 %) could recognise symptoms of the disease on infected potato plants. However, they had very limited knowledge of the disease including its causal agent, spreading mechanisms, and management methods. All of the farmers were unaware of the causal agent of the disease and there were significant incongruences between scientific explanations and farmers' understanding of the disease. The farmers provided different explanations and confused a causal agent of the disease with various factors, such as water shortage, insects, planting seed potato with high moisture content, and waterlogging. Further, the majority of the farmers (60 %) did not know spreading mechanisms of the disease. There was also no statistically significant association between farmers' knowledge of the disease and the category of the farmers.

Farmers' knowledge of recommended management methods for bacterial wilt was also limited. The study further showed that practices of farmers have striking implication for spreading of the disease instead of controlling it. Previous agricultural extension efforts did not have the desirable effect on farmers' knowledge and practices. Therefore, to bridge the gap between scientific knowledge and farmers' understanding and practices, farmers need to learn about the disease and how to manage it through appropriate learning approaches that foster innovations in their local context.

Keywords: Bacterial wilt, disease management, farmers' knowledge, farmers' practices, potato disease

Contact Address: Shiferaw Tafesse, Wageningen University and Research, Centre for Crop Systems Analysis, Wageningen, The Netherlands, e-mail: shiferawtafesse.gobena@wur.nl

How Agroforestry Systems May Impact Pests and Diseases in Robusta Coffee in Ecuadorian Amazonia

KEVIN PIATO[1,3], LINDSEY NORGROVE[1], CRISTIAN SUBÍA GARCÍA[2], FRANÇOIS LEFORT[3]

[1] *Bern University of Applied Sciences, School of Agricultural, Forest and Food Sciences, Switzerland*

[2] *National Agricultural Research Institute, Cocoa and Coffee Research Program, Ecuador*

[3] *Institute Land Nature Environment, HES-SO University of Applied Sciences and Arts Western Switzerland, Life Sciences, Switzerland*

Coffee agroforestry systems could potentially reconcile agricultural, social and environmental objectives, especially in tropical regions, where farmers are generally poor. We assessed how agroforestry and different types and levels of intensification affected pest and disease development on *Coffea canephora* (robusta coffee) trees in Joya de los Sachas, in the Ecuadorian Amazon. The five coffee shading methods assessed were: 1) full sun (no shade); or *Musa* spp. AAB (plantain) at 333 plants ha^{-1} combined with trees of 2) *Myroxylon balsamum*; 3) *Inga edulis*; 4) *Erythrina* spp; or, 5) *Erythrina* spp. and *Myroxylon balsamum*. The four coffee farming practices assessed were: conventional farming at either 1) moderate or 2) intensified input or organic farming at 3) low or 4) intensified input. The experiment was an RCBD thus with 20 treatment combinations, replicated three times. A pyranometer was used to assess shade cover above the coffee in each plot. Infestation of pests and disease incidence of the following were evaluated monthly: the brown twig beetle (*Xylosandrus morigerus*), the coffee leaf miner (*Leucoptera coffeella*), the coffee berry borer (*Hypothenemus hampei*), the anthracnose disease (*Colletotrichum* spp.), the thread blight (*Pellicularia koleroga*) and cercospora leaf spot (*Cercospora coffeicola*). Furthermore, the anthracnose disease severity was assessed with the help of ImageJ.
Agroforestry with *Inga edulis* reduced brown twig beetle infestation by 9 %, compared with in the full sun treatment. Both brown twig beetle and coffee leaf borer infestation were both reduced by 12 % in the intensified organic treatment compared with the intensified conventional treatment. The anthracnose disease severity was found to be only 3 % greater within the intensified organic farming in comparison to the intensified conventional treatment. We conclude that both shade tree treatments and management strongly influence pest infestation levels and diseases incidence and therefore should be considered when selecting optimum management strategies for coffee cultivation.

Keywords: Disease, Ecuadorian Amazon region, pests, pyranometer, robusta coffee, shade

Contact Address: Kevin Piato, Bern University of Applied Sciences, School of Agricultural, Forest and Food Sciences, Route de Bertigny 28, 1700 Fribourg, Switzerland, e-mail: kevin.piato@students.bfh.ch

Rural Female-Headed Households' Perception, Strategies, and Practices of Pest Management in Amhara Region, Ethiopia

BIRTUKAN ASMARE[1], BERNHARD FREYER[1], JIM BINGEN[2], MERKUZ ABERA[3]

[1] *University of Natural Resources and Life Sciences (BOKU), Div. of Organic Farming, Austria*

[2] *Michigan State University, Dept. of Community Sustainability, United States of America*

[3] *Bahir Dar University, Plant Science, Ethiopia*

Studies on farm pest management perception, strategies, and practices with a specific focus on female headed households (FHHs) (single, divorced, widowed) was found to be very scarce. A mixed method approach was applied, integrating diverse qualitative and quantitative surveys including key informant interviews, focus group discussion (FGD), household survey, field observation and photo documentation. We used data from 334 randomly selected respondents from 4 districts of the Amhara Region of Ethiopia. The results reveal that crop pests, disease and weeds are among the constraints for food security with varied intensity and severity across agro-ecologies. As part of FHHs agricultural activities, crop pests, diseases and weeds are managed through agro-chemicals, following the mainstream and socially accepted agricultural practices. Integrated Pest Management (IPM) as a more environmentally friendly, economically viable and socially acceptable approach is uncommon, and rarely promoted by advisory services. The same is with alternative organic farming practices. An excessive and indiscriminate use of pesticides is a cumulative effect of the adoption of various inputs in the 1990s, driven by the Green Revolution (GR), lack of effective alternative methods of pest management, inadequate extension advisory services, and limited institutional control of local pesticide markets. As part of dealing with agro-chemicals, pesticide handling was made in violation of safety practices; with unsafe storage, with little or no use of personal protective equipment (PPE), unsafe disposal and dependence on less expensive but more acutely toxic and environmentally persistent pesticides such as DDT. By applying the knowledge, attitude and practice approach, however, we identified knowledge and perception have detrimental role in influencing pesticide-handling practices. However, high knowledge and risk perception does not always translate into good behavior, rather influenced by socio-economic, institutional and climatic contexts. Therefore, it is vital to understand the various inter-related factors for identifying consequences and recommend more gender sensitive programs and interventions. Therefore, this study suggests the need for action for the reduction of pesticide risks through research, advisory and policy.

Keywords: Environment, Ethiopia, exposure, female headed households, human health

---

Contact Address: Bernhard Freyer, University of Natural Resources and Life Sciences (BOKU), Div. of Organic Farming, Gregor Mendel Straße 33, 1180 Wien, Austria, e-mail: Bernhard.Freyer@boku.ac.at

Effect of *Mentha piperita* and *Thymus vulgaris* Essential Oils on Seed Germination of *Zea mays*

SULEIZE ROCHA TERRA[1], LEANDRO RODRIGUS[1], LETÍCIA BATISTA LACERDA[1], CECÍLIA JOSÉ VERÍSSIMO[2], LUCIANA MORITA KATIKI[3], GERMANO SCHOLZE[4], WALDSSIMILER TEIXEIRA MATTOS[1]

[1]*Institute Animal Science (IZ/APTA), Laboratory of Seed, Brazil*

[2]*Institute Animal Science (IZ/APTA), Lab. of Animal Production Biotechnology, Brazil*

[3]*University of São Paulo, Inst. of Animal Science, Brazil*

[4]*HYG Systems, HYG Systems, Brazil*

Brazil is the world's second largest exporter of corn, surpassed only by the United States (US). The product is recognised for its good quality and for guaranteeing supplies in several countries exactly during the US off-season. To achieve this high production it is necessary to guarantee a high index of vigorous seeds, and this can be reached through seed stimulants. Essential oils can act as germination stimulant. Therefore, the objective of this study was to evaluate the effect of *Mentha piperita* and *Thymus vulgaris* essential oils, at concentrations of 1 %, 5 % and 10 %, on germination of maize seeds. The experiment was conducted with four replicates of 50 seeds each, totaling 200 seeds evaluated. The essential oils were diluted in organic solvent to form the doses. The corn seeds were immersed in the oil solutions for 1 minute, and after drying, they were wrapped in sheets of paper towel. The rolls of paper were packed in plastic bags. The tests were carried out in a germinating chamber, type B.O.D., with photoperiod of 12 hours for a period of 8 days with controlled temperature of 25 ± 2 °C. The evaluations were performed on the 4th and 8th days after the test installation. The results were expressed as a percentage, where it was observed that the *Mentha piperita* oil at the 1 % dose presented the best value with 99 % for normal seedlings, while the percentages of 5 and 10 % presented 92 and 88 % respectively. For the *Thymus vulgaris* oil, it was observed that concentrations 1 and 5 % presented the best values for normal seedlings (100 and 99 %, respectively) in relation to the 10 % dose (88 %). The control treatment showed 100 % of normal seedlings in both tests. It was concluded that 1 % *Mentha piperita* oil and 1 and 5 % *Thymus vulgaris* oil can be used as germination stimulants, since vigorous plants were observed in these treatments, compared to the control group. However, the doses of 10 % for both plant oils are not feasible, since they hinder germination and development of the seedlings.

Keywords: Corn, essential oil, germination seeds

Contact Address: Suleize Rocha Terra, Institute Animal Science (IZ/APTA), Laboratory of Seed, Rua Heitor Penteado 56, 13460000 Nova Odessa, Brazil, e-mail: suleize.terra@gmail.com

Optimising Conidial Concentration of *Isaria cateniannulata* for Management of Common Cutworm *Spodoptera litura* (Fabricius)

PIYACHAT BOONYAPRAPA[1], CHEN BIN[2], PATCHARIN KRUTMUANG[1]
[1]*Chiang Mai University, Entomology and Plant Pathology, Thailand*
[2]*Yunnan Agricultural University, College of Plant Protection, China*

Spodoptera litura (Fabricius) (Lepidoptera:Noctuidae) is a common polyphagous pest infesting several crop species as groundnut, cotton, tomato, cabbage, cauliflower and causing heavy crop losses. To manage this pest in an eco-friendly manner, an experiment was conducted to study effective conidial concentrations of *Isaria cateniannulata* against common cutworm *Spodoptera litula* (Fabricius) larvae. Common cutworm were collected from fields in Kunming province, China. Collected individuals were placed in insect rearing glass tube in laboratory at temperature is 25±1 °C. Six concentrations of conidia (10^3, 10^4, 10^5, 10^6, 10^7 and 10^8 conidia ml^{-1}) along with distilled water as a control were evaluated. Tween 80 was used as adjuvant and the larval populations were dipped in conidial suspensions for 5 seconds. Though all the concentrations tested could infected the larval populations of cutworm, highest mortality of larvae was recorded when larvae were treated with a concentration of 10^8 conidia ml^{-1} of *I. cateniannulata in vitro*. The common cutworm that received 10^7, 10^6, 10^5, 10^4, 10^3 conidia ml^{-1} also showed mortality of the larvae but at a lower level. The mortality of the common cutworm increased with increase in spore concentration and exposure time. The results suggest that *I. cateniannulata* has a potential to be developed into an entomopathogen and further field evaluations and followed by effective formulation development. Furthermore, pest management strategies employing microbial agents offers the environmental advantages. Microbial control of insects is the concerted use of insect-specific pathogens for the biological control of insects. Microbial pesticides have a number of advantages over conventional chemical pesticides. Although the advantages of microbial pesticides are numerous, some of their characteristics are regarded as disadvantages.

Keywords: Cutworm, eco-friendly, Entomopathogens, *Isaria cateniannulata*, *Spodoptera litura* (Fabricius)

Contact Address: Patcharin Krutmuang, Chiangmai University, Entomology and Plant Pathology, 39, 50200 Chiangmai, Thailand, e-mail: patcharink26@gmail.com

Allelopathic *Pseudomonas* Consortium: A Sustainable Weed Control Approach in Wheat (*Triticum aestivum* L.)

Abubakar Dar[1], Zahir Ahmad Zahir[2], Hafiz Naeem Asghar[2], Rashid Ahmad[2], Thomas Hilger[1], Frank Rasche[1]

[1] *University of Hohenheim, Inst. of Agric. Sci. in the Tropics (Hans-Ruthenberg-Institute), Germany*

[2] *University of Agriculture, Faisalabad, Inst. of Soil & Environmental Sciences, Pakistan*

Weeds are notorious biological pests reducing crop production by sharing soil nutrient pools, water and space, and through allelopathy. Global annual crop yield losses by weed infestation range from 20–30 % while in Pakistan *Triticum aestivum* L. production loss due to weeds was estimated up to 24 % annually. Environmental pollution, residual effect, weed resistance, health issues in humans and animals are some limitations of using chemicals. This situation calls for adopting sustainable and eco-friendly approaches (biological weed control). *Pseudomonas* bacteria have the potential to minimise yield losses in wheat due to the production of phytotoxic metabolites (cyanide, phenolics, antibiotics and overproduction of Indole-3-acitic acid), environment friendliness, no residual effects and no chances of weed resistance which makes them a suitable candidate for weed control. The present study aimed to use the potential of pre-isolated and characterised (for secondary metabolites production and growth promoting traits) strains i.e. *Pseudomonas fluorescence* (B11), *Pseudomonas fulva* (T19), *Pseudomonas thivervalensis* (T24) and *Pseudomonas fulva* (T75) for *Avena fatua* L., *Phalaris minor* Retz., and *Rumex dentatus* L. suppression in wheat. Three pot studies were conducted in order to check the weed suppression potential of selected combinations i.e. C9 (B11xT24xT75) and C11 (B11xT19xT24xT75) from axenic studies using two application methods (seed coating and fertiliser coating) following completely randomised design replicated thrice in wire house. The results of the experiments depicted significant weed suppression in terms of reducing germination, SPAD contents, photosynthesis, respiration and stomatal conductance, and wheat growth promotion in terms of SPAD contents, photosynthesis, respiration, stomatal conductance, and grain yield by both combinations over their respective un-inoculated controls. The combination C9 performs significantly better for weed suppression and wheat growth promotion than C11 under seed coating treatment while the results under fertiliser coating of both combinations were statistically at par. The use of these combinations to develop a bioherbicidal product for biological control of *A. fatua*, *P. minor* and *R. dentatus* can be a viable option to meet both food security and the sustainable development goals of the UN.

Keywords: *Avena fatua*, consortium, food security, *Phalaris minor*, *Pseudomonas*, *Rumex obtusifolius*, sustainability

---

Contact Address: Abubakar Dar, University of Hohenheim, Inst. of Agric. Sci. in the Tropics (Hans-Ruthenberg-Institute), Garbenstraße 13, 70599 Stuttgart, Germany, e-mail: abubakardar3206@gmail.com

Effect of some Botanical Oils on the Control of the Cotton Jassid on Eggplant, Sudan

FAIZA SALAH

University of Gezira, Dept. of Crop Protection, Sudan

Eggplant, *Solanum melongena* L., is one of the major vegetable crops in the Sudan. Eggplant is infested by species of insect pests. However, the cotton jassid (*Jacobiasca lybica*) is the major one. This study was conducted to investigate the effects of botanical oils, castor bean seed oil, cotton seed oil and sunflower seed oil on the population of the cotton jassid in eggplant, cultivar Wizzo. Two field experiments were executed during season 2015/16 at two sites, the experimental farm in the Faculty of Agricultural Sciences, University of Gezira, and at Elalafoon on the eastern Bank of the Blue Nile, in Khartoum state. The experiments in both sites were arranged in Randomised Complete Block Design (RCBD) with four replications. The botanical seed oils at a concentration of 5 % were applied at 7 days interval with four sprays. Post-treatment observations on number of jassid were taken after 2 days from the application. The results indicated that all oil treatments were apparently effective against the cotton Jassid compared to the untreated control. The study found that the cotton seed oil was more effective in reducing the number of jassid on eggplant (64 insects/100 leaves) followed by sunflower seed oil (132 insects/100 leaves), castor bean seed oil (140 insects/100 leaves) and the untreated control (312 insects/ 100 leaves) at Wad Medani site while, at Khartoum Site the cotton seed oil recorded high reduction of jassid population (92 insects/100 leaves) followed by castor bean seed oil (140 insects/100 leaves), sunflower oil (208 insects/100 leaves) and the untreated control (608 insects/ 100 leaves). This study suggests that, cotton seed oil can be used to reduce cotton jassid infestation on eggplant crop.

Keywords: Biological pest control, eggplant, insect pest, plant oils

Contact Address: Faiza Salah, University of Gezira, Dept. of Crop Protection, P.O. Box20 Nishishiba, 111111 Wad Madeni, Sudan, e-mail: faizaruba2@gmail.com

Enhancing Potato Production in Kenya through Resistance Based Management of Potato Cyst Nematode

JAMES MWANGI[1], GRACE MWANGI[2], SEBASTIAN KIEWNICK[1]

[1]*Julius Kühn-Institut, Federal Research Centre for Cultivated Plants, Institute for Plant Protection in Field Crops and Grassland, Germany*

[2]*Ghent University, Belgium*

Potato cyst nematode (PCN), *Globodera pallida* and *G. rostochiensis* are potato pests accountable for significant reduction in crop yield. They are common pests in temperate region, however in the last few decades the pests have spread and adapted to tropical environment where they are causing high damage. Symptoms associated with PCN infestation are often mistaken for other abiotic and biotic factors. Therefore the presence of the nematodes in the field may go undetected for a long time. If introduced in a potato field, PCN are difficult to control since they form survival structures known as cysts which protect the eggs and enable them remain viable in the soil for over a decade. *Globodera rostochiensis* and *G. pallida* were first reported in Kenya in 2015 and 2018, respectively. The presence of these nematodes is a major threat to food security considering that potato is the second most important food crop in Kenya. Use of resistant potato cultivars in the management of PCN has the potential to reduce damage and maintain nematode density below the damage threshold. A study was done to investigate the viability of resistant potato cultivars in the management of the Kenyan PCN. The virulence of the Kenyan PCN populations was tested on selected potato cultivars. PCN populations had significantly lower reproduction on resistant cultivars but higher fitness on susceptible cultivar. There was no difference in hatching and penetration of susceptible and resistant cultivars. However investigations revealed that the populations lacked obligatory diapause that is common with all temperate PCN populations. This has a high implication on the management of the pests. In addition, the pathotype(s) of the Kenyan populations was tested in a glasshouse experiment. Similarly, the biology of the populations was studied in resistant and susceptible cultivars and compared with the standard reference population *G. rostochiensis* Ro1 (Ecossee). The study showed that resistance potato cultivars have the potential of reducing the reproduction of the Kenyan PCN and hence boost potato production.

Keywords: Fitness, *Globodera pallida*, *Globodera rostochiensis*, virulence

Contact Address: James Mwangi, Julius Kühn-Institut, Federal Research Centre for Cultivated Plants, Institute for Plant Protection in Field Crops and Grassland, Messeweg 11/12, 38104 Braunschweig, Germany, e-mail: james-maina.mwangi@julius-kuehn.de

Evaluation of Entomopathogenic Fungi Isolates for Control of Sweetpotato Weevils (*Cylas formicarius* Fab.)

Vang Thi Tuyet Loan[1], Tran Van Hai[2], Trinh Thi Xuan[2]

[1]*An Giang University, Crop Science Department, Vietnam*

[2]*Can Tho University, Plant Protection Department, Vietnam*

The objective of this work is to identify five entomopathogenic fungi isolates collected from Hungary using molecular techniques and evaluate their virulence to sweetpotato weevils (*Cylas formicarius* Fab.) in laboratory conditions. Molecular identification was made by sequencing the ITS - rDNA region. Pathogenicity study was based on mortality of adults of *Cylas formicarius* inoculated with *Conidia dasoge* 10^8 conidia ml^{-1}. The results of sequencing of the ITS - rDNA region indicates that the MHU1, MHU2 and MHU3 isolates belong to the clade of *Metarhizium anisopliae*, which was identified as *Metarhizum anisopliae* with the sequences of ITS – rDNA region similar from 99 % to 100 %. Other two isolates MHU4 and MHU5 showed a high degree of siminarity with *Pencillium variabile* (99 %). Laboratory bioassays were conducted to assess the pathogenicity of 4 isolates of *Metarhizium anisopliae* (MHU1, MHU2, MHU3 and MVN1) and 2 isolates of *Penicillium variabile* (MHU4 and MHU5) toward sweetpotato weevils (*Cylas formicarius* Fab.). The *Metarhizium anisopliae*-MVN1 that was isolated from sweetpotato weevils death in Vietnam used as control. The pathogenicity tests show that three *Metarhizium anisopliae* isolates (MHU1, MHU3 and MVN1) caused early adult sweetpotato weevils adults mortality rate from 44.4 % to 50 % at four days after treatment. At six days after treatment, two *Metarhizium anisopliae* (MHU1 and MHU3) isolates were highly virulence on adult sweetpotato weevils adults with the mortalities 92.5 % and 94.2 %, repectively, and same the virulence degree of *Metarhizium anisopliae*-MVN1 isolate (100 % mortality rate). Four isolates *Metarhizium anisopliae* (MHU1, MHU2, MHU3 and MVN1) reached 100 % mortality rate as the same the rate of colony growth on death adult sweetpotato weevils adults at 10 days after treatment. However, two isolates of Penicillium variabiles were not pathogenic to *Cylas formicarius* with from 2.2 % to 3.3 % mortality rate. It is proved that two *Metarhizium anisopliae* isolates (MHU1 and MHU3) have great potential as biological control agents against sweetpotato weevils.

Keywords: ITS - rDNA molecular, *Metarhizium anisopliae*, sweetpotato weevil

Contact Address: Vang Thi Tuyet Loan, An Giang University, Crop Science Department, 18 Ung Van Khiem Street Dong Xuyen Wrad, 90116 Long Xuyen, Vietnam, e-mail: vttloan@agu.edu.vn

Gender Specific Factors Associated with Hazards of Pesticide Usage among Cocoa Farmers in Nigeria

OLAMIDE OLOWOYO[1], OLANIKE DEJI[2], WAHEED SULAIMON[1], EBUNOLUWA FANIYI[1]

[1]*Obafemi Awolowo University, Dept. of Agricultural Extension and Rural Development, Nigeria*

[2]*Humboldt-Universität zu Berlin, Dept. of Gender Research in Rural Areas, Germany*

This study assessed the gender specific factors associated with health and environmental hazards of pesticide usage among cocoa farmers in Ekiti State, Nigeria a region known with predominance in cocoa production. It specifically described the socio-economic characteristics of male and female cocoa farmers and isolate factors associated with health and environmental hazards of pesticide usage in cocoa production. 120 male cocoa farmers and 120 female cocoa farmers were selected through multi stage purposive sampling technique in five local government areas of the state, to make a total of 240 respondents for the study. The study adopts the use of inferential and descriptive statistics to analyse the data from the survey. The mean age for male and female cocoa farmers was 52.5 ± 9.82 and 45.2 ± 8.49 respectively while the mean years of cocoa farming experience for male and female cocoa farmers was 27.9 ± 9.23 and 22.3 ± 9.68 respectively. Varimax factor rotation pattern was used to isolate six factors associated with health and environmental hazards for male cocoa farmers which include: 'information source factor', 'extension contact factor', 'experience factor', 'knowledge-pesticide usage factor', 'group membership factor' and 'family factor' and six factors associated with health and environmental hazards for female cocoa farmers which include: 'knowledge factor', 'information source factor' 'group membership factor' 'extension contact factor' 'family factor' and 'pesticide usage' were isolated. Awareness of the gender specific factors to understanding hidden gender issues associated with hazards of pesticide usage among cocoa farmers was recommended. Hence, this is to ensure sustainable pesticide usage and to boost environmental productivity for rural development.

Keywords: Cocoa farmers, gender specific factors, pesticide hazards, pesticide usage

Contact Address: Olamide Olowoyo, Obafemi Awolowo University, Dept. of Agricultural Extension and Rural Development, Ile-Ife, Nigeria, e-mail: meedey4real@yahoo.com

Information needs and decision support to increase productivity

---

Oral Presentations

Posters

The Role Crop Modelling Can Play in Supporting Diversification of Cropping Systems in Southern Africa

MUNIR HOFFMANN[1], CORRIE SWANEPOEL[2], WILLIAM NELSON[3], DANIE BEUKES[2], MICHAEL VAN DER LAAN[4], REIMUND P. RÖTTER[3]

[1]*Leibniz Centre for Agric. Landscape Res. (ZALF), Landscape Res. Synthesis, Germany*

[2]*Agricultural Research Council (ARC), Inst. for Soil, Climate and Water, South Africa*

[3]*Georg-August-Universität Göttingen, Dept. of Crop Sciences; Tropical Plant Production and Agricultural Systems Modelling (TROPAGS), Germany*

[4]*University of Pretoria, Dept. of Plant Production and Soil Science, South Africa*

Diversifying cropping practices is perceived as an important part of sustainable and productive and resource efficient farming systems in southern Africa. However, there are few field experiments to evaluate the long-term effects (10 years) of the different farming systems (such as fertiliser application, residue management, tillage practices and cropping systems). Given that each practice interacts with the environment (soil and climate), process-based crop modelling could be a useful tool, but has been scarcely applied to investigate long-term effects of cropping practices in southern Africa.

On this background, we first evaluated the widely used crop model APSIM against field data from two environmentally different sites in South Africa: Zeekoegat (heavy clay soil) and Buffelsvlei (sandy soil). Experiments were conducted for six and eight years, respectively, comparing a wide range of cropping practices: fertiliser rate (low, high), system (intercropping, rotation), tillage (reduced and conventional), and crop (cowpea, soy, maize, pearl millet, sunflower, and oats). Field measurements included grain yield, dry matter, N_{min} and soil organic carbon (SOC). APSIM was capable of reproducing the crop growth dynamics at both sites. At Zeekoegat, high yields in the first two years were associated with good rainfall and high available N_{min} resulting from a fallow period prior to the trial. After the two years, unfavourable rainfall patterns and soil compaction reduced yields substantially. At Buffelsvlei, yields were higher, as the sandy soil allowed deeper rooting and more efficient water use than the clay soil at Zeekoegat. Simulated and observed SOC at both sites did not differ strongly from the initial conditions.

Secondly, we used the evaluated model to investigate the effect of increased diversification in cropping practices on profitability, yield, water use, N_{min}, and SOC at both sites over a 10-year period. Simulations showed high diversification is most beneficial in dry years, in particular with pearl millet. While legume integration improved soil fertility and the productivity of the following crop, it may reduce the profitability of cropping at current price levels. These few findings already illustrate the usefulness of crop models to improve management decisions, particularly when data availability is limited.

Keywords: APSIM, intercropping, legumes, maize, rotation

Contact Address: Munir Hoffmann, Leibniz Centre for Agric. Landscape Res. (ZALF), Landscape Res. Synthesis, Eberswalderstr. 84, 15374 Müncheberg, Germany, e-mail: munir.hoffmann@zalf.de

Decision Support Tool to Aid Wetlands Policy Making in East-Africa

ERICK OMWANDHO OPIYO, MATTHIAS LANGENSIEPEN, MATHIAS BECKER

Univeristy of Bonn, Inst. Crop Sci. and Res. Conserv. (INRES), Germany

Wetlands provide good source of natural water, which is available throughout the year. Research conducted in the past, have shown that wetlands are being converted into agriculture productions sites at an alarming rate, this could be attributed to the unproductive state of upland agricultural soils; due over cultivation, unfavorable climate, land segmentation and population pressure. Majority of East-Africa countries are at different stages of formulating wetland policy that will guide wetland conservation and wise use. However, there is inadequate information and knowledge about the extent and scale of wetlands and how they are being utilized. This can be attributed to the lack of a clear universal accepted definition of what wetlands are, since wetlands naturally differ in size, scale, elevation, soil physical and chemical properties, seasonality and vulnerability status; that is the extent to which wetland have been used and converted to other landforms to the detriment of losing their original state. With the limited definition of what wetlands are, some countries are considering broader wetland policy that will cover all possible types of wetlands without putting into consideration, the unique nature of wetlands in relation to their typology. There is also limited input from stakeholders, who might relay directly and indirectly on wetlands as a source of income.

In the GlobE Wetlands Project (www.wetlands-africa.de), a Decision Support Tool (DST) was developed, based on input from wetlands policy makers and stakeholders in the four East-Africa countries; Kenya, Tanzania, Rwanda and Uganda. The DST development process, incorporated views from wetland policy makers and stakeholders, provided open access information on wetlands and other data that directly or indirectly affected wetlands conservation and use. This information was presented in the DST using a decision tree concept, which was developed with input from wetlands experts, discussed and approved by wetland policy makers and stakeholder. It summarized decision making steps in a way that prioritized wetlands conservation and also looked at wise use in the context of biophysical and socio-economical nature of wetlands and its surrounding by factoring in humanitarian, alternative livelihood, typology, land use and land cover changes into the decision making process.

Keywords: Decision making, decision support, East Africa, policy making, software engineering, wetlands

Contact Address: Erick Omwandho Opiyo, Univeristy of Bonn, Inst. Crop Sci. and Res. Conserv. (INRES), Nussallee 1, 53773 Bonn, Germany, e-mail: opiyo@uni-bonn.de

Model Choice Matters – Sensitivity of Chill Metrics to Future Warming in Mediterranean and Temperate Environments

EDUARDO FERNANDEZ, CORY WHITNEY, EIKE LUEDELING

University of Bonn, Inst. Crop Sci. and Res. Conserv. (INRES) - Horticultural Science, Germany

Winter chill, which temperate trees require to overcome dormancy, is expected to decrease substantially in the future in most deciduous fruit tree growing areas. Several mathematical models have been applied in different regions around the world to quantify this chill requirement. While the 'Dynamic' model has emerged as the most plausible and reliable model, chill models are still widely reported to be inadequate for some areas. We compared the outputs from 13 agricultural and forest chill models using past and projected weather data for nine sites in Chile, Tunisia and Germany. We used a weather generator calibrated with 45 years of temperature data to produce chill distributions for 100 synthetic years for multiple climate scenarios. Chill was computed for 10 past scenarios and projected for future scenarios (for 2050 and 2085 according to greenhouse gas concentration scenarios RCP4.5 and RCP8.5). Results show that models' estimations differ substantially for the same site and scenario. The 'North Carolina', 'Utah', 'Modified Utah' and 'Low Chill' models indicated negative chill levels for past and future scenarios in Tunisia. These models, together with the 'Positive Utah' model, projected the greatest chill increases for locations in Germany. The 'Chilling Hours' model and the 'Chilling Rate' function showed high sensitivity across regions in future scenarios. Just two models projected chill decreases in all sites. In Mediterranean climate areas (central Chile and Tunisia) the 'Dynamic' model and 'Positive Utah' model forecasted similar chill reductions in both year and RCP scenarios, whereas in temperate locations (Germany) the 'Dynamic' model forecasted lower chill increase compared with 'Utah' and 'Positive Utah' models. Despite the 'Dynamic' model and the 'Positive Utah' model showing similar performance among climates, the 'Dynamic' model appears to be the best current option, due its more physiologically credible approach. However, further research is needed to develop or identify models that are valid across wide climatic gradients. Such models are crucial for the development of quantitatively appropriate climate change adaptation strategies for temperate orchards.

Keywords: Bud burst, *Prunus* sp., sub-tropics, temperature increase, warm winters

Contact Address: Eduardo Fernandez, University of Bonn, Inst. Crop Sci. and Res. Conserv. (INRES) - Horticultural Science, Auf Dem Hügel 6, 53121 Bonn, Germany, e-mail: efernand@uni-bonn.de

Can Intercropping Increase Climate Resilience of Smallholder Dryland Cropping Systems? Insights from Experimentation and Modelling

William Nelson[1], Munir Hoffmann[2], Vincent Vadez[3], Reimund P. Rötter[1], Marian Koch[1], Anthony M. Whitbread[3]

[1]*University of Goettingen, Dept. of Crop Sciences; Tropical Plant Production and Agricultural Systems Modelling (TROPAGS), Germany*

[2]*Leibniz Centre for Agric. Landscape Res. (ZALF), Landscape Res. Synthesis, Germany*

[3]*International Crops Research Institute for the Semi-Arid Tropics (ICRISAT), India*

With the potential threat of more frequent climate extremes putting semi-arid crop production in jeopardy, there is a need to establish more resilient cropping systems. Intercropping is often practised by farmers in such regions, but to what extent and how can it be a viable option for future conditions? As field testing complex adaptation strategies in well-controlled environments is often difficult, we opted for a different approach: combining experimentation and modelling. A field trial was run in semi-arid India over a two-year period (2015 & 2016) in the dry season. These trials tested a split-plot designed experiment with four replicates, assessing the performance of sole versus intercropped stands, with two densities (30 cm & 60 cm between row spacing), and three drip irrigation treatments (severe stress, partial stress, and well-watered). Under low rainfall conditions, results showed that total grain yields were in-line with the irrigation treatments applied. Intercropping pearl millet led to a significantly lower total grain yield in comparison to the sole equivalent. Pearl millet achieved 1.1 t/ha when intercropped and 2.5 t/ha when grown as a sole crop. Cowpea yielded 0.8 t/ha when intercropped, and 0.8 t/ha as a sole crop. From this study, we can conclude that even when temperatures exceed 43°C crops produce reasonable yields when irrigated. In terms of pearl millet production, sole as opposed to intercrop cultivation could be more suitable. Such experiments during the dry season are arguably an opportunity for testing cropping strategies under extreme but real-world conditions.

Subsequently, we used the above-described detailed data in conjunction with the agro-ecosystem model APSIM. Linking such experimental data with models is important to evaluate models that can be applied to explain and quantify the performance of such systems. Model performance was satisfactory and reproduced the effects of density, irrigation and year on the variables chosen. Plant height proved to be crucial for model evaluation. Simulation experiments were conducted to further evaluate plant densities, as well as genetic traits. Our combined approach is capable of improving intercropping strategies, through understanding the processes that determine interactions between specific environments and management practices.

Keywords: Cereal-legume intercropping, climate-smart, model improvement, sensitivity analysis

Contact Address: William Nelson, University of Goettingen, Dept. of Crop Sciences; Tropical Plant Production and Agricultural Systems Modelling (TROPAGS), Griesebachstrasse 6, 37077 Göttingen, Germany, e-mail: wnelson@gwdg.de

Agronomic Intelligence: Data Management, Modelling and ICT to Deliver Site-Specific Recommendations to Smallholder Farmers

MEKLIT CHERNET TARIKU[1], KODJOVI SENAM EZUI[2], CHRISTINE KREYE[3], STEFAN HAUSER[3], THERESA AMPADU-BOAKYE[2], VERONICA N.E. UZOKWE[4], BERNARD VANLAUWE[1], PIETER PYPERS[1]

[1]*International Institute of Tropical Agriculture (IITA), Kenya*
[2]*African Plant Nutrition Institute, African Cassava Agronomy Initiative, Kenya*
[3]*International Institute of Tropical Agriculture (IITA), Nigeria*
[4]*International Institute of Tropical Agriculture (IITA), Tanzania*

Blanket recommendations for agronomic interventions often bring along substantial risks for smallholder farmers, as responses vary due to heterogeneity in soil conditions, weather, management practices and input-output prices. By considering this heterogeneity, however, site-specific recommendations can be delivered that maximise the profitability and reduce the risk of investments in these technologies. Developing site-specific recommendations at scale requires identifying best agronomic interventions, testing how these perform in field experiments across conditions within the target intervention area, and then using advanced data analysis techniques and crop models to predict crop responses to these interventions. This is then packaged in a convenient format as decision support tools (DSTs) to deliver recommendations to extension agents and farmers. The African Cassava Agronomy Initiative (ACAI) has developed tailored recommendations for cassava growers in Nigeria and Tanzania, responding to demands identified by stakeholders in the cassava value chain. To overcome the limitations of conventional agronomy research, ACAI developed a smart agronomy data management system (SandMan) by combining barcode identifiers and digital data collection forms, a cloud-based server to host data centrally and R scripts for data curation and analysis. This approach allowed managing thousands of field trials and assess the impact of the environment on the efficacy of agronomic interventions. Prediction models were developed to relate responses to tillage, cassava plant density, fertiliser application, intercropping and varied planting and harvest dates to GIS soil and weather layers using a combination of crop models and machine learning techniques. Field-scale recommendations are then generated accounting for crop produce and fertiliser prices, the farmer's investment capacity and risk attitude. Recommendations are packaged in diverse formats tailored to the dissemination strategies of last-mile delivery partners. DST formats include paper guides and maps, as well as electronic tools: mobile apps, IVR, USSD and SMS. Currently, a pilot study involving over 2,000 households and 200 extension agents is ongoing to evaluate the accuracy and return-on-investment of the recommendations, and the user experience of the various formats to anticipate the reach and impact of the recommendations delivered.

Keywords: African Cassava Agronomy Initiative (ACAI), crop models, decision support tools, digital data collection

---

Contact Address: Meklit Chernet Tariku, International Institute of Tropical Agriculture (IITA), Nairobi, Kenya, e-mail: m.chernet@cgiar.org

Validating a Decision Support Tool for Cassava Maize Intercropping in Southern Nigeria

CHRISTINE KREYE[1], CHARLES CHIGEMEZU[2], PIETER PYPERS[3], ADEYEMI OLOJEDE[2], MUTIU BUSARI[4], MARK TOKULA[2], FLORENCE OLOWOKERE[4], ADEMOLA ADEBIYI[1], SHOLA EJALONIBU[2], MEKLIT CHERNET TARIKU[3], STEFAN HAUSER[1], JOHAN SIX[5]

[1]*International Institute of Tropical Agriculture (IITA), Nigeria*
[2]*National Root Crops Research Institute (NRCRI), Nigeria*
[3]*International Institute of Tropical Agriculture (IITA), Kenya*
[4]*Federal University of Agriculture Abeokuta (FUNAAB), Nigeria*
[5]*ETH Zurich, Dept. of Environmental System Science, Switzerland*

Cassava is a major staple food in Nigeria and often intercropped with maize and/or other crops. However, average yields in Nigeria are low, about 10 t ha^{-1}. Thus, farmers request advice on improving cassava-maize cropping. Based on more than 150 researcher and/or farmer managed trials, planted in 2016 and 2017, in Anambra, Benue, Cross River, Ogun and Oyo states, we developed a first version of a decision support tool (DST). It advised on simultaneous planting of both crops, variety choice (erect growth type for cassava and early maturing maize varieties (90 to 95 days), planting density (12500 cassava plants ha^{-1}, 40000 maize plants ha^{-1}) and fertiliser application. The height of the previous maize crop was a proxy for soil fertility. Through statistical modelling, we estimated the number of additional maize ears due to fertiliser application. Fertiliser was recommended when the expected income from the additional maize was twice the estimated fertiliser cost. In 2018, farmers and their extension agents planted 143 validation trials in all 5 states. The extension agent used the DST to derive a specific recommendation. Next, they implemented, 3 plots, testing fertiliser application and maize planting density. Based on the maize harvest from 109 fields, we evaluated the performance of the DST. In 45 % of the cases, we gave a correct recommendation, for 19 % it was wrong with added costs for the farmer (NB: input costs were borne by the project). In 36 %, the recommendation was too conservative: the farmers missed the opportunity of additional profit through fertiliser application. Maize height alone was a poor indicator that was difficult to handle for farmers who use a range of different varieties. Thus, the current revised version of the DST advises planting maize at high density only at sites of high soil fertility or when fertiliser application is recommended. Fertiliser recommendation is coupled with an assessment of height and appearance (including greenness) of a would-be maize crop without fertiliser, fertiliser costs, price expectations for the maize produce and the farmer's risk assessment of the investment into fertiliser.

Keywords: African Cassava Agronomy Initiative (ACAI), agronomy, fertiliser, growth duration, planting density, risk, variety choice

---

Contact Address: Christine Kreye, International Institute of Tropical Agriculture (IITA), PMB 5320, Oyo Rd., Ibadan, Nigeria, e-mail: c.kreye@cgiar.org

Evaluating the African Cassava Agronomy Initiative's Framework for Site-Specific Fertiliser Recommendation

Kodjovi Senam Ezui[1], Meklit Chernet Tariku[2], Veronica N.E. Uzokwe[3], Joy Adiele[4], Adeyemi Olojede[4], Mutiu Busari[5], Mark Tokula[4], Florence Olowokere[5], Rebecca Enesi[6], Ademola Adebiyi[6], Christine Kreye[6], Stefan Hauser[6], Shamie Zingore[1], Bernard Vanlauwe[2], Pieter Pypers[2]

[1]*African Plant Nutrition Institute, African Cassava Agronomy Initiative, Kenya*
[2]*International Institute of Tropical Agriculture (IITA), Kenya*
[3]*International Institute of Tropical Agriculture (IITA), Tanzania*
[4]*National Root Crops Research Institute (NRCRI), Nigeria*
[5]*Federal University of Agriculture Abeokuta (FUNAAB), Nigeria*
[6]*ETH Zürich, Dept. of Environmental Systems Science, Switzerland*

The African Cassava Agronomy Initiative (ACAI) is set to develop decision support tools (DST) to provide advise on site-specific fertiliser recommendations to extension agents and farmers to sustainably intensify and increase cassava production with a focus on commercial farmers. A crop modelling framework comprising two complementary models, LINTUL (Light Interception and Utilisation) and QUEFTS (Quantitative Evaluation of the Fertility of Tropical Soils) was used. QUEFTS is known for understanding N, P and K nutrient interactions and effects on crop production. However, as a static model, QUEFTS is limited in capturing the effect of seasonal weather variability on root yield, which is effectively handled by mechanistic models like LINTUL. Using daily historical weather data from CHRISP (rainfall) and NASA-POWER (solar radiation, wind speed, minimum and maximum temperature), and soil grid data from ISRIC (International Soil Reference and Information Centre), water-limited yields of cassava roots were simulated with LINTUL for a series of planting dates over the year. With the simulated water-limited yield as maximum attainable yields, the site-specific fertiliser rates to achieve target yields were generated with QUEFTS considering indigenous soil fertility, fertiliser price and the cassava root price in order to maximise net returns. Model evaluations using field experiment data from Nigeria in 2016 and 2017 indicated good performance of QUEFTS on the one hand. LINTUL on the other hand performed sub-optimally, especially for planting dates later than July, by overestimating water stress during the long dry season, which lasts from late November to end of March. With additional data collected in 2017 and 2018 from researcher-managed experiments, and from validation trials to test the recommendations, the current efforts are focussing on enhancing the predictions' accuracy and precision for year-round planting dates. Nigerian farmers have moved to stretch planting and harvesting over longer periods to provide roots to the processing industry throughout the year and to benefit from temporal price peaks. With improved fertiliser recommendations, profitable and higher yields can be attained in addition to using off season marketing opportunities.

Keywords: African Cassava Agronomy Initiative (ACAI), decision support tool, LINTUL, QUEFTS, water stress

Contact Address: Kodjovi Senam Ezui, African Plant Nutrition Institute, African Cassava Agronomy Initiative, ICIPE Compound, Kasarani, Nairobi, Kenya, e-mail: g.ezui@apni.net

Cassava Root Yield Response to Tillage Intensity, Planting Density and Fertiliser across South Western Nigeria

Stefan Hauser[1], Rebecca Enesi[1], Felix Salako[2], Omolara Onasanya[2], Mutiu Busari[2], Florence Olowokere[2], Thanni Bolaji[2], Johan Six[3], Pieter Pypers[4], Christine Kreye[1], Bernard Vanlauwe[4]

[1]*ETH Zürich, Dept. of Environmental Systems Science, Switzerland*
[2]*Federal University of Agriculture Abeokuta (FUNAAB), Nigeria*
[3]*ETH Zurich, Dept. of Environmental System Science, Switzerland*
[4]*International Institute of Tropical Agriculture (IITA), Kenya*

Cassava is growing in importance in Nigeria as food security and industrial crop. Current yields are low while production costs are high. Tillage, weed control and fertiliser are the main cost factors. Furthermore, the influence of planting density on yield has not been investigated. To develop decision support tools (DST) targeting a reduction of production costs, effects of tillage intensity, fertiliser application and increased plant density were tested in about 50 farmers' fields in south western Nigeria over two years. Tillage treatments were zero, single and double disc ploughing, followed by ridging versus leaving the soil flat. Fertiliser application was NIL versus 75:20:90 kg ha^{-1} NPK. Plant density was 10000 versus 12500 ha^{-1}. Cassava was harvested 1 year after planting. Root yields are fresh mass. In the first year, double plough (10.34 Mg ha^{-1}) had no advantage over zero plough (10.32 Mg ha^{-1}), single plough produced 12.61 Mg ha^{-1} ($p < 0.0047$). Ridging increased yields by 2.29 Mg ha^{-1} across plough treatments but interacted with ploughing such that in zero plough, ridging added 4.05 Mg ha^{-1}, in single plough 2.84 Mg ha^{-1}, yet no change in double plough. Increasing plant density had a non-significant positive effect (+0.28 Mg ha^{-1}). Fertiliser application increased yields by 4.42 Mg ha^{-1} across tillage treatments with largest increments in zero (+5.31 Mg ha^{-1}) and single plough (+5.58 Mg ha^{-1}), yet no significant plough or ridge × fertiliser interaction was found. In the second year, double plough and low plant density were dropped and herbicide based weed control introduced. Ploughing increased yield by 2.26 Mg ha^{-1}, ridging by 3.31 Mg ha^{-1} and fertiliser application increased root yields by 2.31 Mg ha^{-1}. Using herbicides instead of manual weeding had no effect on cassava yields, yet reduced costs. Root yield had a significant ploughing × ridging interaction with increases when ridged after ploughing being 0.71 Mg ha^{-1} (ns), yet adding 5.92 Mg ha^{-1} when not ploughed and attaining 16.63 versus 16.28 Mg ha^{-1} when ploughed and ridged. Planting on untilled soil produced 10.71 Mg ha^{-1}. Yield increments of +2 (plough) and +1.5 (ridging) Mg ha^{-1} were included into the DST.

Keywords: African Cassava Agronomy Initiative, fertiliser application

Contact Address: Stefan Hauser, International Institute of Tropical Agriculture (IITA), Root & Tuber Cropping Systems Agronomy, Oyo Road, 200001 Ibadan, Nigeria, e-mail: s.hauser@cgiar.org

Exploring Larger Planting and Harvest Phases to Increase Cassava Productivity in South-West Nigeria

REBECCA ENESI[1,2], STEFAN HAUSER[2], PIETER PYPERS[3], JOHAN SIX[1]

[1]*ETH Zürich, Dept. of Environmental Systems Science, Switzerland*

[2]*International Institute of Tropical Agriculture (IITA), Root & Tuber Cropping Systems Agronomy, Nigeria*

[3]*International Institute of Tropical Agriculture (IITA), Kenya*

Cassava (*Manihot esculenta* Crantz) is a perennial crop grown predominantly for its starchy roots in the humid and sub-humid tropics. Seasonal rainfall determines cassava planting dates and farmers have preferred harvest windows based on crop maturity and soil moisture conditions. Cassava being perennial and drought tolerant, the planting and harvesting windows can be expanded to enable year-round production. However, starch accumulation and remobilisation are regulated by physiological processes depending on environmental conditions, nutrient availability and varietal attributes. For sustainable cassava production, adequate supply of nitrogen and potassium is important to support aboveground biomass growth, root starch accumulation and improved tolerance to drought. We investigated: Growing time required to produce profitable root and starch yields; varietal suitability for late planting; NPK effects on starch yield; planting and harvesting dates with best responses to fertiliser. Trials were conducted at 3 sites, with 2 varieties (TMS980581 and TME419), planted in April, July and September, each planting harvested at 9, 11 and 13 months after planting (MAP) and 4 fertiliser levels: control and NPK at 75:20:90, 75:20:135 and 75:20:180 kg ha^{-1}. Starch and root yields were significantly affected by planting and harvest dates ($p < 0.01$) with lowest yields at 9 MAP. Root yields for first planting were 15.2, 22.7 and 26.6 Mg ha^{-1} for 9, 11, and 13 MAP, respectively; for second planting 17.7, 27.7 and 27.5 Mg ha^{-1} for 9, 11, and 13 MAP, respectively and for third planting 20.6, 33.6 and 38.6 Mg ha^{-1} for 9, 11, and 13 MAP, respectively. Starch yields were 2.8, 4.6 and 6.3 Mg ha^{-1} for 9, 11, and 13 MAP, respectively, for first planting, 3.2, 6.6 and 6.6 Mg ha^{-1} for 9, 11, and 13 MAP, respectively for second planting and 4.3, 8.8 and 9.1 Mg ha^{-1} for 9, 11, and 13 MAP, respectively, for third planting. Root yields of TMS980581outperformed TME419 but starch content was higher in TME419. Fertilised plots attained higher root yields than the control across all plantings, without significant effects of fertiliser on starch yield for second and third plantings. Strongest positive yield response to fertiliser application was found at 13 MAP.

Keywords: African cassava agronomy initiative, cassava, fertiliser, variety

Contact Address: Rebecca Enesi, ETH Zürich, Dept. of Environmental Systems Science
current address: International Institute of Tropical Agriculture (IITA), Ibadan, Nigeria, e-mail: R.Enesi@cgiar.org

Sustainable Management Options for Improved Cassava-Maize Intercropping System Productivity and Resource Utilisation in Southeastern Nigeria

CHARLES CHIGEMEZU[1], CHRISTINE KREYE[2], MAGDA NECPLOVA[3], PIETER PYPERS[4], ADEYEMI OLOJEDE[1], SHOLA EJALONIBU[1], STEFAN HAUSER[2], JOHAN SIX[5]

[1]*National Root Crops Research Institute (NRCRI), Nigeria*

[2]*International Institute of Tropical Agriculture (IITA), Nigeria*

[3]*Swiss Federal Institute of Technology, Switzerland*

[4]*International Institute of Tropical Agriculture (IITA), Kenya*

[5]*ETH Zurich, Dept. of Environmental System Science, Switzerland*

Cassava-maize intercropping is a popular, traditionally practised system in southern Nigeria where both crops are staples. Maize offers food and income early in the season (within 3 months) before the cassava harvest (9 months later). However, both crops produce low yields (cassava <10 t ha^{-1} and maize <1 t ha^{-1}) in farmers fields relative to their attainable yields (>48 and >5 t ha^{-1}, respectively). To increase the system's productivity, two intensification options were investigated: i) two N:P:K inorganic fertiliser rates (90:20:40 (F1) vs 75:20:90 (F2)), and ii) increased maize population densities (20000 vs 40000 plants ha^{-1}) on-farm in three agroecologies in Nigeria. F1 and F2 regimes were 300 kg ha^{-1} N:P:K 15:15:15 (basal), two equal splits of urea 3 and 5 weeks after planting (WAP), and 100 % P (TSP basal), three equal splits of urea and MoP at 4, 11 and 17 WAP, respectively. The objectives of the study were to elucidate the effects of: i) F1 and F2 on the growth, development and root yield of cassava and maize cobs, ii) the increased maize population density and fertilisation on solar radiation capture and soil moisture, and iii) the rates of inorganic fertilisers (N:P:K 90:20:40 vs 75:20:90) on cassava root quality (starch yield). Cassava (height, canopy dimension, leaf production) and maize (height, leaf number, visible leave collar per plant) growth were increased by fertiliser application; no effect of maize density was observed. Correspondingly, maize and cassava intercrop yields increased with fertiliser application and were highest at high maize density (>3.5 tha^{-1}) with F1 for maize, and 20 tha^{-1} of cassava roots with F2. Nitrogen uptake was significantly correlated with maize yield and yield components. Highest incident PAR and soil moisture (3 months data) was intercepted and retained, respectively under the two fertilisation regimes with high density maize. F2 was superior over F1 and zero fertilisation in starch yield (>158 kg ha^{-1}) across locations. The use of 40000 maize plants ha^{-1} and inorganic N:P:K fertilisers at the rates and regimes used here was a viable option to increase productivity of cassava-maize cropping in farmers' fields.

Keywords: African Cassava Agronomy Initiative (ACAI), agroecology, fertiliser, soil moisture, solar radiation, sustainable intensification, yield

Contact Address: Charles Chigemezu, National Root Crops Research Institute (NRCRI), Umudike, Nigeria, e-mail: nwokorocc@gmail.com

Improving Cassava Root Yields through Supplementary Dry Season Irrigation and Fertiliser Application

VALENTINE ELETA[1], STEFAN HAUSER[2], EMMANUEL AJAV[1], ADEMOLA AREMU[1], MARIA EGWAKHIDE[2]

[1] *Pan African University, Institute of Life & Earth Sciences, Nigeria*

[2] *International Institute of Tropical Agriculture (IITA), Root & Tuber Cropping Systems Agronomy, Nigeria*

The growing global demand for cassava cannot be met by the current production practices. In sub-Saharan Africa cassava is grown mainly on small holdings by low-income farmers, using little or no external inputs or irrigation. Most of Nigeria experiences a 4–5 months dry season during which cassava sheds leaves and growth ceases. With the onset of rains cassava mobilises starch from roots and stems to form a new canopy, leading to root and starch yield depression, which to compensate for requires additional growing time. Trials were established at IITA Ibadan to assess the effects of supplementary irrigation through the dry season, on the fresh root yield and the response to fertiliser. Two two-factorial experiments were planted with the first factor irrigation regime: I0 (no irrigation) versus I1 (2 mm/day), applied as furrow irrigation; second factor was fertiliser application: F0 (Nil) versus F1 (75:20:90 kg ha $^{-1}$ N:P:K). Irrigation started on December 1st about 1 month after rains stopped. Water was applied every other day by pouring the required amount (equiv. 4 mm) into the furrows between cassava ridges. Trial one was harvested at 12, trial two at 15 months after planting (MAP). Irrigation increased the number of plants reaching harvest, the number of stems and branches and the fresh root yield significantly in both trials. Fertiliser application increased the branch number and the fresh root yield in both trials. Adding 3 months of growing time increased the stem yield and the fresh root yields. Lowest yields were attained at 12MAP without irrigation (16.1 Mg ha $^{-1}$), irrigation added 5.3 Mg ha $^{-1}$ at 12MAP. Harvesting at 15MAP increased root yield to 22.6 Mg ha $^{-1}$ (+6.5 Mg ha $^{-1}$) without irrigation and from 21.4 to 30.3 Mg ha $^{-1}$ (+8.8 Mg ha $^{-1}$) when irrigated. This indicates that maintaining water supply through the dry season affects root yields positively even after the onset of rains. Fertiliser application had no effect when harvested at 12MAP but added 6 Mg ha $^{-1}$ when harvested at 15MAP. Generally a longer growing time increased effects on root yield of irrigation and fertiliser application.

Keywords: Cassava, dry matter, fertiliser application, irrigation, root yield

Contact Address: Stefan Hauser, International Institute of Tropical Agriculture (IITA), Root & Tuber Cropping Systems Agronomy, Oyo Road, 200001 Ibadan, Nigeria, e-mail: s.hauser@cgiar.org

A Researcher Friendly Digital Sample Tracker for Cost Effective Sample Processing

Ademola Adebiyi[1], Meklit Chernet Tariku[2], Christine Kreye[1], Stefan Hauser[1], Veronica N.E. Uzokwe[3], Kodjovi Senam Ezui[4], Pieter Pypers[2], Ouma Turry[2]

[1]*International Institute of Tropical Agriculture (IITA), Root & Tuber Cropping Systems Agronomy, Nigeria*

[2]*International Institute of Tropical Agriculture (IITA), Kenya*

[3]*International Institute of Tropical Agriculture (IITA), Tanzania*

[4]*African Plant Nutrition Institute, African Cassava Agronomy Initiative, Kenya*

Collecting plant and soil samples is an integral part of agricultural research. Large scale research projects collect soil and plant samples from a large number of trials to be processed in different ways, analysed at different laboratories which may result in loss of entire samples or losing track of samples eventually causing data gaps. The African Cassava Agronomy Initiative (ACAI) is introducing an efficient sample tracker system that records a bar code based digital link between samples, plots, and trials, thus is crucial for success of the project. ACAI is developing decision support tools (DST) at scale with its partners in Nigeria and Tanzania and has been collecting thousands of soil and plant samples over the last three years. The integrated sample tracker system creates a digital identity for samples and keeps record of the associated plot and trial metadata. This is facilitated by a barcoding system, which tags each trial, plot, plant and sample with linked bar codes to provide every sample with a unique identifier. The sample tracker consists of 2 web-based forms hosted by ONA. The first an enketo web-form, is designed to i) log in all new samples by composing sample batches containing a group of similar samples; ii) log and record the fate of each sample during its life cycle in the project: reception as part of a batch from partner stations, selection of samples from a batch for processing. The second web-based form is a shinyapp (an interactive web app built from R) in which the decision is made to discard, store or process for analysis. These decisions are based on relevant information about the sample from the project database: trial type, location, sampling dates, validity of trials, sample quality, etc. Thus, this sample tracker tool enables researchers to know where the sample is located and at which stage of processing or analysis the samples are, at any moment. As such the system helps the project to save on handling and processing labour, time and costs by limiting all processing and analyses to only those samples relevant to creating the decision support tools.

Keywords: African Cassava Agronomy Initiative (ACAI), barcoding, database, decision support tools, enketo webform, identifier, metadata, shinyapp

Contact Address: Ademola Adebiyi, International Institute of Tropical Agriculture (IITA), Root & Tuber Cropping Systems Agronomy, Oyo Road, 200001 Ibadan, Nigeria, e-mail: a.adebiyi@cgiar.org

Impact of Planting and Harvest Dates on Root and Starch Yield and Soil Physical Properties

OLOGUNDE OLANREWAJU[1], OLUWASEGUN ADEBAYO[1], MUTIU BUSARI[2], STEFAN HAUSER[1], CHRISTINE KREYE[1], FELIX SALAKO[2], PIETER PYPERS[3]

[1]*Federal University of Agriculture Abeokuta, Soil Science and Land Management, Nigeria*

[2]*Federal University of Agriculture Abeokuta (FUNAAB), Nigeria*

[3]*International Institute of Tropical Agriculture (IITA), Kenya*

Availability of cassava roots for human consumption or the processing factories is controlled by the vagaries of the harvesting time. This study was carried out to examine the effect of varying planting and harvesting dates on cassava growth parameters, root yield, starch content and soil properties. The experiment was a split-split-plot design with planting periods (April, June and August), varieties (TME 419 and TMS 30572) and harvesting dates (9, 11 and 13 months after planting (MAP) as main plot, sub plot and sub-sub plot, respectively, replicated three times. The fertilised plots received NPK 15-15-15, urea and MOP fertiliser at rates equivalent to 75 kg N, 20 kg P, and 90 kg K ha^{-1}. TME 419 produced a mean root yield of 15.2 t ha^{-1}, about 29.7 % higher than 13.0 t ha^{-1} from TMS 30572. The starch content was 20.05 % in TME 419 and 18.45 % in TMS 30572. Cassava root yield and starch content increased significantly ($p < 0.05$) with later harvest dates. The highest root yield and starch content were recorded at 13 MAP. At 13 MAP TME 419 produced 23.95 t ha^{-1} and TMS 30572 produced 21.35 t ha^{-1} compared the root yield at 11 MAP and 9 MAP harvests. At 13 MAP harvest, the starch contents was 20.84 %. This was higher than 20.58 % and 16.32 % at 9 MAP and 11 MAP, respectively. Though not statistically different, the lowest soil bulk density (1.32 g cm^{-3}) and highest total porosity (50.10 %) was found at 9 MAP harvest. Soil under TMS 30572 had higher saturated hydraulic conductivity than soil under TME 419 with mean of 34.58 (cm hr^{-1}) and 32.17 (cm hr^{-1}), respectively. Likewise, the Clay Dispersion Ratio (CDR) decreased with harvesting periods with 13 MAP harvest having the lowest CDR of 2.41 %. Therefore, harvesting of cassava at 13 MAP is considered more productive, irrespective of variety. The selection of the variety to be cultivated should be guided by the intended cassava root sales to fetch higher prices based on root yield or starch content. Also, changes of soil properties should be considered to support soil conservation measures and avoid rapid fertility and physical degradation.

Keywords: African Cassava Agronomy Initiative, extended planting, harvesting date, root yield, starch content

Contact Address: Ologunde Olanrewaju, Federal University of Agriculture Abeokuta, Soil Science and Land Management , Abeokuta, Nigeria, e-mail: ollylarry@gmail.com

Influence of Fertiliser Types and Rates on Soil Nutrients, Growth and Yield of Cassava

Blessing Oyebade[1], Azeez Jamiu[1], Kodjovi Senam Ezui[2], Joy Odedina[1], Mutiu Busari[1], Stefan Hauser[3], Felix Salako[1], Florence Olowokere[1], Pieter Pypers[4]

[1] *Federal University of Agriculture Abeokuta (FUNAAB), Nigeria*

[2] *African Plant Nutrition Institute, African Cassava Agronomy Initiative, Kenya*

[3] *International Institute of Tropical Agriculture (IITA), Root & Tuber Cropping Systems Agronomy, Nigeria*

[4] *International Institute of Tropical Agriculture (IITA), Kenya*

Cassava yield response to fertiliser is often erratic. A trial was conducted at the Federal University of Agriculture, Abeokuta, Ogun State, to evaluate the effect of fertiliser rates on growth and yield of cassava and the changes in soil chemical properties. The experiment was a randomised complete block design with 2 soil sampling depths (0–20 cm and 20–50 cm) and 11 fertiliser rates: Control (no fertiliser), full rate of NPK, ½ NPK, NPK + Micro-N, NP, NK, PK, ½ N + PK, ½ P + NK, ½ K + NP and ½ K + P replicated thrice. Cassava variety TME 419 was planted at 1 m $\times$ 0.8 m spacing. Fertiliser was applied in split, with basal application of 100 % P, while N and K were applied at 4, 10 and 16 weeks after planting, respectively. Soil samples were taken with an auger to determine: ammonium, nitrate, phosphorus and pH. Simultaneously, plant height, stem girth, canopy length, canopy width and number of leaves were measured and recorded. Data obtained were subjected to Analysis of Variance (ANOVA) using Genstat and significant means were separated using Duncan's Multiple Range Test at 5 % probability level. Ammonium concentration of plots receiving ½ P + NK was significantly ($P \leq 0.05$) higher than other fertiliser treated plots. Nitrate concentrations were significantly ($P \leq 0.05$) increased in plots treated with full rate of NPK. Significant ($P \leq 0.05$) decreases in soil pH were recorded mostly across weeks with respect to fertiliser application. Fertiliser treatments had significant ($P \leq 0.05$) effect on stem girth, canopy length and width. Highest number of leaves was produced on plots treated with ½ NPK. There was a significant ($P \leq 0.05$) increase in dry weight of leaves in plots treated with P and K while least was produced on plots treated with ½ K + NP. Dry weight of stems was significantly ($P \leq 0.05$) increased in plots treated with full rate of NPK and ½ K + P while least was produced on plots treated with NPK + micro-nutrients. Dry weight of roots was significantly higher ($P \leq 0.05$) in plots treated with ½ P + NK.

Keywords: African Cassava Agronomy Initiative, fertiliser rates, root yield

Contact Address: Blessing Oyebade, Federal University of Agriculture Abeokuta, Dept. of Soil Science and Land Management, No 10, Ire-akari avenue, Osiele, Abeokuta, Nigeria, e-mail: oyebadeblessing@gmail.com

Tracking the African Cassava Agronomy Initiative Project Theory of Change to Establish the Actual Change

THERESA AMPADU-BOAKYE[1], SABURI ADEKUNBI[2], PIETER PYPERS[1]

[1]*International Institute of Tropical Agriculture (IITA), Monitoring and Evaluation, Kenya*

[2]*International Institute of Tropical Agriculture (IITA), Monitoring and Evaluation, Nigeria*

The African Cassava Agronomy Initiative (ACAI) seeks to improve benefits of cassava farmers through generation of knowledge and development of decision support tools (DST), packaged in formats suitable for extension agents and farmers to apply site specific recommendations in Tanzania and Nigeria. To enable the measurement, the integration of feedback, and to overcome the common MEL approach of focusing on outputs and the time lag between measuring outputs and impacts, a well-structured, gender-inclusive monitoring, evaluation and learning (MEL) system has been developed. The MEL system has two components: i) routine data collection to obtain the necessary information to support steering the project, and ii) impact evaluation, focusing on changes effected by project activities. To ensure attribution of results, a baseline study was conducted in Tanzania and Nigeria, in which unique ID cards were given to participating households in the project target areas. These ID cards enabled easy identification of the households for subsequent panel data collection, serving to identify critical entry points to achieve impact before the end of the project. In Nigeria and Tanzania, a total of 2113 households (40 % female) were given the unique ID card. A Knowledge, Attitude, Behaviour and Practices (KABP) survey was conducted among 334 extension agents (EAs) to identify their current knowledge status regarding the DST's recommendations and relationships with farmers. Data collection focuses on five issues; i) awareness creation, ii) insights gained by farmers, iii) uptake-use of the DSTs, iv) adoption-changes in practices and behaviour and v) ultimate benefits/impacts). Data collection is done routinely on awareness creation (reach) and feedback on insights gained, while the panel survey is bi-annual, focusing on uptake and adoption. To have quick feedback from users (farmers and EAs) on understanding of the recommendations, quick feedback questions have been integrated in the various formats of the DSTs (paper-based and electronic). All MEL data are collected using Information and Communication Technologies (ICT) with programmed data collection tools directly linked to an approved ICT platform. This system already established reference points for project indicators, provides data for reporting and follow up systems on beneficiaries.

Keywords: African Cassava Agronomy Initiative (ACAI), attribution, baseline, feedback, MEL

Contact Address: Theresa Ampadu-Boakye, International Institute of Tropical Agriculture (IITA), Monitoring and Evaluation, Nairobi, Kenya, e-mail: T.Ampadu-Boakye@cgiar.org

Water Availability and its Impact on Cropping Intensity Patterns of Rice-Based Systems in Southeast Asia

CHRISTIN HECKEL, CARLOS ANGULO

University of Bonn, Inst. Crop Sci. and Res. Conserv. (INRES) - Plant Nutrition, Germany

In the frame of the BMBF funded RICH-3P project coordinated by the University of Bonn, agricultural changes in cultivation patterns of rice-based systems in Southeast Asia have been investigated. Six (representative) sites were considered: in Cambodia Takeo and Svay Rieng, in Myanmar Meiktila and Mawgyun and in the Philippines Pangasinan and Muñoz. Among other changes, a shift from single rice (wet season) towards double rice cropping (wet and dry season) per year was found at five out of six sites. For instance, in the Central Dry Zone in Meiktila-Myanmar 108 farmers (out of 160 interviewed) indicated to produce rice twice a year in the present (2018) whereas in the past (baseline 2000) none of them practiced double rice cultivation. This development was made possible through increasing water availability (amount and timing) in the respective regions. The present study (work in progress) aims to evaluate the external pressures and the (farming system) internal drivers leading to the sustained change of single to double rice cropping. On the one hand we consider factors such as improving or installing irrigation public infrastructure, cooperative work on communal irrigation systems, extension campaigns and aid in the implementation of swallow dwells of pumps, etc., which might explain the increase of water availability for the farmers. On the other hand we investigate the thresholds of water amount and availability leading the farmers to definitely implement a second rice crop. For these purposes, apart from the evaluation of secondary sources, our work focuses on the acquisition and evaluation of optical satellite data from 1990 onwards (Landsat, Sentinel-2) with the support of radar data (Sentinel-1) in the corresponding study areas. A field-wise evaluation (polygon- and pixel-wise) of spatial and temporal water cover patterns during the land preparation and pre-flowering stages in the study areas will serve as proxy for the assessment of water availability and the determination of thresholds that have induced (or are inducing) farmers to move from single to double rice cropping in the region.

Keywords: Agricultural change, irrigation, remote sensing, rice

Contact Address: Christin Heckel, University of Bonn, Inst. Crop Sci. and Res. Conserv. (INRES) - Plant Nutrition, Bonn, Germany, e-mail: christin.heckel@uni-bonn.de

Using Crowdsourcing and Machine Learning for Predicting the Spatial Distribution of Banana-Based Cropping Systems in Uganda

DENNIS OCHOLA[1], GODFREY TAULYA[2], GERRIE VAN DE VEN[1], KEN GILLER[1]

[1] *Wageningen University & Research, Plant Production Systems Group, The Netherlands*

[2] *International Institute of Tropical Agriculture (IITA), Uganda*

Uganda is the leading global producer of highland bananas (*Musa* spp. AAA) endemic to East Africa. For decades, expert opinion has been the source of information on the spatial distribution of banana-based cropping systems in Uganda. Lack of accurate and reliable spatial data undermines strategic planning and sustainable intensification at various scales. This study uses 18,956 crowdsourced presence-absence data coupled with geospatial data from 71 covariates (21 climatic, 19 edaphic, 19 vegetation, 6 topographic and 6 socio-economic) to predict the spatial distribution of banana-based cropping systems using the machine learning algorithms Random Forests (RF), Gradient Boosting Machines (GBM) and Neural Networks (NNET). Performance of RF and GBM was better than NNET in terms of accuracy, receiver operating characteristic (ROC) and sensitivity. But, NNET performed better with regards to Cohen's kappa and specificity. The ensemble model aggregat ing outcomes of RF, GBM and NNET performed better (AUC = 0.881) compared to the logistic regression model (AUC = 0.852). Spatial predictions revealed that banana-based cropping systems occupied 9.6% of the total land area of Uganda. The probability of banana presence was greater (> 0.6) in the western (i.e. Ankole, Toro and foothills of Mt Rwenzori), central (i.e. Buganda in Kooki and Buddu) and eastern (i.e. foothills of Mt Elgon), and least (< 0.2) in the northern region. Geographic shifts are defined by declines in the eastern (-13.4%), stagnation in the central (-4.3 %) and expansion in the western (+17.3 %). Although machine learning can iteratively search and filter through covariates to achieve high prediction accuracy, including redundant covariates in the best-fit model may not explicitly describe prediction outcomes. Thus, hypothesis-based selection of covariates with known influence on banana growth and agronomic management is a better option for identifying the drivers of geographic shifts of banana-based cropping systems in Uganda.

Keywords: Banana, banana-based cropping systems, crowdsourcing, geographical shift, machine learning, Uganda

Contact Address: Dennis Ochola, Wageningen University & Research, Plant Production Systems Group, Wageningen, The Netherlands, e-mail: dennis.ochola@wur.nl

Early Detection of Rice and Cotton in Irrigated Indus Basin Using Optical and Synthetic Aperture Radar (SAR) Data

TALHA MAHMOOD, MUHAMMAD USMAN, NIMA AHMADIAN, CHRISTOPHER CONRAD

University of Wuerzburg, Geography Department / Remote Sensing Unit, Germany

Accurate information on areal and yield mapping is very important during the current growing season for effective agricultural monitoring. The aim of this study is to map accurate information on rice and cotton in Lower Chenab Canal Command (LCC) area during the Kharif cropping season. Rice and cotton are two major crops of the region in Kharif season which contribute to a lot of agricultural exports and raw material to the textile industry. The study presents the integrated use of multitemporal optical and Synthetic Aperture Radar (SAR) data with Random Forest (RF) for early mapping of rice and cotton. Sentinel 1 and Sentinel 2 data was used for optical and SAR data respectively. The optical data involves normalised vegetation Index (NDVI) and all bands except bands 1,9 and 10 which are not important for crop mapping. The SAR features include Sigma0 VV and VH intensity and Gray level co-occurrence (GLC) texture. The incremental classification with RF was employed which performs classification at every new instance (availability of new satellite image). The classification performance of the crops at every classification was assessed with F1 scores value. F1 score of 0.80 was used as threshold for early mapping of rice and cotton. Kharif season ranges from end April or early May to end October. The results reveal that the rice area can be map as early as July with 0.82 F1 scores. Similarly, cotton crop area can be mapped accurately with 0.81 F1 scores in mid-September. This early mapping of major crops in the area can be very useful for decision making and planning. Further investigations should be carried out with more features and object-oriented classification to map these crops at earliest.

Keywords: Early mapping, Indus basin, random forest, synthetic aperture radar

Contact Address: Talha Mahmood, University of Wuerzburg, Geography Department / Remote Sensing Unit, Zollrain 5a, 06124 Halle, Germany, e-mail: talha.mahmood@stud-mail.uni-wuerzburg.de

The Use of Aerial Photographs as a Tool for Landscape Analysis in an Oil Palm Plantation

ADRIANA GOMEZ[1], STEPHANIE DOMPTAIL[1], RAINER WALDHARDT[2], SANTIAGO CHIQUITO[3]

[1]*Justus Liebig University Giessen, Inst. of Agricultural Policy and Market Research, Germany*

[2]*Justus Liebig University Giessen, Research Centre for Biosystems, Land Use and Nutrition, Germany*

[3]*Photographer and consultant, Colombia*

In Colombia, the oil palm sector is growing as an important economic sector. Its challenge is to ensure the growth of the palm oil production and maintain or improve the environment and the local landscape. The Macondo oil palm plantation is located in in Mapiripán, Meta, and is 5800 ha large. Landscape analysis is done to show the diversity of natural elements, the complexity and connectivity of the oil palm plantation. The aim of this paper is to explore the use of aerial pictures in landscape analysis. Using the aerial pictures to reduce time in ground-truthing, and simultaneously, collecting useful information for landscape analysis. Besides, it allows us to obtain "real time" pictures in spatio-temporal scales. The landscape analysis was based on a GIS analysis of the land cover maps of 2009. Using landscape metrics, we applied a raster of squares of 500x500 meters over the plantation area. The squares were classified using the cluster analysis according to similarities in land cover into 9 groups. Four squares per group were selected to obtain 36 squares. Using a DJI drone, Mavic Pro 2, camera 20 mpx, RGB, we took over 10800 aerial photophraphs at a height of 100 meters to create the aerial map for each selected square. The landscape analysis showed that the plantation as a landscape can be divided into 9 different groups. They differ in terms of diversity in natural elements and land cover. The pictures realised by the drone enabled us to conduct a quick and effective "real-time" ground-truthing. Furthermore, using the aerial maps allowed to observe the horizontal structure of the landscape for 2019 and contrast information of 2009. Furthermore, working with good resolution 3D models and landscape indices it could give significant results on the vertical structure. Thus, using drone-made pictures is relevant for science since it provides landscape detailed information in a reduced time range. Therefore, the pictures can help with landscape monitoring within the plantations, and identify the related ecosystem services. This characterisation allows the comprehension of the landscape and the local ecosystems.

Keywords: Aerial photographs, cluster analysis, landscape analysis, landscape metrics

Contact Address: Adriana Gomez, Justus Liebig University Giessen, Inst. of Agricultural Policy and Market Research, 35390 Giessen, Germany, e-mail: adriana.gomez@agrar.uni-giessen.de

A Decision Tree for Reconciling Human Needs with Conservation in East-African Wetlands

MATTHIAS LANGENSIEPEN[1], ERICK OMWANDHO OPIYO[1], ALICE KAUDIA[2], MATHIAS BECKER[1], RICHARD KYAMBADDE[3], JULIUS K. NINGU[4], DANIEL P. ASHITIVA[5], THEOGENE NGABOYAMAHINA[6]

[1]*University of Bonn, Inst. Crop Sci. and Res. Conserv. (INRES), Germany*

[2]*Ecoentrepreneurs Ltd, Environment Restoration and Climate Change, Kenya*

[3]*Ministry of Water and Environment, Wetland Department, Uganda*

[4]*Tanzania Tobacco Board, Director General, Tanzania*

[5]*National Environment Management Authority, Principal Wetland Officer, Kenya*

[6]*Rwanda Environment Management Authority, Rwanda*

Hunger and undernourishment are still widespread in Africa and demographic growth is currently the highest in the history of the continent thereby drastically increasing food demand in the coming decades. Inappropriate agricultural practices have lead to degradation of upland soils and farmers are seeking alternative areas for crop production. African wetlands have therefore been encroached and converted to farmland in recent years. This is due to their favourable ecological production conditions in spite of environmental legislation that actually aims at preventing agricultural use. Large wetland areas of Africa will be destroyed if that trend continues in the future. Land-use planning and policy action should be adapted to prevent further destructions of wetlands. Decisions to protect or use wetlands for agriculture must necessarily be knowledge-driven. A collaborative research project GlobE Wetlands (wetlands-africa.de) developed an integrated understanding of wetland characteristics, functions and importance across East-Africa during the past five years. A decision tree has been developed in cooperation with environmental and agricultural policy makers across East-Africa to take informed decisions about future wetland uses in this region. They include assessments of general country-wide conservation issues (Level 1), identification of wetlands located in hotspots of hunger, negative food supply-demand synchrony and poverty (Level 2), exploration of alternative livelihood options making wetland use unnecessary (Level 3), a method for building a wetland typology for protecting fragile wetlands and selecting those which are potential candidates for usage (Level 4), comprehensive ecosystem-services assessments for each identified wetland-type (Level 5) and planning guidelines for defining use and protection zones for wetlands which may be considered for agricultural protection (Level 6). The tree supports the identification of socio-ecological niches in East-African wetlands.

Keywords: Agricultural policy, agriculture, conservation, decision tree, East-Africa, environmental policy, food-demand, hunger, socio-ecological niches, wetlands

---

Contact Address: Matthias Langensiepen, University of Bonn, Inst. Crop Sci. and Res. Conserv. (INRES), Karlrobert-Kreiten-Str. 13, 53115 Bonn, Germany, e-mail: matthias@langensiepen.net

Maize Yield Gains from Seasonal Forecasts Using the CCAFS Regional Agricultural Forecasting Toolbox (CRAFT)

Andree Nenkam Mentho[1], P. C. Sibiry Traoré[1], Vakhtang Shelia[2], Sridhar Gummadi[3], James Hansen[4], Anthony M. Whitbread[5], Gerrit Hoogenboom[2]

[1]*International Crops Research Institute for the Semi-arid Tropics (ICRISAT), Mali*

[2]*University of Florida, Agricultural and Biological Engineering Department, USA*

[3]*International Rice Research Institute (IRRI), Philippines*

[4]*International Research Institute for Climate and Society, United States of America*

[5]*International Crops Research Institute for the Semi-arid Tropics (ICRISAT), India*

The 'maizification' of Sudan Savannah associated with agricultural intensification increases smallholder farmers' exposure to inter-annual rainfall fluctuation related risk, hence rising their dependence on agricultural inputs, as well as on weather and climate information. However, agricultural input allocation remains a serious issue because farmers do not have access to the required quantity at the right time, thus jeopardising crop productivity and consequently food security. The integration of weather and climate data and crop modelling with the CCAFS Regional Agricultural Forecasting Toolbox (CRAFT) offers new insights into regional yield forecasting for maize and presents a great opportunity to optimise resource allocation and thus improve food security. CRAFT is a multi-model gridded framework that provides access to regional agricultural production at approximately 10 km resolution with up to three administrative levels for the region of interest. CRAFT is being piloted in Southern Mali to assess the sensitivity of the maize cropping system to recent climate variability, to evaluate its potential for maize regional yield forecasting and to assess its benefit for agronomic decision-making. The study uses one-year farmers' observed management practices and 26 years (1990–2015) observed yield aggregated to the third administrative unit of the region of study to calibrate and evaluate the crop models employed in the framework. CRAFT performance was tested for four fertiliser management scenarios. Our results show high variability in predicted yield with NRMSE of 27 %, 25 %, 13 % and 17 % kg ha^{-1}, respectively. This result further suggests that fertiliser application should be based on seasonal rainfall amount for its optimal use when there is sufficient rain for crop growth and associated fertiliser requirements. The performance of CRAFT for maize yield forecasting is still being explored. Nonetheless, CRAFT has the potential to improve decision making at regional level through the optimisation of agricultural resource input allocation and food security improvement when yield forecasts show that food demand will not be met.

Keywords: CRAFT, gridded crop modelling, maize, Mali, yield forecast

Contact Address: Andree Nenkam Mentho, International Crops Research Institute for the Semi-arid Tropics (ICRISAT), Icrisat Bp 320, 91093 Bamako, Mali, e-mail: a.nenkam@cgiar.org

Seed Selection Strategies in a Sparse Social Network in Rural Zambia: An Empirical ABM Approach

BEATRICE NÖLDEKE, STEVEN GRONAU, ULRIKE GROTE, ETTI WINTER
Leibniz University Hannover, Institute for Environmental Economics and World Trade, Germany

Identifying the optimal set of seeds (farmers who obtain information initially) in social networks poses an important question for policy-makers and organisations in developing countries where information is commonly spread through word-of-mouth communication. Widespread provision of adequate information to farmers is particularly important in the context of innovation adoption which has high potential to improve farmers' productivity and adaptation abilities. This paper systematically evaluates different strategies for seed selection with the aim of optimising the seed set to improve knowledge diffusion in a sparse social network in a case study area of rural Zambia. The seed selection strategies include random, hierarchy (village heads), betweenness, closeness, degree, and eigenvector based choice. In addition, the effect of the number of seeds on the diffusion process is investigated. To test for robustness, the study includes the assessment of interaction effects between seed size and seeding strategy. An agent-based model adjusted to a case study area in rural Zambia is applied.
Key findings show that degree based seed selection performs best out of the selection strategies in terms of speed and reach of the diffusion in the sparse network. Also farmers with high betweenness centrality who function as bridges to connect network components enhance the diffusion. Eigenvector and closeness based seed selection do not result in a widespread reach. Further simulations shows that higher seed size increases reach, but this effect is more pronounced for small seed sizes and in the short run. These results highlight the importance of taking the time frame into account when planning diffusion of information. While for random, degree, and betweenness based seed selection the results are robust, the simulation shows that closeness and Eigenvector based seed selection significantly improves if the seed size is enlarged. All in all, the findings indicate that seed selection according to degree centrality is a good starting point to improve the diffusion process and that seed sizes should not be too small to ensure widespread reach, but increasing large seed sizes further will only result in marginal benefits.

Keywords: Agent-based modelling, information diffusion, seeding, sparse social networks

Contact Address: Beatrice Nöldeke, Leibniz University Hannover, Institute for Environmental Economics and World Trade, Am Königsworther Platz 1, 30167 Hanover, Germany, e-mail: noeldeke@iuw.uni-hannover.de

Canopy Cover Evolution of Banana Plantations: Drought Effects, Diurnal Patterns and Leaf Area Index Relationships

BERT STEVENS[1], JAN DIELS[2], ELINE VANUYTRECHT[2], PATRICK NDAKIDEMI[3], ALLAN BROWN[4], ALVIN RUJWEKA[3], RONY SWENNEN[5]

[1]*KU Leuven, Dept. of Biosystems (BIOSYST), Belgium*

[2]*KU Leuven, Dept. of Earth and Environmental Sciences, Belgium*

[3]*Nelson Mandela African Institution of Science and Technology, Dept. of Sustainable Agriculture and Biodiversity Conservation (SABE), Tanzania*

[4]*International Institute of Tropical Agriculture, Tanzania*

[5]*KU Leuven / IITA / Bioversity International, Belgium*

Water shortage is considered the most limiting abiotic factor to banana production, but reliable physiological indicators for stress and irrigation necessity are lacking for banana plants. Canopy Cover (CC) can be used as an indicator of growth, but CC research in banana plantations is lacking. This paper is the first to study the effect of irrigation on CC in banana plantations. Two cultivars (cv), Mchare Huti Green (AA) (HG) and Cavendish-Grand Naine (AAA) (GN), were studied over a full crop cycle for two irrigation treatments: full irrigation (FI) as control, and deficit irrigation (DI) as drought. Soil moisture was measured daily in every cv-treatment plot using TDR sensors. CC was determined through monthly drone images obtained at hourly intervals (between 8 AM and 16 PM) of every cv-treatment plot. Linear mixed models were used to determine the effect of time and treatment on CC, CC growth curves where obtained through non-linear logistic regression. Monthly leaf area index (LAI) values were related to CC to obtain CC-LAI curves. Drought reduced growth rates (r) and CC_{max} values. CC_{max} values for FI were similar in both cultivars (89 %), but time to reach CC_{max} was longer in HG (50 WAP) than in GN (32 WAP). CC differences between FI and DI ranged from 2.22 % until 17.67 % for HG, while differences ranged from -2.5 % to 4.5 % for GN. Soil moisture values differed significantly in the vegetative stage for HG, explaining this divergence. Diurnally, CC drops occurred on days characterised with a high evaporative demand. CC had a maximum in the morning, dropped to a minimum around noon, and climbed again in the afternoon, with drops being more significant in DI than FI plots, when moisture values differed significantly. Relating CC and LAI, showed CC-LAI curves followed an exponential curve reaching maxima CC at LAI 4 for both cvs. CC curves are cultivar dependent, and CC is reduced by a lack of soil moisture. Comparison with optimal CC curves may pinpoint stress periods. On a daily basis, relative CC drops cannot be used to assess drought stress, as its severity is more influenced by weather parameters.

Keywords: Agronomy, canopy cover, drought stress, growth model, leaf area index

Contact Address: Bert Stevens, KU Leuven, Dept. of Biosystems (BIOSYST), Willem de Croylaan 42 - Bus 2428, 3001 Leuven, Belgium, e-mail: bert.stevens@kuleuven.be

Monsoon Crop Biomass Estimation Using Terrestrial Hyperspectral Imaging

SUPRIYA DAYANANDA, THOMAS ASTOR, JAYAN WIJESINGHA, MICHAEL WACHENDORF

University of Kassel, Grassland Science and Renewable Plant Resources, Germany

India's majority (60%) of the population depend on the agriculture sector for their livelihood. With the agriculture facing major challenges, remote sensing can be an effective tool in monitoring crop production and estimating the yield. It can lead to better planning and policies to ensure food security. This study was conducted with the main objective of predicting the fresh matter biomass (FMB) using the spectral reflectance extracted from hyperspectral images. Three monsoon crops (lablab, maize and finger millet) were grown simultaneously in each of these two experiments rainfed (R) and drip irrigated (I) at University of Agricultural Sciences, Bengaluru, India. The images from full frame hyperspectral camera UHD-185 was used along with destructive biomass sampling to measure the FMB in t ha^{-1}. A total of 11 sampling dates in the monsoon season of 2016 to 2018 were sampled. The spectral data was used with random forest regression model to estimate the FMB in rainfed, irrigated experiments and generalised (data sets of R and I combined) condition. The prediction accuracies based on the relative error (rRMSEP) was found to be lower in generalised condition with 13.9 % for lablab (R^2 = 0.53), 18 % for finger millet (R^2 = 0.46) and 18.7 % for maize (R^2=0.53). Overall, the results show that the FMB prediction model is not specific to rainfed and irrigated experiments as it performed better in the generalised condition. In future, it must be tested to predict the FMB on a larger scale using the sensor on unmanned aerial vehicles.

Keywords: Biomass prediction, hyperspectral imaging, machine learning, multitemporal

Contact Address: Supriya Dayananda, University of Kassel, Grassland Science and Renewable Plant Resources, Steinstrasse 19, 37213 Witzenhausen, Germany, e-mail: supriyaD@uni-kassel.de

Towards a Cropping System Sustainability Tool (CROSST) Evaluating Performance of Green Manure Cover Crops in Benin and Kenya: A Pilot Study

JESSICA MUKIRI[1], RODRIGUE V. CAO DIOGO[2], SÈNAMI G. M GBEDJISSOKPA[2], MICHAEL KINYUA[1], REIN VAN DER HOEK[3], ROLF SOMMER[4], BIRTHE PAUL[1]

[1] *International Center for Tropical Agriculture (CIAT), Tropical Forages Program, Kenya*

[2] *University of Parakou, Dep. of Sci. and Techn. of Animal Prod. and Fisheries, Benin*

[3] *International Center for Tropical Agriculture (CIAT), Central America, Nicaragua*

[4] *World Wide Fund for Nature (WWF), Germany*

Farming practices in sub-Sahara Africa have resulted in declining soil fertility. Hence, Green Manure Cover Crops (GMCC) are promoted for soil improvement and protection. For farmers, adopting GMCCs and making formed decisions to integrate them in their cropping systems, they require a good understanding of the multi-dimensional impacts of these crops. We therefore developed the Cropping Systems Assessment Sustainability Tool (CROSST) which can assess the performance of different cropping systems with and without the integration of GMCCs. CROSST is an Excel based tool that assesses both agro-environmental and socio-economic impacts of GMCC technologies. The tool looks at gross margin, productivity (yield), soil health (N and P balances, soil structure and soil organic carbon), labour hours and the trade-offs between these. It adopts a static rule-based framework and follows a three-step approach (i) identification of alternative crop rotations, (ii) selection of agronomic, environmental and socio-economic parameters and (iii) assessing and comparing the different crop rotations. The tool was pilot tested in Benin and Kenya under the BMZ-GIZ programme on 'Soil Protection and Rehabilitation for Food Security'. Data were collected through literature reviews, focus group discussions and key expert interviews. The crop rotations were selected and designed by experts with in-depth knowledge on the local cropping systems of North and South of Benin and Western Kenya. First results indicate that GMCCs improve soil structure/soil organic matter as well as soil N balances in both countries. However, investing in soil improvement can result in loss of profitability especially when a crop that produces grain for consumption or sale is swapped for a crop that produces biomass for soil amendment only. CROSST still needs further data refinement with recent official census as well as independent field measurements. Once validated it can serve as a decision-support tool for development agencies, implementing partners and local stakeholders when designing sustainable cropping systems that integrate GMCCs.

Keywords: Cover crops, green manure, impact assessment tool, soil fertility, sub-Saharan Africa

---

Contact Address: Jessica Mukiri, International Center for Tropical Agriculture (CIAT), Tropical Forages Program, 823, 00621 Nairobi, Kenya, e-mail: j.mukiri@cgiar.org

Evaluating Groundwater Management Options in a Semi-Arid and Rapidly Urbanising Area Using a Socio-Hydrological Model

JOHANNES WEGMANN

University of Goettingen, Department of Agricultural Economics and Rural Development, Germany

In areas where water resources are under stress due to high population growth and related increasing agricultural production, water resource management becomes increasingly important. Understanding water user behaviour thereby is the basis to analyse potential outcomes of different management options to ensure sustainably managed resources. Socio-hydrological models which combine natural processes with human decision making to evaluate policy designs have been improved considerably in recent years. However, these models often rely on data which is either at a highly aggregated level for the natural processes or human decision making is not calibrated with micro-level data. In this study, a socio-hydrological model is elaborated which represents observed farmer irrigation practices and local (geo)hydrological conditions. The model is then calibrated with observed farmer behaviour and hydrological and soil information in the study area. Data comes from 600 agricultural households and 6 experimental plots in the vicinity of the Indian megacity Bengaluru. The region is characterised by a drastic drop in groundwater levels which is partly a consequence of intensive irrigated agriculture. The socio-hydrological model is used to evaluate changes in climate condition and in irrigation behaviour on groundwater resources and welfare implication for rural households. Preliminary results show that the modeled groundwater level approximates the stated groundwater level of the farmers in the survey. Moreover, a change in irrigation behaviour is more dramatically on the resource than changing climatic conditions. In a next step, different management option will be incorporated and evaluated.

Keywords: Groundwater management, India, policy evaluation, socio-hydrological model

Contact Address: Johannes Wegmann, University of Goettingen, Department of Agricultural Economics and Rural Development, Platz der Göttinger Sieben 5, 37073 Göttingen, Germany, e-mail: johannes.wegmann@agr.uni-goettingen.de

Assessing Impacts of Crop Expansion and Overstocking in Kenyan Savannah Rangelands Using the LUCIA Model

BENJAMIN WARTH, CARSTEN MAROHN, FOLKARD ASCH

University of Hohenheim, Inst. of Agric. Sci. in the Tropics (Hans-Ruthenberg-Institute), Germany

There is potential to increase agricultural productivity by converting Guinea Savannah rangelands to cropland on a large scale, but such land use change is expected to have negative effects on ecosystem functions. Conversion of pasture to cropland and tillage affect seasonal plant cover, soil organic matter replenishment, and soil water storage. Additionally, herd mobility becomes more limited and remaining pastures are overgrazed. This can cause shifts in vegetation composition towards higher plant cover and abundance of trees and bushes at the cost of grasses. Our aim is to assess the impacts of land use change in savannah rangelands on vegetation and soil as well as to investigate major processes involved. We studied a savannah watershed in Laikipia, Kenya, by simulating different crop expansion- and herd movement patterns using the mechanistic and spatially explicit Land Use Change Impact Assessment tool (LUCIA). In order to represent the processes described above, we amended the agronomic LUCIA model for functions on plant regeneration after grazing, as well as on tree seed dispersal and seedling recruitment depending on grass competition. The new functions allow simulating pasture degradation with growth reserve depletion, as well as bush encroachment due to enhanced seedling establishment under reduced grass competition. We, thus, build a bridge between agricultural modelling and simulation of plant community ecology, which allows assessing ecological sustainability of agricultural actions in landscapes. Scenario simulations show how ecosystem functions such as soil organic carbon storage, water retention, plant productivity, and fodder provision are affected by different cropping systems (e.g. inter-cropping,) and management regimes (e.g. zero tillage), as well as their expansion rates and -pattern into the savannah landscape. On rangelands, grazing intensity and timing affect biomass regrowth and grass productivity, which influence grass-tree competition and vegetation composition. We discuss model improvements representing important processes involved in land use change impact pathways in savannah ecosystems by newly implemented ecological model routines. We expect, that upcoming LUCIA coupling with livestock- and farmer's decision models will enhance the understanding of major processes and feedbacks within complex socio-ecological savannah systems that are under pressure of land use change and agricultural intensification.

Keywords: Ecosystem functions, pasture degradation, spatially explicit modelling

Contact Address: Benjamin Warth, University of Hohenheim, Inst. of Agric. Sci. in the Tropics (Hans-Ruthenberg-Institute), Garbenstr. 13, 70593 Stuttgart, Germany, e-mail: benjamin.warth@uni-hohenheim.de

Investigating Farmers' Knowledge about Climate Change in Iran

MASOUD YAZDANPANAH[1], YOUSOF AZADI[2], NOZAR MONFARED[3]

[1]*Agricultural Sciences and Natural Resources University of Khuzestan, Agricultural Extension and Education, Iran*

[2]*University of Zanjan, Dept. of Agricultural Extension, Communication and Rural Development, Iran*

[3]*Ministry of Jihad-e-Agriculture, Institute of Applied Scientific Higher Education, Iran*

Climate change is a reality that poses severe threats to agriculture and rural livelihoods in developing countries. Mounting evidence has revealed that farmers can effectively manage negative impacts by adapting their farming practices to climate change. However, it is a common belief that more information and knowledge about climate change will lead to a better understanding of the phenomenon and adaptation options. Therefore, farmers need to refresh and continually update their knowledge about climate change to raise their potential capacity and output in the facing of climate change impacts. Therefore, the aim of this study was to investigate factors affecting knowledge of wheat growers about climate change and the associated impacts in Kermanshah County in western Iran. To achieve this goal a quantitative study (survey methodology) was used, applying a multi-stage random sampling technique and selecting 350 farmers. Data were collected through a questionnaire with confirmed internal reliability and validity. Descriptive analysis revealed that the age of the participants ranged from 25 to 84 with a mean value of 48.7 years (SD 12.47). The sample consisted of 13 female farmers (3.7 %) and 337 male farmers (96.3 %). The majority of the participants (26.6 %) had a high school equivalent degree. The mean farm size for rain fed and arable farmland farmers, respectively, was 7.5 and 2.6 hectares. The mean scores of self-efficacy, risk attitude and knowledge with respect to climate change were 2.05 out of 5 (SD 0.72), 3.57 out of 5 (SD 0.64) and 3.64 out of 5 (SD 0.77), respectively. Pearson's correlation test revealed a significant relationship between 'knowledge of climate change' and all other variables, including 'environmental attitude', 'trust', 'self-efficacy' and 'risk attitude'. Structural equation modelling revealed that environmental attitude ($\beta = 0.31$, $p < 0.001$), risk attitude ($\beta = 0.18$, $p < 0.002$) and trust ($\beta = 0.14$, $p < 0.013$) were significant predictors of farmers' knowledge about climate change. The present study provides a justification for programs that intend to encourage farmers' adaptation behaviour in view of climate change impacts.

Keywords: Agriculture, climate change, knowledge, structural equation modelling

---

Contact Address: Masoud Yazdanpanah, Agricultural Sciences and Natural Resources University of Khuzestan, Agricultural Extension and Education, Mollasani, 744581 Ahvaz, Iran, e-mail: masoudyazdan@gmail.com

Soils and soil fertility

Sustainable land use - the soil-plant interface

Oral Presentations

Posters

The Effect of Soil Conservation Practices on Soil Carbon and Yields in Smallholder Oil Palm Plantations

NINA HENNINGS[1], KATRIN RUDOLF[2], MICHAELA DIPPOLD[3], MEIKE WOLLNI[4]

[1]*University of Goettingen, Dept. of Soil Science of Temperate Ecosystems, Germany*

[2]*University of Goettingen, Dept. of Environmental and Resource Economics, Germany*

[3]*University of Goettingen, Dept. of Biogeochemistry of Agroecosystems, Germany*

[4]*University of Goettingen, Dept. of Agricultural Economics and Rural Development, Germany*

The progressive expansion of oil palm plantations in South East Asia leads to a conversion of tropical rainforests. This transformation reduces soil organic carbon (SOC) contents significantly. Since a reduction in SOC levels is also linked to a decrease in many important soil ecosystem services, such as water and nutrient regulation which positively affect yields, it might negatively affect the income of oil palm farmers.

One option to sustain and enhance SOC contents is the use of soil conservation practices such as mulching or cover crop application. These practices might affect yields via various channels: First, via an increase in soil carbon content, but also via a direct fertilisation or nutrient competition effects.

Despite its theoretical relevance for farmers' income, the agronomic effects of SOC on oil palm yields, especially in smallholder plantations, have rarely been analysed. Therefore, we study the effect of soil conservation practices on soil indicators and yields, and the effect of SOC on yields to distinguish between the pathways through which soil conservation practices can affect oil palm yields.

Our study was conducted in Jambi Province, Indonesia, in 2017. We focus on the application of empty fruit bunch (EFB) mulching and cover crops (CC) as soil conservation practices in smallholder plantations. Smallholder plantations are chosen because of their growing importance in the Indonesian oil palm sector.

Using a stratified random sampling procedure, we selected 142 independent oil palm growing smallholders. Per household, one plantation was selected. Soil samples were collected for SOC content and bulk density measurements.

Our results hint to a positive effect EFB application on soil carbon, but the effect of CC is insignificant. While the direct effect of SOC on per hectare oil palm yields is insignificant, application of EFB is associated with statistically higher oil palm yields in comparison to CC application and the control group. This suggests that the fertilising effect of EFB mulching might be more important than the effect via improved soil fertility.

Keywords: Oil palm yields, smallholder farmers, SOC, soil conservation practices

---

Contact Address: Katrin Rudolf, University of Goettingen, Dept. of Environmental and Resource Economics, Goettingen, Germany, e-mail: krudolf@gwdg.de

Effects of Emerging Crop Rotations and Changing Soil Aeration Status on B and Zn Availability and Vegetable Responses in Nepal

Shova Shrestha[1], Mathias Becker[1], John Lamers[2]

[1] *University of Bonn, Inst. Crop Sci. and Res. Conserv. (INRES), Germany*

[2] *University of Bonn, Center for Development Research (ZEF), Germany*

Since recently, the traditional rice-wheat rotation systems in Nepal are subject to drastic changes. Water shortages drive the replacement of the water-consuming lowland rice by maize, and progressing urbanisation drives a replacement of wheat by high-value vegetables. The changes in soil aeration status by integrating maize and the higher nutrient demand of vegetables compared to wheat are associated with changing demands of and responses to key micro-nutrients such as boron (B) and zinc, (Zn) which are highly deficient in the soils of Nepal. We compared the status of soil B/Zn availability and the responses of rice and maize (hot wet season) as well as of cauliflower, tomato and wheat (cold dry season) to added B and Zn in a greenhouse trial as well as in field validation trials at two sites in Nepal. The shift from rice to maize and concomitant changes in soil aeration status tended to increase the responsiveness of both vegetable crops to added B and Zn at both sites. The shift from wheat to high-value vegetables increased the overall demand for B and to a lesser extent for Zn, irrespective of the soils and sites. Thus, the addition of 4 kg ha^{-1} each of B and Zn increased crop uptake by cauliflower significantly. While biomass accumulation and yield of wheat were largely unaffected by B/Zn additions, both vegetables responded with 2-4-fold yield increases. At these effects lowland sites were generally more pronounced than at the mid-hill site, and response tended to be more in maize (aerobic) than with rice (anaerobic) based systems. It is concluded that in the inherently B and Zn deficient soils in much of Nepal, the application of B and Zn fertilisers improves the performance of the traditional rice-wheat rotations. Therefore, with the shift towards maize and high-value vegetable-based systems, B and Zn applications become essential. The demand for such nutrients as fertilisers is foreseen to dramatically increase with progressing urbanisation and associated system shifts.

Keywords: Boron, cauliflower, *Oryza sativa*, soil fertility, tomato, zinc

Contact Address: Shova Shrestha, University of Bonn, Inst. Crop Sci. and Res. Conserv. (INRES), Karlrobert-Kreiten-Strasse 13, D-53115 Bonn, Germany, e-mail: shovashrestha@gmail.com

Combining Land-Based Organic and Landless Food Production: A Concept for a Circular and Sustainable Food Chain for Africa in the Year 2100

Gerold Rahmann[1], Daniel Grimm[1], Anja Kuenz[2], Engel Hessel[2]

[1]*Thuenen Institute of Organic Farming, Germany*

[2]*Thuenen Institute of Agricultural Technology, Germany*

Agriculture in tropical countries has to adapt not only to climate change and resource depletion but also to population growth – especially in Africa, where around 80 % of the growth until 2100 is predicted to happen. The medium estimate of the United Nations predicts an increase from currently 1.2 to 4.4 billion people by the end of the century – but no strategic plans exist beyond the year 2050. Since Africa is a continent with large deserts, arid and semi-arid areas, only about 400 m^2 of agricultural land will be available per person according to these predictions.

In the LandLessFood project, we are trying to design an agricultural system, based on the principles of circular economy, in which it is possible to feed everyone on this little space, focusing on Nigeria as an example.

The main idea is the combination of "landless" food systems such as microalgae production and mushroom cultivation, with organic, "land-based" food production. Microalgae are very spatially efficient, can be produced independent of fertile soils, could be used as a pathway of nutrient recycling in the sewage system and can produce large amounts of oils or starch, to fulfil the energy-need of the population. Land-based food production should be used to fulfil the need for "quality food" (vitamins, nutrients, minerals) of the population. Mushroom cultivation is a form of "selective composting" in which food is produced. While mushrooms themselves are an excellent meat-substitute, spent mushroom substrate can be used as animal feed, ideally for invertebrates, such as earthworms, which are good chicken, pig and fish feed.

Keywords: Africa, bioreactor, circular economy, food security 2100, landless food, organic agriculture

Contact Address: Daniel Grimm, Thuenen Institute of Organic Farming, Landlessfood, Trenthorst 32, 23847 Westerau, Germany, e-mail: daniel.grimm@thuenen.de

Phytoremediation of Petroleum Hydrocarbon-Contaminated Soils with *Jatropha curcas* and *Vetiveria zizanioides*

BERTRAND NERO

Kwame Nkrumah University of Science and Technology, Dept. of Forest Resources Technology, Ghana

Accidental oil spills during oil exploration and mineral mining operations have deleterious consequences on soils and overall ecosystem health. Environmentally benign methods of reducing hydrocarbon contamination levels in soils and ecosystems are often encouraged. However, there is scanty data on the effects of vegetation on petroleum hydrocarbon concentration in crude oil contaminated soils. The present study investigated the growth performance of two plant species on hydrocarbon contaminated soils and the effects of these plants on hydrocarbon concentration levels in soils. A $2 \times 2 \times 3$ factorial arrangement of treatments in a completely randomised design with 3 replicates was adopted. Two plant species, 2 soil contamination levels, and 3 soil amendments constituted the major treatment factors. Soil total oil and grease (TOG), total petroleum hydrocarbon (TPH), plant height, collar diameter, and number of leaves or tillers were monitored over a sixteen week period at Ghana Manganese Company Ltd.

Amending the soil with compost, *Jatropha curcas* (JC) caused up to 78.8 and 82.2 % decline in soil TPH and TOG concentrations, while *Vetiveria zizanioides* (VZ) caused 51.1 and 39.7 %, respectively, after 16 weeks. Compost amendments significantly reduced TOG and TPH concentrations compared to fertiliser and no amendments in both JC and VZ ($p < 0.001$). However, the effect of species on TOG and TPH concentrations were not statistically significant ($p = 0.081$).

Growth in height, collar diameter and number of leaves in JC were significantly higher in the compost amendment compared to the fertiliser and no amendment treatments ($p < 0.001$). Number of leaves ($p = 0.009$) and collar diameter growth ($p = 0.010$) were significantly lower in contaminated soils compared to non-contaminated soils but no significant differences were observed between the two with respect to height growth. Furthermore, only the number of tillers in Vertivar was significantly influenced by the soil amendments ($p = 0.003$) and the soil hydrocarbon contamination levels ($p = 0.048$).

It is concluded that phytoremediation, particularly with *Jatropha curcas*, is an alternate means to reducing soil hydrocarbon concentration levels. However, soils must be amended with compost for effective remediation and enhancement of rapid, vigorous early growth of plants.

Keywords: Collar diameter, compost, hydrocarbon, soil amendment

Contact Address: Bertrand Nero, Kwame Nkrumah University of Science and Technology, Dept. of Forest Resources Technology, DFRT FRNR KNUST, Kumasi, Ghana, e-mail: nerbefest@yahoo.com

The Impact of Long-Term Tropical Grassland Conversion on Soil Quality and Soil Carbon Stocks

John Kormla Nyameasem[1], Carsten Malisch[1], Thorsten Reinsch[1], Charles Yaw Fosu Domozoro[2], Friedhelm Taube[3], Iraj Emadodin[1], Esther Marfo-Ahenkora[2]

[1]*University of Kiel, Grassland and Forage Science / Organic Agriculture, Germany*

[2]*CSIR Animal Research Institute, Natural Resources Management and Environmental Health, Ghana*

[3]*University of Kiel, Institute of Crop Science & Plant Breeding, Germany*

Enhancing the capacity of livestock systems to sequester carbon is an important measure to tackle soil degradation and climate change. Therefore, accurate assessments are required, particularly at farm level, to estimate soil carbon sinks under different land use systems by available predictors. This study evaluated the impact of different long-term land-use scenarios on soil carbon storage and ascertained the impact of condensed tannins (CT) and soil chemical properties on soil C dynamics in grasslands of southern Ghana. Soil samples were taken (0–30 cm depth) from 50 years old food crop fields, seeded grazing fields, monoculture fields of fodder grass, legume herbs, legume browse, non-legume browse species and native grassland. CT concentration in the forages ranged from 4–67 mg/kg dry matter and were higher ($p < 0.01$) in browses compared to herbs. Nitrogen (N) levels were highly correlated with soil carbon stocks and were significantly higher ($p < 0.01$) for fodder grass fields and legume herbs fields. C:N ratio in soils was not significantly affected by the land use system ($p > 0.05$). Plant available phosphorus and potassium represented highest ($p < 0.01$) values in food crop fields. Soil pH varied only with a significant rate ($p < 0.001$) between food crop fields and seeded gazing fields. Soil carbon stocks ranged from 16.6 — 64.1 t C ha^{-1} (mean ± s.e: 33.1 ± 1.13 t C ha^{-1}) across land use systems and were lower ($p < 0.01$) for grazed seeded-pasture fields and herbaceous legume plots compared to the other land use systems. Conversion of the natural grassland resulted in a mean loss of 480 kg C ha^{-1} year^{-1}. There was significant ($p < 0.05$) positive correlation between the long-term changes in soil C stock and CT fractions (r = 0.33–0.49). Also, correlation tests showed positive relationships between change in soil C-stock and soil chemical traits (r = 0.043–0.91). The current case study indicated that a multiple linear regression equation with N, CN and K as principal factors could explain 98 % of the long-term changes in SOC stock.

Keywords: C4 grasses, carbon sequestration, condensed tannins, soil nutrients, sub-Saharan Africa

Contact Address: John Kormla Nyameasem, University of Kiel, Grassland and Forage Science / Organic Agriculture, Hermann-Rodewald Str. 9, Kiel, Germany, e-mail: jnyameasem@gfo.uni-kiel.de

Optimisation of Mineral Fertilisers of Cowpea (*Vigna unguiculata*) in Benin

FIRMIN ANAGO[1], GUSTAVE DAGBENOMBAKIN[2], EMILE AGBANGBA[3]

[1] *University of Abomey-Calavi, Laboratory of Soil Sciences of Crop Production School, Benin*

[2] *Ministry of Agriculture, Breeding and Fisheries, National Institute of Agronomic Research of Benin, Benin*

[3] *University of Abomey-Calavi, Laboratory of Applied Biology Research and Study, Benin*

Cowpea is an important legume crop in the dry savannahs of Africa, especially West Africa because of its important role as a primary source of protein. Its productivity in Benin is deeply affected by many biotic factors. Among this factors, soil poverty is the mainly factor that limited cowpea productivity. The aim of this study was to determine the optimal fertiliser rates. A field experiment was carried out in growing season 2018 in centre Benin to assess the effect of mineral fertilisers (N, P, K, Mg and Zn) on growth and yield of cowpea (*Vigna unguiculata*). The experiment was conducted in Box Behnken design with five factors. This design was repeated on eight experimental sites. Each factor had three levels: N (0, 25, 50 kg ha $^{-1}$, P_2O_5 (0, 75, 150 kg ha $^{-1}$, K_2O (0, 40, 80 kg ha $^{-1}$, MgO (0, 7.5, 15 kg ha $^{-1}$ and Zn (0, 7.5, 15 kg ha $^{-1}$. Mixed-effects linear models and response-surface method was used to analyse the data with R software version 3.5.1. The results showed that only nitrogen influenced significantly the height of cowpea while nitrogen, phosphorus, potassium and magnesium influenced significantly the number of leaves and ramification by plant of cowpea. Likewise, the number of pods/plant, seeds/pod, pod yield and seed yield was significantly influenced by the five factors (N, P_2O_5, K_2O, MgO, Zn). However, only nitrogen and phosphorus influenced significantly the aerial biomass and root biomass yield. The mineral fertiliser rates which allowed to achieve a higher seed yield were 25.3 ± 1.51 kg ha $^{-1}$ of N, 70.0 ± 20.32 kg ha $^{-1}$ of P_2O_5, 43.2 ± 5.58 kg ha $^{-1}$ of K_2O, 7.1 ± 1.14 kg ha $^{-1}$ of MgO and 7.9 ± 0.86 kg ha $^{-1}$ of Zn.

Keywords: Cowpea, mineral fertiliser, optimal rate

Contact Address: Firmin Anago, University of Abomey-Calavi, Laboratory of Soil Sciences of Crop Production School, Abomey-Calavi, Benin, e-mail: firmin.anago@gmail.com

Influence of Compost, Lime and NPK on Performance of three Cassava Varieties

Maria Egwakhide[1], Stefan Hauser[1], Olugbenga Adeoluwa[2]
[1]*International Institute of Tropical Agriculture (IITA), Root & Tuber Cropping Systems Agronomy, Nigeria*
[2]*University of Ibadan, Soil Fertility, Nigeria*

Soil fertility is one of the major limitations to productivity of Cassava (*Manihot esculenta* Crantz). The use of different soil amendments such as fertilisers could improve the productivity of cassava and also increase the fertility of the soil. Thus, the potential of brewery-based compost, lime and NPK 15:15:15 were evaluated on the growth and yield of three cassava varieties (TMS01/1393, TMS1980581 and TMS101/0040). The study was conducted at the International Institute of Tropical Agriculture, Ibadan, Nigeria. The experiment was a 4 factorial randomised complete block design in a split plot arrangement with three replicates. The first factor was compost application at 2 levels nil versus 5 Mg ha^{-1}; second factor was NPK 15:15:15 at 2 levels: nil versus 500 kg ha^{-1} (equivalent to 75:33:62 kg ha^{-1} N:P:K); third factor was lime at 2 levels: nil versus 500 kg ha^{-1}; fourth factor was the cassava variety at 3 levels: TMS01/1393, TMS1980581 and TMS101/0040, with the input combinations nested within varieties. Cassava was planted at 1 by 0.5 m distance and harvested after 12 months.
TMS01/1393 had the highest stem yield with an average of 22.65 t ha^{-1} followed by TMS1980581 20.60 t ha^{-1} and TMS101/0040 14.30 t ha^{-1}. Compost + lime combination gave the highest stem yield (42.54 t ha^{-1}, $p \leq 0.0001$), significantly different from other treatments.
Root yield across all treatments of TMS01/1393 was 36.96 t ha^{-1} fresh mass (FM) and 10.29 t ha^{-1} dry mass (DM), for TMS1980581 root fresh yield was 25.13 t ha^{-1} and 7.01 t ha^{-1} DM, and TMS101/0040 produced 28.00 t ha^{-1} FM being 6.33 t ha^{-1} DM. Combination of compost + lime had the highest yield in TMS01/1393 at 42.32 t ha^{-1} FM (10.97 t ha^{-1} DM) which was not statistically different from the control at 40.29 t ha^{-1} FM (9.40 t ha^{-1} DM). There was a significant increase over the control yield 22.14 t ha^{-1} (5.14 t ha^{-1} DM) of TMS101/0040 in the following combinations: compost +lime + fertiliser 36.06 t ha^{-1} FM (8.20 t ha^{-1} DM), compost + fertiliser 36.60 t ha^{-1} FM (8.55 t ha^{-1} DM), lime + fertiliser 34.85 t ha^{-1} FM (8.36 t ha^{-1} DM).

Keywords: Cassava, compost, lime, NPK, soil fertility

Contact Address: Maria Egwakhide, International Institute of Tropical Agriculture (IITA), Iita Oyo Road, 200285 Ibadan, Nigeria, e-mail: m.egwakhide@cgiar.org

Comparative Nitrogen Use Efficiency of Urea, Manure and Different Mulch Types in Horticulture in Semi-Arid Bolivia

LAURA KUONEN[1], NOEMI STADLER KAULICH[2], LINDSEY NORGROVE[1]

[1] *Bern University of Applied Sciences, School of Agricultural, Forest and Food Sciences, Switzerland*

[2] *Center for Andean Agroforestry, Cochabamba, Bolivia*

Onion (*Allium cepa*) is an important horticultural crop in Bolivia, where the main production region is in the semi-arid region of Cochabamba, in the inter-Andean valleys. Here, soils are degraded, due to overgrazing and agricultural intensification. Soil erosion and water availability as well as nitrogen (N), limits crop growth. We hypothesised that using locally available mulch might improve growth and finally yield more than the equivalent N application as either inorganic fertiliser or manure as it might also release nutrients more in synchrony with the developing crop as well as suppress weeds, increase soil moisture and reduce soil erosion. We also hypothesised that growth improvement due to adding nutrients would be less in the more shaded blocks where growth would be light-limited.

We tested three mulches *Dodonaea viscosa, Melinis repens*, and *Chamaecytus proliferus* (application rate 4 t DM ha^{-1}, equivalent to 44, 26 and 39 kg N ha^{-1}) versus farmyard manure (5 t DM N ha^{-1} equivalent to 165.5 kg total N ha^{-1}); two urea treatments (40 and 80 N eq ha^{-1} and an unfertilised, unmulched control in an onion monoculture, planted in January 2019 and manually weeded at 60 days after planting (DAP). We used a randomised complete block design (n=4) with shade and competition from a tree boundary as the blocking factor. Plant height, circumference, number of green and of newly produced leaves were evaluated every two weeks.

At 12 weeks after planting, onions were taller in the urea treatments (40 and 80 kg N eq ha^{-1}) than in the control with the 80 kg N eq ha^{-1} urea treatment having the most weed biomass. In the least shaded block, weed biomass was lower than in the most shaded block (10.1 t fresh weight ha^{-1} compared to 3.9 t fresh weight ha^{-1}). None of the mulch treatments significantly affected growth.

Keywords: *Allium cepa*, horticulture, mulch, nitrogen efficiency, semi-arid climate

Contact Address: Laura Kuonen, Bern University of Applied Sciences, School of Agricultural, Forest and Food Sciences, Länggasse 85, 3052 Zollikofen, Switzerland, e-mail: laura.kuonen@bfh.ch

Sustaining Intercropped Maize and Peanut Yields through Controlled Organic Fertiliser Usage in Muna Regency, Indonesia

LA KARIMUNA, RACHMAWATY HASID, HALIWU HALIM
University of Halu Oleo Kendari, Dept. of Agrotechnology, Indonesia

The worldwide demand of staple food crops for various needs has increased doubled for the last 10 years, especially for Indonesia government, import of such basic foods as rice, maize, peanut and soybean has to be taken but this policy can't be kept in the long term. The farmers in Java, agricultural crops are commonly cultivated in arable land, so the crop yield planted is sufficient to fulfil the community needs. However, recently most of favoured lands have been changed to other land uses for industrial areas, housing and other economic purposes. Southeast Sulawesi has tremendous role for the development acceleration of Indonesian economy to improve community welfare. Soil limiting factors identified can be overcome properly by the application of controlled organic fertiliser to increase soil fertility sustainably and sufficient yields of agricultural crops can be achieved. The main objective of this paper was to analyse the current results of research carried out in sustaining improvement of intercropped maize and peanut yields through controlled usage of organic fertiliser technology in Muna regency. The research work was carried out in three villages, in a split plot experimental for two factors of various doses of bokashi plus fertiliser. Local peanut was cultivated with maize in an intercropping system. Variables measured were plant height (cm), stem diameter (cm) and Leaf Area Index (LAI) at 14, 28, 42 and 56 DAP, cob length (cm), bean dry weight 100 g at 14 % moisture content, Land Equivalent Ratio (LER), and dry beans (t ha^{-1}). All variables were analysed using analyses of variances (ANOVA) followed by Duncan's Multiple Range Test (DMRT) with 95% confidence level. The results showed that the higher the doses of bokashi plus fertiliser applied, the higher the growth and yields of intercropped maize and peanut obtained. Land use efficiency was better indicated by a LER value of 1.46, and the highest yield of maize and peanut was 7.6 and 2.43 t ha^{-1}, respectively. This finding confirms the significant contribution of bokashi plus fertiliser usage to not only maintain high soil fertility but also to improve crop productivity

Keywords: Bokashi, mineral fertiliser, intercropping system, maize, marginal lands, peanut, poverty

Contact Address: La Karimuna, University of Halu Oleo Kendari, Dept. of Agrotechnology, Terong Street, 93232 Kendari, Indonesia, e-mail: lkarimuna@hotmail.com

Growth and Yield Response of two Mungbean (*Vigna radiata* L.) Genotypes to Inoculation with N_2-fixating and P- and K-Solubilising Bacteria

REZA MIRZAEITALARPOSHTI[1], SAEID SOUFIZADEH[1], JAFAR NABATI[2], AFSANEH YOUSEFI[3]

[1]*Shahid Beheshti University, Environmental Sciences Research Institute, Iran*

[2]*Ferdowsi University of Mashhad, Research Center for Plant Sciences, Iran*

[3]*Ferdowsi University of Mashhad, Fac. of Agriculture; Dept. of Agrotechnology, Iran*

Economic problems as well as environmental concerns related to the excessive application of agricultural chemicals have shifted the attention to application of biological fertilisers in many agroecosystems. In order to evaluate the effect of free-living nitrogen fixating and phosphorus and potassium solubilising bacteria on yield and nitrogen fixating root nodes of two mungbean (*Vigna radiata* L.) genotypes, a two-year factorial experiment was carried out as a randomised complete block design with three replications. Two mungbean genotypes (Dezfouli and Indian) were planted under six fertilisation systems at the Agricultural Research Station of Ferdowsi University of Mashhad, Iran, in 2018. Fertilisation treatments were free-living nitrogen fixating bacteria (NFB), phosphate solubilising bacteria (PhSB), potassium solubilising bacteria (PSB), NFB+PhSB+PSB, chemical nitrogen fertiliser (NF), and no fertiliser application as control (C). The crop growth parameters including leaf area index (LAI), crop growth rate (CGR), accumulation dray matter (DM) as well as its yield and its components were assessed every year and reported in a two-year average, as there were no significant differences between two growing years. According to the results, the highest LAI (3.85), CGR (27.14 g m^{-2} d^{-1}) and DM (796 g m^{-2}) were obtained for Indian genotype planted under NFB+PhSB+PSB treatment. Similarly, the values recorded for LAI, CGR and DM for Dezfuli genotype were 3.84, 26.27 g m^{-2} d^{-1} and 836 g m^{-2}, respectively, under NFB+PhSB+PSB treatment. The number of nitrogen fixating root node was affected by treatment significantly increasing from 5.83 in C treatment to 25.33 per plant in NFB+PhSB+PSB treatment. The highest biomass and grain yield obtained also under NFB+PhSB+PSB treatment by the value of 7621 kg ha^{-1} and 1828 kg ha^{-1}, respectively. All the investigated parameters decreased markedly in both genotypes under C treatment as compared to either chemical and/or biological treatments. In conclusion, application of nitrogen fixating bacteria as well as phosphate and potassium solubilising bacteria can be an appropriate and environment friendly strategy for mungbean cultivation in areas with low nutrient availability.

Keywords: Crop growth rate, leaf area index, Mungbean

Contact Address: Reza Mirzaeitalarposhti, Shahid Beheshti University, Environmental Sciences Research Institute, Tehran, Iran, e-mail: rezamirzaei57@gmail.com

Gross Nitrogen Transformation Rates Do not Support Previously Described BNI Capacities of Selected *Brachiaria* Genotypes

Eduardo Vazquez[1], Nikola Teutscherová[2], Mirjam Pulleman[3], Michael Dannenmann[4], Klaus Butterbach-Bahl[4], Paul Töchterle[4], Jacobo Arango[5]

[1] *Universidad Politécnica de Madrid, Dept. of Agricultural Production, Spain*

[2] *Czech University of Life Sciences Prague, Fac. of Tropical AgriSciences, Dept. of Crop Sciences and Agroforestry, Czech Republic*

[3] *Wageningen University and Research, The Netherlands*

[4] *Karlsruhe Institute of Technology, IMK-IFU, Germany*

[5] *International Center for Tropical Agriculture (CIAT), Colombia*

Nitrification is one of the key processes leading water contamination and greenhouse gas emissions in the form of N_2O in pasture systems. As vast areas of tropical pastures are considered nitrogen (N) limited, grasses from the *Brachiaria* genus have adapted to reduce N losses and increase N use efficiency by releasing substances capable of biological nitrification inhibition (BNI) in the rhizosphere. Although the release of BNI compounds and its impact on N_2O emissions and net nitrification rates in soil have been studied, the impact of BNI on gross nitrogen transformation rates have not been addressed, despite its relevance to mechanistic understanding of this phenomena. Using intact soil cores and ^{15}N dilution technique we evaluated gross N transformation rates in five *Brachiaria* genotypes, including high (CIAT-679, CIAT-16888 and Bh08–1149) and low BNI (Mulato hybrid and CIAT-26146) accessions. Two experimental plots, one established 14 and one 5 years ago were used, located in areas with contrasting soil types of Colombia were used.

Contrary to our expectations, gross nitrification was not lower in soil covered by the high-BNI than in low-BNI genotypes. Surprisingly, gross nitrification rates in the long-term plot were higher under CIAT-16888 compared to Mulato. However, in the long-term plot, the high-BNI CIAT-16888 soil exhibited the highest gross ammonification rates and immobilisation of both ammonium and nitrate, when compared to other genotypes. Similarly, in the medium-term plot, CIAT-16888 and Bh08–1149 (considered high-BNI) showed higher ammonium immobilisation rates than both low-BNI genotypes (Mulato and CIAT-26146). Nevertheless, the relative microbial N retention, i.e. the ration of microbial immobilisation and microbial production of inorganic N, was not affected by *Brachiaria* genotype at any of field locations studied.

Our results suggest, for the first time, that BNI capacity may not be due to a suppression of the gross nitrification rates but to higher N immobilisation rates, which could equally explain lower net nitrification rates and N_2O emission previously described in fields under high-BNI genotypes. The N immobilisation could lead to temporal N storage and reduce N availability to leaching or gaseous losses.

Keywords: N immobilisation, N mineralisation, nitrification, tropical grassland

Contact Address: Eduardo Vazquez, Universidad Politécnica de Madrid, Dept. of Agricultural Production, Av. Puerta de Hierro, 2, 28040 Madrid, Spain, e-mail: eduardo.vazquez@upm.es

A Novel Method of Vermicompost Placement in Organic Tomato Production

EHSAN EBRAHIMI[1], PETER VON FRAGSTEIN UND NIEMSDORFF[2]

[1] *University of Kassel, Soil Science, Germany*
[2] *University of Kassel, Organic Farming and Cropping Systems, Germany*

The need to produce more crops for feeding a growing population world-wide raises questions of more efficient uses of agricultural inputs. In agricultural fields, composts are commonly applied to the field soil by broadcast. This method does not ensure that nutrients as part of applied compost are available to the plant roots at the right time and at the right quantity. In this field study, we compare the effect of compost placement (in two different methods) to compost broadcast on yield attributes and nutrient uptake for organic tomato field. Three methods of placement of the vermicompost (VC) were used in a two-year field trial in north east Iran: 1. VC placed in a row on the soil surface with incorporation, behind the plantation lines (R), 2. Broadcast on the field (B), and 3. in the transplant hole, under the root (U) which is our novel method for this study. As a second factor, VC was applied at three different rates of application, namely 3, 6 and 9 t ha^{-1} for R and B, and 2, 4 and 6 t ha^{-1} for U. In both years, the different rates and placement methods had no significant effect on the fresh yield of tomatoes. However, in treatments with a higher rate and using the U placement increased the dry matter yield of the plants by up to 50 % (8.4 t ha^{-1}) in the second year. This suggests that our novel method (U) could reduce the amount of compost application without a significant change in yield production. Treatments with U placement method showed 23 % higher nitrogen uptake (156 kg ha^{-1}) compared to B method (121 kg ha^{-1}). The U method of placement seems to be a suitable alternative to the B method for field-grown vegetables such as the tomato. Considering the high price and quality of vermicompost compared to the other types of compost, this novel method can be economically considered by farmers.

Keywords: Fertiliser placement, nutrient uptake, organic tomato, soil conditioners, vermicompost

Contact Address: Ehsan Ebrahimi, University of Kassel, Soil Science, Nordbahnhofstr. 1a, 37213 Witzenhausen, Germany, e-mail: eh.ebrahimi@uni-kassel.de

Enhancing Smallholder Wheat Production in the Highlands of Ethiopia through Improved Legume Residue Management Systems

Amsalu Nebiyu Woldekirstos[1], Jan Diels[2], Pascal Boeckx[3]
[1]*Jimma University, Plant Sciences, Ethiopia*
[2]*KU Leuven, Dept. of Earth and Environmental Sciences, Belgium*
[3]*Ghent University, Isotope Bioscience Laboratory (ISOFYS), Belgium*

Faba bean - wheat rotation is a traditional cropping system in most parts of the Ethiopian tropical highlands to maintain adequate N and P for crop growth. However, net contribution of the legume to soil nutrient balance is determined by the extent to which crop residue is removed from the field. We conducted a two-phase on-farm experiment to test whether high N_2-fixing and P-efficient faba beans could benefit subsequently grown wheat. In the first phase, three improved faba bean varieties (Degaga, Moti, Obse) were grown at four levels of P fertilisation (0, 10, 20 and 30 kg P ha^{-1}) along with local faba bean and reference wheat but without fertilisation. N_2-fixation, soil N balance and P-uptake were determined for the faba beans. In the N balance, two scenarios were examined. Scenario-I assumed that all the aboveground biomass is exported from the fields and scenario-II assumed that all the aboveground biomass except grains and empty pods is returned to the soil. In the second phase, the N and P benefits of faba beans to rotational wheat were assessed. In the first phase, N_2-fixed ranged from 258 $\pm$ 17 to 387 $\pm$ 15 kg N ha^{-1}. Scenario-I gave negative net N balance (kg N ha^{-1}) in the range of -86.5 $\pm$ 5.8 (Degaga) to -9.4 $\pm$ 8.7 (Moti) with significant differences between varieties. Scenario-II showed that all balances were significantly ($p < 0.01$) improved and the varieties were positively contributing N (kg ha^{-1}) to the system in the range 50.6 $\pm$ 13.4 (Degaga) to 168.3 $\pm$1 3.7 (Moti), which is equivalent to 110 to 365 kg N ha^{-1} in the form of urea (46 % N). In the second crop phase, biomass and grain yield of wheat grown after faba beans were improved significantly by 112 and 82 %, respectively, compared to wheat-wheat rotation. The incorporated legume residues likely played a role in improving wheat yield through positive effects of N and P sparing via biological N fixation and straw-P addition. The study demonstrates the prospects of improved faba bean germplasm and management as a key component for sustainable wheat based cropping systems in the tropical highlands.

Keywords: Faba bean—wheat rotation, P addition, N balance, N_2 fixation, wheat yield

Contact Address: Amsalu Nebiyu Woldekirstos, Jimma University, Plant Sciences, College of Agriculture, POB1316 Jimma, Ethiopia, e-mail: anebiy@yahoo.com

The Integration of Organic Farm Waste in Degraded Smallholder Banana-Coffee-Farming Systems in the Kagera Region, Tanzania

ANIKA REETSCH[1,2], KARL-HEINZ FEGER[2], KAI SCHWÄRZEL[1], CHRISTINA DORNACK[2], GERALD KAPP[2]

[1] *United Nation University (UNU), Inst. for Integrated Management of Material Fluxes and Resources (FLORES), Germany*

[2] *Technische Universtät Dresden, Dept. of Forest Sciences, Germany*

Smallholder farming account to more than 90 % of the agricultural production in highland perennial-banana-coffee-based farming systems intercropped with beans, maize, cassava, and other annual crops in the Kagera region in NW Tanzania. Densely grown banana-coffee-based farming systems developed over eight centuries and have been severely degraded in less than half a century. For at least one century, traditional composting and mulching techniques have played a crucial role to replenish soil nutrients, combat soil erosion, and soil drying. However, increased biomass production and deforestation of a growing local and refugee hosting population have accelerated soil nutrient depletion and could not be sufficiently replenished through farm waste management. This situation has been worsened through the weakening of farm households after the outbreak of the HIV/AIDS epidemic accompanied by reduced labour, orphanhood, and interrupted knowledge transfer of traditional farm waste practices. Our research shows the current use of organic farm waste in 150 surveyed farm households and how farm waste management is related to biomass production and thus the livelihood. We further present the results of five focus group discussion with 22 lead farmers who have trained 700 households in organic soil fertility, farm waste, and pesticide management fostering sustainable agroforestry. The results compare non-trained with trained farm households. Among the non-trained households, not all households use organic farm waste in composting and mulching. Further, the land size determines biomass up to one hectare. Households either concentrate on crop farming with integrated livestock keeping or livestock keeping and forestry. The majority of the most vulnerable households is female-headed. In comparison, trained households differently succeed in applying the skills they have learnt. The 'wealthiest' group of households use all organic farm waste in strategically applied *in-situ*, ring-hole, pit or drainage composting, whereas those who do not follow the main instructions have a lower biomass production and cannot leave the poverty trap. Soil fertility and biomass production can be significantly improved by training organised by local organisations. However, these organisations need governmental support and funding because they significantly contribute to combat poverty and climate change through soil and organic waste management.

Keywords: Agroforestry, banana-coffee-based farming systems, composting, farm waste management, farmer field training, food availability, gender, Kagera region, Karagwe and Kyerwa district, mulching, smallholder agriculture, Tanzania

---

Contact Address: Anika Reetsch, United Nation University (UNU), Inst. for Integrated Management of Material Fluxes and Resources (FLORES), Ammonstr. 74, 01067 Dresden, Germany, e-mail: reetsch@unu.edu

The Impact of Integrated Soil Fertility Management Practices (ISFM) on Dry Grain Yields of Teff, Wheat, Barley, Maize, Sorghum and Faba Bean of Small Scale Farmers in the Ethiopian Highlands

HAILE DERESSA[1], STEFFEN SCHULZ[1], ABIYOT MEKONNEN[2], ELIAS KADIRO[3], ASFAW NIGUS[4], BORAN ALTINCICEK[5], ANN-KATHRIN LICHTNER[1]

[1]*Deutsche Gesellschaft für Internationale Zusammenarbeit (GIZ) GmbH, Intergrated Soil Fertility Project Ethiopia, Ethiopia*

[2]*Amhara Bureau of Agriculture, Crop Production and Protection Directorate, Ethiopia*

[3]*Oromia Bureau of Agriculture and Matural Resources, Soil Fertility Improvement Directorate, Ethiopia*

[4]*Tigray Bureau of Agriculture and Rural Development, Horticulture and Crop Directorate, Ethiopia*

[5]*Consultant for Research and Data Analysis, Ethiopia*

Ethiopia's food security and economy depends on its agricultural production. Crop yields are, however, below expected values, particularly on the high degraded soils in the highlands. Cereals account for about 75 % of the total area cultivated. We investigated the increased production of selected cereals along with that of the legume faba bean by using integrated soil fertility management practices restoring soil fertility and thereby increasing productivity. To reduce the impact of environmental factors, such as terrain elevation/slope, weather/climate or soil's water/nutrition availability, we used a paired data approach by splitting fields of small-scale farmers into a treatment and control plot growing the same crop with varying treatment elements.

This study examined the yield response of teff, wheat, maize, and faba bean to the individual and combined effects of multiple technologies of integrated soil fertility management practices (ISFM) under small scale farming conditions. Dry grain yields of crops in defined micro-watersheds were recorded for three years along with the information on the use of improved seed, line seeding, Lupine-based green manure, lime, organic fertiliser, blended fertiliser, urea-split dressing and Rhizobia-based biofertiliser for legumes. Yields in kg per hectare were on average for control and treatment fields for teff 1346 and 2220, wheat 2839 and 4552, maize 4879 and 8224, and faba beans 1895 and 3354, respectively. Using linear mixed models (LMM) with field code, year and woreda as random factors, we found that treatment had a significant effect on increased yields of on average +1500 kg ha-1 (s.e., 186.37; d.f., 1760.9, t-value, 8.05, $p < 0.01$). Since crops showed significantly varying yields alone and in combination with treatment, we analysed the various treatment practices along with total N mineral application, disaggregated for teff, maize, wheat and faba bean. For

Contact Address: Ann-Kathrin Lichtner, Deutsche Gesellschaft für Internationale Zusammenarbeit (GIZ) GmbH, Intergrated Soil Fertility Management Project Ethiopia, Addis Ababa, Ethiopia, e-mail: ann-kathrin.lichtner@giz.de

analysed crops we found that one or more integrated soil fertility management practices enhanced resulting crop yields significantly.
Our results indicate that practicing integrated soil fertility management can significantly drive yield increases of crops of small scale farmers in the highlands of Ethiopia while contribute to soil rehabilitation and sustainably improve soil health for stable yields.

Keywords: Ethiopia, integrated soil fertility management, iSFM, productivity increase, small-scale farming, soil fertility

Socio-Economic Impact of Organic Matter Management on Smallholder Crop Farmers in the Rwenzori Region, Uganda

DEOUS MARY EKYALIGONZA, BERNHARD FREYER, JUERGEN FRIEDEL

University of Natural Resources and Life Sciences Vienna (BOKU), Dept. of Sustainable Agricultural Systems, Austria

To react on ongoing decrease of yields and farm economy, the aim of this study is to test farming system specific best soil fertility and cropping practices, their economic performance, and farmers potential to adapt. Two activities serve for fulfiling this objective: to analyse the impact of cropping systems with application of forage legumes, grain legumes, alley crop leaves and farmyard manure on biomass and yield improvement; to understand the conditions enabling farmers to adopt the cropping systems. An experiment (randomised complete Block Design) was conducted in Rwenzori region of Uganda with five treatments running for two seasons: short rains (SR; March-June 2018) and long rains (LR; August- December 2018). The treatments included T1: SR: maize; LR: maize with DAP (18–46-0) at a rate of 50 kg ha^{-1} (N: 9 kg ha^{-1}; P: 23 kg ha^{-1}); T2: SR: cowpea (*Vigna unguiculata*); LR: maize. T3: SR: cowpea + farmyard manure (2.5 t ha^{-1}; FYM): LR: maize and *Mucuna pruriens*. T4: SR: cowpea and *Faidherbia albida*; LR: maize and *M. pruriens*. T5: SR: cowpea and F. albida + FYM; LR: maize and *M. pruriens*. Grain weight and total biomass were significantly higher in T1, T3, T4 and T5 in the LR and T1 in the SR. The highest average input costs (114.29 US$ ha^{-1}) were observed in T4 and T5 in the SR although the costs reduced by 50.28 % in the LR. The average input price for T1 increased by 67 % in the LR. In the SR, T3 and T5 involved the highest average labour costs of 400 US$ ha^{-1} and 414.29 US$ ha^{-1}, respectively, which reduced by 10.7 % and 10.4 %, respectively, in the LR. The highest average revenue was observed in T2 (1924.72 US$ ha^{-1}) and T3 (1687.25 US$ ha^{-1}) in LR although the highest average grain revenue of 83.4 % and 78.4 % was observed in T3 and T5, respectively. T3 and T5 also increased the average gross margin by 111.1 % and 172.8 %, respectively, in the LR. Farmers mentioned that T3 and T5 could be adopted since they can increase yield and in the meantime reduce the costs in comparison to T1 with the fertiliser treatment.

Keywords: Alley crops, biomass, crop yield, cropping systems, farmyard manure, legumes, organic matter management, Rwenzori region, smallholder farmer, socio-economic, Uganda

---

Contact Address: Deous Mary Ekyaligonza, University of Natural Resources and Life Sciences Vienna (BOKU), Division of Organic Farming, Gregor-Mendel-Straße 33, 1180 Vienna, Austria, e-mail: mary.ekyaligonza@students.boku.ac.at

Economic Valuation of Micronutrients Soil Reviving Programme on Ragi and Groundnut Crops in Karnataka

HAMSA K R, UMESH K B, VEERABHADRAPPA BELLUNDAGI
UAS GKVK Bengaluru, Dept. of Agricultural Economics, India

Globally rainfed areas are hotspots of poverty, malnutrition and degradation of natural resources. There are 7 billion people in the world today, with an estimated 925 million or 14% hungry. In India, out of 142 million ha, about 60% are under rainfed agriculture. A similar situation exists in Karnataka: the total cultivated area of 12.4 million ha, about 70% are under rainfed cultivation. A majority of these are small and marginal farmers with landholding of less than 2 ha, responsible for nearly half the food production in the state. Inappropriate soil, water and crop management practices are depleting soil nutrient reserves and further degrading land resources which results in low crop productivity. The Government of Karnataka in association with various developmental agencies has undertaken many initiatives, has converged all the Government schemes through a mission mode project called "Bhoochetana" meaning "reviving the soils" to benefit dryland farmers. The study was conducted in Tumakuru district of Karnataka known for a high deficiency of micronutrients (gypsum, zinc, boron) in the soil. Multiple regression analysis was employed to evaluate the marginal effect of the soil reviving programme among the adopters and non-adopters. Field information was gathered from 120 farmers covering adopters and non-adopters of the programme. The application of micronutrients was the major intervention of this programme and it had significant impact on income and yield of of selected crops. In groundnut the application of the micronutrients by adopters resulted in yield increase of 1.23 quintals extra over and above the non-adopters (5.9 quintals). In case of ragi (finger millet), the adopters yielded 2.02 quintals extra over and above the non-adopters (6.6 quintals). Therefore, the policy focus must be on creating awareness and adoption of new technologies through developmental programmes in rural areas to reach millions of small farmers.

Keywords: Adopters, micro nutrients, multiple regression analysis, non-adopters, soil reviving programme

---

Contact Address: Hamsa K R, UAS GKVK Bengaluru, Dept. of Agricultural Economics, 560065 Bengaluru, India, e-mail: hmmshamsa@gmail.com

Systematic Review on the Social-Ecological Impacts of Urbanisation on Agricultural Systems

PRAMILA THAPA[1], MARIO TORRALBA[1], ANDREAS BUERKERT[2], CHRISTOPH DITTRICH[3], TOBIAS PLIENINGER[1]

[1]*University of Kassel / University of Goettingen, Social-Ecological Interactions in Agricultural Systems, Germany*

[2]*University of Kassel, Organic Plant Production and Agroecosystems Research in the Tropics and Subtropics, Germany*

[3]*University of Goettingen, Inst. of Geography: Human Geography, Germany*

Though urban land covers only around 5 % of the Earth's land area, urbanisation often leds to significant changes in land use and land cover through the conversion of natural or semi-natural ecosystems to urban ecosystems. Urban land conversion often threatens agrobiodiversity and modifies the provision of multiple ecosystem services. However, specific knowledge on the social and ecological impacts of the conversion on agricultural systems is largely fragmented across a multitude of local-level studies. This paper explores the social-ecological impacts of urbanisation in agricultural systems based on a global systematic review of the scientific literature to identify the positive and negative impacts. Our review follows the established guidelines called Preferred Reporting Items for Systematic Reviews and Meta-Analyses (PRISMA). A scooping exercise using different combinations of key words on ecological and social dimensions of urbanisation is performed using Web of Science and CABI Direct. Out of 7904 articles hit in the scooping, 1850 articles are selected from the titles and the key words screening. The number is further reduced to 559 after reading the abstracts, and finally to 110 after reading the full-text. The following three inclusion criteria are considered in the process: (1) a study must cover the topic urbanisation; (2) the study must be based on an empirical fieldwork; and (3) the study must present original findings. Data on positive and negative impacts of urbanization on variables such as water quality, soil fertility, income, and food security are extracted. How these impacts are mediated by the explanatory variables (such as continents, household wealth, ecological degradation, and population) is also assessed. The robustness of the impacts is assessed by using a scale from 1 (poor evidence) to 5 (best evidence). Fisher's exact test is conducted to examine significant differences of reported impacts of urbanisation. Different types of synergies and trade-offs between social and ecological outcomes are categorized. Identifying positive and negative social-ecological impacts of urbanisation in agricultural systems contextualizes the regional focus of the research unit with a global perspective and allows a meaningful interpretation and upscaling of scientific insights.

Keywords: Agricultural systems, social-ecological impacts, systematic review, urbanisation

Contact Address: Pramila Thapa, University of Kassel / University of Goettingen, Social-Ecological Interactions in Agricultural Systems, Steinstr.19, 37213 Witzenhausen, Germany, e-mail: thapa.prami@gmail.com

Resilient *Coffea arabica* Production Strategies for Smallholder Family Farms in the Rwenzori Mountain, Uganda

THADDEO TIBASIIMA KAHIGWA[1], BERNHARD FREYER[1], JUERGEN FRIEDEL[1], ANDREAS MELCHER[2]

[1] *University of Natural Resources and Life Sciences (BOKU), Dept. of Sustainable Agricultural Systems, Austria*

[2] *University of Natural Resources and Life Sciences (BOKU), Dept. of Center for Development Research, Austria*

The Rwenzori Mountain in Uganda is inhabited by smallholder (0.25–0.33 ha) farmer families producing *Coffea arabica* at 1400–2300 ma.s.l with 1800 mm annual rainfall. In this area, soil erosion is high (37.5 t ha^{-1} yr^{-1}) associated with the steep slopes (20°-25°). Due to land degradation and climate change effects such as drought, higher temperature and erratic rains, even the suitable *Coffea arabica* for this area yields 1267.84 kg ha^{-1} (dry cherry) which is below the optimum yield of 3500 kg ha^{-1}. Resilience against degradation can be attained through soil erosion control via organic matter addition and micro-climate regulation but adoption is constrained due to high labour needs, high financial investment and high technical skills required by current practices. This study aimed at developing and selecting feasible strategies for smallholder farmers to build resilience of degraded *Coffea arabica.* An experiment was conducted to reset the cropping system of the *Coffea arabica* using undersown *Mucuna pruriens* (T1), agroforestry *Millettia dura* Dunn (T2), combination of T1&T2 (T3) and; a control (TC), at 26 sites. Household interviews of 26 family farmers were conducted to better understand how to enable adoption of the coffee system. Results indicate that T1 attains full ground cover in three months, accumulating 52.4 t FM ha^{-1} (24.7 DM t ha^{-1}). T1 increases coffee production by 1327.5 kg ha^{-1} dry cherry and attains profit of USD 887.6 ha^{-1} within one year. T3 was not significantly different from T1 except that T3 operates at USD 100.3 ha^{-1} loss for the first year due to costs of resetting the cropping system. We found that smallholder farmer families rent land in less degraded areas to compensate and supplement the low incomes generated from *Coffea arabica.* This extra activity constrains farmers from providing the time and labour for adopting resilience building strategies into their degraded *Coffea arabica* fields. We conclude that T1 and T3 are feasible resilience building strategies for smallholder farmers producing mountain *Coffea arabica.* We recommend a strategy of adopting T1 during the first year of resetting the cropping system, then transition into T3 in the second year of adoption.

Keywords: *Milletia dura*, *Mucuna pruriens*, resilience, Rwenzori Mountain

Contact Address: Thaddeo Tibasiima Kahigwa, University of Natural Resources and Life Sciences (BOKU), Division of Organic Farming, P.O Box 716 Fort Portal Uganda, N/A Fort Portal, Uganda, e-mail: thaddeo.tibasiima@students.boku.ac.at

Land Use Effects Sediment Dynamics in the Headwaters of a Tropical Montane Forest in Western Kenya

Jaqueline Stenfert Kroese[1], Suzanne Jacobs[2], Lutz Breuer[3], John Quinton[1], Mariana Rufino[1]

[1]*Lancaster University, Lancaster Environment Center, United Kingdom*

[2]*Justus-liebig University Giessen, Centre for Intern. Development and Environmental Research (ZEU), Germany*

[3]*Justus-Liebig University Giessen, Inst. for Landscape Ecology and Resources Management (ILR), Germany*

Land use effects sediment yield in the headwaters of the Sondu River basin in Kenya's largest remaining tropical montane forest, the Mau Forest Complex. Streams containing sediments originating from eroding cultivated land or unpaved roads can lead to on- and off-site effects that have economic and social impacts. The Sondu River drains into Lake Victoria, the second largest fresh water lake in the world, which is an important water resource for five countries and the source of the River Nile. East African streams are often characterised by high concentrations of suspended sediment, however, the availability of sediment yield data is severely limited. Within Kenya, there are only a few studies. Here, we report on a unique, four-year, high temporal resolution turbidity monitoring suspended sediment dynamics data set (10.2014–12.2018) which we use to assess temporal and spatial variability within three catchments (27–35 km^2) under distinct land uses (natural forest, smallholder agriculture and tea/ tree plantations). As a surrogate for total suspended sediments, turbidity was calibrated by using an ex-situ suspended sediment increment suspension method. The outlet of the natural forest catchment had the lowest suspended sediment concentrations (34.5±76.4 mg L^{-1}) followed by the tea/ tree plantation (46.9±88.6 mg L^{-1}) and highest concentrations were observed at the outlet of the smallholder agricultural catchment (121.0±180.6 mg L^{-1}). Annual rainfall varied from 1554, 1730 to 1842 mm yr^{-1} among the three catchments of the tea/ tree plantations, smallholder agriculture and natural forest, respectively. Sediment catchment yields showed seasonal variations in all three catchments with highest seasonal yields (4.2 – 24.1 t km^{-2} $month^{-1}$) during the long rains (Apr - Jul) and lowest yields (0.4 - 0.5 t km^{-2} $month^{-1}$) during the dry season (Jan - Feb), closely followed by discharge and rainfall patterns during the four year observation period. The high-resolution data set showed that the catchment under smallholder agriculture had the highest sediment yields when compared to the catchments under tea/ tree plantation and natural forest (106.0, 40.3 and 20.4 t km^{-2} $year^{-1}$, respectively). The study indicated the importance of keeping natural forest ecosystems and managing agricultural systems to protect water resources from sediment pollution.

Keywords: Land use, suspended sediment yield, temporal variation, ropical mountain catchments, spatial variation

---

Contact Address: Jaqueline Stenfert Kroese, Lancaster University, Lancaster Environment Center, LA1 4YQ Lancaster, United Kingdom, e-mail: j.stenfertkroese@lancaster.ac.uk

Influence of Farm Size and Management Practices on Landscape Trajectories in Southern Mexico

CAROLINA BERGET MORENO[1,2], GERARD VERSCHOOR[2], FRANS BONGERS[1], EDUARDO GARCIA FRAPOLLI[3]

[1] *Wageningen University & Research, Forest Ecology & Management, The Netherlands*
[2] *Wageningen University & Research, Sociology of Developm. & Change, The Netherlands*
[3] *Universidad Nacional Autonoma de Mexico, Economia Ecologica, Mexico*

The Marqués de Comillas region in Southern Mexico is embedded in the Selva Lacandona which has been designated as a priority area for conservation because of the important ecosystem services it provides. This research concentrates on comparing two adjacent communities (Loma Bonita and Chajul) in the region, by describing and analysing how the land use history, farm size, management practices and perceptions towards the land, has influenced the landscape trajectories of 40 farmers' plots since their colonisation around forty years ago. Qualitative and quantitative data were derived from 40 farms' visits (where farms were mapped manually and with GPS) and respective semi-structured interviews with owners (20 in each community). Interviews consisted of questions about the history, characteristics, management and perceptions of each of the land uses found in each plot. The results show two contrasting case studies with different land use histories, in which Chajul is characterised by more diversified landscape trajectories: large-tracts of conserved forest, pastures, agriculture, reforestations, oil palm and rubber plantations; while Loma Bonita presents only three landscape trajectories: pastures, agriculture, and some small patches of forest. These trajectories are related to differences in farm sizes and consequently different land use practices in each community. Chajul is a bigger community in which the farm size per household ranges from 50–150 hectares. Loma Bonita is smaller and farms per household are only 20 hectares. Bigger farms in Chajul allowed farmers to diversify their land-use management portfolio. Smaller farms in Loma led to less land use practices and therefore less trajectories. We conclude that a major finding of this research is the relationship between farm size and landscape change. Generally a link is made between "large" = "bad" (in terms of forest cover), but, at least in our case study, the relationship is the other way around: "large" = "more forest cover". Other important factors physically influencing landscape change have been the land management practices of the farms, the reasoning behind these practices, and the perceptions towards the land.

Keywords: Farm size, forest, land-use change, landscape trajectories, management practices

Contact Address: Carolina Berget Moreno, Wageningen University & Research, Forest Ecology & Management, Gerdesstraat 85, 6701 AH Wageningen, The Netherlands, e-mail: berget.carolina@gmail.com

Effects of Land Use Change in the Hydrophysical Properties in Vertisols in Northeastern Mexico

María Inés Yáñez Díaz, Israel Cantú Silva, Humberto González Rodríguez

Autonomous University of Nuevo Leon, Fac. of Forest Sciences, Mexico

The impact of the use of natural resources associated with anthropogenic activities has increased evidently, mainly through land use changes which have altered hydrophysical properties of soils. We hypothesised that, in the same soil type (Vertisol), four types of land use system (grassland, agriculture, Eucalyptus plantation and thornscrub) and seasonal variation can modify the soil hydrophysical properties. Results showed significant differences between land use systems and seasons for hydraulic conductivity (K), infiltration capacity (fp), and cumulative infiltration (fc). There were no seasonal differences in soil penetration resistance (SPR), bulk density (ρd) and total porosity (φP). Grassland presented higher values for ρd (1.2 g cm $^{-3}$) and SPR (5.3 kg cm^{-2}) and lower for K (0.8 × 10–5 cm s^{-1}) and φP (53 %), unlike thornscrub. Agriculture presented lower SPR (0.4 kg cm^{-2}), while plantation showed similar values when compared to the thornscrub. Kostiakov infiltration model was fitted to land use infiltration curves, showing differences between land use and season. The values for fp oscillated between 53.6 and 548.8 mm h^{-1} and fc ranged from 105.3 to 1061 mm. The order of the infiltration values goes as follows: agriculture $>$ plantation $>$ thornscrub $>$ grassland. Land use changes in Vertisols induced modification of soil physical properties affecting processes like permeability, soil compaction, and water availability. It can be concluded that although the edaphic factor is dominated by the characteristics of the interaction of factors at the time of its formation, the anthropogenic activities will determine the functioning of the soil-system, contributing to changes that modify the physical properties, affecting ecological stability and economic continuity.

Keywords: Bulk density, hydraulic conductivity, infiltration, land use system, porosity, soil penetration resistance, Vertisol

Contact Address: María Inés Yáñez Díaz, Autonomous University of Nuevo Leon, Dept. of Agroforestry, Carretera Nacional km 145, 67755 Linares, Mexico, e-mail: inesyd@gmail.com

Assessing the Effect of Different Spatial Resolutions in Soil Erosion Modelling - Case Study in a Highland Tropical Watershed in Southeast Mexico

SERGIO NARANJO[1], GEORG CADISCH[1], FRANCELINO RODRIGUES[2]

[1]*University of Hohenheim, Inst. of Agric. Sci. in the Tropics (Hans-Ruthenberg-Institute), Germany*

[2]*International Maize and Wheat Improvement Center (CIMMYT), Sustainable Intensification Program,*

Cuauhtemoc watershed, subject of this study, is located in the Santa Catarina Tayata municipality in the Mixteca Alta region in southeast Mexico. The municipality is characterised by low population density, smallholder land tenure (bienes comunales), and diverse land cover (e.g., mature forest, grass-dominated areas, arable land, eroded land). The region combines culturally diverse history, complex geology, diverse topographical relief, and moderate precipitation regime. Some areas in the region, experience a soil erosion problem termed as "ecological disaster" as a consequence of dynamic interactions of these conditions coupled with a long history of anthropogenic influences.

Aiming to predict soil erosion at the study unit / landscape level, a watershed of 2.5 km^2 was chosen. Sediment yield per precipitation event was measured using a collection station at the outlet of five study units under different land cover (i.e. forest -SUFO-, maize cultivation -SUM1 and SUM2-, fallow -SUFA-, and eroded lands - SUEL-). Soil properties, biological and topographical influences, and precipitation from May to September 2017 were measured, together with a Digital Surface Model (DSM) derived from an unmanned aerial vehicle (UAV) flight and a Digital Elevation Model (DEM) derived from public sources (INEGI).

High resolution UAV´s DSM (0.2 m) was the basis for study unit level modelling. This dataset provided the most realistic topographical conditions given its level of detail. Study unit parameters (e.g. soil properties, biological influences) were the input to OpenLISEM, a physically-based soil erosion model. Using measured soil loss, a validation set was derived and model performance parameters evaluated for adequacy of the model.

The availability of high-resolution DSM provided the opportunity to assess the effects of six different spatial resolutions on hydrologic / soil erosion processes. Validated parameters derived for the 0.2 m resolution set previously derived, were the input for other five resampled resolutions (0.4, 1.0, 4.0, 8.0, and 15.0 m) sets to assess the effect of different spatial resolution in soil erosion modelling at the study unit level.

Keywords: Deposition, detachment, eroded land, forest, infiltration, runoff

---

Contact Address: Sergio Naranjo, University of Hohenheim, Inst. of Agric. Sci. in the Tropics (Hans-Ruthenberg-Institute), Stuttgart, Germany, e-mail: naranjosa@gmail.com

Investigation of Soil Quality Index at Different Land Management Practices in North of Iran

REZA MIRZAEITALARPOSHTI, HOUMAN LIAGHATI
Shahid Beheshti University, Environmental Sciences Research Institute, Iran

Several different soil management-based approaches have been developed for assessing changes in soil quality caused by land management practices. Most Iranian soils have been severely exhausted since many decades because of over utilisation and mismanagement. Because sustainability in agricultural production systems is closely related to soil quality, its assessment has received a great attention. The current study was conducted to evaluate soil quality under different land use management on Talar Sub-Basin, Northern Iran. Sixty-five soil sites were sampled under (i) native forest (NF), (ii) pasture (P), (iii) cropland converted to orange (O), (iv) double-cropping rice, followed by winter fallow (DR), (v) rice followed by winter faba-bean (RF) and (vi) single rice followed by winter fallow (RF). We adopted 10 physico-chemical and biological soil indicators to evaluate overall soil quality index and its chemical, physical, and biological sectors using the Soil Management Assessment Framework (SMAF). Land use change effects were efficiently described by SMAF scores representing soil quality over different land management practices. Native forest showed the highest overall SQI for its chemical, physical, and biological sectors by scores ranging from 0.83 to 0.93. The overall SQI of native forest soil was 0.91 showing a major proportion of functioning of their potential capacity. SQI values decreased significantly when native forests were converted to pasture and agriculture production systems. The SQI values ranged from 0.69 at DR to 0.81 at O and RF production systems. Generally, conversion from cropland to perennial trees like orange and adopting legumes in rice-based cropping systems led to significant improvement in SQI. A positive linear correlation was obtained between SQI scores and soil organic carbon representing 73 % of the overall SQI variability. SMAF enabled to combine the individual soil indicators into an overall index for assessing land-use management effects on soil quality and functions. Evaluated SMAF scores could identify the principle soil limitations and therefore can be used to adopt priorities for specific management practices. Therefore, appropriate farming management practices must be developed to improve soil quality and the sustainability of agroecosystems in Northern Iran.

Keywords: Agroecosystems, forest, land management, SMAF, soil quality index

Contact Address: Reza Mirzaeitalarposhti, Shahid Beheshti University, Environmental Sciences Research Institute, Tehran, Iran, e-mail: rezamirzaei57@gmail.com

Improving the Model Prediction of Soil Temperature under Rubber Plantations

XIAOMIN YU[1], GERHARD LANGENBERGER[2], GEORG CADISCH[2], SERGEY BLAGODATSKY[2]

[1]*University of Hohenheim, Inst. of Farm Management, Germany*

[2]*University of Hohenheim, Inst. of Agric. Sci. in the Tropics (Hans-Ruthenberg-Institute), Germany*

The expansion of natural rubber plantations in the tropics has been the major driving force for deforestation and forest degradation in mountainous Southeast Asia, coupled with losses of ecosystem functions. The observed losses of soil organic carbon due to organic matter decomposition is mainly influenced by soil temperature, which is a key variable in Land Use Change Impact Assessment (LUCIA) model. However, measured records of site-specific soil temperature are often not available, especially in regions where data are hard to acquire. This has led to a number of studies on estimating and modelling soil temperature with easy-to-get data.

The current study aims at calibrating and further developing the existing Kang soil temperature model, to make it more suitable for the application in tropical rubber plantations. The model is a hybrid model based on heat transfer physics and empirical relationships between air and soil temperature, considering the effects of canopy and ground litter on heat attenuation. Input data needed for the model include daily air temperature and leaf area index (LAI). Our model development includes: (i) adding an additional coefficient to the ground litter term, followed with parameter optimisation; and (ii) developing a litter decomposition function calibrated for rubber plantations, and integrating this function into the existing soil temperature model.

The study is done within the framework of SURUMER project and data used for model development were collected from two locations in Naban Nature Reserve in Yunnan, southwest China. As outcomes, two model versions were validated and can be provided to users, depending on their data availability. Package A needs measured air temperature and LAI on a daily basis, and Package B additionally needs litter dynamics data. Both models outperformed Kang's model significantly, with increased Nash-Sutcliffe efficiency and decreased mean absolute error at 10 cm, 20 cm and 40 cm soil depths.

Keywords: Modelling, rubber plantations, soil temperature

Contact Address: Xiaomin Yu, University of Hohenheim, Inst. of Farm Management, Schwerzstraße 42, 70593 Stuttgart, Germany, e-mail: xiaomin.yu@uni-hohenheim.de

Functional Diversity of the Soil Macro Fauna in Different Ecosystems

ONELIO FUNDORA[1], EMMA PINEDA[2], ILIÁ LUGO[2], GEORGINA GÁLVEZ[3], BETTINA EICHLER-LOEBERMANN[4]

[1] *University of Santa Clara, Fac. of Agriculture, Cuba*

[2] *Institut of Sugar Cane Research, Cuba*

[3] *Central University of Las Villas, Cuba*

[4] *University of Rostock, Fac. of Agricultural and Environmental Sciences, Germany*

Cropping systems strongly affect biological soil characteristics. In order to determine long-term effects (30 years) of cropping systems on the diversity of soil macro fauna a study was carried out on Vertic Haplustept soils. Four sites with different cropping types were included in this study: i) annual crops, ii) perennial crop (monoculture of sugar cane), iii) forest, and iv) natural grassland. Different indicators of the macro fauna in the soil were determined, highlighting their function (soil engineers, detritivores, herbivores and predators) and considering their impact on soil fertility. The number of individuals per taxon (orders and classes) and per function was determined in an area of 25 cm $\times$ 25 cm and a soil depth of 0 to10 cm and 10 to 20 cm. The results showed that in all cropping systems investigated indicators of the macro fauna were related to the soil fertility (soil organic matter, bulk density, and nutrient contents). The forest showed the highest functional biodiversity and a relative high abundance of taxa belonging to soil engineers (Haplotaxida and Hymenoptera) and detritivores (Dictyoptera, Isoptera, Pulmonata, and Spirobolina), which are also decisive for physical properties of soils and quality of soil organic matter. For the grassland and the sugarcane sites fewer individuals in comparison to forest were found. In tendency the grassland was superior to the sugarcane site in terms of functional biodiversity and soil fertility. The site with annual crops was found to have the lowest numbers of individuals. Presumably, the relative low amount of crop residues, periods without soil coverage, and higher frequency and intensity of tillage resulted in negative effects on these soil characteristics. Following, the management of cropping systems should also focus on aspects of soil biology and the diversity of soil organisms.

Keywords: Biodiversity, cropping systems, soil biology

Contact Address: Bettina Eichler-Loebermann, University of Rostock, Fac. of Agricultural and Environmental Sciences, J. von Liebig Weg 6, 18059 Rostock, Germany, e-mail: bettina.eichler@uni-rostock.de

Soil fertility and nutrient cycling

Oral Presentations

Posters

Towards Integrated Soil Fertility Management (ISFM) Practices to Increase Cocoa Productivity in Cameroon

Urcil Papito Kenfack Essougong[1], Lotte Woittiez[1], Precillia Ijang Tata Ngome[2], Wouter Vanhove[3], Philippe Boudes[4], Syndhia Mathé[5], Maja A. Slingerland[1]

[1]*Wageningen University and Research, Plant Production Systems, The Netherlands*

[2]*National Institute of Agricultural Research Development (IRAD), Cameroon*

[3]*Ghent University, Dept. of Plant and Crops - Lab. for Tropical Agronomy, Belgium*

[4]*Agrocampus Ouest, Physical Environment, Landscape Architecture and Territorial Development, France*

[5]*CIRAD, UMR Innovation, France*

Africa supplies most of the world cocoa, but yields remain low. ISFM practices are required to increase cocoa productivity in Africa. Peoples' perceptions, closely coupled with their knowledge, are key in explaining their behaviour. Little is known on how cocoa farmers understand and manage soil fertility, and view its importance for achieving high yield. We address this gap by providing data from a survey of 120 Cameroonian cocoa farmers equally divided over two cocoa production basins with distinct predominant vegetation and mean annual rainfall: Mbam-and-Inoubou (low-shrub-savannah and 1300 mm), and Mvila (humid dense forest and 1650 mm). We found that most farmers agreed or strongly agreed that high yields (90 %), dark colour (85 %), easy ploughing (82 %), specific plant species presence (80 %), and diverse-and-abundant vegetation (79 %) indicate a fertile soil. Respectively 30 % and 13 % of farmers in Mbam-and-Inoubou and Mvila consider soil fertility low or very low. 100 %, 40 %, 34 %, 16 % and 14 % of farmers respectively used crop residues, tree planting, mineral fertilisers, compost and manure. More farmers in Mbam-and-Inoubou than in Mvila used manure (20 % and 7 %, $p = 0.03$) and mineral fertilisers (47 % and 20 %, $p = 0.002$). To justify the non-use of mineral fertilisers, limited accessibility (52 %), "satisfying soil fertility" (23 %), and lack of knowledge (15 %) were cited. Regarding organic fertilisers (manure and compost), lack of knowledge, labour requirement, limited accessibility, and "satisfying soil fertility" were cited by farmers respectively 42.5 %, 24.5 %, 23.5 % and 9.5 % of times. Furthermore, farmers view SFM as the least important cocoa farming practices for high yields; SFM practices scoring on average 2.92 compared to 4.27 out of 5 for the other. The findings suggest that soil fertility perceptions, inputs access, and local habits influence farmers' SFM practices. Farmers combine different practices, but lack knowledge on the optimal practices combination for high yield. The authors recommend to raise farmers' awareness on the need to 'continuously feed cocoa trees'; move towards ISFM practices by combining fertilisers uses and good agricultural practices taking into account local conditions. Furthermore, it is essential to think about measures that improve inputs access and returns on ISFM investment.

Keywords: Adoption, Africa, cocoa, good agricultural practices, indicators, intensification, knowledge, soil fertility

---

Contact Address: Urcil Papito Kenfack Essougong, Wageningen University and Research, Plant Production Systems, Wageningen, The Netherlands, e-mail: urcil.kenfackessougong@wur.nl

10 m x 10 m Map of Soil Organic Carbon and Major Nutrients: Towards Plot Level Soil Fertility Management

POWELL MPONELA[1], GRACE VILLAMOR[1], LULSEGED TAMENE[2], SIEGLINDE SNAPP[3], QUANG BAO LE[4], CHRISTIAN BORGEMEISTER[1]

[1]*University of Bonn, Center for Development Research (ZEF), Ecology and Natural Resources Management, Germany*
[2]*International Center for Tropical Agriculture (CIAT), Agroecosystems and Sustainable Landscapes Research Area, Malawi*
[3]*Michigan State University, Plant Soil and Microbial Sciences, United States of America*
[4]*International Center for Agricultural Research in the Dry Areas (ICARDA), Jordan*

Much as the declining trends of soil fertility is recognised as a major threat to food production for the rural smallholder farmers in Africa; the knowledge of the status and gaps at the management scale of a managed plot is limited. This study uses measured soil attributes collected from 219 locations within 10 $\times$ 10 km land degradation surveillance site in rural Malawi in 2013 and 2018 to predict major plant nutrients and carbon in soils. We used randomForest regression model in R with rain and dry season Sentinel2 imagery, Shuttle Radar Topography Mission (STRM) terrain attributes, geo-coordinates, and soil lithological classes as predictors. the prediction accuracy using out of the bag error was around 90 %. Findings suggest that there are pronounced spatial variations with most soils deficient in soil organic carbon (SOC) and total nitrogen (TN) but have low to adequate phosphorus (P) and potassium (K). The predicted mean$\pm$sd for SOC and TN of 1.06$\pm$0.27 % and 0.07$\pm$0.01 % are lower than the critical levels of 2.0 and 0.15 whilst for P and K, 35.99$\pm$33.04 mg kg^{-1} and 132.87$\pm$23.15 mg kg^{-1} are within the low to high ($>$11 and $>$17) and deficient to moderate ($<$125 and $>$190), respectively. The C:N and C:P stoichiometry and structural stability index show that the limited SOC could potentially affect retention and availability of both N and P but also lowers soils physical stability. Hence, organic input sources should be integrated in the currently inorganic fertiliser dominated soil management to minimise soil degradation risk and ensure agricultural sustainability.

Keywords: Plot level, predictive soil mapping, randomForest, soil fertility, soil nutrient limits, soil structural stability

---

Contact Address: Powell Mponela, University of Bonn, Center for Development Research (ZEF) - Ecology and Natural Resources Management, Genscherallee 3, 53113 Bonn, Germany, e-mail: powell.mponela@uni-bonn.de

Soil Phosphorus and Crop Phosphorus Use Efficiency as Affected by Residue Return on Weathered Soils of West Africa

EEUSHA NAFI[1], HEIDI WEBBER[2], ISAAC DANSO[3], JESSE B. NAAB[4], MICHAEL FREI[1], THOMAS GAISER[1]

[1] *University of Bonn, Inst. Crop Sci. and Res. Conserv. (INRES), Germany*

[2] *Leibniz Centre for Agricultural Landscape Research (ZALF), Germany*

[3] *CSIR - Oil Palm Research Institute, Ghana*

[4] *West African Science Service Center for Climate Change and Adapted Land Use (WASCAL), Burkina Faso*

Most soils in West Africa are acutely deficient in Phosphorus (P); therefore, sustainable P management is important in order to restore their capacity to supply adequate P, improve crop P use efficiency, and sustain crop production. Thus, we investigated the impacts of alternative management practices on soil available phosphorus (Pcal), soil P balance (Pb), and crop P use efficiency (Pe) under maize-cotton rotation system on four weathered soils [Ferric Lixisol (FL), Eutric Plinthosol (EP), Haplic Lixisols (HL), and Plinthic Lixisol (PL)] of West Africa. To this end, on-farm trials were set up in a strip-split plot layout, where 2 levels of tillage (contour ridge tillage, and reduced tillage) were considered as main-plot factor, and sub-plot factors included 2 levels of crop residue management (with and without), and 2 levels of N fertiliser doses (control and recommended dose). After 5 cycles of maize-cotton rotation (2012–2016), Pcal in the topsoil layer were significantly affected by single and interactive effects of soil and residue management. The highest content of Pcal was recorded on HL (64.4 kg ha^{-1}), while both FL and EP had the lowest content (27.1 kg ha^{-1}). Returning crop residues caused an 34.8 % increase in Pcal when averaged across soil types. Also, both Pe and Pb of cotton and maize varied according to soil and crop residue management and their interactions. Pe trends of both cotton and maize were ranked as: HL > PL > EP > FL. As expected, Pb showed the reversed pattern. With the retention of crop residues on surface soils, Pe of cotton and maize improved by 8 % and 8.5 %, respectively. We concluded that soils derived from volcanic materials (high clay content) attributed to poor topsoil Pcal and crop Pe compared to those derived from sandstone (low clay content) materials. Additionally, retention of crop residues led to improved Pcal and crop Pe by acting as an additional source of soil P and protecting surface soil from erosion. Hence, we suggest that returning residues can be a viable measure to manage soil P and sustain crop production in weathered soils of West Africa.

Keywords: Cotton-maize rotation, crop residue, phosphorus use efficiency, soil available phosphorus, West Africa

Contact Address: Eeusha Nafi, University of Bonn, Inst. Crop Sci. and Res. Conserv. (INRES), Katzenburgweg 5, 53115 Bonn, Germany, e-mail: eeusha.nafi@uni-bonn.de

Management Effects on Soil Microbial Stoichiometry and Fungal-Bacterial Ratio of Rural-Urban Interfaces in Bengaluru

Virna Estefania Moran Rodas[1], Rainer Georg Joergensen[1], Chickadibburahalli T. Subbarayappa[2], V.C. Suvarna[3], Hanumanthappa Dimba Chowdappa[2], Z. Mudalagiriyappa[2], Christine Wachendorf[1]

[1] *University of Kassel, Soil Biology and Plant Nutrition, Germany*

[2] *University of Agricultural Sciences, Dept. of Soil Sci. and Agricultural Chemistry, India*

[3] *University of Agricultural Sciences, Department of Agricultural Microbiology, India*

Soil microbial communities perform functions essential to agricultural production, and to regional and global element dynamics. The specific element ratios (stoichiometry) of these soil communities are directly linked to organic matter decomposition rates and may allow to predict patterns of nutrient mineralisation or immobilisation. Such stoichiometry is affected by the type and intensity of land management such as governed by urbanisation, nitrogen fertilisation and irrigation. However, the underlying pathways and their variation across agroecosystems are difficult to predict, due to the many factors involved. To contribute in filling this knowledge gap we analysed the elemental ratios (C, N and P) and microbial communities, including their fungal and bacterial portions, pH, water content, texture and water-holding capacity of typical Nitisol top-soils exposed to different management intensities in the rapid-growing city of Bengaluru, India. Samples were collected in plots that differed in water supply (intensively irrigated and rain-fed), low versus high N-fertilisation intensity and two major crops (maize and millet). Our results showed that differences are largest between the two experiments. Organic C and total N, microbial biomass C and N, the ratio of organic C to total N and the contribution of the fungal biomarker, ergosterol to microbial biomass were higher in the irrigated site. In contrast the microbial biomass ratio of C:N and microbial biomass phosphorus were higher in the rain-fed experiment. Treatment effects within each experimental site are less pronounced, and the stoichiometric differences between crop types seemed to be larger than among N fertilisation intensities. The different stoichiometric patterns observed in this study reflect management effects on microbial carbon use efficiency and, therefore, in nutrient cycling and organic matter stabilisation.

Keywords: Irrigation, microbial stoichiometry, nitrogen fertilisation, soil organic matter, soil stoichiometry

---

Contact Address: Virna Estefania Moran Rodas, University of Kassel, Soil Biology and Plant Nutrition, Nordbahnhofstr. 1a, 37213 Witzenhausen, Germany, e-mail: virna.moran@uni-kassel.de

Eating the Soil – Environmental Effects on Nutrient Concentrations in Food Produced on Soils of Different Fertilities

Sahrah Fischer[1], Thomas Hilger[1], Irmgard Jordan[2], Jeninah Karungi[3], Georg Cadisch[1]

[1] *University of Hohenheim, Inst. of Agric. Sci. in the Tropics (Hans-Ruthenberg-Institute), Germany*

[2] *Justus-Liebig University Giessen, Center for International Development and Environmental Research, Germany*

[3] *Makerere University, Department of Agricultural Production, Uganda*

Very little is known about the connection between food nutrient composition and environmental factors. The main aim of this study is therefore to elucidate the importance of environmental factors on the variance of food nutrient concentration in different types of food crops; specifically (i) whether soils of different fertilities produce foods with significantly different nutrient concentrations; (ii) do environmental or anthropogenic factors have a significant influence on the nutrient concentration of different food crops and different soil types?

Maize grain (*Zea mays*, n=31) and cassava tuber (*Manihot esculenta*, n=27) samples were collected in Teso South, Kenya (low fertility) and maize grain (n=30) and matooke fruit (*Musa acuminate*, n=54) samples collected in Kapchorwa, Uganda (higher fertility) paired with soil samples. All samples were measured using a portable X-Ray Fluorescence Spectrometer for the total concentration of macro (Mg, P, S, K, Ca) and micronutrients (Fe, Zn, Mn, Cu). Soils were evaluated for eCEC, N and C content, pH and texture. Yields per field as well as anthropogenic factors (fertilisation, distance to household, crop species richness and diversity) were collected. Canonical Correspondence Analysis (CCA) coupled with permutation ranking tests per crop and region were done.

Soil of higher fertility produced foods with significantly higher food nutrient concentrations and yields. Although the largest food nutrient difference between the two areas was micronutrients, the largest deficiencies in both areas were macronutrients. Variance in food nutrient concentration was described best by full CCA models (Maize >80%; Cassava 76%; Matooke 39 %). Surprisingly, pH was not a significant factor affecting nutrient concentration in the acidic low fertility region. Agrobiodiversity had no effect in the low fertility area and a significant positive effect in the higher fertility region. A natural dilution effect observed in the higher fertility region becomes important when planning fertiliser strategies, as a higher nutrient input could exacerbate this effect. Generative storage organs (grain) of annual plants were affected most by environmental effects compared to storage organs (tuber) of perennial crops, or generative organs of perennial crops (fruits). Soil fertility as well as cultivation practices can affect the quantity and quality of produced foods.

Keywords: Agriculture, canonical correspondence analysis, environment, nutrients, plant nutrition

Contact Address: Sahrah Fischer, University of Hohenheim, Inst. of Agric. Sci. in the Tropics (Hans-Ruthenberg-Institute), Garbenstr. 13, 70599 Stuttgart, Germany, e-mail: sahrah.fischer@uni-hohenheim.de

Modelling Crop Diversification Strategies for Sustainable Intensification in Dryland Cropping System of Semi-Arid Southern India

RANGANATH LAKSHMAIAH, ERNST-AUGUST NUPPENAU, STEPHANIE DOMPTAIL

Justus-Liebig University Giessen, Inst. of Agric. Policy and Market Res., Germany

Agriculture intensification in dryland cropping systems has relied upon adoption of crops with high yielding varieties and external chemical fertilisers. This intensification further coupled with declining land size has led to increase in monocropping and has completely disregarded potential of crop diversification as soil fertility management strategy. Legume-based crop diversification with appropriate input management strategy have proven to increase the productivity of the soil based on long-run field level experiments. However, in short-run, farmer adopting a legume-based rotation must forgo returns from relatively high remunerative cereal crop for less remunerative legume crop. Any recommended crop diversification strategy must inform farmer on this trade-off in short-run and appropriate input management strategies in cropping system for sustainable long-run benefit. In this study, we model adoption of legume based cropping patterns as a part of field level sustainable intensification strategy for dryland agriculture from economic point in semi-arid region of southern India. Sustainable intensification of dryland cropping systems is defined as efficient allocation of external inputs (non-renewable resources) and temporal choice (long-run) of crops over finite period for given output, factor prices and crop yield levels.

A plot level data of 838 dryland plots collected from 198 family households for the year 2013–14 and 2014–15 is used for analysis. Maize, finger millet and groundnut are major crop in the region that are grown individually as monocrop or two crop rotation (finger millet and groundnut) or three crop rotation (finger millet, maize and groundnut). Optimal input allocation strategy is evaluated using yield response function and alternative input management strategies (with only high fertiliser, fertiliser and organic matter and low chemical fertiliser) observed at field level for each crop in five different crop patterns. Grid point approximation technique is used obtain optimal input. Finally, long term optimal cropping pattern is evaluated using dynamic linear programming technique by imposing different level of yield penalty for monocropping based on long term field experiments results. This model further adds value to experimental station results by identifying possible ways to achieve sustainable intensification through optimal choice of cropping pattern and input management strategies at farmers' field.

Keywords: Cropping system, soil fertility management, sustainable intensification

Contact Address: Ranganath Lakshmaiah, Justus-Liebig Universität Gießen, Inst. for Agricultural Policy and Market Research, unterhof, 35392 Gießen, Germany, e-mail: ranga.laki@gmail.com

Organic Carbon and Micro Nutrients Distribution in Agricultural Systems along Rural-Urban Interface of Southern Transect Bengaluru

ALTAF KUNTOJI, CHICKADIBBURAHALLI T. SUBBARAYAPPA, GURURAJ KOMBALI

University of Agricultural Sciences, Dept. of Soil Science and Agricultural Chemistry, India

Urbanisation is a global trend rapidly transforming the biophysical and socioeconomic structures of metropolitan areas. Bengaluru being one of the megacities in the world, expanding rapidly over time. Urbanisation is witnessing consumer demand for vegetables, fruit trees and flowers which induced major shift in agricultural land use towards an intensive irrigated multi cropping system from conventional crops. Intensification of agriculture has an adverse effect on physical, chemical and biological properties of soils, and their potential to synchronise nutrient supply and demand in agricultural cropping systems. This change in cropping system made farmers to use inorganic fertilisers indiscriminately. This monopoly in the use of inorganic fertilisers (N fertilisers in particular) has led to decline in organic carbon content of agricultural lands and their by imbalance of micronutrients in the soil and tends to decline in the yield levels. Hence, study was conducted to know the distribution of organic carbon and micro-nutrient agricultural systems along rural-urban interface of southern transect of Bengaluru. The sampling was done across villages of southern transect based on survey stratification index (SSI), which is worked out by taking distance and buildup area into an account. The transect was devided into urban, transition and rural areas. Different cropping systems were considered while taking the samples from each urbanisation gradients. The mean organic carbon content of soil varies from 0.69, 0.474 and 0.38 per cent in rural, transition and urban areas, respectively. Nutrient index values for organic carbon is found to be low across transect. However, available Fe, Zn, Cu, Mn and B content of soils varied significantly along urbanisation gradient. Available Fe, Zn, Cu, Mn and B were 7.28, 0.83 0.60, 5.41 and 0.75 ppm in rural areas, 11.44, 0.56, 1.12, 8.29 and 1.06 ppm in transition and 4.16, 0.54, 0.44, 3.44 and 0.44 in urban areas, respectively. Nutrient index values for all the micronutrients are found to be medium in rural areas, high in transition and low in Urban areas of Bengaluru.

Keywords: Agricultural systems, Bengaluru, micronutrients, organic carbon

Contact Address: Gururaj Kombali, University of Agricultural Sciences, Bengaluru, Dept. of Soil Science and Agricultural Chemistry, UAS, GKVK, 560065 Bengaluru, India, e-mail: rajck.kakol@gmail.com

Leaching of Soil Nutrients from Tropical Cropping Systems at Different Levels of Agricultural Intensity in Bengaluru, Southern India

Renuka Suddapuli Hewage[1], Beate Madan[1], Andrea Mock[1], Hanumanthappa Dimba Chowdappa[2], Chickadibburahalli T. Subbarayappa[3], Ellen Hoffmann[1], Andreas Buerkert[1]

[1]*University of Kassel, Organic Plant Production and Agroecosystems Research in the Tropics and Subtropics, Germany*

[2]*University of Agricultural Sciences, AICRP on Agroforestry, India*

[3]*University of Agricultural Sciences, Dept. of Soil Sci. and Agricultural Chemistry, India*

Emerging mega cities with their rapidly growing urban population exert increasing pressure on the available arable land to satisfy burgeoning food demand. Meeting these demands requires intensified input use in multiple cropping systems as can be studied along the rural-urban interface in Bengaluru, South India. Under deeply weathered soil and tropical monsoon conditions, high rates of inorganic inputs into the soil improve the soil nutrient stock temporally, but may also cause high leaching losses which so far are poorly quantified. This study therefore aimed to measure seasonal leaching losses of nitrogen (N), phosphorus (P), potassium (K) and sulphur (S) at two crop specific inorganic N-fertiliser intensities (low and high), and two different water regimes (rainfed with life-saving supplementary irrigation and drip irrigated) in a factorial on-station experiment with a rotation of cabbage, eggplant, tomato in the dry (*rabi*) season followed by maize, finger millet, lablab in the wet (*kharif*) season. Leaching of soil water was determined at weekly intervals throughout the growing using micro-lysimeters. Analysis of collected data showed cumulative water losses by leaching to be 35 % higher in the dry season than in the wet season 2017. As an example, for the variation of nutrient losses among different crops, the concentration of N (as NO_3^- and NH_4^+) in the leachate were higher in tomato than in eggplant and cabbage in the dry season, while in the wet season, lablab showed the lowest concentrations compared to finger millet and maize. In addition to those initial results, ongoing data analysis in a multi-annual study should give an even deeper understanding of the effects of inter- and intra-seasonal variations on nutrient leaching in intensive high inputs cropping systems under the influence of the monsoon climate condition in India.

Keywords: Agricultural intensification, India, leaching, lysimeters, mega city, soil nutrients

---

Contact Address: Renuka Suddapuli Hewage, University of Kassel, Organic Plant Production and Agroecosystems Research in the Tropics and Subtropics, Steinstrasse 21, 37213 Witzenhausen, Germany, e-mail: r.suddapuli@uni-kassel.de

Estimation of the Optimal Nitrogen Dose in a *Brachiaria humidicola* – Corn Rotation System in the Colombian Eastern Plains

Brenda Muñoz, Karen Enciso, Stefan Burkart
International Center for Tropical Agriculture (CIAT), Trop. Forages Program, Colombia

Improving nitrogen use efficiency (NUE) by optimising the N fertiliser application dose is one way to reduce greenhouse gas emissions in agriculture and livestock production, especially in higher demanding crops such as corn. Taking a *Brachiaria humidicola* (Bh)-corn rotation system in the Colombian Eastern Plains as a case study, we seek to determine both the optimal economic dose (OED) and the optimal technical dose (OTD) of N, which allow to maximise income at producer level and minimise environmental impacts. This particular rotation system was chosen as research subject given the presence of the residual effect of Biological Nitrogen Inhibition (BNI) in permanent lots of Bh (established for >10 years), which has positive impacts on corn production such as increased yields and better N efficiency. The data for this study was obtained from trials conducted between 2013 and 2017, where corn production in a Bh-corn rotation system (with residual BNI effect) was compared with conventional corn production (without residual BNI effect). For determining the OED and OTD of N, three response models will be applied: a pseudo-quadratic model, a continuous curvilinear model and a discontinuous rectilinear model. The comparison of these models will also allow to identify the most suitable model for obtaining the best OED and OTD estimates of N. The results of this study will be key for providing recommendations to primary producers on the correct doses of N in a Bh-corn rotation system. This will contribute to improving both efficiency in production and profitability, and help to avoid the excessive and unnecessary application of nitrogen fertilisers and its associated negative effects on the environment.

Keywords: Agro-pastoral systems, bovine livestock, climate change, Latin America, sustainable intensification

---

Contact Address: Stefan Burkart, International Center for Tropical Agriculture (CIAT), Trop. Forages Program, Km 17 Recta Cali-Palmira, Cali, Colombia, e-mail: s.burkart@cgiar.org

Impacts of Integrated Soil Fertility Management on Yield and Household Income: The Case of Tamale (Ghana) and Kakamega (Kenya)

IVAN ADOLWA[1], STEFAN SCHWARZE[2], ANDREAS BUERKERT[3]

[1]*Kenya Agricultural and Livestock Research Organization (KALRO), KALRO Food Crops Research Centre, Alupe, Kenya*

[2]*Thünen Institute, Institute of Rural Studies, Germany*

[3]*University of Kassel, Organic Plant Production and Agroecosystems Research in the Tropics and Subtropics, Germany*

Integrated soil fertility management (ISFM) has been widely promoted by research and philanthropic organisations as well as governments to increase crop yields and improve livelihoods of smallholder farmers in Africa. Therefore, it is surprising that there is still scant information on its impact on crop yields and household income.
Taking the case of maize farmers in Tamale (northern Ghana) and Kakamega (western Kenya) we estimate a yield and income model using inverse-probability-weighted regression adjustment, which is considered a powerful estimator for observational data as it is more robust to misspecification. Farm household data has been collected between July 2014 and February 2015 from 285 farmers in Tamale and 300 in Kakamega.
The analyses reveal that ISFM adoption leads to an increase in maize yields by up to 27 % in Tamale and 16 % in Kakamega. Increasing the number of ISFM components, however, does not improve yields. Despite the positive effect on yields, adoption of ISFM does not increase total household incomes at both locations, which might be due to increased cost of production, family labour reallocation, and the low share of maize income in total household income. Cost of production for ISFM adopters are much higher because they apply more chemical fertiliser and use improved varieties. Our results suggest that ISFM is not a particularly attractive choice from the farmers' point of view when considering just private costs and benefits. At national and global levels, however, this might change, because ISFM produces positive environmental externalities such as carbon sequestration and the reduction of nutrient mining through building-up of soil organic matter. Some implications for future research are also discussed.

Keywords: Impact assessment, inverse-probability-weighted regression adjustment, maize yield, sub-Saharan Africa, total household income

Contact Address: Stefan Schwarze, Thünen Institute, Institute of Rural Studies, Bundesallee 64, 38116 Braunschweig, Germany, e-mail: stefan.schwarze@thuenen.de

Enhancing Livelihoods through Integrated Soil Fertility Management in the Highlands of Ethiopia

Jim Ellis-Jones[1], Ann-Kathrin Lichtner[2], Tesfay Halefom[2], Haile Deressa[2], Steffen Schulz[2], Rodger White[3]

[1]*Agriculture-4-Development, United Kingdom*

[2]*Deutsche Gesellschaft für Internationale Zusammenarbeit (GIZ) GmbH, Intergrated Soil Fertility Management Project Ethiopia, Ethiopia*

[3]*Rothamsted Research, Centre for Computational and Systems Biology, United Kingdom*

Many studies have shown that land degradation and declining productivity are serious problems in the Ethiopian Highlands, the major cause being a decline in soil fertility resulting from topsoil erosion, soil nutrient and organic matter depletion and often increased soil acidity. At the same time Ethiopia is being adversely affected by climate change weather patterns.

Integrated Soil Fertility Management (ISFM) aims to improve soil fertility and productivity of small-scale farmers by promoting locally adopted combinations of various ISFM technologies. These include the use of improved seed, organic fertiliser (improved compost, manure, vermi-compost) in combination with inorganic fertiliser, legumes with bio-fertiliser, line seeding, and lime application on acidic soils. Crop cuts show that ISFM technologies increase crop yields substantially, but farmers are also interested to know if their investment is also financially viable. Therefore, a cost benefit analysis has been undertaken to evaluate the financial effect of ISFM under small scale farm conditions. These have been based on farmer-led demonstrations during the 2016 cropping season, comparing ISFM and farmer practice for four major crops - maize, faba bean, teff and wheat in Amhara, Oromia and Tigray regions. Gross margins, returns to labour and benefit-cost ratios were calculated to compare the benefits with an increase of costs of purchased inputs (seed, fertiliser and lime) and an increase in labour (compost production, line seeding and applying inputs). Results show that benefits considerably exceed the costs, hence it can be financially attractive for small scale farmers to invest in soil fertility enhancing technologies. At the same time, many of the components of ISFM can also be described as "Climate Smart", meaning a win-win situation for both the environment and improving livelihoods. Nevertheless, the availability of inputs, finance and labour as well as knowledge about ISFM remain critical to scaling up and long-term sustainability.

Keywords: Climate smart agriculture, cost benefit analyses, integrated soil fertility management, ISFM, livelihood security

Contact Address: Ann-Kathrin Lichtner, Deutsche Gesellschaft für Internationale Zusammenarbeit (GIZ) GmbH, Intergrated Soil Fertility Management Project Ethiopia, Addis Ababa, Ethiopia, e-mail: ann-kathrin.lichtner@giz.de

Reducing Crop Yields Gap in Acidic Soils of Northern Tanzania Using Local Liming

DEODATUS STANLEY KIRIBA[1], KELVIN MARK MTEI[2], FLORIAN WICHERN[3], PATRICK ALOIS NDAKIDEMI[1]

[1]*The Nelson Mandela African Institution of Science and Technology (NM-AIST), Sustainable Agriculture and Biodiversity Conservation (SABEM), Tanzania*

[2]*The Nelson Mandela African Institution of Science and Technology (NM-AIST), Water and Environmental Science and Engineering (MEWES), Tanzania*

[3]*Rhine-Waal University of Applied Sciences, Fac. of Life Sciences, Germany*

Soil acidity (pH<5.4) is a widespread problem negatively affecting crop production in all agro-ecological zones of Tanzania including the northern zone. The use of lime in combination with organic and/or inorganic fertiliser has been reported to sufficiently rehabilitate very poor or depleted soils by increasing pH, reducing Al toxicity, enhancing Ca and Mg, contributing to soil structure improvement and increase both, P uptake in high P fixing soils and the plants root system. Despite the availability of the liming materials in different parts of Tanzania and their potentiality to alleviate soil acidity, there is limited use in agricultural production by smallholder farmers in the country, as there are no established liming recommendations (effective rates) of these locally available liming materials and it is even unknown, how effective local liming materials are for soil acidity management in the northern zone of Tanzania. The objectives of this research project are; i) to characterise soil fertility status (acidity levels, physical and chemical properties) of soils from Mbulu district, Northern Tanzania, ii) to characterise the quality of locally available liming materials collected from different parts of Tanzania and iii) to establish liming requirements (effective rates) of the locally available liming materials and determine their effectiveness on soil pH changes, selected soil nutrients and yield performance of maize crop under greenhouse and field experiments at the Nelson Mandela African Institution of Science and Technology (NM-AIST) and two sites in Mbulu district, northern Tanzania. The research project outline will be presented and first results will be shown.

Keywords: Buffer methods, lime, lime requirement, soil acidity, soil fertility

Contact Address: Deodatus Stanley Kiriba, The Nelson Mandela African Institution of Science and Technology (NM-AIST), Sustainable Agriculture and Biodiversity Conservation (SABEM), P.O. Box 447, +255 Arusha, Tanzania, e-mail: kiribad@nm-aist.ac.tz

Agro-Ecology Determines Farm Typology Effect on Soil Fertility Variability in Small Holder Farmers; Ethiopia

BIRHANU AGUMAS[1], GENEROSE NZIGUHEBA[2], GEORG CADISCH[1], FRANK RASCHE[1]

[1]*University of Hohenheim, Inst. of Agric. Sci. in the Tropics (Hans-Ruthenberg-Institute), Germany*

[2]*International Institute of Tropical Agriculture (IITA), Kenya*

The aim of this study was to investigate inter-related effects of agro-ecology and farmers' resource endowment (i.e., "wealthy", "medium wealthy", and "poor" farmers) on the spatial variability of soil fertility in Central and Western Ethiopia. Using a mid-infrared spectroscopic research approach, coupled to partial least squares regression analyses (midDRIFTS-PLSR), a prediction model was developed to assess soil fertility indicators across a regional scale, including various agro-ecological zones: "dega" (D) and "high dega" (HD) (2500–3500 m.a.s.l., temperature $\leq$9°C, rainfall 938 mm), "weina-dega" (WD) (1500–2500 m.a.s.l., 15 to 27°C, 1376 mm), and "kola" (K) (<1500 m.a.s.l., 15 to 27°C, 2037 mm). Furthermore, midDRIFTS peak area analysis of selected spectral frequencies (2930, 1620, 1159 cm^{-1}) was applied to characterise functional groups of soil organic carbon (SOC) and to calculate the SOC stability index 1620:2930, which were used as proxies of soil quality. Total carbon (TC) content in soils was predicted accurately ($R^2 = 0.92$, RPD = 3.46), whereas prediction of total nitrogen (TN) ($R^2 = 0.86$, RPD = 2.71) and pH ($R^2 = 0.89$, RPD = 3.02) was acceptable. Predictions of available phosphorous (Pav) and potassium (Kav) were not successful; hence, wet chemistry was used instead. Higher contents of Kav and TN (K) as well as higher TC (HD) were found in fields of wealthy compared to poor farmers. Highest and lowest areas of peak 2930 cm^{-1} were found on fields of wealthy farmers in D and K, respectively ($p < 0.05$). On the contrary, highest and lowest areas of peak 1620 cm^{-1} and SOC stability index were found on fields of poor farmers in D and K, respectively ($p < 0.05$). We conclude that for the observed soil fertility variability inter-related effects of agro-ecology and farmers' resource endowment were stronger than the individual factors. In this context, the mid-DRIFTS approach allowed a comprehensive insight into the spatially heterogeneous soil fertility across a regional scale.

Keywords: Agro-ecology, midDRIFTS, soil organic carbon

Contact Address: Birhanu Agumas, University of Hohenheim, Inst. of Agric. Sci. in the Tropics (Hans-Ruthenberg-Institute), Garbenstrasse 13, 70599 Stuttgart, Germany, e-mail: birhanuagumas1980@gmail.com

Effect of Changing Climate on Erosion in Oke-Oyi Dam Agricultural Watershed, Ilorin, Nigeria

HENRY AHAMEFULE, JOHN OLANIYAN, RIDWAN TAIWO
University of Ilorin, Nigeria, Agronomy, Nigeria

Soil loss is one of the major problems resulting from water erosion in watersheds. Agricultural lands are constantly degraded by erosion and sometimes completely lost yet not much information on these occurrences is documented particularly for agricultural watersheds in Kwara State, Nigeria. Documented information of this nature could be a fundamental resource for land-use planning and commercial crop production in kwara state to ensure food security in Nigeria. Therefore a study was conducted in Oke-Oyi Dam watershed in Ilorin, Kwara State to estimate annual soil loss using the Revised Universal Soil Loss Equation (RUSLE) model with the aid of remote sensing and GIS techniques. Data on rainfall, soil erodibility, topography, cover management and support practice were collected. The rainfall erosivity of the study area is estimated to be 83.48 MJ mm ha^{-1} $y^{-1}$and projected to reach 94.17MJ mm ha^{-1} y^{-1} in the next decade. Erodibility of the soil was found to be high and results indicated that the annual soil loss ranged from 0–1272 t ha^{-1} y^{-1}. Soil loss in about 46 % (345 ha) of the study area was observed to be very low, less than 32 % (236 ha) of the study area falling within low and moderate soil loss class whereas 22 % (163 ha) of the study area suffered severe soil loss. The soil loss in the watershed fall far above the maximum tolerable limit of 12 t ha^{-1} y^{-1} set by the Food and Agriculture Organisation (FAO) for sustainable land use. The total annual soil loss for the whole watershed is about 30,000 t ha^{-1} y^{-1} with a mean soil loss of about 40.02 t ha^{-1} y^{-1}, this figure is expected to continue to rise if adequate soil management measures are not expediently instituted. The precursors of soil loss in the study area were found to be low vegetation cover, unfavourable topographic factor and poor erosion support practice. In the long term, if the present erosion control practices of contouring is substituted with strip-cropping or terracing, soil loss will be reduced by 50.00 % and 79.99 % respectively.

Keywords: ArcGis, erosion hotspot, Oke-Oyi dam, soil loss, USLE, watershed

Contact Address: Henry Ahamefule, University of Ilorin, Nigeria, Agronomy, Tanke Akata, Ilorin, Nigeria, e-mail: ahamefule.he@unilorin.edu.ng

Towards Sustainable Land and Water Management in the Farm Level in Iran: Case of Honam Sub Catchment of Karkheh River Basin

M.R. Balali[1], Sina Mallah Nowkandeh[1], K. Eftekhari[1], H. Rezaei[1], K. Bazargan[1], V. Nangia[2], F. Moshiri[1], S. Ghalebi[1], N. Navidi[1], M. Panahi[1], F. Rejali[1], M.R. Emdad[1], N. Davatgar[1], S. Omidvari[1], M. Sepahvand[3], A. Momvandi[3], M. Matinkia[3], M. Azizolahi[3]

[1]*Soil and Water Research Institute (SWRI), Iran*

[2]*International Center for Agricultural Research in the Dry Areas (ICARDA), Jordan*

[3]*Agriculture - Jahad of Lorestan province, Iran*

To improve the production systems and livelihoods resilience of dryland area's habitats of the country, CGIAR Research Program on Dry land System (CRP-DS) methodology has taken into account to shift from linear to spiral paradigm and system approach. Institutionalization of this innovative, inter- and multi-disciplinary approach needs changing dominant culture of top down approach to bring together all stakeholders to cope with the complex issues which dry areas are confronted. Regarding this, Soil and Water Research Institute (SWRI) setup the project follow of Challenge Program on Water and Food (CPWF) during 2004 to 2009 in Karkheh River Basin (KRB) by consideration of CRP-DS concerns to be implemented in the farm level since 2012. To this end, through review of lessons learned, it has been tried to consider the experiences of CPWF1 by the team of project. Also by collaboration among international, national and local experts, the conceptual framework of project was prepared in a participatory way. Honam sub catchment (142 km^2) in Lorestan province which was one of the four pilot areas worked in the CPWF has selected to implement the project. The goal of project is to prepare soil and water packages which could be applicable on farm level. Therefore, different activities have been conducted in parallel which can be divided into two parts including stakeholders' inclusion and best bet practices and also determination of Soil and Water Homogenous Management units (SWHMUs). Data needed for determining SWHMUs which is the basis for out- and up-scaling of prepared packages contain soil profiles, surface physical and chemical properties, land use type, water allocation, supply and demand management, water consumption were collected. Soil map, soil limitation map, land use type in the farm level (1:25,000) were prepared, processed and finalized. Modifying fertility capability soil classification system (FCC) of Sanchez et al. (1982) used as a basis to separate 30 FCC for rainfed and 23 SWHUs for irrigated landuses and recommended packages (RP) was determined per homogeneous units. To validate, modify and develop RP, best bet practices has been carried out in the selected pilots through technology fair.

Keywords: Capability classification, soil homogeneous unit, sustainable management, water homogeneous unit

Contact Address: M.R. Balali, Soil and Water Research Institute, Soil Chemistry, Fertility and Plant Nutrition Department, Soil and Water Research Institute Imam Khomeini Blvd, 3177993545 Karaj, Iran, e-mail: mrbalali68@gmail.com

Partial C and N Balances of Small-Scale Farms in a River Oasis of Western Mongolia

SVEN GOENSTER-JORDAN[1], GRETA JORDAN[1], BAIGAL ULZIISUREN[2], ANDREAS BUERKERT[1]

[1]*University of Kassel, Organic Plant Production and Agroecosystems Research in the Tropics and Subtropics, Germany*
[2]*National University of Mongolia, Biology Department, Mongolia*

During the last decades, Mongolia's national policy fostered the expansion and intensification of crop and forage production. Non-sustainable agricultural management practices, however, may have led to disequilibria in soil surface matter balances of agricultural plots jeopardising the scarce and susceptible agroecological resources of river oases.
The objective of this study therefore was to calculate partial carbon (C) and nitrogen (N) balances in order to assess the bio-physical sustainability of current management strategies for agricultural plots of small-scale farms in the Mongolian river oasis Bulgan sum centre as a representative example. Based on a baseline survey, six small-scale farms were selected. After the quantification of main C and N input and output fluxes, partial balances were calculated for hay, carrot and rye plots in each farm for the growing seasons 2013 and 2014.
All C and N balances were strongly negative (-1366 kg C ha^{-1} $season^{-1}$ and -44 kg N ha^{-1} $season^{-1}$ on average). Gaseous emissions accounted for 67 % of total C losses for all agricultural products on average, while the removal of biomass through harvest and weeding was a major N output pathway (90 % on average). Although fertilisers were an important input flux particularly for N (47 % of total N and 12 % of total C input on average), the application was with an average of 569 kg C ha^{-1} $season^{-1}$ and 22 kg N ha^{-1} $season^{-1}$ low. C-fixation by plants was the main C input flux for all agricultural products (87 % of total C input) whereas N-fixation was only relevant for hay plots (67 % of total N input).
Possible measures to lower disequilibria of soil balances include the increase of C and N input through alfalfa cultivation and the intensification of organic fertilisation (sufficient organic manure could be provided by livestock from agropastoral households) whereby fertiliser should be incorporated into the soil to minimise gaseous emissions. Our study underlines the need of corrective measures to sustain soil fertility and thus agro-pastoral livelihoods in the floodplain of the Mongolian river oasis of Bulgan sum centre and similar ones in the (semi)arid areas of East Asia.

Keywords: Altai Mountains, carbon losses, nitrogen turnover, soil fertility

---

Contact Address: Sven Goenster-Jordan, University of Kassel, Organic Plant Production and Agroecosystems Research in the Tropics and Subtropics, Steinstr. 19, 37213 Witzenhausen, Germany, e-mail: goenster@uni-kassel.de

Effects of Different Stocking Densities on Soil Properties in Montane Pastures of the Chinese Altai Mountains

MARIKO INGOLD[1], SVEN GOENSTER-JORDAN[1], RAMIA JANNOURA[2], ANDREAS BUERKERT[1], RAINER GEORG JOERGENSEN[2]

[1]*University of Kassel, Organic Plant Production and Agroecosystems Research in the Tropics and Subtropics, Germany*

[2]*University of Kassel, Soil Biology and Plant Nutrition, Germany*

Overgrazing is one of the major threats for extensive terrestrial ecosystems, such as the Central Asian steppe, which belongs to the largest continuous rangeland. The constantly increasing livestock pressure in the Chinese Altai Mountains and changes of traditional mobile pastoralism to paddock-based livestock management are likely to affect soil properties, but little data exists to substantiate this claim. A grazing experiment was conducted on a summer pasture at 2400 m a.s.l. in the Chinese part of the Altai Mountain range, with average annual precipitation of 174 mm and average monthly temperatures ranging from -26 to 30°C. Sheep were kept in paddocks arranged in a completely randomised block design for a period of 56 days per year at densities of 0, 8, 16, and 24 sheep ha^{-1} in four replicates. After two grazing periods, six samples per paddock were collected from the topsoil and analysed for physical, chemical and microbial soil properties. After two years the sheep density statistically significantly affected clay content and metabolic quotient (qCO_2) at a significance level of $p < 0.1$, while soil organic C (SOC), total N and other microbial indices were unaffected. The clay content decreased from 23.6 % to 21.8 % at sheep densities > eight sheep ha^{-1}. From zero grazing to eight sheep ha^{-1}, the qCO_2 decreased from 13.5 to 12.9 mg CO_2-C g-1 microbial biomass C d-1 and remained relatively constant at about 12.5 mg CO_2-C g-1 microbial biomass C d-1 with further increase of sheep density. As expected, SOC strongly correlated with microbial indices such as microbial biomass C and N, ergosterol and basal respiration (Pearson's r = 0.6 to 0.9) and with the average seasonal aboveground biomass (r = 0.6). The gradually decreasing clay content may indicate commencing soil degradation by an increase of grazing pressure, whereas other chemical and microbial soil parameters remained unaffected. The short duration of the experiment and the high spatial variability of measured parameters may explain the marginal effects of sheep density. The results nevertheless underline the significance of SOC for preserving soil quality of summer pastures in Central Asia.

Keywords: Exclosure experiment, metabolic quotient, microbiological indicator, soil degradation, soil fertility

Contact Address: Mariko Ingold, University of Kassel, Organic Plant Production and Agroecosystems Research in the Tropics and Subtropics, Steinstrasse 19, 37213 Witzenhausen, Germany, e-mail: ingold@uni-kassel.de

Reclaiming a Technosol Using *Ficus thonningii* and Domestic Wastes

Dora Neina[1], Sibylle Faust[2]

[1] *University of Ghana, Department of Soil Sience, Ghana*

[2] *University of Kassel, Soil Biology and Plant Nutrition, Germany*

The incorporation of native tree species in mine land reclamation has been promoted over the years. However, high-value multipurpose species for local people' livelihoods has received less attention. Also, human excreta (HE), comprising fecal matter (FM) and human urine fertiliser (HUF), a product of ecological sanitation and kitchen wood ash (WA) have been least explored in land reclamation. Yet, these could be indispensable in areas with limited amounts of organic plant nutrient sources coupled with a high demand for natural resources for livelihood support. This study evaluated the effects of HE and WA on the growth and establishment of *Ficus thonningi*i Blume (called ficus) in a pegmatite-rich tantalum Technosol. Hardwood cuttings of ficus from homesteads in the Gatumba Mining District of Western Rwanda were planted in 20.4×19 cm diameter pots containing 5 kg forest soil (FS) and 6 kg Technosol. Five treatments including No amendment; HUF alone (100 mL/pot); HUF+WA (100 mL + 60 g/pot); FM (200 g/pot); and FM+WA (200 g + 60 g/pot) prepared in ten replicates each. After five and seven months of planting, plant height, number of leaves, shoot and root biomass were determined. The HE and WA treatments significantly increased ficus height ($P = 0.003$) ranging from 39 – 42 cm in the FS and 31 – 34 cm in the Technosol after seven months of planting. Shoot biomass weights ranged from 17 – 21 g in the FS and 10 – 16 g in the Technosol. Root: shoot ratios generally decreased with time in all the treatments except for the non-amended soils of both soils and ranged from 0.8 – 1.7 and 1.0 – 1.7 in the FS and Technosol, respectively. Ficus exhibited an efficient rooting system that stabilised the loose particles of the Technosol. This suggests the potential of using ficus, human excreta, and wood ash in degraded mine soil reclamation and provides an opportunity for further research in different conditions of soil degradation with low metal toxicity, different multipurpose tree species, and repeated applications throughout the period of growth and establishment.

Keywords: Ecological sanitation, fecal matter, human urine fertiliser, root and shoot biomass, wood ash

Contact Address: Dora Neina, University of Ghana, Department of Soil Sience, P.O. Box Lg 245, School of Agriculture, University of Ghana, Accra, Ghana, e-mail: dneina@ug.edu.gh

Effect of Soil Aeration and Genotype on the Response of Rice to Increasing Arsenic in Mining Soils from Ghana

PETER BADU, VARUNSEELAN MURUGAIYAN, MATHIAS BECKER
University of Bonn, Inst. Crop Sci. and Res. Conserv. (INRES) - Plant Nutrition, Germany

Arsenic pollution of soils is typically associated with gold mining activities in Ghana. Opportunities for reclaiming former mining sites for crop production depend on the level of As pollution, the soil aeration status (dominance of As-3 vs. As-5 species) and genotype-specific tolerance mechanisms (mainly related to P efficiency). Soils were collected from the strongly polluted centre and a largely unaffected field site at the fringe of a former artisanal gold mine in Ghana. The samples represent five positions along a gradient with soil As contents of 0, 19, 36, 57 and 76 ppm As. Potted soils were transplanted with rice genotypes with contrasting P-efficiencies (IR 64 and Mudgo) and incubated under aerobic (field capacity) and anaerobic (flooded) conditions in a greenhouse. After 6 weeks of growth, rice plants were harvested and analysed for As and P contents. Increasing soil As from 0.2 to 76 ppm negatively affected rice growth, irrespective of the soil aeration status or genotype. However, increasing soil As concentrations enhanced As uptake up to >70 ppm in case of the P-inefficient IR 64 but only up to 40 ppm in case of the P-efficient Mudgo. Increasing As uptake resulted in a concomitant reduction in plant P contents from 0.3 to <0.1 %. The As-induced decline in rice P accumulation was more in aerobic (predominance of arsenate – As^{5+}) than in flooded soil (predominance of arsenite – As^{3+}), and rice P uptake by the P-efficient genotype at moderate soil pollution of 19 to 40 ppm As was nearly twice that of the inefficient one. As a result, plant As and P showed a highly significant negative correlation. However, the P/As ratio was higher (more favourable) in anaerobic soil and in the P-efficient genotype. We conclude that for reclaiming as-polluted mining areas it may be advisable to cultivate P-efficient lowland rice rather than non-P-efficient upland crops.

Keywords: Arsenate, arsenite, *Oryza sativa*, P transporters, soil reclamation

Contact Address: Peter Badu, University of Bonn, Inst. Crop Sci. and Res. Conserv. (INRES) - Plant Nutrition, Bonn, Germany, e-mail: pierro55gh@yahoo.co.uk

Nutrient and Water Uptake of Rice in Response to Day and Night Root Zone Temperatures under Different Vapour Pressure Deficits

DUY HOANG VU[1], SABINE STÜRZ[2], FOLKARD ASCH[2]

[1] *Vietnam National University of Agriculture (VNUA), Dept. of Cultivation Sci., Vietnam*

[2] *University of Hohenheim, Inst. of Agric. Sci. in the Tropics (Hans-Ruthenberg-Institute), Germany*

With climate change, air temperature is predicted to increase, particularly during the night. Likewise, temperature of the soil, and thus of the root system will rise. In flooded rice fields, the water layer can buffer peak temperatures, but with the progressive introduction of water saving irrigation technologies in many parts of the world, root zone temperature (RZT) will vary more widely. Moreover, increasing vapour pressure deficit (VPD) caused by climate change is also challenging rice production in many rice growing regions by increasing evaporative demand. In the study, we investigated the effects of day and night RZT on water and nutrient uptake under low and high vapour pressure deficit. Two rice varieties, IR64 (international check) and NU838 (hybrid, widely used in Vietnam) were grown hydroponically at three root temperature levels (19°C, 24°C, 29°C). Twice per day during seven days, the fresh weight of the plants, nutrient (NH_4^+, NO_3, $PO_4{}^{3-}$, K+) and water uptake were measured both at the end of the day and at the end of the night. Results showed that water uptake of the plants was strongly influenced by VPD, but not by RZT. In contrast, nutrient uptake was not influenced by VPD and did not correlate with water uptake, but strongly increased with RZT for both varieties in the observed temperature range. This increase was larger during the day than during the night, but indicates a temperature optimum for nutrient uptake above 30°C both during day and night. Furthermore, the higher N uptake at high RZT led to an increased concentration of free amino acids in the leaf tissue, implying a beneficial effect of higher temperatures on the plant's nutrient status and growth, which was also reflected in dry biomass of plants.

Keywords: Nitrogen assimilation, nutrient uptake, *Oryza sativa*, temperature, VPD

Contact Address: Duy Hoang Vu, Vietnam National University of Agriculture (VNUA), Dept. of Cultivation Sci., Ngo Xuan Quang - Gialam, Hanoi, Vietnam, e-mail: vdhoang87@gmail.com

Non-Additive Effects of Mixing Rice Straw and Groundnut Stover Alter Decomposition

WIMONSIRI PINGTHAISONG, PATMA VITYAKON

Khon Kaen University, Dept. of Soil Science and Environment; Soil Organic Matter Management Research Group, Thailand

Rice straw (RS) is a residue remained in high quantity in rice-based agricultural systems. However, RS incorporated singly into a sandy soil decomposes rapidly leading to C loss which is not conducive to soil organic C accumulation. The objective of this study was to compare decomposition rate of RS mixed with groundnut stover (GN), to RS and GN only. A microcosm experiment with three replicates was initiated and RS (low N, high cellulose; CL) was mixed GN (high N and moderate lignin; L), at a ratio of 1:1 w/w , and compared with RS and GN applied singly in a sandy soil. The litter bag technique was used and decomposition was investigated at 3, 7, 14, 28 and 56 days after incorporation. Groundnut and the mixture had higher mass losses than RS especially from day 14 onwards. Microbial biomass nitrogen (MBN) (mg N kg^{-1}) peaked at 7 days in GN (23.16), while it was at 14 days in the mixture (24.18). During 14–28 days, MBN in the mixture significantly declined below that of GN, while RS had the lowest MBN. The mixture showed non-additive effects of synergistic type for mass loss throughout decomposition. Microbial biomass N showed antagonistic type of non-additive effects except for day 14. It appears that GN component in the mixture enhanced decomposition of the mixture relative to that of RS alone. The enhanced decomposition of the mixture corroborated the significantly lower C/N ratio of the mixture than that of the RS alone throughout the decomposition. Temporal changes in MBN and soil mineral N (NH_4^+ -N + NO_3 -N) concentrations were examined to explain the non-additive effect of MBN. Temporal patterns of change of MBN and soil mineral N concentrations were similar in the mixture and GN alone treatments. However, both parameters were lower in the mixture than those of the GN alone except for MBN at day 14 which was comparable in both treatments. The results showed that mixing RS with GN altered decomposition as indicated by non-additive effect of MBN which was related to altered pattern of mineral N release.

Keywords: Decomposition in residue mixture, litter bag technique, microbial biomass nitrogen, Northeast Thailand, soil mineral nitrogen

Contact Address: Patma Vityakon, Khon Kaen University, Dept. of Soil Science and Environment; Soil Organic Matter Management Research Group, 123 Mitraparp, 40002 Khon Kaen, Thailand, e-mail: patma@kku.ac.th

Effects of Biochar Prepared from Agricultural Residue on Water Retention Indices in Different Suctions

BIJAN KHALILIMOGHADAM, SHIMA GHAZIANI

Agricultural Sciences and Natural Resources University of Khuzestan, Dept. of Soil Science, Iran

The aim of the present study is to investigate the impact of biochar prepared from agricultural residues as a modifier of soil physical properties, including aggregate stability, water retention, porosity, and air and water capacity indices on sugarcane agro-industries in Khuzestan province, Iran. To this end, in each of the three agro-industries of Amirkabir, Haft Tapeh, and Karun, a factorial experiment in a completely randomised design with the factors; soil texture at three levels, and biochar type at three levels (sugar cane bagasse, wheat straw, and sugar cane leaves) was conducted with three replicates. Subsequently, soil pH, soil electrical conductivity, mean weight diameter of soil aggregates, and soil moisture curve in suctions with pF of 0, 0.4, 1, 1.8, 2, 2.52, and 4.17 were measured. The results showed that by increasing biochar, mean weight diameter of soil aggregates decreased significantly in Amirkabir and Haft Tapeh agro-industries. However, increasing biochar did not affect mean weight diameter of soil aggregates in Karun agro-industry. Furthermore, adding biochar to soil improved soil electrical conductivity and pH. However, its effects on soil salinity and pH in each agro-industry varied according to the type of biochar. Adding different amounts of biochar to soil increased soil moisture in all suctions, but wheat biochar increased soil moisture significantly. Biochar had different effects on other soil physical properties such as porosity, air capacity, plant available water, and relative field capacity. The results showed that biochar increases soil moisture and decreases plant available water, relative field capacity, porosity, and air capacity. In general, sugarcane biochar, compared to wheat biochar, results in lower soil moisture and higher plant available water.

Keywords: Biochar, porosity, sugarcane agro-industries, water retention

Contact Address: Bijan Khalilimoghadam, Agricultural Sciences and Natural Resources University of Khuzestan, Dept. of Soil Science, Mollasani, Ahvaz, Iran, e-mail: moghaddam623@yahoo.ie

Nitrogen Dynamics from Seasonal and Perennial Legume Residues in Mushinga South-Kivu Eastern, DR Congo

ISAAC BALUME[1], SVEN MARHAN[2], FRANK RASCHE[1]

[1]*University of Hohenheim, Inst. of Agric. Sci. in the Tropics (Hans-Ruthenberg-Institute), Germany*

[2]*University of Hohenheim, Inst. of Soil Science and Land Evaluation, Germany*

The use of leguminous organic residues in resource-prone smallholder farming systems is widely acknowledged. A central knowledge gap remains, however, to what extent a change of biochemical quality (e.g. similar C/N ratio, different lignin and ([polyphenol+lignin]/N (PP+L/N) ratio) of organic inputs as a result of different legume types shapes nitrogen dynamics in agricultural soils. The aim was thus to assess effect of legume type i.e. differences in biochemical quality of organic inputs, on soil N dynamics during a defined decomposition period. Organic inputs of the perennial legume *Calliandra calothyrsus* (CC) (C/N ratio: 13.80, lignin: 14.86 and PP+L/N: 8.00) *Leucaena leucocephala* (LL) (14.96, 10.63, 6.44) versus seasonal *Phaseolus vulgarus* (PV) (11.80, 8.82, 10.55) (10 ton ha^{-1} dry weight) were incorporated into Ferralsol (soil pH 4.79; TC 3.8; TN 0.35; P 5.5; K 0.12; Ca 5.9; Mg 0.7; Clay 34; %Sand 48; %Silt 18). A control treatment without residues was included. Soil samples were obtained at 15, 45, 75, 105, 165, 255 and 347 days after residue application. Samples were analysed for different N sources, including ammonia (NH_4^+), nitrate (NO_3) and dissolved organic nitrogen (DON). Generally, under residues treatment, increasing NH_4^+ ($p < 0.001$) and NO_3 ($p < 0.01$) contents were observed respectively at the beginning and at the end sampling points. Input type and decomposition time showed a significant interaction for NH_4^+ ($p < 0.001$), which leveled out with decomposition time while NO_3 increased ($p < 0.001$). NO_3 was negatively correlated with soil pH (r=0.5, $p < 0.0001$) and positively correlated with soil moisture (r=0.2, $p < 0.05$). No correlation was found between NH_4^+ with soil pH and soil moisture, similar for DON. The study demonstrated that the N dynamics were influenced by biochemical quality of residues obtained from differently legume types, which was mainly attributed to the increase in NO_3 concentration under seasonal residue (PV), a phenomenon that may have been explained by the low C/N ratio and lignin contents of used residue types. Results showed that changes in biochemical quality of organic inputs is a determinant of soil nitrification and has to be considered when using contrasting organic residues for adapted soil fertility management in resource-prone smallholder agricultural systems.

Keywords: Legume residues, nitrogen dynamic, residue quality

Contact Address: Isaac Balume, University of Hohenheim, Inst. of Agric. Sci. in the Tropics (Hans-Ruthenberg-Institute), Garben Str 13, 70599 Stuttgart, Germany, e-mail: isaac.balume@uni-hohenheim.de

Residue Quality and Agro-Ecological Effect of Microbial Prokaryotes Abundance in Agricultural Soils; Ethiopia and DRC

Claudia Afua Asare-Bediako, Isaac Balume, Birhanu Agumas, Mary Musyoki, Frank Rasche

University of Hohenheim, Inst. of Agric. Sci. in the Tropics (Hans-Ruthenberg-Institute), Germany

Microbial communities have been acknowledged to play major role in geochemical cycling of soil nutrients in agricultural systems. Organic residue management has been reported to improve soil fertility, enhance soil productivity and quality. However, effects of biochemical quality such as C/N ratio and complex compounds (phenols (P), lignin (L) and cellulose (CL)) of legumes (*Leucaena leucocephala* (LL), *Calliandra calothyrus* (CC) and *Acacia decurrens* (AD)) on the abundance of total (bacterial and archaeal 16SrRNA genes) and nitrifying (bacterial and archaeal amoA genes) microbial communities under different agro-ecologies is still unknown. The objective of this study was to investigate the response of bacterial and archaeal 16SrRNA and amoA gene abundance to biochemical quality (C/N ratio, phenols, lignin, cellulose) of different perennial residues during decomposition. Residues of LL (C/N ratio (11.61), L (1.71), CL (12.54), P (12.08)) and AD (C/N ratio (23), L (13.57), CL (16.73), P (15.09)) were incorporated in fields of Ethiopia (Injebara and Koga) and LL (C/N ratio (4.96), L (10.63), CL (9.2), P (9.11)) and CC (C/N ratio (13.84), L (14.86), CL (12.28), P (11.4) in DRC (Mushinga and Murhesa). Soil samples were subjected to quantitative PCR analysis; copy numbers of target genes were measured for all residue treatments. Results showed an interactive effect of both residue quality and agro-ecology on archaeal 16SrRNA ($p < 0.001$) with AD recording higher copy numbers under relatively stable soil moisture conditions. This interaction also had an effect on archaeal 16SrRNA and bacterial amoA genes ($p < 0.05$) in DRC. Over time, in Ethiopia, residue quality showed an effect on total 16SrRNA (bacterial & archaeal) and archaeal amoA genes ($p < 0.05$) but not on bacterial amoA ($p > 0.05$). For DRC, a time-residue quality interaction was observed for all investigated genes ($p < 0.05$). The findings of this study indicate that both residue quality and agro-ecology have an effect on abundance of archaeal 16SrRNA and amoA genes (bacterial and archaeal) for both countries. This might be attributed to lignin and C/N ratio differences of residues and agro-ecology (temperature and rainfall differences). Further studies should include more samples and sampling times to give a more conclusive result.

Keywords: Agro-ecology, gene abundance, organic residue quality

Contact Address: Claudia Afua Asare-Bediako, University of Hohenheim, Inst. of Agric. Sci. in the Tropics (Hans-Ruthenberg-Institute), Garbenstrasse 13, 70599 Stuttgart, Germany, e-mail: claudiaafua.asarebediako@uni-hohenheim.de

Effect of Residue Rate, Forward Speeds and Tillage Methods on the Soil Unsaturated Hydraulic Conductivity

Bijan Khalilimoghadam[1], Zahra Asgharian Dehkordi[1], Mahmoud Ghaseminejad Raeini[2]

[1]*Agricultural Sciences and Natural Resources University of Khuzestan, Dept. of Soil Science, Iran*

[2]*Agricultural Sciences and Natural Resources University of Khuzestan, Agricultural Machinery Engineering and Mechanization, Iran*

Inappropriate use of land and lack of knowledge of proper management operations leads to soil erosion and degradation, loss of soil organic carbon, low soil performance and crop productivity. This experiment was conducted to investigate the effect of tillage methods, residue rate and forward speed on the soil unsaturated hydraulic conductivity using tension infiltrometer at Agricultural Sciences and Natural Resources University of Khuzestan in 2016. The experiment was a split-split-plot design. The treatments included: conventional tillage (moldboard plough+disc), and two conservation tillages - combination tillage and chisel packer+plough, with three residue rates at surface of 0, 40 and 80 %, and three machine forward speeds of 4, 7 and 10 km hr $^{-1}$ in three replications. Unsaturated hydraulic conductivity at matric tensions (1, 4, 14 and 20 cm), soil bulk density, organic matter content, sorptivity, macroscopic capillary length, number of effective macropores, mesopores and micropores per unit area and their contribution to total saturated flux and yield were measured. The results showed that after one year of tillage management, soils under conventional tillage compared to combination tillage (with the lowest bulk density) had 6.1 % greater bulk density. High travel speed of 10 km hr^{-1} lowered stress by reducing soil pressure and decreased soil bulk density at a rate of 3.1 % compared to the low travel speed 4 km hr $^{-1}$. The treatments had no significant effect on hydraulic conductivity. Unsaturated hydraulic conductivity of 1 metric tension increased by 25.4 % and 18.7 % under conventional tillage and conservation tillage, respectively. Residue rates of 80 and 40 percent were in the suction capacity of 14 and 20 cm, respectively, with the highest values of hydraulic conductivity (0.121 and 0.0297 cm min $^{-1}$, respectively). In general, the effect of tillage on soil physical property characteristic is dynamic and the resulting effects over time are moderate and change their trend.

Keywords: Residue, soil unsaturated hydraulic conductivity, tillage

Contact Address: Bijan Khalilimoghadam, Agricultural Sciences and Natural Resources University of Khuzestan, Dept. of Soil Science, Mollasani, Ahvaz, Iran, e-mail: moghaddam623@yahoo.ie

The *CEP1* Peptide Modulates Cluster Root Morphology in P-Deficient *Lupinus albus*

UPAMA SARKER, YAPING ZHOU, UWE LUDEWIG, GÜNTER NEUMANN
University of Hohenheim, Inst. of Crop Science: Nutritional Crop Physiology, Germany

White lupin (*Lupinus albus*) has an adaptive mechanism to aid the acquisition of phosphorous from the soil. In phosphorous stress condition, white lupin forms proteiod (cluster roots) roots which mediate the secretion of citric acids, malic acids, protons, phenolics and acid phosphatases which influence releasing of phosphorous. *CEP1* is a member of *CEP* signaling peptide that has a short, conserved 15-amino-acids domain near the C terminus, a feature that has common to several post-translationally modified small signal peptides in plants. *CEP1* was investigated to reduce the lateral root formation and slow down the primary root growth in *Arabidopsis* and *Medicago*. However, it has a positive correlation with nodules formation in *Medicago*. In this study, the effect of *CEP1* gene on cluster roots morphology changes in white lupin plants was examined by *CEP1* gene overexpression via *A. rhizogenes* transformation. These transgenic plants were initially grown in agar plates and transferred into hydroponics without phosphorus supply after the formation of hairy roots. Five-weeks old plants were harvested and changes in root morphology were observed, such as number of cluster roots, rootlet density, rootlet length, root hair length and the number of root primordia of the first lateral root. Number of cluster roots and number of root primordia were documented throughout the culture period. Compared with the control plants (*agrobacterium* control *A. rhizogenes* infected plant without construct) and vector control (empty vector transformed), no significant difference in the number of cluster root and root primordia was found in *CEP1* overexpressed plants. However, *CEP1* overexpressed plants always showed a trend of reduction of cluster roots number in the entire cultural period. On the other hand, a significant reduction in root length and root hair length were observed in *CEP1* overexpressed plants compared to control plants. Taken together, our data suggested that the *CEP1* overexpression has a decreased trend on cluster root formation, and decrease in rootlet length and root hair length was observed in Pi-deficient white lupin plants. Hence, *CEP1* act as a negative growth regulator in cluster roots growth and development under phosphorus deprived condition via a direct or indirect way.

Keywords: *CEP1* overexpression, cluster roots, negative plant growth regulator, P deficiency

Contact Address: Upama Sarker, University of Hohenheim, Inst. of Crop Science: Nutritional Crop Physiology, Landhaustrasse, 68, 70190 Stuttgart, Germany, e-mail: upama.sarker@yahoo.com

Livestock

Resource efficiency in ruminant husbandry

Challenges, Opportunities and Strategies to Increase the Contribution of Urban Livestock Production to Sustainable Resource Management

REGINA RÖSSLER

University of Kassel / University of Goettingen, Animal Husbandry in the Tropics and Subtropics, Germany

The purpose of this contribution is to outline the challenges, opportunities and strategies related to the sustainability of livestock development in urban production systems in developing countries. Sustainability is thereby especially discussed in the context of the management of animal genetic and feed resources. Population growth and urbanisation have led to an increasing demand for animal-sourced food in developing countries. This resulted in an expansion of livestock production in and around urban centres in the last few decades. Notwithstanding the higher associated knowledge and financial needs, an intensification of urban livestock production systems has been successfully encouraged in South and Southeast Asia, mainly for monogastric animals, but also for sheep and cattle. The importation of exotic livestock breeds and concentrate feeds enhanced production outputs in many Asian countries, and the market share of the intensive commercial livestock sector has been steadily increasing. However, continuously rising costs of livestock feed call the long-term sustainability of these intensive landless livestock production systems into question. In addition, the concentration of livestock production in Asia raises concerns of environmental pollution. In sub-Saharan Africa (SSA), on the other hand, many countries have difficulties to meet the increasing demands for food, and the overall increase in animal productivity has been minimal. The lower adaptability of imported breeding stock and indiscriminant crossbreeding often do not allow animals to express their full genetic potential. Restricted availability of sufficient livestock feeds of adequate quality and possibilities to provide external feed inputs further limit the required intensification of livestock production in SSA. The major challenge for livestock development in developing countries is therefore to increase the production outputs without having much additional natural resources to utilise, while ensuring that environmental resources (land, water and air) and animal genetic diversity are sustained. Solutions are needed that utilise the potential of existing animal genetic and locally available feed resources to sustainably develop urban livestock production in developing countries.

Keywords: Animal genetic resources, livestock feed resources, sustainable livestock development, urban livestock production

Contact Address: Regina Rößler, University of Kassel / University of Goettingen, Animal Husbandry in the Tropics and Subtropics, Göttingen, Germany, e-mail: regina.roessler@agr.uni-goettingen.de

Modelling Heat Stress Effects on Milk Production in a Tropical Environment Using Test-Day Records and Random Regression Models

Jackson Mbuthia, Manfred Mayer, Nobert Reinsch
Leibniz Institute for Farm Animal Biology (FBN), Genetics and Biometry, Germany

In the tropical environments, dairy cattle production is constrained by several factors among them being climate. The seasonal loss of milk due to heat stress is a recurring challenge for many dairy producers. The objective of this study was to detect heat stress thresholds and individual animal variations from test-day milk records using random regression models for dairy cattle in Kenya. Data were obtained from the Kenya Livestock Breeders Organisation for the years 2000 to 2017 and merged with meteorological data. After editing, the records comprised of 50743, 44899 and 35958 test-day records for first, second, and third lactations, respectively, for the four main dairy breeds: Friesian (68.1 %), Ayrshire (21.0 %), Jersey (7.6 %) and Guernsey (3.3 %). Variance components were estimated using Restricted Maximum Likelihood in ASREML software. Random regression models with heterogeneous residual variance were applied. Legendre polynomials of order 3 were fitted to the average and individual lactation curves and the reaction norms. The heat stress indicators studied were; daily average temperature (Tave), daily maximum temperature (Tmax), Temperature Humidity Index (THI) and Heat Load Index (HLI). Considering a one day lag, the estimated heat stress thresholds for the different indices were about 22°C, 69 and 81 for Tave, THI and HLI, respectively. Tmax did not adequately capture the response of milk yield to increasing temperature. On average, the milk yield loss for each unit increase in Tave from 22°C to 26°C was -0.47, -0.71 -0.98 and -1.41 kg for first lactation milk yield. The loss was -0.51, -0.78, -1.09 and -1.58 kg for second lactation, and -0.61,-0.85, -1.18 and -1.69 kg for third lactation respectively. The temperature and THI thresholds obtained in this study are important in management of dairy production systems in Kenya and in the tropics with similar climatic conditions.

Keywords: Dairy cattle, heat stress indicators, reaction norm

Contact Address: Jackson Mbuthia, Leibniz Institute for Farm Animal Biology (FBN), Genetics and Biometry, Wilhelm-Stahl-Allee 2, 18196 Dummerstorf, Germany, e-mail: mbuthia@fbn-dummerstorf.de

Breed Improvement Using Sexed-Semen for Filling Gaps and Removing Traps for Commercial Cattle Farming in Nepal

LOK NATH PAUDEL[1], RESHMI RAJ GNYAWALI[2]

[1] *Ministry of Agriculture and Livestock Development, Department of Livestock Services, Nepal*

[2] *Sandhikharka Municipality, Province Number 5, Livestock Section, Arghakhanchi, Nepal*

Livestock is an integral part of Nepalese agricultural system that contributes about 12 % to the national Gross Domestic Production (GDP) and 30 % to the agricultural GDP. Livestock enterprise is becoming semi-commercial to commercial especially after launching the Dairy Cattle and Buffalo Genetic Improvement Program (DCBGIP) and subsequently artificial insemination (AI) mission programme in Nepal. The AI mission programme (2011–19) has increased the number of animals inseminated to around 0.7 million in 2018 as compared to 0.15 million before seven years. This programme has been adding more than 0.2 million crossbred calves every year in the country. Because of religious and cultural value systems, cattle are not allowed to slaughter in Nepal. Therefore, yearly production of around 0.1 million male-calves, other than needed for natural services and AI, are becoming a burden to the nation. To mitigate this problem, an effort has been made to introduce sexed semen that could result in more than 90 % female-calves.

A pilot research was conducted by introducing 600 doses of sexed semen obtained from the Improved Seed for Farmers Programs (ISFP) project procured from the USA. This semen was used in cattle farming pocket areas of Arghakhanchi district of Nepal. Out of the 600 doses semen, 570 were used to inseminate the cross bred cattle of smallholders farmers, covering 198 households from February 2017 to March, 2018. The study revealed that the use of sexed semen in cross bred cattle resulted in a conception rate of 58.5 % and female-calves born were 90.2 %. These results proved that the use of sexed semen in cattle produce significantly ($p < 0.01$) higher number of female-calves. The survey with the farmers and key informants also revealed that the farmers who had used the sexed semen were very happy and explained that their burden of rearing male-calves, which have no monetary importance, has been reduced significantly. It can be concluded that the use of sexed semen would be a cornerstone for commercialisation of cattle farming. This technology would be an exemplary work for filling gaps and removing traps for sustainable cattle farming in Nepal.

Keywords: Cattle, commercialisation, gaps, religion, sexed semen, sustainability, traps

Contact Address: Lok Nath Paudel, Ministry of Agriculture and Livestock Development, Department of Livestock Services, Hariharbhawan, Lalitpur, Nepal, e-mail: paudelln2015@gmail.com

Year Round Feed and Fodder Availability in Smallholder Dairy Farms Across High and Low Altitude Areas in Eastern Africa

Patrick Ongadi Mudavadi[1], Ben Lukuyu[2], Emmanuel Abaraham Mpolya[3], Minn Wang[4], Mizeck Chagunda[5], Derseh Malkamu[6], Adie Aberra[6], Endalkachew Woldemeskel[6]

[1]*Nelson Mandela African Institution of Science and Technology, Sustainable Agriculture, Biodiversity and Ecosystems Management, Tanzania*

[2]*International Livestock Research Institute (ILRI), Feeds and Forages Program, Uganda*

[3]*Nelson Mandela African Institution of Science and Technology, Health and Biomedical Sciences, Tanzania*

[4]*Institute of Subtropical Agriculture, The Chinese Academy of Sciences, Animal Nutrition, China*

[5]*University of Hohenheim, Inst. of Agric. Sci. in the Tropics (Hans-Ruthenberg-Institute), Germany*

[6]*International Livestock Research Institute (ILRI), Ethiopia*

An understanding of seasonal variation in availability of feeds and fodder resources is important in future planning and development of appropriate technologies to ensure resilience of smallholder dairy systems to seasonality driven milk fluctuations. It is against this context that this study was carried out to: 1) evaluate the current pattern of seasonal variation in feed and fodder availability in smallholder dairy farms across high and low altitude areas of Kenya and Tanzania in Eastern Africa; and 2) assess seasonal variation in year-round feeds and fodder based feeding strategies in smallholder dairy farms across the high and low altitude areas of Kenya and Tanzania in Eastern Africa. Data was collected from a purposive representative sample of 400 smallholder dairy farmers using the Feed Assessment Tool (FEAST) through cross sectional survey and observational study from 2016-2018 to capture the season's effect (wet and dry). Data was analyzed using the general linear model procedure of SPSS version 21.0 and FEAST Version 2.21. Results showed that location (country), agro-ecological zone and season were significant ($p \leq 0.05$) on rainfall variability throughout the year. Further, country, agro-ecology, seasons, production systems and their interaction were significant ($p \leq 0.05$) on year-round availability and utilisation of concentrate feeds, green and dry crop residues, improved fodder, natural grass and legume forage. Rainfall variability was crucial in determining year-round variation in availability and utilisation of feeds and fodder. Correlation between the feeds and fodder resources revealed highly significant ($p \leq 0.001$) positive relationships across the two countries, pointing further to the dynamics of seasonality change effects. In conclusion, different seasonality driven site/region/country specific year-round feeding/supplementation strategies could be applied depending upon the type, availability of feeds and fodder to overcome seasonal milk fluctuations in smallholder dairy farms in Eastern Africa.

Keywords: Agro-ecology, feeds and fodder, seasonal variation, utilisation, year-round

Contact Address: Patrick Ongadi Mudavadi, Nelson Mandela African Institution of Science and Technology, Sustainable Agriculture, Biodiversity and Ecosystems Management, P.O. Box 447, 0206 Arusha, Tanzania, e-mail: pmudavadi@gmail.com

A Social-Ecological Framework for Managing and Adding Value to the Lagune Cattle Breed in Benin

COSSI MAURICE AHOZONLIN, MAHAMADOU DAHOUDA, LUC HIPPOLYTE DOSSA

University of Abomey-Calavi, Fac. of Agricultural Sciences, School of Sciences and Technics of Animal Production, Benin

The West African Shorthorn Lagune cattle breed of Benin is known for its adaptability to humid regions and resistance to parasitic diseases: But since the 1970s', its population size, estimated to about 37000 in 2004, is sharply decreasing. Yet, past initiatives of conserving this small-bodied breed were limited to its characterisation out of its production system context, paying little attention to interacting ecological, social, cultural and economic factors. Consequently, the results are not translated into policies and legal institutions favourable to a large motivation of the main stakeholders for its conservation and sustainable use. We explore how a social-ecological framework can be used to better understand the relationships between the different social, economic and ecological factors affecting the sustainable use of this breed. We carried out, from February to May 2017, questionnaire-based interviews of 312 Lagune cattle keepers in the two agro-ecological zones (AEZs) of Oueme Valley and Pobe. Owing to its cultural value, the Lagune breed was used for sacrifices and social events in both AEZs. However, the opportunities for its *in situ* conservation varied between AEZs, being greater in Pobe where breed admixture was almost inexistent. Similar to farmers' breeding practices, feed constraints and farmers' resilience strategies varied ($p < 0.001$) between AEZs. In contrast to Pobe, seasonal river flooding in the Oueme Valley prompted collective actions related to Lagune cattle herd management. The results of a logistic regression analysis revealed AEZ, socio-cultural factors (main income source and ethnicity) and cattle herd size as the main determinants of farmers' decision to keep Lagune cattle in purebred herds. Furthermore, on-farm sale accounted for more than 95 % of all Lagune cattle sales. This might be explained by the absence of a physical cattle market in both zones which probably prevents farmers to upgrade their lagune cattle with zebus. Nevertheless, because of their proximity to urban areas, few keepers (17.7 %), produced or expressed their willingness to produce crossbred cattle to satisfy the urban market demand for large-sized cattle. Using the social-ecological framework can further assist in the identification of sound actions and policies towards a better management and promotion of this local breed.

Keywords: Agro-ecology, animal genetic resources, collective actions, conservation, multi-stakeholder approach, socioeconomic factors

Contact Address: Luc Hippolyte Dossa, University of Abomey-Calavi, Fac. of Agricultural Sciences, School of Science and Technics of Animal Production, 01 BP 526, Cotonou, Benin, e-mail: dolhip@yahoo.com

Vegetation Dynamics and Livestock Productivity in a Kenyan Savannah Rangeland: The Relevance of Herding Efforts

Anna Seidel, Folkard Asch, Benjamin Warth, Carsten Marohn
University of Hohenheim, Inst. of Agric. Sci. in the Tropics (Hans-Ruthenberg-Institute), Germany

In Laikipia County, Kenya, landowners involved in both commercial ranching and wildlife conservation hire pastoralists to herd and tend to cattle, sheep, goats and camels on a daily basis. The distribution of livestock in relation to the palatable forage density and the palatable forage offered to animals over grazing itineraries have effects on sustainable rangeland management and animal performance. The objective of this study was to examine vegetation dynamics and livestock productivity in response to grazing itineraries around a night resting place and identify factors influencing daily herding decisions of hired pastoralists.

On a large-scale ranch in semi-arid Laikipia County, the vegetation neighbouring a night resting place enclosing one herd of camels, one mixed herd of sheep and goats and two herds of cattle was studied between June and September 2018. Through participant observation, the activities of hired pastoralists and herds were monitored, and by means of a hand-held global positioning system receiver, grazing ranges were measured. Georeferenced data of grazing itineraries were diurnally recorded with 20 s tracking intervals. All animals were weighed monthly to determine their live weight gain. Vegetation dynamics were monitored on six regularly spaced transects around the night resting place with sampling points at 50 m, 150 m, 250 m and 350 m distance from the night resting place. Herbaceous biomass classified into monocotyledonous and dicotyledonous plants and litter was collected at each sampling point every fourth day. Samples of herbaceous biomass pooled by species and month were analysed for their nutrient contents.

The employment of both qualitative and quantitative methods offered insights into the heterogeneous utilisation of rangeland resources by hired pastoralists and the resulting relevance of their daily herding efforts. The results indicate that the herding efforts of hired pastoralists strongly influenced the distribution of livestock and thus vegetation dynamics. The relevance of their herding decisions to vegetation dynamics and livestock productivity and the resultant potential implications for a sustainable use of rangeland resources will be discussed.

Keywords: Daily grazing itineraries, herding decisions, hired herders, sustainable rangeland management

---

Contact Address: Anna Seidel, University of Hohenheim, Inst. of Agric. Sci. in the Tropics (Hans-Ruthenberg-Institute), Garbenstr. 13, 70599 Stuttgart, Germany, e-mail: AnnaSeidel1@gmx.de

Livestock Mobility in the Context of Changing Land Tenure on Community Lands of Kenya

HUSSEIN WARIO[1], BRIGITTE KAUFMANN[2]

[1] *Center for Research and Development in Drylands, Kenya*

[2] *German Institute for Tropical and Subtropical Agriculture (DITSL), Germany*

Kenya is on the path of transforming land governance and tenure by putting in place various legal and institutional reforms. A significant shift in the land reforms is the proposed changes in ownership and management of the community lands in the arid and semi-arid areas, which were previously managed by government as trusts on behalf of the communities. In these arid and semi-arid areas, livestock production, through extensive pastoral systems, remains the main economic stay, where the practice of mobility is the main strategy in accessing feed and water resources that are variable in distribution over space and time. Mobility of livestock already faces a number of constraints, however within the context of the changing land tenure on communal lands the constraints may well be heightened. This is particularly so due to the fact that the need for securing resource tenure rights and the requirement for a spatially flexible resource use system for successful pastoral production are contradictory. This presentation therefore explores the possible challenges posed to mobile resource use systems by the implementation of the community land act in Kenya. The presentation further recommends efforts that need to be considered during the implementation of the community land act to maintain livestock mobility. It assesses the opportunities in application of different internationally acknowledged legal and policy guidelines such as the Voluntary Guidelines on Governance of Tenure (VGGT), strengthening of pastoral institutions and the use of experiences from other regions in order to understand how nested tenure rights can be protected under formalized arrangements. The presentation concludes that unless deliberate measures are taken to maintain mobile resource use systems, the implementation of the community land act can exacerbate conflicts along communal resource borders, curtails livestock mobility further and increases the vulnerability of pastoral populations to climate related disasters, which is a polar opposite of what is envisioned in the land reform agenda.

Keywords: Community land, flexible resource use, livestock mobility

---

Contact Address: Hussein Wario, Center for Research and Development in Drylands, Kenya, Hospital Road, 60300 Marsabit, Kenya, e-mail: husseinwario76@gmail.com

Yak Herding Strategies on High Altitude Rangelands of Gilgit Baltistan, Pakistan

Asif Hameed[1], Muhammad Tariq[2], Andreas Buerkert[3], Eva Schlecht[1]

[1] *University of Kassel / University of Goettingen, Animal Husbandry in the Tropics and Subtropics, Germany*

[2] *University of Agriculture, Dept. of Livestock Management, Toba Tek Singh, Pakistan*

[3] *University of Kassel, Organic Plant Production and Agroecosystems Research in the Tropics and Subtropics, Germany*

To study different herding strategies of yak grazing on communal high altitude rangelands in the Gilgit-Baltistan region of Pakistan, individual interviews with 90 herders were conducted in three valleys (Hoper, Phander, Shimshal) and 10 group interviews (5–8 herders per group) in three additional valleys (Chipurson, Haramosh, Khaplu) from May to November 2018. Individually interviewed herders had a herding experience of 25 years and kept 10 yaks on average. Yak herding practices varied between valleys and seasons and were influenced by labour availability and frequentation of an area by tourists. Collective yak herding by a group of herders was widespread in Shimshal (practised by 70 % of interviewees), Chipurson and Khaplu (66 %), but rare in the other valleys. This system of herding continued for full seasons in Shimshal but was only used during a few weeks per season in the other valleys. Individual herding, largely by male family members (>90 % overall) was widespread in Phander (93 %) and Hoper (78 %), intermediate in Chipurson and Khaplu (66 %), of low relevance in Shimshal (37 %) and non-existent in Haramosh. Family herding was found where yaks were milked, namely in Phander (both partners: 10 %, other family members: 54 %) and Shimshal (both: 11 %, other members: 31 %), while there was mixed species herding in Hoper (both: 8 %, other members: 24 %). Employed herding in which the community arranges for herders was common in Chipurson (53 %) and Phander (20 %), particularly from late spring throughout summer when farmers needed to work on their farmland. Year-round unsupervised grazing was predominant in Haramosh (100 %), Khaplu (67 %) and Hoper (77 %) but restricted to autumn in Chipurson (67 %). The different herding practices across valleys and seasons led to spatially very patchy and sometimes extremely concentrated utilisation of already depleted rangelands, herd size fluctuations through mortality, and in some areas created problems for wildlife. Moreover, support of yak-keeping families by developing local milk and wool markets, and maintenance of the genetic diversity of yak herds is challenged by the very heterogeneous herding strategies in general and the high prevalence of unsupervised grazing in particular.

Keywords: Alpine rangelands, Gilgit-Baltistan, herding, yaks, summer season

Contact Address: Eva Schlecht, University of Kassel / University of Goettingen, Animal Husbandry in the Tropics and Subtropics, Steinstraße 19, 37213 Witzenhausen, Germany, e-mail: tropanimals@uni-kassel.de

Drinking Water Intake of Beef Cattle in Pasture-Based Systems of Brazil

MARIANA PEREIRA[1], JESSICA WERNER[1], MANUEL C. M. MACEDO[2], ROBERTO G. ALMEIDA[2], UTA DICKHOEFER[1]

[1] *University of Hohenheim, Inst. of Agric. Sci. in the Tropics (Hans-Ruthenberg-Institute), Germany*

[2] *EMBRAPA Beef Cattle, Integrated Production Systems, Brazil*

Beef cattle has a large water requirement per kg of liveweight (LW). Although water intake accounts for a minor proportion of this requirement, even small changes may impact on reducing this demand. Water intake (WI) of Nellore heifers was measured in three pasture-based systems with three paddocks per system: integrated crop-livestock-forestry (ICLF), integrated crop-livestock (ICL) and continuous pasture (CP) in Brazil. ICLF and ICL were a spatial-temporal integration of soybean and *Brachiaria brizantha*. For ICLF, there were *Eucalyptus urograndis* trees in array 22×4 m. In CP cattle were grazing *Brachiaria decumbens*. Heifers (n = 36) were randomly allocated to the systems, individually weighed at the beginning and at end of measurements (mean LW 317 ± 36.1 kg). Forage allowance (kg DM 100 kg $^{-1}$ LW) was 3.2 in ICLF, 7.1 in ICL and 4.4 in CP. Ambient air temperature, relative air humidity, precipitation, and radiation were measured in ICLF and CP by Tinytag dataloggers, rain gauges, and Accupar LP-80 Ceptometer, respectively, during rainy season from January to February 2019. Evaporation was calculated from class A pan. Drinking fountains were equipped with water meters that were read every day at 3 p. m. for 27 days, corrected for precipitation and evaporation. Data were grouped per system (n = 9), subjected to analysis of variance, means were compared by t-test. Total precipitation in 37 days was 346 mm. Temperature-humidity index (mean ± standard deviation) calculated was 77 ± 2.0 in ICLF, and 76 ± 1.6 in CP. Mean radiation (μmol m^{-2} s^{-1}) was 789 ± 245 in ICLF, whereas 1518 ± 327 in CP. Daily WI (L 100 kg^{-1} LW) was lower ($p = 0.0135$) in ICLF (3.58 ± 0.4) than in CP (5.55 ± 0.9), but similar in ICL (4.29 ± 0.8) compared to the other two systems ($p > 0.05$). Daily WI (L animal^{-1}) did not differ ($p = 0.3141$), it was 12.5 ± 1.6, 14.8 ± 2.6 and 14.6 ± 2.5 for ICLF, ICL and CP, respectively. Integrated systems reveal potential to decrease drinking water requirement. However, herbage intake should also be considered to explain the results.

Keywords: Agrosilvopastoral, beef cattle, daily water intake, grazing system

Contact Address: Mariana Pereira, University of Hohenheim, Animal Nutrition and Rangeland Management in the Tropics and Subtropics, Garbenstrasse 13, 70599 Stuttgart, Germany, e-mail: mariana.pereira@uni-hohenheim.de

Responses of Crossbred Lactating Cows to Heat Stress and its Alleviation under a Tropical Environment

Joaquín Castro-Montoya[1], Elmer Edgardo Corea[2]

[1] *University of Hohenheim, Animal Nutrition and Rangeland Management in the Tropics and Subtropics, Germany*

[2] *University of El Salvador, Dept. of Animal Production, El Salvador*

Rising temperatures in the world are an increasing concern for the welfare and production of livestock. Dairy cows are particularly sensitive to heat stress (HS) due to their metabolic heat production from milk synthesis. Heat stress decreases milk yield (MY) beyond decreases in dry matter intake (DMI), by impairing the carbohydrate and protein metabolism. In high yielding cows, multiparous are more susceptible to HS than primiparous, due to their greater MY (~10 kg) and body weight (BW) (~100 kg) compared with primiparous. However, in the tropics, smaller size, lower production, and long-term exposure to HS may influence the cows' response to HS and its alleviation. To prove this hypothesis, 12 cows, six primiparous and six multiparous (¾ Holstein, ¼ Brahman, 520±35 kg BW, 16.6± 1.2 kg d^{-1} MY) were exposed to four 15-days-long periods, two controls (no cooling), alternated with two cooli ng periods (fans and sprinklers in 1-hour cycles at 10:00, 12:00, 14:00 and 16:00 h). Cows were monitored for rectal temperature (RT) and respiration rate (RR) at 09:00, 11:00, 13:00, 15:00 and 17:00 h, and for MY.

Contrary to high yielding cows, primiparous were more susceptible to HS than multiparous, with higher RT (39.46 vs. 39.16°C) and RR (76 vs. 66 breath min^{-1}), likely due to smaller differences in MY (δ=1.5 kg d^{-1}) and body weight (δ=35 kg) between primiparous and multiparous compared with those in high yielding animals. Cooling effectively reduced RT and RR and increased MY in both primiparous (+9.7 %) and multiparous (+6.5 %). However, cooling did not improve feed efficiency (0.87 and 0.88 kg milk kg^{-1} DMI, with or without cooling, respectively). This additional discrepancy with high yielding cows may be related to the lower level of production, smaller size, and better adaptation to HS by tropical cows. But it could also indicate that HS was not alleviated to the point of improving nutrients metabolism, as evidenced by an overall low feed efficiency, and RT and RR not differing with or without cooling before the first cooling cycle of the day. These results highlight different responses of cows in tropical environments to HS and its alleviation, a phenomenon worth of further research.

Keywords: Dairy cattle, heat stress, multiparous, primiparous

Contact Address: Joaquín Castro-Montoya, University of Hohenheim, Animal Nutrition and Rangeland Management in the Tropics and Subtropics, Fruwirthstrasse 31, 70599 Stuttgart, Germany, e-mail: jcm@uni-hohenheim.de

Methane Emissions in Dairy Cattle in Dependency of Rural-Urban Gradients

ANA PINTO[1], TONG YIN[1], MARION REICHENBACH[2], RAGHAVENDRA BHATTA[3], EVA SCHLECHT[2], SVEN KÖNIG[1]

[1]*Justus-Liebig-University Giessen, Inst. of Animal Breeding and Genetics, Germany*

[2]*University of Kassel / Georg-August-Universität Göttingen, Animal Husbandry in the Tropics and Subtropics, Germany*

[3]*National Institue of Animal Nutrition and Physiology, Inst. of Animal Breeding and Genetics, India*

Greenhouse gas (GHG) emissions from livestock farming have being criticised for being a main contributor to climate change. Methane (CH_4) is one of the most important GHGs in dairy farming as it is 25 times more potent than carbon dioxide. Different strategies (nutrition, genetic, management) have been researched to decrease this emission. The aim of the present study was to assess the environmental impact on dairy cattle CH_4 emissions, individually recorded using a mobile laser methane detector (LMD). CH_4 emissions were measured along the rural-urban gradient of the rising megacity of Bengaluru, in the south of India, from June 2017 to April 2018. A simplified survey stratification index (SSI) was calculated based on building density and distance to the city centre. According to this index, three districts were defined, urban (SSI < 0.3), mixed (0.3 – 0.5) and rural (SSI > 0.5). Individual overall CH_4 mean, CH_4 eructation and CH_4 respiration were calculated based on 2-minutes CH_4 emissions from 452 cows. The basic statistical model considered fixed effects of lactation number, days in milk, breed, fasting duration prior CH_4 measurement, and temperature humidity index. Random effects were cow, farm, and residual. Effect of location, SSI, access to pasture, and milk yield group were tested stepwise. For the overall mean and respiration CH_4, cows from urban areas responded stronger than cows from mixed and rural areas. Cows kept on pasture emitted less CH_4 than cows mainly or only kept in indoor systems. Highly productive cows had a significantly higher CH_4 output (overall mean, respiration and eructation) compared to cows with medium productivity. Hence, a mixture of cow associated factors and social-ecological descriptors contribute to individual CH_4 output, and in further consequence, to resource efficiency. In a next step, we will consider genetic aspects, in order to infer genotype by social-ecological interactions.

Keywords: Dairy cattle, methane emission, rural-urban farms, survey stratification index

Contact Address: Ana Pinto, Justus-Liebig-University Giessen, Inst. of Animal Breeding and Genetics, Ludwigstr. 21b, 35390 Giessen, Germany, e-mail: ana.pinto-garcia@agrar.uni-giessen.de

Methane Emissions of Zebu Steers Fed Tropical Forages of Contrasting Nutritional Value

Xiomara Gaviria Uribe[1], Diana María Bolivar Vergara[1], Ngonidzashe Chirinda[2], Jacobo Arango[2], Rolando Barahona[1]
[1]*National University of Colombia, Dept. Animal Production, Colombia*
[2]*International Center for Tropical Agriculture (CIAT), Colombia*

Methane emissions from livestock contribute significantly to climate change and diet plays a fundamental role in the emissions generated. Research in animal nutrition should aim at identifying ways to increase production efficiency with the least possible environmental impact and manipulation of diet quality becomes one of the most viable options to both mitigate emissions and increase animal productivity. In this study, methane emissions of diets based on tropical forages differing in nutritional value and level of voluntary intake was evaluated. Methane emissions were measured in four animals per evaluated diet using the polytunnel technique. Forage voluntary intake was calculated as the difference of forage offered and rejected by each animal. Treatments evaluated corresponded to five different diets: T1: low quality *Brachiaria* hybrid cv. Cayman; T2: high quality *Brachiaria* hybrid cv. Cayman; T3: *Brachiaria* hybrid cv. Cayman + *Leucaena leucocephala*; T4: *Brachiaria* hybrid cv. Cayman + *Leucaena diversifolia*; T5: *Dichantium aristatum* hay. There was a tendency for the different evaluated parameters to improve when Leucaena was included. Treatments with Leucaena had higher protein, digestibility and dry matter (DM) intake (T3 and T4). The digestibility in the treatments varies in a range between 47.9 % (T1) and 61.8 % (T5), being the treatments with the lowest values. Methane emission per kg of dry matter intake was lowest in diets of better nutritional value (T3=27.5 g, T4=19.8 g) in contrast with treatments T1 (60.34) and T5 (35.9) with lower nutritional content. Diets with greater DM intake had the highest methane emissions (g day^{-1}). However, when emissions are expressed per kg of DM intake and/or digested, this relationship changes and it is the treatments with higher DM intake that have lower emissions per kg DM consumed and/or digested. The results of this study show that methane emissions decrease significantly when the nutritional value of the diet increases. Likewise, DM intake increases and therefore higher production parameters are also potentially attained.

Keywords: Dry matter intake, enteric methane emissions, polytunnel technique

Contact Address: Xiomara Gaviria Uribe, National University of Colombia, Dept. Animal Production, Carrera 57 56-31, 51040 Copacabana, Colombia, e-mail: xgaviri0@unal.edu.co

Climate Smart Livestock: The Interaction between Mitigation, Adaptation and Efficient Livestock Management Practices in Ecuador

PAMELA GERMANIA SANGOLUISA RODRÍGUEZ, ARMANDO RIVERA, JUAN MERINO, JONATHAN TORRES

Food and Agriculture Organization of the United Nations, Ecuador

Livestock supply chain greenhouse gas emissions account for 14.5 % of anthropogenic gases worldwide, which indicate that the sector is an important GHG contributor. In addition, this sector plays a significant role for income and employment generation. At the same time, livestock productive activity is directly influenced by the impacts of climate change. For such reasons, identifying and implementing management practices that promote sustainable production in a climate change context, becomes a challenge. In Ecuador, the livestock production systems are evaluated from a climate-smart perspective by considering three objectives: 1. sustainably increasing productivity; 2. adapting to climate change; and 3. reducing greenhouse gas emissions whenever possible. The Climate Smart Livestock Project (CSLP) quantified GHG emissions from beef and dairy cattle in the country using local information to derive the estimates. For 2016, the preliminary results show a value of 16547 Gg of CO_2-eq from direct emissions, being 76.92 % CH_4 form enteric fermentation, 18.12 % NO_2 from manure in pastures, 2.66 % CH_4 and 2.30 % NO_2 from manure management. Regarding climate risk, the CSLP conducted a study to compare the relationship between intense rains, droughts, heat waves and frosts; with the exposure in three dimensions (environmental: pasture area, socioeconomic: number of animals per producer, governance: established productive associations). The results show that the increase of extreme events in drought and intense rains constitute threats with high impacts on livestock systems. Based on the preliminary results, the CSLP compiled 126 good livestock practices that contribute to a climate-smart management, a efficient production systems that are adapted to climate change and aim to reduce GHG. A customized set of practices were implemented on 165 pilot farms, being pasture and water sources management the most relevant factors for the development of intervention strategies at field level. The impact of implementing the practices in the pilot farms is monitored through two web apps, developed by the CSLP, applying algorithms to estimate emissions and climate risk using herd management (number of animals, production, reproduction, weights, etc.) and farm data (area, conservation area, pastures, infrastructure, etc.).

Keywords: Adaptation, climate change, climate risk, Ecuador, greenhouse gases

---

Contact Address: Pamela Germania Sangoluisa Rodríguez, Food and Agriculture Organization of the United Nations, Quito, Ecuador, e-mail: psangoluisa@yahoo.com

LivestockPlus: Supporting Low Emissions Development in the Latin American Cattle Sector

JACOBO ARANGO[1], IDUPULAPATI RAO[1], TODD ROSENSTOCK[2], JEIMAR TAPASCO[1], JENNIFER TWYMAN[1], ROLANDO BARAHONA[3], CRISTOBAL VILLANUEVA[4], MAURICIO CHACÓN[5], NELSON VIVAS[6], NGONIDZASHE CHIRINDA[1]

[1]*International Center for Tropical Agriculture (CIAT), Colombia*

[2]*World Agroforestry Centre (ICRAF), Kenya*

[3]*National University of Colombia, Dept. of Animal Production, Colombia*

[4]*Tropical Agricultural Research and Higher Education Center (CATIE), Costa Rica*

[5]*Ministry of Agriculture and Livestock, Costa Rica*

[6]*University of Cauca, Colombia*

The LivestockPlus project was created in 2015 with the aim to provide technical assistance and raise up baseline information for the development and implementation of the Nationally Appropriate Mitigation Actions (NAMAs) for the livestock sector in Latin-American countries. The first pilot regions were Colombia and Costa Rica, where the process of developing sectorial NAMAs was in early to intermediate stage. This concept demonstrates how improved forages, when properly managed, can lead to the sustainable intensification of mixed crop-forage-livestock systems in the tropics, while recognising the multiple social, economic and environmental objectives. Minimizing trade-offs, LivestockPlus emphasises the synergy between soils, plants, animals, people and the environment.

This initiative helped to determine technical options for low emissions pasture development in Latin America through intensification at the three levels: (i) genetic, to provide a wide range of forage/feed options; (iii) ecological, to generate multi-dimensional benefits and minimise trade-offs; and (iii) socio-economic, to promote widespread use of improved forages.

As a result, Costa Rica and Colombia have made significant progress in the development and early implementation of their respective NAMAs for the cattle sector, and their experience have played an important role for the scaling up of these initiatives to other regions through the NAMAs and other policies. The project facilitated synergies that enabled overcoming barriers to the adoption of improved management practices and, consequently, the achievement of low emission development by: (i) fostering partnerships among relevant stakeholders; (ii) identifying and evaluating best-fit mitigation options; and (iii) evaluating measurement reporting and verification (MRV) systems for the cattle sector in the target countries.

Keywords: Climate change mitigation, nationally determined contributions

Contact Address: Jacobo Arango, International Center for Tropical Agriculture (CIAT), A A 6713, Cali, Colombia, e-mail: j.arango@cgiar.org

Integral Assessment of Productive and Environmental Parameters of a Forage-Based Silvopastoral System

Jacobo Arango[1], Mauricio Sotelo[1], Jhon Fredy Gutierrez[2], Belisario Hincapie[1], Eduardo Vazquez[3], Nikola Teutscherová[4], Xiomara Gaviria[5], Isabel Molina[6], Rolando Barahona[7], Mirjam Pulleman[8], Ngonidzashe Chirinda[1], Michael Peters[1]

[1]*International Center for Tropical Agriculture (CIAT), Colombia*

[2]*University of Cauca, Dept. of Agricultural Sciences, Colombia*

[3]*Universidad Politécnica de Madrid, Dept. of Agricultural Production, Spain*

[4]*Czech University of Life Sciences Prague, Fac. of Tropical AgriSciences, Dept. of Crop Sciences and Agroforestry, Czech Republic*

[5]*National University of Colombia, Colombia*

[6]*Autonomous University of Yucatan, Dept. of Animal Nutrition, Mexico*

[7]*National University of Colombia, Dept. of Animal Production, Colombia*

[8]*Wageningen University and Research, The Netherlands*

To quantify the productivity and environmental benefits of improved forage-based silvopastoral systems, a six hectares field trial was established in 2013 at the headquarters of the International Center for Tropical Agriculture (CIAT) using improved forages developed at CIAT and the International Livestock Research Institution (ILRI). A randomised complete block design with three replications was used with three treatments: T1 grass alone, *Brachiaria* hybrid cv. Cayman; T2: Grass + herbaceous legume, *Canavalia brasiliensis* and T3: Grass + herbaceous legume + tree legume *Leucaena diversifolia.* In order to measure animal productivity, a group of steers was fattened (from 200 to 450 kg) and subgroups of animals were assigned to the three treatments (T1, T2 and T3) in a rotational grazing system (stocking rate: 4 animal per hectare). Animal subgroups were kept in each treatment for comparison purpouses. Live weight gain per animal was recorded monthly and large differences were found between treatments with an average of 240, 744 and 819 kg ha^{-1} yr^{-1} for T1, T2 and T3, respectively. Environmental parameters evaluated in the three treatments included soil chemical and physical parameters as well as soil macrofauna, as an indicator of soil health. A positive impact of forage diversification on key soil properties and the abundance and diversity of soil macroinvertebrates was found with best results in T3 compared to the other two treatments. The taxonomic abundance and functional diversity of soil macrofauna were found to correlate positively with forage diversification and the level of inclusion of legumes in a grass-alone system. Finally, the enteric methane emissions of animals grazing T1 and T3 were measured. Average amounts per treatment (n=4), were 30.1 and 19.9 g CH_4 kg^{-1} of dry matter intake for T1 and T3, respectively. Our results evidence the benefits of silvopastoral systems in supporting sustainable intensification of livestock systems and improving livestock productivity per unit area, while delivering ecosystem services and simultaneously reducing the environmental footprint of the animals.

Keywords: Climate change, greenhouse gases, soil quality, tropical forages

Contact Address: Jacobo Arango, International Center for Tropical Agriculture (CIAT), A A 6713, Cali, Colombia, e-mail: j.arango@cgiar.org

Sustainable Feed Management through Fodder Preservation and Feed Rationing in Kenyan Dairy Farms

BONIFACE SAKWA[1], OGHAIKI ASAAH NDAMBI[2], JAMES ONDIEK[1], ANTHONY KINGORI[1]

[1]*Egerton University, Dept. of Animal Science, Kenya*
[2]*Wageningen University and Research, Animal Science Group, The Netherlands*

The year-round availability of quality forage is a major challenge of dairy farmers in Kenya leading to low milk yields, low milk solids content and high production costs. Use of conserved forages and proper ration formulation has a great potential to bridge the gap in dairy nutrition and reduce seasonal variations in milk yield. The SNV Kenya Market-led Dairy Program (KMDP), funded by the Embassy of the Kingdom of The Netherlands, introduced some interventions on fodder conservation and ration formulation on dairy farms in Kenya. This study aimed to determine the effects of these interventions on dairy farm performance. Twelve farms each using one of four interventions (use of maize train/baled silage, production of silage with support from Service Provider Enterprises (SPEs), ration formulation aided by Rumen8 software, and feed balancing without software support) were compared with 12 control farms without interventions, making a total of 60 purposively sampled farms. The data was analysed using Rumen8 software and a linear regression model was used to assess the relationship between independent variables (fodder interventions) and dependent variables (dairy performance). The results show that farms using maize train/baled silage had a better performance with an average daily milk yield of 20 kg/cow than those using silage from SPEs (15 kg/cow) and control farms (8 kg/cow). Ration formulation aided by Rumen8 software resulted in highest daily milk yield (27 kg/cow) while feed balancing without software led to an average milk yield of 20 kg/cow. The average income was KES 1000 (10 USD) per cow (maize train) compared to 7.50 USD/cow (SPEs) and 5.50 USD/cow (control), while average feed costs per farm were 205, 181, and 245 USD per ton for maize train, SPEs and control farms, respectively. Milk-related income on Rumen8 software farms was 17.20 USD/cow while farms not using feed rationing software achieved 11.45 USD/cow. Feed costs were 260 and 226.5 USD/ton for Rumen8 software farms and farms without feed rationing software, respectively. In conclusion, use of maize train/baled silage and Rumen8 software aided ration formulation provide ideal and sustainable fodder management and feed rationing measures for dairy farms by reducing seasonal milk yield fluctuation, enhancing year-round fodder availability and increasing farm income.

Keywords: Dairy performance, feed interventions, maize train silage, Rumen8 software

Contact Address: Oghaiki Asaah Ndambi, Wageningen University and Research, Animal Science Group, Droevendaalsesteeg 1, 6780 PB Wageningen, The Netherlands, e-mail: asaah.ndambi@wur.nl

Intake, Apparent Digestibility, and Nitrogen Balance in Steers Fed with different Grasses Grown alone or Intercropped with *Lablab purpureus*

PEDRO ALAN SAINZ-SANCHEZ[1], DANIEL KORIR[2], ASEP INDRA MUNAWAR ALI[3], LUTZ MERBOLD[2], JOHN GOOPY[2], EVA SCHLECHT[3], UTA DICKHOEFER[1]

[1] *University of Hohenheim, Animal Nutrition and Rangeland Management in the Tropics and Subtropics, Germany*

[2] *International Livestock Research Institute (ILRI), Kenya*

[3] *University of Kassel / Georg-August Universität Göttingen, Animal Husbandry in the Tropics and Subtropics, Germany*

Dry matter intake (DMI) and nutrient digestibility are major limiting factors for cattle productivity and nutrient use efficiency in tropical livestock systems. There is a lack of quantitative *in vivo* information on voluntary DMI, nutrient conversion and nitrogen (N) balance in tropical cattle fed tropical forage grasses grown alone or intercropped with forage legumes. The objective of this experiment was to investigate the effects of feeding three tropical grasses grown alone or intercropped with *Lablab purpureus* on DMI, apparent total tract digestibility (ATTD) and N balance in tropical cattle. Eighteen Boran steers of 230±26.9 (mean±standard deviation) kg of live weight (LW) were used in a randomised block design with six experimental periods of 21 days (14 days of adaptation and 7 days of total urine and feces collection). Animals were stratified by LW and randomly allocated to four treatments: *Bracharia brizantha* alone (BA); *B. brizantha* intercropped with Lablab (BL); *Pennisetum purpureum* alone (PA); *P. purpureum* intercropped with Lablab (PL); *Chloris gayana* alone (CA) and *C. gayana* intercropped with Lablab (CL). Data (n=36) was analysed using the mixed procedure of SAS; the statistical model included grass species and intercropping as fixed factors as well as their interaction, and animal as random factor.

Daily DMI (BA=98; BL=100; PA=96; PL=98; CA=101; CL=111 g kg^{-1} LW 0.75) did not differ between treatments ($p > 0.05$). The ATTD of the diet consumed by the steers was similar ($p > 0.05$) for grasses fed alone or intercropped (BA=614.2; BL=609.4; PA=625.3; PL=632.2; CA=597.1; CL=609.9 g kg^{-1}. Daily N balance was positive for all animals and tended to be higher ($p = 0.07$) when the forage grasses where intercropped with Lablab (BA=50.11; BL=65.2; PA=54.9; PL=63.8; CA=54.5; CL=57.4 % of N intake).

A positive N balance indicates that the evaluated forage grasses satisfied the protein requirement of the steers. Feeding tropical forage grasses intercropped with forage legumes is a promising strategy to improve the nitrogen balance in tropical cattle.

Keywords: Forage legume, intake, nitrogen balance, tropical cattle

Contact Address: Pedro Alan Sainz-Sanchez, University of Hohenheim, Animal Nutrition and Rangeland Management in the Tropics and Subtropics, Fruwirthstraße 31, 70593 Stuttgart, Germany, e-mail: alan.sainz@uni-hohenheim.de

The Evaluation of Dry Matter and Nutrient Intake Adequacy in Dairy Cattle in Indonesia

IDAT GALIH PERMANA, ZAHARA KHAIRANI, DESPAL

Bogor Agricultural University, Dept. of Animal Nutrition and Feed Technology, Indonesia

The scale of ownership of lactation dairy cows in Indonesia is relatively low with an average of 3–4 heads. This study aims to determine and evaluate adequacy nutrient and nutrient utility in milk production for dairy cows based on the scale of ownership of farmers. This research was carried out in dairy cattle villages on the island of Java, observing 144 lactationg cows from 56 farmers. The parameters measured in this study were the amount of feeding, nutrient intake, body weight, (BCS) Body Condition Scoring, milk production, milk quality and nurient adequacy. Observation data were analysed by using a T-test based on the scale of ownership. The results of this study indicate that the performance of dairy cows in Indonesia is still not optimal. The average body weight was between 394.54 ± 37.26 and 424.38 ± 43.33 kg, and BCS ranged from 2.51 ± 0.27 to 2.60 ± 0.26. The consumption of dry matter (DM) in high-production cows was higher compared to medium-production cows and low-production cows: the average consumption of DM was 12.8 kg, 12.37 kg and 11.22 kg per head, respectively, with a total digestible nutrient content varying from 57.1 to 67.5 %. However, based on the analysis of nutrient adequacy, there were still cows fed with no standard feeding, namely 51.8 % of high-production cows, 37.2 % of medium-producing cows and 38.6 % of low-production cows. Milk production was largely determined by the scale of ownership. The average milk production on the scale of ownership of 1–3 heads, 4–5 heads and > 6 heads, respectively, was 12.42 ± 5.07, 11.81 ± 3.62, and 16.00 ± 2.00 kg per head with milk fat varying from 4.2 ± 1.73 % to 4.8 ± 1.73 %. Based on these results it was concluded that the feeding of lactating dairy cows in Indonesia is still below the standard requirements so that it does not support high milk production.

Keywords: Dairy cattle, milk production, milk quality, nutrient intake

Contact Address: Idat Galih Permana, Bogor Agricultural University, Dept. of Animal Nutrition and Feed Technology, Jl. Agatis Kampus IPB Darmaga, 16680 Bogor, Indonesia, e-mail: permana@apps.ipb.ac.id

Evaluation of a Modified Livestock Simulator (LIVSIM) Model

CHRISTIAN ADJOGO BATEKI, UTA DICKHOEFER
University of Hohenheim, Animal Nutrition and Rangeland Management in the Tropics and Subtropics, Germany

Models can play an important role in identifying and removing traps for sustainable resources-use in tropical livestock production systems (TLPS). However, simply adopting procedures for estimating nutrients and energy requirements or the dry matter intake (DMI) developed for cattle in temperate regions may limit models' capacity to accurately quantify resources-use in TLPS. Therefore, we modified a selected model developed for TLPS to account for differences between cattle in (sub-)tropical and temperate regions and evaluated its accuracy to predict dairy cattle performance using data from a published study.
The LIVestock SIMulator (LIVSIM) model was selected upon review of different livestock models suitable for TLPS. It is dynamic and simulates animal performance based on availability and nutritional quality of feedstuffs. We modified the LIVSIM by simplifying and modifying the British metabolisable energy (ME) and protein requirement system for dairy cattle. Using literature values, maintenance ME requirements were increased from 0.488 to 0.631 MJ ME kg^{-1} metabolic live-weight (LW) and age-sex specific ME and protein requirements for gain (MJ ME kg^{-1} LW) were defined for dairy cattle in TLPS. Next, DMI was estimated as the lower of either the physically regulated (using the animal's capacity of neutral detergent fibre (NDF) intake (g kg^{-1} LW)) or physiologically regulated (using the animal's total ME requirements (MJ d^{-1})) estimated DMI. Then, ME intake was partitioned to favour milk production rather than LW gain during early lactation. The original and modified LIVSIM versions were evaluated using data from a published study with tropical stall-fed dairy cattle, and the mean bias (MB), root mean squared error-of-prediction (RMSEP), and relative prediction error (RPE) used to assess their accuracy.
Both models under-estimated (i.e. positive MB) DMI and over-estimated (i.e. negative MB) animals' LW changes. Yet, the modified LIVSIM was more accurate (i.e. low RMSEP and RPE) than the original LIVSIM in predicting DMI, final LW, and average daily milk yield over a 140-d-period.
Adapting underlying procedures adopted from temperate regions when simulating DMI, nutrient partitioning, and performance improves accuracy of livestock models and their ability to predict resources-use of cattle in TLPS.

Keywords: Dairy cattle, LIVestock SIMulator, models, resources-use

---

Contact Address: Uta Dickhoefer, University of Hohenheim, Animal Nutrition and Rangeland Management in the Tropics and Subtropics, Fruwirthstr. 31, 70599 Stuttgart, Germany, e-mail: aninutrop@uni-hohenheim.de

Breeding Objectives and Practices in Three Cattle Production Systems in Burkina Faso

DOMINIQUE OUÉDRAOGO[1], ALBERT SOUDRÉ[2], SALIFOU OUÉDRAOGO-KONÉ[3], BIENVENUE.L ZOMA[1], BERNADETTE YOUGBARE[1], NEGAR KHAYATZADEH[1], PAMELA BURGER[4], GÁBOR MÉSZÁROS[1], AMADOU TRAORÉ[5], ALI MWAI OKEYO[6], MARIA WURZINGER[1], JOHANN SÖLKNER[1]

[1]*University of Natural Resources and Life Sciences Vienna (BOKU), Dept. of Sustainable Agricultural Systems, Austria*

[2]*Université Norbert Zongo, Unité de Formation et de Recherche en Sciences et Technologies, Burkina Faso*

[3]*Université Nazi Boni, Institut de Développement Rural, Burkina Faso*

[4]*University of Veterinary Medicine Vienna (Vetmeduni), Dept. of Integrative Biology and Evolution, Austria*

[5]*Institute of Environment and Agricultural Research (INERA), Burkina Faso*

[6]*International Livestock Research Institute (ILRI), Kenya*

This study was conducted in South–west Burkina Faso, the traditional habitat of trypanotolerant Baoulé cattle, to identify farmers preferences for cattle breeding traits. A structural questionnaire was administered to 194 heads of households randomly sampled. Own herd ranking method was additional used to detect the most important criteria for breeding cows' selection; bulls were excluded due to their low numbers in herds. Pearson's chi-square test was employed to test the independence and to compare the categorical variables. Arithmetic and least square means of continuous variables were calculated and compared by Tukey-test or Wilcoxon pairwise test. The importance of criteria of selection of breeding animals by farmers was calculated through index of ranking. According to breed and herd mobility, three production systems were defined: sedentary pure Baoulé, sedentary crossbred and Zebu, and transhumant Zebu and crossbred systems. Cash income, social reasons and good wealth status were the main reasons of keeping cattle in the three production systems while milk for home consumption was reported specifically in transhumant system. Baoulé females were perceived to be significantly older at first mating, first calving and with longer reproduction life time. Breeding bulls were selected among young males in herd. The majority of transhumant farmers reported castration of no selected males contrary to pure Baoulé farmers. The most important common criteria of selection between the production systems were body size for both males and females, growth for male and calf growth for female. In addition to body size, the survey showed that females' mothering ability was preferred by pure Baoulé owners while milk yield and udder state were highly considered by transhumant people. The ranking revealed that

Contact Address: Dominique Ouédraogo, University of Natural Resources and Life Sciences Vienna (BOKU), Dept. of Sustainable Agricultural Systems, Gregor-Mendel-Straße 33, A-1180 Vienna, Austria, e-mail: ouedom@gmail.com

a cow is considered best according to its fertility in pure Baoulé system, and according to its milk yield in sedentary crossbred and Zebu and transhumant systems. Bulls were furthermore selected according to docility in pure Baoulé system and dam in crossbred and Zebu systems. These findings suggest taking into account farmers and their specific characteristics and practices in the design and implementation of successful breeding programs to improve and conserve locale cattle breed in this area.

Keywords: Baoulé, breeding practices, crossbred, traits preference, West Africa, Zebu

Morphometric Characterisation of Purebred Baoulé and their Crosses in the Southwestern Region of Burkina Faso

BERNADETTE YOUGBARE[1], ALBERT SOUDRÉ[2], DOMINIQUE OUEDRAOGO[1], BIENVENUE.L ZOMA[1], ARNAUD S.R. TAPSOBA[3], MOUMOUNI SANOU[3], SALIFOU OUEDRAOGO-KONÉ[4], PAMELA BURGER[5], MARIA WURZINGER[1], NEGAR KHAYATZADEH[1], HAMIDOU H. TAMBOURA[3], AMADOU TRAORÉ[3], JOHANN SÖLKNER[1], GÁBOR MÉSZÁROS[1]

[1]*University of Natural Resources and Life Sciences, Vienna (BOKU), Dept. of Sustainable Agricultural Systems, Austria*

[2]*Norbert Zongo University, Burkina Faso*

[3]*Institute of Environment and Agricultural Research (INERA), Animal Production Department, Burkina Faso*

[4]*Nazi Boni University, Inst. of Rural Development, Burkina Faso*

[5]*University of Veterinary Medicine Vienna (Vetmeduni), Dept. of Integrative Biology and Evolution, Austria*

The cattle population of Burkina Faso is currently around 8.5 million providing draft power, meat and milk to the country as well as foreign exchange from exports. The various natural resources of the country from semi-arid Sahelian North to sub-humid Sudanese South yielded a diverse local cattle populations, with comparatively large and strong indicine (zebu) cattle in the North and physically very small taurine cattle in the disease ridden south. The overall objective of the LoCaBreed project is to contribute to livelihood improvement and a better understanding of Burkina Faso local cattle breeds.

It is this context that a morphometric characterisation study has been conducted to explore the prospects for improvement and conservation of the Baoule cattle breed in southwestern region of Burkina Faso. A total of 421 cattle including 266 Baoulé × zebu crossbreds and 155 purebred Baoulé cattle were characterised for 24 quantitative and 20 qualitative traits.

The analyses of quantitative traits confirmed the notable differences in size between Baoulé and their crosses with zebu cattle. This difference was significant in all of the 24 measured quantitative traits. In fact, crossbreeding with zebu increases body size, which is considered as an improvement because larger animals are preferred by the breeders. The qualitative traits, such as coat colour, and horn shape showed large variety within Baoulé and crossbreds, likely due to local preferences of livestock keepers rather than breed identity. The combined genotype-location effect showed that facial length, distance point to point horn, ear length, chest depth, colour of muzzle, head tache and horn shape differ between purebred Baoulé and crossbred populations. The differences of qualitative traits are a result of the selection by the breeders, rather than due to environment. These results will serve as basis for further characterisation, conservation and improvement strategies for Burkina Faso Baoulé breed.

Keywords: Cattle breeds, characterisation, crossbreds, linear model, morphological traits

Contact Address: Bernadette Yougbare, University of Natural Resources and Life Sciences, Vienna (BOKU), Dept. of Sustainable Agricultural Systems, Gregor-Mendel-Strasse 33, 1180 Vienna, Austria, e-mail: yougbernadette@hotmail.fr

Spatial and Socio-Economic Factors Shaping the Genetic Diversity of Indigenous Cattle Breeds in Benin

Sèyi Fridaïus U. S. Vanvanhossou[1], Carsten Scheper[1], Luc Hippolyte Dossa[2], Kerstin Brügemann[1], Sven König[1]

[1] *Justus-Liebig-University Gießen, Inst. of Animal Breeding and Genetics, Germany*
[2] *University of Abomey-Calavi, Fac. of Agricultural Sciences, School of Science and Technics of Animal Production, Benin*

Do ecological factors shape the genetic structure of the Beninese cattle population? Are the indigenous taurine (Somba and Lagune) and hybrid (Borgou, Pabli-Kerou) breeds in Benin endangered by ongoing indicine introgression? What is the role of transhumance movements in this context? These are critical questions for the development of effective management policies for the conservation of indigenous cattle genetic resources in Benin, a country regularly influenced by transboundary transhumance movements of zebu herds.
We developed a balanced sampling design based on the distribution of the agro-ecological zones and the socio-economic factor transhumance to study the genetic diversity of the four indigenous Beninese cattle breeds. 461 animals (Borgou 181, Pabli-Kerou 61, Lagune 157, Somba 62) were phenotyped for conformation traits based on FAO guidelines and genotyped with a 50k SNP chip. SNP genotypes were used for basic measures of genetic diversity, unsupervised k-means clustering in the framework of a discriminant analysis of principal components (DAPC) and estimation of ancestry coefficients using DAPC, TESS and ADMIXTURE. The diversity analysis is complemented with linear regression-based genome-wide association studies (GWAS) for conformation traits and selection signature analysis (SSA).
Unsupervised clustering using DAPC identified 6 genetic clusters in the dataset. The two hybrid breeds Borgou and Pabli-Kerou built two nearby mixed clusters reflecting a close relationship and admixture between the two breeds. The taurine Somba breed built a highly homogeneous distant cluster, whereas the Lagune breed is more diverse with 2 separated heterogeneous clusters reflecting local transhumance effects. A separate sixth cluster reflected potential crossbreeding with non-indigenous breeds of unknown origin. The clustering results were supported by the basic measures of diversity and the admixture analysis. The main drivers of genetic diversity are spatial and socio-economic factors such as transhumance. The multi-breed GWAS revealed 9 and 23 highly conserved SNPs associated with Heart Girth and Hip Width. Similarly, we identified several significant loci across breeds, and breed specific loci for the 10 other phenotypic traits investigated. GWAS and SSA shed light on functional gene-networks shaping the genomic architecture of body conformation traits in light of environmental adaptation and artificial selection.

Keywords: Agro-ecological zone, animal genetic resources, conservation, taurine, transhumance, zebu introgression

Contact Address: Sèyi Fridaïus U. S. Vanvanhossou, Justus-Liebig-University Gießen, Inst. of Animal Breeding and Genetics, Ludwigstraße 21b, 35390 Gießen, Germany, e-mail: seyi.f.vanvanhossou-2@agrar.uni-giessen.de

Farmers' Willingness to Conserve the Endangered Sheko Cattle Breed in Benchi Maji Zone, Ethiopia

Emnet Gashaw[1], Tewodros Tefera[2], Regina Rössler[3], Simret Weldenegodguad[4]

[1] *Bonga University, College of Agriculture, Ethiopia*
[2] *Benefit Program of Partnership Office, Ethiopia*
[3] *University of Kassel / University of Goettingen, Animal Husbandry in the Tropics and Subtropics, Germany*
[4] *Hawassa University, College of Agriculture, Ethiopia*

The Sheko cattle breed has a promising potential to form the genetic backbone to cope with new tropical diseases, unpredictable future climate changes and food insecurity in Ethiopia. Despite its recognised trypanotolerance and adaptation to environmental stressors, the breed currently faces clear risk of extinction. In order to support the establishment of cost-efficient conservation plans, this study aimed at assessing farmers' willingness to conserve the Sheko breed in its natural breeding tract of Benchi Maji zone, Ethiopia. Choice experiments were employed on 400 cattle keepers to assess their preferences for important cattle attributes. Preferences for cattle attributes are used as indicators for farmers' willingness to conserve the local Sheko cattle breed. The study revealed that farmers' generally preferred cattle breeds with a high trypanotolerance, low feed requirements and aggressiveness, and high milk yield. However, preferences for cattle attributes differed according to agro-ecological zone (AEZ). Farmers in the midland AEZ were willing to pay 603 Birr (21.60 €) for a cow with a high milk yield. On the contrary, their counterparts in the lowland AEZ were willing to pay 2.3 and 1.9 times more for a cow with low feed requirements and high trypanotolerance than farmers in the midland AEZ. This suggests that farmers, in particular in lowland AEZ, valued adaptation traits more than a high milk performance. Furthermore, farmers who were born in the Sheko community were more likely to prefer aggressive cattle, while a longer experience in keeping the Sheko cattle breed increased the probability that farmers preferred cows with a high milk yield. Similarly, a positive interaction was found between satisfaction with the veterinary service and a high milk yield, as well as high feed requirements and trypanotolerance. Consequently, conservation programs for the Sheko cattle breed in Benchi Maji zone of Ethiopia will imply compensation payments of the farmers due to the high feed requirements and aggressiveness of this breed. It is recommended to implement conservation strategies in the midland AEZ by involving farmers who originate from the Sheko community or have experience in keeping Sheko cattle, and to improve veterinary services to reduce compensation costs.

Keywords: Choice experiments, Ethiopia, in-situ conservation, preference heterogeneity, Sheko cattle breed, trait preferences

---

Contact Address: Regina Rößler, University of Kassel / University of Goettingen, Animal Husbandry in the Tropics and Subtropics, Göttingen, Germany, e-mail: regina.roessler@agr.uni-goettingen.de

The Role of ICT Based Extension Services on Dairy Production in Kenya: A Case of iCow Service

Erick Marwa[1], John Mburu[1], James Rao[2], Ali Mwai Okeyo[2]

[1] *University of Nairobi, Dept. of Agricultural Economics, Kenya*
[2] *International Livestock Research Institute (ILRI), Kenya*

There is a consensus that agricultural extension service play an important role for growth and transformation of the agricultural sector in Sub Saharan Africa. Benefits such as high productivity, quality of produce, reduction of diseases and pest, and increased income among smallholder farmers are attributed to access to quality extension service. Despite of these contributions of extension services, smallholder farmers have the challenge of accessing timely and credible agricultural information which limits them in realising maximum farm output. Use of ICT as an agricultural extension tool by smallholder farmers has the potential to reverse the scenario and improve farmers' access to credible agricultural information. The current study aimed at determining the impact of iCow service (ICT- based extension tool) on milk production, household income.

The study randomly sampled a total of 457 small-household dairy farmers in three counties of Kenya, of which 209 farmers regularly used the iCow services and 248 farmers were non-users. Further, the study used Propensity Score Matching (PSM) model to estimate the impact of iCow on milk production, household income.

Descriptive analysis shows that there are significant differences in the means between regular users of iCow and non-users with respect to milk production and household income. Regular iCow users realised higher average annual milk production and annual household income than non-users. The results of PSM model indicates that use of iCow services among dairy farmers had a positive and significant effect on milk production and income. Specifically, the figures reveal that use of iCow services led to increased milk production by 1138.95 litres and earned Ksh. 89,043 more annually. These figures can be considered as an opportunity cost of not using iCow service.

The positive impact shows the potential role of ICT-based extension in rural poverty reduction through increased household incomes. The positive correlation of use of phones in getting timely information among farmers suggest partnership between network providers and research institutes should be encouraged as part of bridging the extension gap occasioned by reduced public expenditure on extension services.

Keywords: Agricultural information, dairy farmers, iCow services, ICT, propensity score matching

Contact Address: Erick Marwa, University of Nairobi, Dept. of Agricultural Economics, 29053, Nairobi, Kenya, e-mail: mwitaerick@gmail.com

Milk Fatty Acid Composition of Camels and Cattle Grazing and Browsing in East-African Rangelands

Paul Leparmarai[1,2], Carmen Kunz[1], David Miano Mwangi[2], Ilona Gluecks[3], Michael Kreuzer[1], Svenja Marquardt[1,3]

[1]*ETH Zurich, Inst. of Agricultural Sciences, Switzerland*

[2]*Kenya Agricultural and Livestock Research Organisation (KALRO), Kenya*

[3]*International Livestock Research Institute (ILRI), Kenya*

Cattle and camels are important livestock species in the semi-arid and arid rangelands of East-Africa. Their diet selection behaviour differs. Cattle are grazers preferring grasses while camels are browsers that rely on shrubs, trees and herbs. Their diets are affected by seasonal changes. The present study investigated the milk fatty acid (FA) profile of camels and two cattle types (Guernsey × Boran crossbreds and local Pokot cattle) kept on savannah rangelands in two seasons (rainy season (RS) and transition period (TP)) in Laikipia, Kenya. Twelve lactating animals per livestock type were used in RS and 12 different animals in TP, respectively (n=72 animals in total). Half of the animals from each livestock type received a urea-molasses supplement at night during confinement. The animals grazed during the day. Direct observations with bite counting were applied on one animal per day (36 observation days/season) and the chemical composition of the most selected plant species was analyzed. Intakes of nutrients and phenols were estimated. Milk samples per animal were collected at the end of each season and were analysed for FA composition using gas chromatography. Camels ingested higher contents of lignin (ADL), crude protein and phenols, with the last two parameters differing from the cattle especially in the TP. Saturated FA were generally higher in proportion in the milk of both cattle types compared to the camel milk, and increased ($p < 0.05$) from RS to TP in the milk of the cattle types, but not of the camels (season × animal type, $p < 0.001$). Proportions of monounsaturated FA were higher in camel milk than cattle milk and did not differ between seasons in camel milk, while their proportions declined in milk of both cattle types from RS to TP (season × animal type, $p < 0.001$). Proportions of polyunsaturated FA in milk decreased from RS to TP in all the three animal types but were higher ($p < 0.05$) in camel milk compared to cattle milk in both seasons. Urea-molasses supplementation had minor effects on a few FA. In conclusion, the milk FA profile of camels and cattle is differently affected by seasonal changes in forage and diet selected.

Keywords: Browser, diet selection, grazer, livestock species, milk fatty acids, season

Contact Address: Svenja Marquardt, International Livestock Research Institute (ILRI), Mazingira Centre, Nairobi, Kenya, e-mail: s.marquardt@cgiar.org

Urbanising Tropical Environments and the Production Gap – The Case of Dairy Production in Bengaluru, India

MARION REICHENBACH[1], ANA PINTO[2], SVEN KÖNIG[2], RAGHAVENDRA BHATTA[3], EVA SCHLECHT[1]

[1]*University of Kassel / Georg-August-Universität Göttingen, Animal Husbandry in the Tropics and Subtropics, Germany*

[2]*Justus-Liebig-University Giessen, Inst. of Animal Breeding and Genetics, Germany*

[3]*National Institue of Animal Nutrition and Physiology, Inst. of Animal Breeding and Genetics, India*

While drivers of change such as urbanisation and population growth have a huge impact on agriculture in (sub-)tropical regions, this is also were gaps in terms of resources use efficiency are the largest. In the emerging megacity of Bengaluru in India, rapid urbanisation is combined with a high demand for dairy products. Yet, neither has the resources use efficiency of the dairy units providing these products been quantified, nor have transition processes that are provoked by urbanisation been considered. Therefore, the aim of this study was to quantify on-farm resources use with a focus on feed efficiency of dairy cattle, in relation to surrounding urbanisation. Twenty-eight dairy units were selected across the rural-urban interface of Bengaluru: 4 urban, 8 peri-urban and 16 rural ones. During one year, each dairy unit was visited at a 6 weeks interval to collected data on herd management, and at animal level growth, offered forages and milk offtake. Temperature and humidity were recorded at the dairy units. Most commonly offered forages were self-cultivated maize and elephant grass, and during the dry season, finger millet straw. Dairy producers also offered grass collected on field margins or public land, and in urban areas, market waste. Cows were sent to pasture by 77% of producers. In densely build-up areas, cows were pasturing in the streets (organic waste) or public green space. Across dairy units and seasons, the daily amount of green forage offered ranged from 0 to 44.4 kg cow^{-1}. In addition, each cow received between 0 and 4.3 kg of concentrate twice a day. Daily milk offtake in the early lactation varied from 5.8 to 18.1 kg cow^{-1}. The data show the high variability in farm resources, feeding strategies and subsequent production (in-)efficiency. On one hand, urban dairy units scored a higher milk production and compensated for the lack of private land by relying on publicly available resources to provide the highly demanded milk directly to consumers. On the other hand, manure management in the city was poor, suggesting that peri-urban dairy units hold the highest potential for efficient resources use as they combine opportunities of rural and urban localities.

Keywords: Dairy production, resources use efficiency, urbanisation, megacity

Contact Address: Marion Reichenbach, University of Kassel / Georg-August-Universität Göttingen, Animal Husbandry in the Tropics and Subtropics, 37213 Witzenhausen, Germany, e-mail: marion.reichenbach@uni-kassel.de

Dairy Producers' Levels of Adaptability Post-Market Shock in Bamenda, Cameroon

JENNIFER PROVOST[1], GABRIEL ROSERO[2], BERNHARD BRÜMMER[2], EVA SCHLECHT[1]

[1] *University of Kassel / University of Goettingen, Animal Husbandry in the Tropics and Subtropics, Germany*

[2] *University of Goettingen, Dept. of Agricultural Economics and Rural Development, Germany*

Formal and informal markets usually complement one another in urbanised centres of developing countries. In Bamenda – the capital of the North-West region in Cameroon – the dairy market consisted of a single private milk factory and a web of informal sellers until mid-2016, when the factory closed its doors. As a result of the sudden departure of the formal market, milk producers made several changes to their farm management. As part of the Urban Food-Plus project, we carried out surveys in 2017 in urban, peri-urban and rural zones of Bamenda with 201 current and 123 former milk suppliers to look at the various strategies undertaken. Four levels of adaptation to post-market shock were observed: producers who remained in the market (Sellers), producers who stopped selling but kept the milk for household consumption (Self-Consumers), producers who stopped selling, stopped breeding, but kept the dry dairy animals (Keepers), and producers who stopped selling and sold all their dairy animals (Quitters). The objectives of this study are to analyse if certain characteristics can predict dairy producers' level of adaptability to market shocks and if there are policy implications to make producers more resilient to such events. We divided the analysis between Mbororo pastoral producers and sedentary Grassfielders because we noticed that the different production systems had separate and distinct adaptation levels. A multinomial logit model was used for the Grassfielders whose adaptation levels varied between Sellers, Keepers, and Quitters, whereas a logit model was used for the Mbororo'en whose levels were limited to Sellers and Self-Consumers. Our results show that Grassfielders are more likely to remain in the market after a shock when they are younger and literate, have bigger households and a land title, and when they use accessible trainings, institutional credit, and market price information. For the Mbororo'en, younger producers with large households who are members of dairy cooperatives and settle more permanently in the rural areas are more likely to remain Sellers. Therefore, policies and development programs should consider the distinct production systems and focus on the aforementioned characteristics in order to strengthen dairy producers' resilience to market shocks.

Keywords: Adaptation, Cameroon, dairy, market shock, milk production, pastoralism

Contact Address: Jennifer Provost, University of Kassel / University of Goettingen, Animal Husbandry in the Tropics and Subtropics, Albrecht-Thaer-Weg 3, 37075 Göttingen, Germany, e-mail: jprovos@gwdg.de

Market Supply of Animals and Animal Products in Northeastern Madagascar – The Influence of the Vanilla Boom

JOËLSON T. HERITIANA[1], STEFANIE KUNZ[2], HENDRIK HÄNKE[3], EVA SCHLECHT[2]

[1] *University of Antananarivo, Ecole Supérieure des Sciences Agronomiques; Section Sci. Animales, Madagascar*

[2] *University of Kassel / Georg-August-Universität Göttingen, Animal Husbandry in the Tropics and Subtropics, Germany*

[3] *University of Goettingen, Dept. of Agricultural Economics and Rural Development, Germany*

In the Sava Region of north-eastern Madagascar, the soaring price of vanilla is accompanied by an improvement in the quality of life of the population. This may boost local investments in agriculture and education. But is the price boom of vanilla also an opportunity for the development of the livestock sector in this region? To answer this question, the present survey targeted key stakeholders in the sector, namely breeders, resellers and butchers of cattle and poultry. Interviewed persons were identified through a snowball sampling approach starting each time at the regular markets of 3 major cities and 3 villages in the Sava Region. First, resellers of animals / products and butchers were interviewed. Subsequently, they were asked to indicate their suppliers. Observations included 30 resellers of animals / products, 21 butchers and 80 breeders. The most important outlets for zebu cattle were the livestock markets of Ampondra in Vohemar district and Antagena-Betsakotsako in Andapa district. The main customers were beef butchers and itinerant resellers. Due to a substantial increase in animal prices, zebu resellers no longer supply animals to local breeders but only to other resellers or butchers. Regions of origin of the resold zebus were Boeny, Sofia and Diana, but not the Sava Region itself. Similar to cattle, the price of poultry increased substantially over the past three years due to growing market demand. Poultry resellers are supplied predominantly from breeders in the Antalaha and Vohemar districts. Poultry resale activities provide income diversification for farm households and birds constitute capital that can be mobilised in case of emergency. Egg production is scarce, despite significant market demand, and eggs sold on regional markets originate from the regions Analamanga and Antsinanana. Intensive pork farming is almost non-existent in the Sava Region due to the arrival of African swine fever. However, butchers noticed an increase in pig meat demand, which they explained by the arrival of immigrants who came to take advantage of the vanilla business. It can be concluded that booming vanilla business in the Sava Region encourages local but also more remote breeders to slightly intensify their livestock activities.

Keywords: Eggs, Madagascar. market demand, meat, supply chain, vanilla production

Contact Address: Eva Schlecht, University of Kassel / University of Goettingen, Animal Husbandry in the Tropics and Subtropics, Steinstraße 19, 37213 Witzenhausen, Germany, e-mail: tropanimals@uni-kassel.de

Development of an Integrated Cattle Farm on an Ex-Coal Mining Area in Indonesia

TEDI YUNANTO[1], FARISATUL AMANAH[1], DODDY HERIKA[2]

[1]*Min. of Energy and Mineral Resources of Republic of Indonesia, Directorate General Mineral and Coal, Indonesia*
[2]*Pt Berau Coal, Post-mining Division, Indonesia*

Coal production plays a significant role for domestic income in Indonesia but its exploitation inflicts environmental and social damages. In order to rehabilitate the ex-mining land and ensure sustainable development, mine reclamation is mandatory and has to be implemented according to the area's function. The Good Mining Practice Implementation Guidelines of the Ministry of Energy and Mineral Resources of Indonesia, specified in Decree No. 1827.K/30/MEM/2018, determine that mine reclamation embraces revegetation to natural succession (particularly in forested areas) and other reclamation to area function including housings, tourism, water bodies (for pit lake or void) and livestock or farming. By this regulation, the government rules how the non-renewable resources contribute to a sustainable development whilst exploitation is over, for example by setting up a livestock (cattle) farming enterprise on ex-mining land. PT Berau Coal has initiated the Integrated Cattle Farm in East Kalimantan Province in 2010, which breeds the Balinese cattle (*Bos javanicus*; since 2010), the *Bos indicus* Donggala cattle breed (since 2011) and Brahman Cross cattle (*Bos indicus*; since 2012). This farm supports the government's project since Indonesia, to supply domestic consumption, imports about 23 % of its beef - mostly from Australia, the United States and New Zealand. From 2010 till date, birth rates are 0.62 for the Brahman Cross, 0.45 for Balinese Cattle and 1.04 for Donggala cattle. According to measurements of weight increments, Brahman Cross gain 0.28 kg day $^{-1}$ of weight in intensive feeding (zero-grazing) and 0.18 kg day $^{-1}$ with semi intensive feeding (i.e., grazing). The nutrients supplied by cover crops and grasses cultivated on on ex-mining land were also tested and found to meet the nutrient requirements of beef cattle. Currently, the Integrated Cattle Farm of PT Berau Coal is not only a training centre for the local community surrounding the ex-mining area but also holds a certain a potential as national beef production facility in Indonesia.

Keywords: Cattle, ex-mining area , integrated farming, reclamation

Contact Address: Tedi Yunanto, Ministry of Energy and Mineral Resources of Republic of Indonesia, Directorate General Mineral and Coal, Jl. Prof. Dr. Soepomo No.10 Rt.1/rw.3 Menteng Dalam Tebet, 12870 Jakarta Selatan, Indonesia, e-mail: genom.tedi@gmail.com

Small ruminants and forage evaluation

Oral Presentations

Posters

Not all Legumes are Created Equal: Different Nutritional Value of Tree, Herb and Shrub Legumes

JOAQUÍN CASTRO-MONTOYA, UTA DICKHOEFER

University of Hohenheim, Animal Nutrition and Rangeland Management in the Tropics and Subtropics, Germany

Legume forages are seen as means to improve the nutritional status of ruminants in tropical regions, where low crude protein (CP) and energy of grasses limit their production. Nevertheless, despite several positive attributes of legumes, their incorporation into the production systems is still limited. The diversity of these forages in the tropics contributes to the difficulty to use them as feed, as their nutritional value strongly varies across species and ways of use. Therefore, the objective of this study was to highlight the differences in the nutritional composition of 743 tropical legumes across 399 in-vivo trials with ruminants depending on their growth habit (tree, herb, shrub) and the form they are fed to the animal (fresh, hay, silage), as well as their comparison with tropical grasses.

Legumes from 107 species representing 53 genera were found in those studies. Tree legumes had higher CP (206 g kg^{-1} dry matter (DM)) than herbs and shrubs (159 g kg^{-1} DM), as leaves and young stems are fed from trees, while herbs and shrubs are fed as whole plants. This reflects in the fiber content, which was lower in trees and higher in shrubs. Lignin concentration was higher for legumes compared with grasses (73 g kg^{-1} DM), particularly for trees and shrubs (101 and 106 g kg^{-1} DM, respectively) a factor that could limit intake of legumes. Accordingly, *in–vitro* DM digestibility and metabolisable energy was higher for herbs, and lowest for shrubs, with those of shrubs and trees being lower than for grasses. Hay and sole-legume-silage CP averaged g kg^{-1} DM less than the fresh legume, highlighting losses when legumes are processed for conservation. Calcium, phosphorus and magnesium were highest in trees and lowest in herbs, but all greater than in grasses. Finally, fiber-bound nitrogen was greater in shrubs and trees than in grasses, particularly legume hays, indicating the need for adjustments in the CP of diets when legumes are included.

The nutritional value of legumes strongly varies with their growth habit, and whether they are fed fresh or as hay/silage. Recognizing and understanding these differences is a step closer to successfully incorporate these forages in ruminants feeding.

Keywords: Legume forages, nutritional value, ruminants, tropics

Contact Address: Joaquín Castro-Montoya, University of Hohenheim, Animal Nutrition and Rangeland Management in the Tropics and Subtropics, Fruwirthstrasse 31, 70599 Stuttgart, Germany, e-mail: jcm@uni-hohenheim.de

Determinants and Impact of Adopting Climate-Smart *Brachiaria* Grass among Dairy Farmers in Kenya

KEVIN MAINA[1], CECILIA RITHO[1], BEN LUKUYU[2], JAMES RAO[3]

[1] *University of Nairobi, Department of Agricultural Economics, Kenya*

[2] *International Livestock Research Institute (ILRI), Feeds and Forages Program, Uganda*

[3] *International Livestock Research Institute (ILRI), Kenya*

Evolving changes such as population growth, urbanisation and a growing middle-income class in Africa are redefining the agro-food systems. Population growth is expected to double the demand for milk and meat products in Africa by 2050. This raises concerns on the capability of African countries to meet the projected demand. It is imperative that farmers seize the opportunities for earning higher and stable income by responding to the new trends and overcoming the constraints tightened by climate change. Access to quality fodder has continued to be the single most important challenge in livestock production systems. The objective of this article is to evaluate the impact of climate-smart *Brachiaria* on feed sufficiency and milk productivity in dairy production in Kenya. A random sample of 237 farmers, 111 adopters and 126 non-adopters of *Brachiaria* was selected in Makueni and Siaya using multi-stage sampling. Data was collected through face to face interviews and Propensity score-matching approach was then employed to evaluate the impact of *Brachiaria* grass on feed sufficiency and milk productivity. In this study, non-adopters of *Brachiaria* were farmers who were using Napier grass as their source of fodder.

The findings reveal that adoption of *Brachiaria* increases milk production by about 27.6 %. This translates to an average increase of about 3 litres daily per animal. Adoption of *Brachiaria* consequently increases feed sufficiency measured by time spent in feed related activities by the primary woman in a household by 31.6 %. Adopters of *Brachiaria* spend 2 hours less in sourcing and preparing feed in dry seasons.

The surplus milk implies more income for the household, improved nutrition from milk consumption and improved wealth status of household. Results on feed sufficiency imply that *Brachiaria* offers an alternative sustainable source of fodder in fodder scarce periods and releases the burden on women in feed related activities. There is also an improvement in the welfare of families. The study concludes that policies and efforts aimed at increasing widespread adoption should address factors that influence adoption. We recommend increasing extension and training on climate-smart fodder and strengthening collective institutions such as farmer groups for sustainable livestock production.

Keywords: *Brachiaria* grass, feed sufficiency, milk productivity, propensity score matching

---

Contact Address: Kevin Maina, University of Nairobi, Department of Agricultural Economics, 30197-00100 Nairobi, Kenya, e-mail: mainakevin.km@gmail.com

Feed Resource Use Strategies and Performances of Small Ruminants in the Peri-Urban Area of Southern Benin

PANINE YASSEGOUNGBE[1], IVAN B. KOURA[2], LUC HIPPOLYTE DOSSA[1]

[1] *University of Abomey-Calavi, Fac. of Agricultural Sciences, Benin*

[2] *National University of Agriculture, School of Management and Exploitation of Livestock Systems, Benin*

A study was carried out in the peri-urban areas of Cotonou in South Benin in order to identify the small ruminant production systems, to understand farmers' feed resource use strategies and assist them to increase their animal performances and farm profitability. A total of 125 small ruminant farms were surveyed, using a semi-structured questionnaire to collect information on the production system and feeding strategies. The two-step cluster procedure was used for classification of surveyed farms in homogenous groups. Cross tabulations with calculation of Chi^2 statistic were used to compare the groups according to key farm qualitative characteristics. Means were calculated for key quantitative variables and the non-parametric test of Kruskal-Wallis was used to assess significance of difference between groups. Subsequently 30 farms, representative of identified systems, were selected for in-depth study. Average daily gain (ADG), body condition scores (BCS), and body measurements including thoracic perimeters (TP) were recorded weekly on young animals during three months. Six small ruminant farm types were identified, as follows: Goat only (G, 46 %), mixed goat-crop (GC, 30 %), mixed sheep-goat-crop (SGC, 16 %), sheep only (S, 15 %), sheep-goat (SG, 11 %) and mixed sheep-crop (SC, 7 %). Free grazing was the main feeding strategy in all farm types. In addition, crop residues (cassava leaf), agro-industrial by-products (corn bran, soybean bran, cassava peels) and tree fodder (from oil palm tree, Moringa) were used for supplementing grazing with significant variations among farm types. ADG, BCS and TP values also varied ($p < 0.05$) over time and according to farm types. ADG was highest in CS farms 28.2 g day^{-1} and lowest in G farms 13.3 g day^{-1}. There was a linear relationship ($p < 0.001$) between ADG and BCS ($R^2 = 0.72$) on the one hand, and between ADG and TP ($R^2 = 0.83$) on the other hand. On-going studies focus on identification of improved feeding and general management strategies for increasing animal productivities and welfare with low environmental impact.

Keywords: Feeding strategies, goat, sheep, urban farming, West-Africa

Contact Address: Ivan B. Koura, National University of Agriculture, School of Management and Exploitation of Livestock Systems, BP 43, Kétou, Benin, e-mail: kouraivan@gmail.com

Assessment of Quality and Rumen Degradability of Mixed Silages of Sugarcane Tops with Marabú Forage

RACIEL LIMA OROZCO[1], IDALMIS BELLO MORALES[2], EINAR ARTILES ORTEGA[3], MIGUEL ÁNGEL ARCE GONZÁLEZ[4], VEERLE FIEVEZ[3]

[1] *Central University of Las Villas (UCLV), Agriculture Research Center (CIAP), Cuba*

[2] *Agriculture Ministry, Animal Health, Cuba*

[3] *Ghent University, Department of Animal Sciences and Aquatic Ecology, Belgium*

[4] *Central University of Las Villas (UCLV), Veterinary Medicine and Zootechnics, Cuba*

The quality and rumen degradability of sugarcane (*Sacharum* spp.) tops with forage of Marabú (*Dichrostachys cinerea* L.) ensiled in combination with either molasses, lactobacillus or fungi as well as their combination were assessed. The labscale silages were made in triplicate from sugarcane crop and *D. cinerea* fields of two years old and harvested after 12 months and 90 d of regrowth, respectively (Santa Clara, Cuba). Sugarcane tops and D. cinerea forage were mixed in a ratio of 60:40, respectively and either or not in combination or not with one of the three following additives or their combination [fungal inoculant [FI; UC1 (*Penicillium sp.)* + UC13 (*Aspergillus sp.*) or *Trichoderma* sp. (L6+R6b)], *Lactobacillus plantarum* (LAB; 3×10^6 colony forming units/g fresh matter (FM)) and molasses (MOL; 39.4 g kg^{-1} FM)]. Both FI were inoculated at three doses (FD; 1.5, 3.0 and 4.5 x 10^5 spores g^{-1} of FM) in order to prepare 28 treatments [2FI×3FD×2LAB×2MOL+4 control treatments (2LABx2MOL)]. The chemical proximate content (CPC) prior and after ensiling, as well as the ensilability and the *in vitro* ruminal degradability (IRD) of the silages were determined. The metabolisable energy (ME) was estimated from CPC. The ensilability parameters (pH, lactate, acetate, ammonia, ethanol) were used to select the best silages by a full factorial design in the GLM of SPSS with FI, FD, LAB and MOL as factors. The CPC, ME and IRD of the best silages were compared by one way ANOVA of SPSS (post-hoc Tukey test was performed when $p < 0.05$). The forages that received simultaneous addition of all additives showed the best silage quality, independently of FI or FD. However, the FD ($p < 0.001$) but not the FI ($p > 0.05$) affected the fiber fraction content and IRD of organic matter (IRDOM). The ADF content was lower ($p < 0.001$) and the ME content and IRDOM were enhanced ($p < 0.05$) in those silages with higher FD. It was concluded that mixed silages of an invasive plant and fibrous by-product inoculated simultaneously with molasses, lactobacillus and fungi at doses of 3.0 and 4.5 x 10^5 spores g^{-1} of FM showed the best silage quality, the lowest ADF content and the highest IRDOM, independently of fungi strain.

Keywords: *Dichrostachys cinerea*, ensiling, fungi, *Lactobacillus*, molasses, rumen degradability, sugarcane tops

Contact Address: Veerle Fievez, Ghent University, Animal Sciences and Aquatic Ecology (Lanupro), Coupure Links 653, Block F, 1st floor, 9000 Gent, Belgium, e-mail: veerle.fievez@ugent.be

Replacement of Egyptian Clover (*Trifolium alexandrinum*) Hay with Tanniniferous Legumes in Sheep Diets: Nutritional Assessment

Hani El-Zaiat[1], Mohamed Moharam[2], Marwa Attia[1], Adibe Luiz Abdalla[3], Sobhy M.A. Sallam[1]

[1] *University of Alexandria, Faculty of Agriculture, Department of Animal Production, Egypt*

[2] *Damanhour University, Fac. of Agriculture; Animal and Poultry Production, Egypt*

[3] *University of São Paulo, Centre for Nuclear Energy in Agriculture, Lab. of Animal Nutrition, Brazil*

In Egypt, a large diversity of tanniniferous legumes are widely grown in abundance all year even during droughts. Thus, the objetives of this study were to assess the partial (50 %) replacement of Egyptian clover (*Trifolium alexandrinum*) hay with tanniniferous legumes under *in vitro* and *in vivo* conditions. Dietary treatments were as follows: CTL: the experimental basal diet (clover hay and concentrate; 50:50 ratio) as control; AS, the basal diet with 50 % *Acacia saligna* replaced clover hay; LL, the basal diet with 50 % *Leucaena leucocephala* replaced clover hay; and AH, the basal diet with 50 % *Atriplex halimus* replaced clover hay. An *in vitro* semi-automatic system was employed to evaluate gas production (GP), degradability and fermentation profile of diets. For the *in vivo* experiment, twelve Barki rams (43.29 kg of BW) were assigned to 1 of 3 treatments in a complete randomised design and housed in metabolic cages for total collection of faeces and urine (21 days as adaptation and 7 days for data collection). Enteric CH_4 emission was measured using 6 open-circuit respiration chambers with some modifications. The net GP was lower ($p = 0.022$) with the AS diet than with CTL, LL and AH diets. Whereas, net CH_4, truly degraded organic matter, ruminal NH_3-N concentrations and total protozoa were decreased ($p < 0.05$) with AS, LL and AH diets than for CTL diet. Acetate and acetate:propionate ratio were lower, while propionate and partitioning factor were increased with AS or LL than with CTL diet ($p < 0.05$). Sheep fed with LL diet lowered digestibility of OM ($p < 0.05$) compared with those fed with CTL or AS diets. Urinary-N was decreased, while faecal N and retained N was increased ($p < 0.05$) for sheep fed with AS or LL compared to CTL diet. Methane emission was higher ($p = 0.021$) from sheep fed with AS or LL than for those fed with CTL diet. Thus, use of such tanniniferous legumes provides a promising source of forages for sheep with positive impact on mitigation of methane emission without adverse effects on animal performance.

Keywords: Feed degradability, methane emission, nutrients digestibility

---

Contact Address: Hani El-Zaiat, University of Alexandria, Fac. of Agriculture, Dept. of Animal Production, Aflaton St. El-Shatby P.O Box 21545, Alexandria, Egypt, e-mail: hani.elzaiat@alexu.edu.eg

Estimating Rumen Undegradable Protein from Proximate Nutrient, Fiber Fractions, and *in-vitro* Fermentation of Tropical Feedstuffs

KHATERINE SALAZAR-CUBILLAS, UTA DICKHOEFER

University of Hohenheim, Animal Nutrition and Rangeland Management in the Tropics and Subtropics, Germany

Estimates of rumen undegradable crude protein (RUP) are a prerequisite in feed evaluation and nutrient recommendation systems. The in-situ method is the reference method used for RUP determination. However, this method is expensive, time-consuming, and requires fistulated animals. Therefore, this study aims to predict the RUP of tropical feedstuffs from proximate nutrients, fiber fractions, and rumen in-vitro feed fermentation characteristics.

A total of 46 tropical feedstuffs, namely, 6 concentrates and by-products, 17 forage legumes, and 23 forage grasses, were used to develop RUP prediction equations. With use of the software Minitab (2018), stepwise linear regression was performed and the determination coefficient (R^2) was calculated for identified RUP equations at rumen passage rates of 2, 5, and 8%/h.

The dataset included information on (arithmetic mean±standard deviation): concentrations (per kg of dry matter [DM]) of organic matter (OM; 902±38.9 g), crude protein (CP; 160±87.3 g), neutral detergent fiber (490±132.1 g), acid detergent fiber (287±86.3 g), acid detergent lignin (48±28.8 g), digested OM (0.6±0.1 g kg^{-1} OM), and metabolisable energy (10±1.9 MJ kg^{-1} DM), as well as cumulative gas production (GP) during 24h of *in-vitro* incubation (32±8.3 ml/200 mg DM) and *in-situ* RUP proportion (per kg CP) at rumen passage rates of 2%/h (228±90.8 g), 5%/h (299±104.0 g), and 8%/h (342±110.6 g).

The CP concentration and GP were good predictors of the proportion of RUP across all feedstuffs. The CP concentration accounted for 60, 72, and 77 % and the GP for 14, 7, and 4 % of the variability in the RUP proportion at rumen passage rates of 2, 5, and 8%/h, respectively. The following models were developed (coefficient±mean standard error):

RUP 2%/h = (40.9±8.5 + 0.3±0.1 × CP – 1.5±0.3 × GP) × 1000/CP (R^2=0.74; $p < 0.01$),

RUP 5%/h = (34.1±10.6 + 0.4±0.1 × CP – 1.4±0.4 × GP) × 1000/CP (R^2=0.80; $p < 0.01$), and

RUP 8%/h = (25.2±12.5 + 0.4±0.1 × CP – 1.3±0.4 × GP) × 1000/CP (R^2=0.81; $p < 0.01$),

where RUP is in g kg^{-1} CP, CP is in g kg^{-1} DM, and GP is in ml/200 mg DM.

In conclusion, the RUP proportions of tropical ruminant feedstuffs can be predicted from CP concentration and GP.

Keywords: *In-vitro* fermentation, tropical feedstuffs, undegradable crude protein

Contact Address: Khaterine Salazar-Cubillas, University of Hohenheim, Animal Nutrition and Rangeland Management in the Tropics and Subtropics, Fruwirthstrasse 31, 70599 Stuttgart, Germany, e-mail: khaterine.salazar-cubillas@uni-hohenheim.de

Comparison of NIRS and Wet Chemistry Methods for Analysing Nutritional Value of Indonesian Local Feedstuffs

DESPAL, LAILA ATIKA, ROBIN CHANDRA, LUKI ABDULLAH, IDAT GALIH PERMANA

Bogor Agricultural University, Dept. Animal Nutrition and Feed Technology, Indonesia

Near Infrared Spectroscopy (NIRS) instrument use for feed analysis is increasingly popular due to its fast, low cost, non-destructive, and no requirements for solvents or reagents. As the used database was developed based on feedstuffs from temperate zones, its utilisation for tropical feedstuffs need to be calibrated. The aim of this study was to compare results of NIRS feed analysis with the result from conventional wet chemistry (CWC) methods. Five types (Napier grass, natural grass, corn leaves, corn husk and rice straw) of mainly used forages in dairy cattle in Indonesia were used in this study. The forages were collected from 5 different dairy cattle production areas. From each area, forage from 4 different farms was collected. In total, 100 forage samples were analysed both using NIRS and CWC to measure proximate compositions (dry matter (DM), ash, crude protein (CP), crude fiber (CF), crude lipid (XL)), and Van Soest cell wall contents (neutral detergent fiber (NDF), acid detergent fiber (ADF) and acid detergent lignin (ADL). The results showed that proximate analyses (DM, ash, CP, CF, XL parameters) were significantly lower when using CWC method than NIRS. While NDF and ADF parameters were higher when using NIRS than CWC. Correlation between the methods was low especially for water content ($r = 0.413$). Therefore, estimation of tropical chemical compositions using NIRS temperate feedstuffs database were biased. Calibration using local database improved the correlation ($r > 0.80$). Validation NIRS based on local data base were also resulted in better result. It is concluded that utilisation of NIRS for analysing tropical forage need to be calibrated with local feedstuffs wet chemistry analysis data.

Keywords: Local forage, NIRS, proximate composition, Van Soest, wet chemistry

Contact Address: Despal, Bogor Agricultural University, Dept. Animal Nutrition and Feed Technology, Fakultas Peternakan IPB Jl. Agatis Kampus IPB Darmaga, 16680 Bogor, Indonesia, e-mail: despal@apps.ipb.ac.id

Livestock Feeds Assessment in Southern Highlands in Tanzania

SOLOMON MWENDIA[1], ANGELLO MWILAWA[2], NZOGELA BEATUS[1], JONAS KIZIMA[2], WALTER MANGESHO[2], ROSE LOINA[2], JULIUS BWIRE[2], AN NOTENBAERT[1]

[1] *International Center for Tropical Agriculture (CIAT), Kenya*

[2] *TALIRI, Pastures, Tanzania*

Importance of livestock in smallholder farms in Africa continue to be significant in households' nutrition, incomes and for socio-cultural reasons. However, the sector faces various challenges including management, diseases and feeding among others. With the estimated doubling of demand for animal source foods especially in sub-Saharan Africa, due to rise in human population, efforts for improving livestock productivity are needed. Amongst the costs involved in cattle production, the feeding component takes 70 % suggesting greater attention is required in feeding interventions for success to be realised. In an effort to contribute to improvement of livestock production in Southern Tanzania highlands, current feeding status was assessed in 3 districts (Mufindi, Njombe, Rugwe) following Feeds Assessment Tool protocol (FEAST https://www.ilri.org/feast). Information captured entailed; land size categories, livestock species reared, feed resources and seasonality, contribution of livestock to household incomes and most suited interventions were elucidated. In the three districts, dairy remains most prevalent amongst livestock species kept, and livestock in general is important for household incomes. Analysis suggest interventions on feeds and forages addressing protein and energy supply would be crucial, including forages suitable under cut and carry context, prevalent under smallholder farmers settings. Dry seasons are characterised by inadequate feed resources. The information will guide selection of forage options to be tested through farmer participation, under Climate Smart Dairy Project (2019-2021) in the three districts. Identified farmer acceptable and productive forages will be recommended for the areas other similar ecologies, and by extension support forage development in Tanzania.

Keywords: Feed availability, forages, livestock

Contact Address: Solomon Mwendia, International Center for Tropical Agriculture (CIAT), Nairobi, Kenya, e-mail: s.mwendia@cgiar.org

Forage Options for Tanzania Southern Highlands: Preliminary Assessment

Solomon Mwendia[1], Angello Mwilawa[2], Angello Kizima[2], Julius Bwire[2], Beatus Nzogela[1], John Mutua[1], An Notenbaert[1]
[1]*International Center for Tropical Agriculture (CIAT), Kenya*
[2]*TALIRI, Forages, Tanzania*

In adequate quality and quantity, forages curtail animal production in sub-Saharan Africa. As such, the potential of livestock agriculture to contribute to household nutrition and incomes is usually compromised, against an increasing consumption of animal source foods. The growing consumption deserve forage interventions that can address the feeding gap, and contribute to improved livestock production, and subsequently livestock keepers' prosperity. Using "Targeting Tools", a web GIS system, we mapped the suitability of a wide array of grass and forage legumes for southern highlands. Using the maps, we selected context-specific varieties and procured available seeds/planting materials for pilot trials. Following farmers' participatory approach, we established forage trials in three districts namely, Mufindi, Njombe and Rugwe of southern Tanzania highlands comprising of fourteen forage treatments. The test forages included; two *Cenchrus purpureus* (Syn. *Pennisetum purpureum*) cultivars, two *Urochloa* (Syn. *Brachiaria*) hybrids and *Chloris gayana*. Where applicable, we intercropped the grasses with three forage legumes- *Lablab purpureus*, *Stylosanthes guianensis*, and *Desmodium intortum*, while *Tripsacum andersonii* (Syn. *Tripsacum laxum*), a grass, was planted as a local check. We observed clear differences amongst the three districts and treatments. Dry matter accumulation (t ha^{-1}) in the districts, and across the various forage treatments was in the order Rugwe>Mufindi>Njombe, even when accumulation kg DM day^{-1} was considered, in the early harvests. Most DM accumulation was by Napier grass intercropped with *Lablab purpureus* that was closely comparable to *Chloris gayana-Desmodium* intercrop. Further observations across more cuttings, farmer' preference rankings and quality analysis are under way to inform reliable conclusions. The results would be applicable elsewhere with similar agricultural context and ecologies.

Keywords: Dry matter yields, forages, participatory evaluation

Contact Address: Solomon Mwendia, International Center for Tropical Agriculture (CIAT), Nairobi, Kenya, e-mail: s.mwendia@cgiar.org

Evaluation of Undigested and Potential Digestible Fiber in Tropical Grasses and Tropical Legumes

RISMA RIZKIA NURDIANTI, UTA DICKHOEFER, JOAQUÍN CASTRO-MONTOYA
University of Hohenheim, Animal Nutrition and Rangeland Management in the Tropics and Subtropics, Germany

Dietary fiber is one of the main determinants for voluntary feed intake (FI) in ruminants, particularly because of its effects on rumen fill. Recently, the undigested neutral detergent fiber (i.e. remaining neutral detergent fiber (NDF) after an infinite period of time under degradation by rumen microorganisms; uNDF) has been presented as the most likely indicator of filling capacity by a forage. Tropical legume forages (TLF) decrease FI when included in large proportions in ruminant diets (i.e. >400 g kg $^{-1}$ dry matter; DM; and the reasons for this may be related to the NDF concentration and its characteristics. Plenty of information is available on the NDF concentration of TLF, which could be as high as that of tropical grasses (TG), but no information is available on the uNDF fraction of TLF. Thus, the objective was to evaluate the uNDF after 240-h *in vitro* incubation (uNDF240) and the potentially digestible NDF (pdNDF) concentrations of TLF as compared with TG. Seven samples of TG and 14 samples of TLF were collected from Indonesia, El Salvador, Peru, and Brazil. The amylase-treated ash-corrected NDF (aNDFom) was measured using an ANKOM200 fiber analyser with addition of sodium sulfite. The uNDF240 was analysed using the modified Tilley and Terry technique after 240-h *in vitro* incubation. The pdNDF was determined by subtracting the uNDF240 fraction from total aNDFom. The results were analysed using descriptive analysis. Values of aNDFom, uNDF240 and pdNDF concentrations varied among different forages and within the same type. TLF had a lower aNDFom concentration than TG (373.9±88.9 and 591.6±31.3 g kg $^{-1}$ DM, respectively), but a higher uNDF240 than TG (239.3±75.4 and 231.3±71.2 g kg $^{-1}$ DM, respectively). Therefore, the proportion of pdNDF was lower in TLF than TG (134.5±53.5 and 360.4±63.2 g kg $^{-1}$ DM, respectively). Furthermore, compared to TG, uNDF240 as a proportion of aNDFom was much higher in TLF (637.6±135.7 vs. 389.5±110.6 g kg $^{-1}$ aNDFom). In conclusion, TLF had high uNDF240 concentrations which could explain the lower FI when grasses are substituted by TLF at high rates in the diet of ruminants.

Keywords: Potential digestible fiber, tropical grasses, tropical legumes, uNDF

Contact Address: Risma Rizkia Nurdianti, University of Hohenheim, Animal Nutrition and Rangeland Management in the Tropics and Subtropics, Fruwirthstraße 31, 70599 Stuttgart, Germany, e-mail: aninutrop@uni-hohenheim.de

Scaling of Feeds and Forages Technologies in East Africa

Uwe Ohmstedt[1], An Notenbaert[1], Michael Peters[1], Solomon Mwendia[1], Stefan Burkart[2]

[1]*International Center for Tropical Agriculture (CIAT), Tropical Forages, Kenya*

[2]*International Center for Tropical Agriculture (CIAT), Trop. Forages Program, Colombia*

Feed for livestock is identified as the limiting factor for livestock productivity and accounts for up to 70 % of the production costs in the Kenyan dairy farms. Year around access to high quality feed determines the productivity and competitiveness of the dairy sector. The actual situation is still characterised by scarcity in quantity and quality with scarcity peaks in the dry season.

CIAT's Tropical Forages meanwhile works for decades on the improvement of forages by breading and selection of *Brachiaria* and *Panicum*. Our private sector partner Papalotla Sementes released 3 hybrids on the Kenyan market and they are commercially available. One more hybrid, 4 cultivars of *Brachiaria* and 1 *Panicum* cultivar will be on the market by mid 2019.

CIAT is convinced that these improved forages can mitigate feed scarcity in quantity and quality and contribute to better feeding practices, increase the productivity of the dairy cows and consequently contribute to higher income generation of dairy farmers. This applies for the high productive rainy season but also shortens the period of low to none production of the classical forages due to the better drought resistence of the introduced *Brachiaria* and *Panicum*.

The scaling approaches (along the value chain, farmer to farmer, partnerships with development organisations) and activities (demonstration plots, farmer field days, trainings, partnerships with public and private sector etc) to reach out to a large number of farmers (>100000 farmers over 5 years) are presented as well as actual achievements (>11000 farmers in 2018) and further planned activities.

The experiences made in Kenya will be the base for further scaling efforts in East African countries which also have a growing dairy sector and face the same challenges for sufficient high value feed.

Keywords: East Africa, forages, livestock, scaling

Contact Address: Uwe Ohmstedt, International Center for Tropical Agriculture (CIAT), Tropical Forages, Icipe Road Off Kasarani Road, Nairobi, Kenya, e-mail: u.ohmstedt@cgiar.org

Agronomic Evaluation of a Collection of *Chloris gayana* for Year Round Forage Supply in a Dry Sub-Humid Environment in Córdoba, Colombia

SERGIO MEJÍA[1], JOSÉ TAPIA[1], MAURICIO SOTELO[2], JACOBO ARANGO[2], MICHAEL PETERS[3]

[1]*Colombian Agricultural Research Corporation (AGROSAVIA), Livestock, Colombia*

[2]*International Center for Tropical Agriculture (CIAT), Colombia*

[3]*International Center for Tropical Agriculture (CIAT), Tropical Forages, Kenya*

Globally about 3400 Mha are being used for grazing livestock, with the large part natural or naturalized pastures with seasonal forage availability limited by water supply. As a result, livestock production is often inefficient and relying on transhumance to seek feed resources. In Colombia and similar regions of the tropics, large areas are prone to extended drought resulting in seasonal feed scarcity. As part of the solution, the inclusion of improved forages tolerant to abiotic stresses, which enhance feed availability in critical periods and overcome seasonality of livestock production, is required. In this contribution we describe the agronomic evaluation of a collection of 20 *Chloris gayana* accessions from the ILRI genebank in Addis Ababa, Ethiopia. The *Chloris* accessions were compared with three reference controls for the region: *Dichantium aristatum, Megathyrusus maximus* cv. Mombasa and *Megathyrsus maximus* cv. Sabanera. The experiment was conducted in the research station of the Colombian Agricultural Research Corporation (AGROSAVIA) in Turipaná in the Córdoba Department of Colombia. Data was collected during 18 months between 2016 and 2017, covering both the wet and the dry season. Parameters evaluated included nutritional quality (e.g. protein content, digestibility, fibre content) and dry matter production. Results are currently being analysed but first insights were obtained already, showing variability in the performance of the *Chloris* accessions evaluated outperforming the controls in at least one parameter. After data analyses will be finished, at least one promising *Chloris* genotype will be identified for further testing and eventual release as cultivar to be utilised in farmers fields.

Keywords: Forage production, forage quality, Latin America, livestock production, sustainable intensification, tropical forages

Contact Address: Michael Peters, International Center for Tropical Agriculture (CIAT), Tropical Forages, ICIPE - Duduville Campus, Nairobi, Kenya, e-mail: m.peters-ciat@cgiar.org

Evaluation of the Feed Quality of Six Dual Purpose Pearl Millet Varieties and Growth Performance of Sheep Fed their Residues in Niger

CLARISSE UMUTONI[1], VINCENT BADO[1], ANTHONY M. WHITBREAD[2], AUGUSTINE AYANTUNDE[3], IBRAHIM ABDOUSSALAM[1], SEYDOU KOROMBE HAMZA[4]

[1]*International Crops Research Institute for the Semi-Arid Tropics (ICRISAT), Dryland Systems and Livelihood Diversification, Niger*

[2]*International Crops Research Institute for the Semi-Arid Tropics (ICRISAT), India*

[3]*International Livestock Research Institute (ILRI), Burkina Faso*

[4]*University Abdou Moumouni, Niger*

Pearl millet (*Pennisetum glaucum* (L.)) is the main staple crop of Niger in the semi-arid region of West Africa. Apart from the use of its grain as a nutritious food for humans, the stover are commonly used to feed ruminants as a basal diet. Historically, crop improvement programs mostly focus on higher grains yields, stover yield, disease resistance and water stress tolerance while crop residue quality is rarely prioritised. Until recently, residue quality (especially digestibility) was not a trait that was screened or selected for the pearl millet improvement programs in West Africa. ICRISAT and its NARES partners have developed many dual-purpose pearl millet varieties as options to produce both food and fodder. This study, funded by the US-AID Livestock Systems Innovation Lab, was conducted to assess the residue quality of five dual-purpose varieties, the feed intake and live weight gains of sheep fed these residues compared to a commonly used local landrace. Data were obtained from 36 sheep (live weight of 27±0.3 kg) fed the pearl millet residues and 600 g of cowpea hay per day as a supplement. Sheep were randomly assigned to the six treatments, confined and fed in individual pens per animal. Over 90 days, the animals fed with variety 167005 had higher ($p < 0.05$) final live weight and average daily live weight gain compared with those fed with cvs. 167006, 167111, 167002, chakti and the local landrace. Feed intake was also higher with cv. 167005 and 167006. The study revealed that the different varieties of pearl millet differ in digestibility and nutrient composition which significantly affects the growth performance of the sheep. Two varieties (167005 and 167006) could be recommended as the best dual purpose crops for mixed crop and livestock systems in Niger.

Keywords: Crop residues, digestibility, dual-purpose millet, feed, growth performance, sheep

Contact Address: Clarisse Umutoni, International Crops Research Institute for the Semi-Arid Tropics (ICRISAT), Dryland Systems and Livelihood Diversification, Niamey, Niger, e-mail: c.umutoni@cgiar.org

Effect of Using Different Levels of Willow Silage on Growth Performance of Jordanian Awassi Lambs

Sami Awabdeh, Rawad Swidan
National Agricultural Research Center, Livestock, Jordan

The objective of this study was to substitute wheat straw as a common forage source in Jordan with different levels of willow silage (*Salix* spp.) to study its effect on growth performance of Awassi lambs. Twenty-six male Awassi lambs were randomly assigned to one of three groups: control (8 lambs) were only wheat straw was used, silage-10 (9 lambs) were both straw and willow silage were used, and silage-20 (9 lambs) were only willow silage was used as forage source.
At the age of 110 days, weaned lambs were caged in separate pins at Khanasri Station for Animal Research in which dry matter intake was measured daily, and lambs body weights were measured weekly. Average initial body weights for control, silage-10 and silage-20 were 22.0, 22.5 and 22.1 kg, respectively. Lambs consumed an isocaloric, isonitrogenous high concentrate diet *ad libitum* (20:80, F:C ratio) for 90 days to study the effect of different levels of willow silage as a substitute of wheat straw on growth performance of fattening Awassi lambs. Metabolisable energy of the diets was between 2.9 - 3.0 Mcal kg^{-1} and crude protein content was 16 % (all on DM basis). Fecal samples were collected after 9 weeks of the experiment to measures the digestibility of crude protein, neutral detergent fiber (NDF) and acid detergent fiber (ADF).
Substitution of different levels of willow silage had no significant effect on dry matter intake or average daily gain compared to the control. Feed to gain ratio for silage-20 was slightly higher than other groups but the difference was not significant. On the other hand, crude protein digestibility was significantly higher ($p < 0.02$) in lambs consumed silage-20 and silage-10 diets (78.3 and 75.5 %, respectively) compared to control group (61.6 %). Lambs consumed silage-20 diet had significantly higher ($p < 0.001$) NDF and ADF digestibility compared to silage-10 and control groups.
In conclusion, using willow silage as a forage source not only increase the digestibility of nutrients but also provide livestock with fresh high quality forage.

Keywords: Akwassi lambs, fattening, growth performance, *Salix* spp., willow silage

Contact Address: Sami Awabdeh, National Agricultural Research Center, Livestock, Jarash Highway, 19381 Ain Basha, Jordan, e-mail: sami_awabdeh@yahoo.com

Bio-Degradation of Water Hyacinth into Value Added Ruminant Feed Using White Rot Fungi

Adejoke Mako[1], Victor Akinwande[1], Oluwanike Abiola-Olagunju[2]
[1]*Tai Solarin University of Education, Dept. of Agricultural Sciences, Nigeria*
[2]*Lead City University, Dept. of Microbiology, Nigeria*

A 40 day experiment was carried out on the biodegradation of water hyacinth (WH) into value added ruminant feed using two white rot fungi *Pleurotus florida* (PF) and *Pleurotus sajor-caju* (PC)) in a solid state fermentation. The chemical composition and *in vitro* gas production degradability of the substrates were determined.
Results revealed that crude protein (CP) increased significantly ($p < 0.05$) from 11.61 in untreated WH to 12.86 % and 14.38 % in WH treated with PF and PC respectively. Same trend was observed for ether extract and ash. However, the crude fibre (CF) decreased significantly from 21.23 % in untreated WH to 18.23 % and 15.25 % in WH treated with PF and PC respectively.
The estimated *in vitro* gas production parameters also ranged significantly ($p < 0.05$), except short chain fatty acids (SCFA) that was not significantly different. The fungal treatment enhanced organic matter digestibility (OMD) and metabolisable energy (ME) compared with untreated WH. The OMD increased from 48.50 % in untreated WH to 52.12 % and 53.89 % in WH treated with PF and PC respectively, while the ME ranged from 5.68 % in untreated WH to 7.56 % and 8.39 % in WH treated with PF and PC respectively. Gas production increased significantly ($p < 0.05$) as the hour of incubation progressed. Methane production decreased significantly from 4.00 ml/200 mg DM in untreated WH to 2.50 and 2.00 ml/200 mg DM in WH treated with PF and PC respectively.
This study revealed that fungal treatment of WH enhanced chemical composition and *in vitro* degradability. Hence, biodegradation of WH will fill in the gab for scarcity of feed during the off season and enhance sustainable ruminant production in water hyacinth endemic areas.

Keywords: Biodegradation, *Pleurotus florida*, *Pleurotus sajor-cajor*, ruminant, water hyacinth

Contact Address: Oluwanike Abiola-Olagunju, Lead City University, Dept. of Microbiology, Toll Gate Area, Ibadan, Nigeria, e-mail: nikkymakky07@yahoo.com

Crude Protein Digestibility of Mixed Legume-Grass Diets Depending on the Quality of the Substituted Grass

JOAQUÍN CASTRO-MONTOYA, UTA DICKHOEFER

University of Hohenheim, Animal Nutrition and Rangeland Management in the Tropics and Subtropics, Germany

Legume forages are promoted in ruminants feeding because of their high crude protein (CP) and an alleged higher digestibility than grasses, especially in tropical regions. It is assumed that increasing the CP supply in the diet via legumes, proportionally increases the amount of digestible CP (digCP). However, evidence exists that diet CP digestibility (CPD) may decrease with increasing proportion of legumes in the diet, particularly if they substitute a basal grass of high quality. Therefore, the objective of this study was to explore the changes in digCP depending on the quality of the grass substituted.

For this, 60 studies where legumes substituted grasses without any additional feedstuff were classified into three categories depending on the CP of the grass: Low (CP<51.0 g kg^{-1} dry matter (DM)), Medium (51.0<CP< 93.1 g kg^{-1} DM) and High (CP>93.1 g kg^{-1} DM). Then within each category digCP and CPD were regressed on the amount (g kg^{-1} DM) of CP from legumes in the diet. The slopes of the regressions were compared among the three categories as indicators of the contribution of legumes to digCP and CPD.

Based on the intercepts of the regressions, grasses supplied 16.2, 38.2 and 68.1 g digCP kg^{-1} DM for the Low, Medium and High category, respectively. On the other hand, slopes increased from 0.432 to 0.493 when legumes substituted grasses of Low or Medium, respectively. This indicates an increased digestibility in the legumes CP when these are combined with a medium quality grass. But for the High category the slope decreased to 0.305, showing that the contribution of legumes to digCP decreases when substituting grasses of High quality. This was confirmed when the CPD was regressed on the amount of CP from legumes, where slopes were 1.13, 1.18 and –0.18 for the Low, Medium and High category. Even though some of these results might be due to the increasing CP concentration in the diet, these findings still highlight the contrasting effects of legumes on the diet's CP digestibility depending on the nutritional value of the substituted grass, an effect that needs further research and consideration depending on the production system targeted for legumes utilisation.

Keywords: Crude protein digestibility, legumes, ruminants feeding, tropical grasses

---

Contact Address: Joaquín Castro-Montoya, University of Hohenheim, Animal Nutrition and Rangeland Management in the Tropics and Subtropics, Fruwirthstrasse 31, 70599 Stuttgart, Germany, e-mail: jcm@uni-hohenheim.de

Nutrient Digestibility and Performance of Sheep Fed Bitter Cassava Leaf Meal Based Diet Supplemented with Cyanide-Degrading Bacteria

SRI SUHARTI, HAFNI OKTAFIA, ASEP SUDARMAN, KOMANG GEDE WIRYAWAN

Bogor Agricultural University, Dept. of Animal Nutrition and Feed Techn., Indonesia

This research aimed to study the effect of inoculating cyanide degrading bacteria on the performance and nutrient digestibility of local sheep fed bitter cassava leaf meal based diet. The experimental design used a randomised block design 5×3 using 15 rams, aged about 12 months with a body weight of 18.8 ± 1.02 kg and placed in individual cages. The treatments were P1= 40 % napier grass: 60 % concentrate, P2 = 40 % napier grass: 45 % concentrate: 15 % bitter cassava leaf meal, P3= 40 % napier grass: 45 % concentrate: 15 % bitter cassava leaf meal + cyanide degrading bacteria, P4= 40 % napier grass: concentrate 30 % : 30 % bitter cassava leaf flour, P5= 40 % napier grass: concentrate 30 % : 30 % bitter cassava leaf meal + cyanide degrading bacteria. Variables observed were feed intake, body weight gain, rumen fermentation characteristics (concentration of ammonia (NH_3), volatile fatty acid (VFA) total and partial) and nutrient digestibility. Data were analysed using analysis of variance. The results showed that innoculation of cyanide degrading bacteria to the sheep fed high level of bitter cassava leaf meal (30 %) did not affect feed consumption, but significantly increased ($p < 0.05$) total VFA production, concentration of NH_3, propionate proportion, body weight gain and feed efficiency. In addition, methane estimation significantly decreased ($p < 0.05$), meanwhile ratio C2/C3 significantly increased. In contrast, the use of bitter cassava leaf meal at the level of 30 % without cyanide degrading bacteria addition significantly decreased body weight gain and feed efficiency of sheep. In conclusion, the addition of cyanide degrading bacteria could detoxify cyanide in bitter cassava leaf meal efficiently to reduce its toxicity for the sheep and enhance rumen fermentation as well as sheep performance.

Keywords: Bitter cassava leaf, cyanide degrading bacteria, digestibility, performance, sheep

Contact Address: Sri Suharti, Bogor Agricultural University, Dept. of Animal Nutrition and Feed Technology, Agatis Kampus Ipb Darmaga, 16680 Bogor, Indonesia, e-mail: sri_suharti@apps.ipb.ac.id

Genetic Diversity of Nubian Ibex in Comparison to other Ibex and Domesticated Goat Species

SIHAM A. RAHMATALLA[1,2], DANNY ARENDS[2], LUBNA M. A. HASSAN[3], MONIKA REISSMANN[2], HENRY REYER[4], KLAUS WIMMERS[4], SUMAIA M. A. ABUKASHAWA[5], GUDRUN A. BROCKMANN[2]

[1]*Humboldt-Universität zu Berlin, Dept. Animal Breeding Biology and Molecular Genetics, Germany*

[2]*Humboldt-Universität zu Berlin, Albrecht Daniel Thaer Inst. for Agricultural and Horticultural Sciences, Germany*

[3]*Wildlife Research Center, Animal Resource Research Corporation, Federal Ministry of Livestock, Fisheries and Rangelands, Sudan*

[4]*Leibniz Institute for Farm Animal Biology (FBN), Institute for Genome Biology, Germany*

[5]*University of Khartoum, Dept. of Zoology, Sudan*

Nubian ibex (*Capra nubiana*) is the only ibex species which has adapted to live in hot, arid regions. Nubian ibex was listed as a vulnerable species in 2008 by the IUCN and is currently considered as critically endangered. The objective of this study was to infer the population structure and genome wide diversity of Nubian ibex (*Capra nubiana*, n = 8) in comparison to Alpine ibex (*Capra ibex*, n = 8), Bezoar ibex (*Capra aegagrus*, n = 4), and domesticated Taggar goats (*Capra aegagrus hircus*, n = 24). Animals were genotyped by using the goat SNP50 BeadChip. The assessment of the genetic diversity in wild breeds by using the SNP chips designed for domesticated breeds is very challenging. Hierarchical clustering using Nei's genetic distances clearly distinguished between the four examined species. Surprisingly, Nubian ibex clusters closer to the Alpine ibex, than to Bezoar ibex. Principal component (PC) analysis confirmed the initial finding from clustering. The principle component 1 (PC1) accounts for 38 % of the total variation and separates the Nubian and Alpine ibex in one compact group from the Bezoar ibex and Taggar goat. The SNPs that contributed strongly to PC1 were used to pinpoint genomic regions contributing to the differentiation between breeds. Such regions are located on chromosomes 3, 6, 8, 10, and 21. Among the genes in the vicinity of those SNPs are genes coding for casein milk proteins on chromosome 6. Variation in milk protein genes likely contributes to the genetic differences between wild and domesticated species and highlights the importance of preservation and genetic analysis of these critically endangered breeds. The first three authors contributed equally to this work.

Keywords: Alpine ibex, Bezoar ibex, genetic diversity, goat, Nubian ibex, Taggar

---

Contact Address: Siham A. Rahmatalla, Humboldt-Universität zu Berlin, Dept. Animal Breeding Biology and Molecular Genetics, Invalidenstraße 42, D-10115 Berlin, Germany, e-mail: rahmatas@hu-berlin.de

Polymorphisms GDF9 and FSHR Genes and its Association with Litter Size in Egyptian Goat Breeds

HELMY METAWI, HEBA ABD EL HALIM
Animal Production Research Institute, Sheep and Goats Research, Egypt

Phenotypic evaluation and culling of candidate animals for traits by applying traditional animal breeding are usually costly tasks which require considerable time to be carried out. Molecular genetics as an alternative method enables animal breeders to select eligible animals for the desirable trait(s) at their earlier ages. Selection based upon markers could result in increasing accuracy as well as selection response of animals. This research was performed for screening polymorphism of growth differentiation factor 9 (GDF9) and follicle stimulating hormone receptor (FSHR) in two goat breeds Zaraibi and Baladi of Egypt. Both breeds are more prolific than other goat breeds found in the country. To find molecular markers to associate with litter size, animals were selected based on single birth (SB) and multiple births (MB) history. Forty samples were collected from each breed. In Baladi goat breed, polymorphic restriction pattern indicate presence of one band with 710 bp among all SB does, and three bands with 710, 600 and 100 bp for MB does. In Zaraibi goats, polymorphic restriction pattern indicate presence of one band with 700 bp among all SB does and two bands with 600 and 100 bp for MB does. These results showed the presence of polymorphic of GDF9 in Baladi goat and monomorphic of GDF9 in Zaraibi MB does. The mutations in the GDF9 gene associate with fecundity were identified only in investigated MB Egyptian goat breeds. On the other hand, the restriction enzyme which was used in PCR-RFLP did not identify any FSHβ gene mutation in exon 1. Aligment of the tested alleles with *Capra hircus* GDF9 sequence from gene bank showed transition in multiple birth does from CCGAGG to GTTCAT and from TT to AG in regions from 52 to 57 and from 61 to 62, respectively. On the other hand, only one transition occurred in FSHB sequence gene from CTGTT to ACAAA in region from 31 to 35. The study indicates possibilities of using these markers for selection for high prolificacy in Egyptian goats.

Keywords: GDF9, goat, FSHR, PCR-RFLP, prolificacy

Contact Address: Helmy Metawi, Animal Production Research Institute, Sheep and Goats Research, 5 Nadi Elsaid Str.Dokki, Giza, 16128 Dokki Cairo, Egypt, e-mail: hrmmetawi@hotmail.com

Farmers' Desired Attributes for Community Owned Persons for Integration in Trypanosomiasis Control Programs in Kenya

ARNOLD MUSUNGU[1], DAVID OTIENO[1], BEATRICE MURIITHI[2]

[1] *University of Nairobi, Dept. of Agricultural Economics, Kenya*

[2] *International Centre of Insect Physiology and Ecology (ICIPE), Social Science and Impact Assessment, Kenya*

Animal trypanosomiasis is a major constraint to livestock production and a threat to food security in most parts of sub-Sahara Africa (SSA). Smallholder farmers are particularly affected because access to veterinary services is limited in most cases. While efforts have been made to control the spread and prevalence of the disease in Kenya, for instance, through the use of traps, repellents and drugs, the disease continues to cause losses to livestock producers. These losses include reduced meat and milk off take, weight loss, reduced traction power of oxen, high calf mortality and overall cost of livestock management. Subsequently, with the current insufficient extension and veterinary services in the country, farmers in endemic areas resort to presumptive and indiscriminate drug administration thus increasing the risks of drug resistance. Integrating communities into tsetse and trypanosomiasis control programs as a way to tackle these challenges has become an important consideration in the development policy agenda in SSA. This approach proposes recruiting and training a pool of community owned resource persons (CORPs) to address the insufficient veterinary and extension services and subsequently reduce cases of drug resistance. Further, CORPs are believed to be vital in facilitating technology transfer to farmers. However, the desired attributes of the CORPs by farmers are not known. This paper applied choice experiment (CE) method to assess the farmers' desired attributes for CORPs in Kwale County, Kenya. A random parameter logit (RPL) model was used to analyse the CE data. The analysis reveals that a CORP must have secondary school education and be a permanent resident of the community. Further, age, marital status, training partner, and recruitment process, are all significant attributes that influence integration of CORPs in trypanosomiasis control programs. The findings call for an integrated approach where CORPs are recruited and trained based of farmers' preferences.

Keywords: Choice experiment, CORPs, farmers, losses, trypanosomiasis

---

Contact Address: Arnold Musungu, University of Nairobi, Dept. of Agricultural Economics, Nairobi, Kenya, e-mail: musunguarnold8@gmail.com

Community Animal Health Workers: Filling Gaps in Animal Health Services in Remote Global South Communities

Maria Victoria Larrateguy[1], Angela R. Schug[1], Margherita Gomarasca[2], Constanze Boenig[1], Antonia Braus[1]

[1] *Vétérinaires Sans Frontières Germany, Germany*

[2] *Vétérinaires Sans Frontières International, Belgium*

Basic animal health services are often insufficient in many rural areas in the global south, such as Eastern Africa, the project region of Vétérinaires sans Frontières Germany (VSFG), leading to certain risks related to poverty, public health and food insecurity and nutrition. Community-Based Animal Health Workers (CAHWs) try to fulfil the previously named gaps. CAHWs are community members who are selected in a participatory way and work in collaboration with private veterinary doctors, veterinary public services and supporting bodies (international programs and NGOs). CAHWs perform a limited range of veterinary health services (such as vaccination, deworming and others), in return of some form of payment (either cash, voucher system or in kind) and disseminate husbandry, farming and natural resources management methods in order to optimise animal production. In 2017, VSFG trained 1345 CAHWs in Ethiopia, Kenya, Somalia, Sudan and South Sudan.
However, CAHWs encounter several challenges related to: i) the lack of a common nomenclature on a global level and even in the same country; ii) no standardised training; iii) inadequate supervision; iiii) their role and status: has often not been officially recognised in regulatory frameworks1,2. CAHWs operate under the One Health approach, since their positive impacts can be measured in animal health and wellbeing, on the people who depend on them (like agro-pastoral communities to whom livestock means food security, income source, savings and employment), and on the environment, through better use and management of natural resources. They play a vital role in epidemiologic surveillance and overall public health matters such as food safety, prevention and control activities of zoonotic diseases (anthrax, brucellosis, rabies etc.) and other Neglected Tropical Diseases. Therefore, larger efforts need to be made in order to overcome the above mentioned challenges and to allow CAHWs to be internationally and nationally recognised and integrated in the Veterinary Health System Legislation, to facilitate and expand their training and supervision; to clarify the roles and responsibilities of each of the actors involved in the local animal health system, to assure the sustainability; and to promote stronger public-private engagements and frameworks to incentivize investments in the system.

Keywords: Agro-pastoral communities, framework, livestock, One Health

Contact Address: Maria Victoria Larrateguy, Vétérinaires Sans Frontières Germany
current address: Alsina 140, 3100 Paraná, Argentinia, e-mail: mvlarrateguy@gmail.com

Potential of Phytogenics in Filling Gaps and Removing Traps for Sustainable Livestock Production without Antibiotics

CHRISTABEL AGHOROWEI, AMAMIEGHAYE EDUM, RUTH TARIEBI S. OFONGO
Niger Delta University, Dept. of Animal Science, Nigeria

In the current era of antibiotic free animal production, medicinal plants and spices are been highlighted to fill gaps and remove traps for sustainable livestock resource development. Secondary plant metabolites such as tannins, saponins, flavonoids, anthraquinones and other compounds with antimicrobial, antibacterial, antifungal, antiviral and immune response inducing potential can successfully fill gaps in sustainable livestock resource development without antibiotic growth promoters. This will ultimately benefit both producers and consumers of animal products. In order to better understand these potentials of phytogenics, qualitative and quantitative analysis of the methanolic extract of *Ocimum gratissimum* (lyn) was carried out. Qualitative analysis for specific phytochemical was carried out as follows: terpenoids (Liebermann – Burchard test); phenolic compound (Lead acetate test); tannins (Ferric chloride test); flavonoids (Shinoda's test) and reducing sugars (Fehling's test). Total flavonoid content was expressed as Rutin equivalents (mg Rutin Equivalents per g extract). The total phenolic content of the extract was estimated using the Folin Ciocalteau reagent method and result obtained reported as Gallic acid equivalent per g of extract. All analysis was carried out in triplicates.

Qualitative analysis revealed the presence of terpenoids, phenolic compounds and tannins – heavily detected (+++) while flavonoids were detected (++). Quantitative analysis indicated reducing sugar concentration of 9.01 mg g^{-1} of extract. Other phytochemicals were; 5.66 mg g^{-1} of extract (tannins). Total phenolic content measured as Gallic acid equivalent was 37.45 mg Gallic acid equivalent per g of extract while total flavonoid content measured as Rutin equivalent was 35.12 mg Rutin equivalent per g of extract.

From the results obtained, it is obvious that *Ocimum gratissimum* has secondary metabolites with potential to fill gaps in sustainable livestock resource development in the absence of antibiotics.

Keywords: Antibiotics, livestock production, *Ocimum gratissimum*, phytogenics

---

Contact Address: Ruth Tariebi S. Ofongo, Niger Delta University, Dept. of Animal Science, Wilberforce Island, P.M.B. 071, Yenagoa, Nigeria, e-mail: tariruth@live.de

How Agro-Ecological Transition Could Sustain Goat Keeping in Nomadic Systems of Iran

FARHAD MIRZAEI

Aryan Rural & Nomads Animal Production Management Association, Iran

Pastoralism is a type of animal production characterised by the use of spontaneous food resources. Since 50 years, the dominant paradigm of agriculture has been based on specialisation and intensification of production systems by maximising the issue of production factors including higher yields, higher inputs, higher dependence on food industry. Its positive effect is increased food safety and its negative effect is environmental impacts and decreasing number of farmers. Five principles of agroecology for livesock are as follows: (i) the integrated management of animal health, (ii) the reduction of inputs by using ecological processes, (iii) the reduction of pollutions by controlling the biological cycles, (iv) the use of diversity in production systems to increase their resilience and (v) the preservation of biodiversity (pastures, landscapes, and local populations) by adapting practices. To define agroecological characteristics for goat keeping in Iran the following items should be considered and developed: animal nutrition, sustainable pasture management, crops and forage practices, disease prevention, breeds and reproduction, animal welfare, food safety and hygiene, marketing and management, conditions of social and economic sustainability, environmental sustainability and societal contribution. The productivity of Iranian goats is low, with a national average estimated at 150 kg of milk and 20 kg of meat per goat per year. The Iranian goat sector is dominated by extensive herds dedicated to the production of goats for meat and milk, which have very seasonal sales concentrated mainly during the feast of sacrifice and Ramadan. The production of goat milk is undergoing an important development, especially in the centre of the country, and is allowing for a significant improvement in the profitability of goat operations.

Keywords: Agro-ecological transition, goat keeping, nomads

Contact Address: Farhad Mirzaei, Aryan Rural & Nomads Animal Production Management Association, Karaj, Iran, e-mail: fmirzaei@gmail.com

Does Transhumance and/or Vegetation Types Affect the Productivity of Natural Rangelands in Benin?

RODRIGUE V. CAO DIOGO[1], LUC HIPPOLYTE DOSSA[2], PÉNÉLOPPE G.T. GNAVO[1], EVA SCHLECHT[3], ANDREAS BUERKERT[4]

[1] *University of Parakou, Dep. of Sci. and Techn. of Animal Prod. and Fisheries, Benin*

[2] *University of Abomey-Calavi, Fac. of Agricultural Sciences, School of Science and Technics of Animal Production, Benin*

[3] *University of Kassel / University of Goettingen, Animal Husbandry in the Tropics and Subtropics, Germany*

[4] *University of Kassel, Organic Plant Production and Agroecosystems Research in the Tropics and Subtropics, Germany*

Transhumance is a way of life and a livestock rearing strategy that greatly contributes to animal production and the livelihood of pastoralists. Although, this production system contributes to livelihoods, it is increasingly being considered as a threat to rangelands. To evaluate this claim and derive recommendation for the sustainable use of rangelands, we investigated the effects of bovine transhumance and of vegetation types on the biomass production and species diversity of plant communities on natural rangelands in Benin. Three distinct agro-ecological zones (AEZ) were studied in north-east, centre and south Benin. Two zones of different intensity of transhumance (ST: strong and WT: weak animal frequentation) and three vegetation types (VT) per site (open forest/woodland savannah, wooded savannah/shrubland savannah and crop field mosaic) were considered. A four-month floristic survey of herbaceous plants was carried out in the three AEZ with a total of 130 phytosociological surveys including 90 surveys in ST and 40 in WT. Total species richness in north-east was 77 in ST compared to 44 herbs in WT. Likewise, in central Benin, 135 herbaceous species were identified in ST against 80 in WT, whereas in south a total of 33 herbaceous species were found in ST compared to 16 species in WT. In all AEZ, forbs dominated as compared to grasses. The diversity of herbaceous species was not significantly affected by transhumance ($p > 0.05$), but by VT and site, with significant interaction ($p < 0.05$). Despite the species richness observed in ST zone, the total plant biomass was neither affected by transhumance nor by vegetation type, but a site (AEZ) effect and a combination of AEZ, VT, and site was determined ($p < 0.05$).

The results reveal that AEZ and VT play a key role in species richness and biomass production, whereas plant diversity and biomass production were not per se affected by the intensity of grazing, most probably due to the combined effect of AEZ and VT. This suggests that pastoral transhumance alone has no direct effect on the floristic composition of the studied natural rangelands and is thus no major driving force of plant biodiversity loss across different AEZ of Benin.

Keywords: Benin, cattle, land degradation, pastoral mobility, plant biodiversity

Contact Address: Rodrigue V. Cao Diogo, University of Parakou, Dep. of Sci. and Techn. of Animal Prod. and Fisheries, 123 Parakou, Benin, e-mail: dcao_bj@yahoo.fr

Effect of Dry Season Supplementary Feeding on Performance of Desert Sheep in North Kordofan State, Sudan

Ibrahim Bushara[1], Norr Eldean Hassan[2]

[1]*University of Kordofan, Fac. of Natural Resources and Environmental Studies, Dept. of Animal Production, Sudan*
[2]*Ministry of Agriculture, Livestock and Fishery, Sudan*

This study was conducted at North Kordofan State, Sudan. The objectives of the study were to investigate the effects of supplementary feeding during mating, late pregnancy and pre-lambing period on ewe productive and reproductive performance for Desert sheep during the dry season. Eighty ewes were selected from the flock of Desert sheep for this study. Ewes were weighed and randomly divided into four groups of 20 ewes each. Group A was supplemented with diet one (40 % sorghum, 35 % groundnut seed cake, 20 % wheat bran, 4 % sheath, 0.25 % salt lick, 0.75 % common salt), group B received diet two (35 % sorghum, 30 % groundnut seed cake, 30 % wheat bran, 4 % sheath, 0.25 % salt lick, 0.75 % common salt), group C received diet three (30 % sorghum, 25 % groundnut seed cake, 40 % wheat bran, 4 % sheath, 0.25 % salt lick, 0.75 % common salt) and group D served as control (un-supplemented with feed intake depending on pasture only, as practice by farmers). Groups A, B and C were offered their respective supplement feed during 30 days before estrus, 30 days after mating, 45 days before lambing and 90 days after lambing, whereby each diet was offered in the evening at 350 grams per ewe and day. Supplementary feeding had a significant ($p < 0.05$) effect on reproductive traits, by increasing conception rate and lambing rate as compared to the control group. Supplementary feeding also significantly ($p < 0.05$) affected abortion rate which was higher in the control group as compared to the supplemented groups. Furthermore, supplementation enhanced prolificacy in the supplemented groups as compared to the control ($p < 0.05$). Supplemented ewes recorded higher ($p < 0.05$) birth weights of lambs (A: 2.32 kg, B: 2.11 kg, C: 2.02 kg, D: 1.80 kg) as well as higher ($p < 0.05$) weaning weights (A: 11.46 kg, B: 10.70 kg, C: 8.82 kg, D: 7.86 kg). In conclusion, supplementation of Sudanese Desert ewes with diet one (group A) during the breeding period most effectively improved the reproductive performance of the ewes.

Keywords: Body change, desert sheep, reproductive, Sudan, supplementation

Contact Address: Ibrahim Bushara, University of Kordofan, Fac. of Natural Resources and Environmental Studies, Dept. of Animal Production, Elobied, Sudan, e-mail: bushara3030@yahoo.com

Village Level Dairy Goat Development Project in Kenya: A Review on Gaps that Affect Sustainability

THOMAS OGOLA[1], ISAAC KOSGEY[2]

[1] *Egerton University, Dept. of Agricultural Economics and Agribusiness Management, Kenya*

[2] *Egerton University, Dept. of Animal Sciences, Kenya*

Many development projects prioritise high-return programme interventions that contribute to mitigate food insecurity, improve nutrition, and increase the incomes of the rural poor. Key to this is the nurturing of innovations and adaptive technologies and techniques that support these outcomes. Human population pressure, ensuing land demarcation, multi-parity and multiple births, and shorter generation interval have stimulated use of dairy goats in rural development efforts in Kenya. This has led to major good production intensification and scaling-up over the last few decades in the Kenyan dairy goat sector driven by the promise of health benefits from the dairy goat milk. Although some areas like central Kenya and Nyanza saw a major expansion, in other areas this has largely stagnated. The industry is struggling to sustain the progresses made by a number of projects. Thus, a review of the existing literature and project experiences related to dairy goat development in Kenya was performed with the aim of identifying gaps that may have contributed to the current situation.

It was found that lack of supportive policy and regulatory environment with regards to milk standards and goat sales, unsustainable breeding strategies, uncoordinated breeding system, uncoordinated milk and goat marketing, lack of technical support, and inadequate stakeholder or industry player's involvement have all contributed to this scenario. In order to bring about sustainable dairy goat development in the country, these factors need to be adequately addressed. The way these identified factors are addressed will determine the success and sustainability of the dairy goat projects. Value addition, promoting quality assurance supporting development of fodder management and feeds, developing breeding strategies and investment in genetics are encouraged.

Keywords: Dairy goats, gaps, Kenya, sustainable development

Contact Address: Thomas Ogola, Egerton University, Dept. of Agricultural Economics and Agribusiness Management, P.O.Box 73538, 00200 Nairobi, Kenya, e-mail: ogolatdo@gmail.com

Challenges to monogastric systems

Biodiversity in Animal Genetic Resources and their Importance for Environmental and Animal Welfare

Rudolf Bühler

Bäuerliche Erzeugergemeinschaft Schwäbisch Hall, Germany

The People's Republic of China does not only host the world's largest numbers of pigs but also the largest pool of animal genetic resources in terms of pig breeds. There are more than 120 autochthonous *Sus scrofa* vitattus breeds (Asian banded pigs) of which several were imported to England at the end of the 18th century and after 1816 also reached the mainland of the European continent. Until then, only *Sus scrofa scrofa* types as partially domesticated wild pigs where found in Europe. The most well-known descendant of the *Sus scrofa vitattus* sub-species is the Schwäbisch-Hällische Landschwein which originates from the Jinhua pig bred Hangzhou province; in 1820 it arrived via England to Stuttgart-Hohenheim in Germany.

Since the 1970s, pig farming in Europe is dominated by mainstream breeds and hybrids that are bred for maximum daily feed intake and optimised conversion of highly digestible feeds such as wheat, barley and soybeans. This implies a misguided luxury consumption of primary agricultural products which should, for reasons of economic and ecological sustainability, be reserved for human nutrition. However, Asian banded pigs and thus also the Schwäbisch Hällische Lanschwein, have the ability to utilise secondary agricultural raw materials in a resource-efficient manner due to physiological characteristics of their digestive tract. Sustainable strategies of pig meat production must account for such parameters that account for high resource efficiency.

Keywords: Breeds, genetic resources, pigs

Contact Address: Rudolf Bühler, Bäuerliche Erzeugergemeinschaft Schwäbisch Hall, Haller Straße 20, 74549 Wolpertshausen, Germany, e-mail: rudolf.buehler@besh.de

Performances and Efficiency of (Peri-)Urban Pig Breeds under Different Production Management in Ouagadougou, Burkina-Faso

SERGE EUGÈNE MPOUAM, REGINA RÖSSLER, EVA SCHLECHT

University of Kassel / Georg-August-Universität Göttingen, Animal Husbandry in the Tropics and Subtropics, Germany

The demand of pork in sub-Saharan Africa is supposed to rise substantially in the coming years due to urbanisation and urban population growth. In this study we compared the production performance and resource use efficiency of local and crossbred pigs under different feeding intensities in six representative (peri-)urban pig farms in Ouagadougou, Burkina Faso, by quantifying input and output every 6–10 weeks over a 16-month period. The overall productivity of pig rearing was low, with mortality rates of 61 % and 14 % in local and crossbred pigs, respectively, and average litter sizes of 5.1±1.9 versus 5.4±2.3 piglets in crossbred and local sows. The inter-farrowing interval averaged 206±42 days in both breeds. The average daily weigh gain (ADG) of growing animals was 110±116 and 70±69 g d^{-1} in crossbred and local pigs, respectively. Suckling crossbred pigs had a 60 % higher ADG than local pigs while their mother sows were losing weight about twice as fast during lactation than the local sows, even though they received about 40 % more feed dry matter per kg metabolic weight and day. Most feeds offered to pigs were protein-rich industrial by-products, and supply of commercial pig feeds was limited to crossbred pigs. There was a high variability in feeding intensities between farms and seasons. In most cases, the provision of metabolisable energy was by far more limiting pig growth than the supply of digestible protein, with severe energy deficits observed in 50 % and 40 % of local and crossbred animals, respectively. A more efficient use of resources and therefore an improvement of production performances can be realised through requirement-based pig feeding, combined with improved animal management that includes the cancelation of scavenging, adoption of proper housing and year-round stall feeding.

Keywords: Feeding intensity, Ouagadougou, pig breeds, production performance, resource use efficiency, West Africa

Contact Address: Eva Schlecht, University of Kassel / University of Goettingen, Animal Husbandry in the Tropics and Subtropics, Steinstraße 19, 37213 Witzenhausen, Germany, e-mail: tropanimals@uni-kassel.de

Pig Production in Uganda - Adapting to Climate Change

Paul Zaake[1], Oghaiki Asaah Ndambi[2], Birthe Paul[3], Karen Marshall[4], An Notenbaert[3], Emily Ouma[5], Michel Dione[5], George Ouma[1]

[1] *University of Nairobi, Institute of Climate Change and Adaptation, Kenya*

[2] *Wageningen University and Research, Animal Science Group, The Netherlands*

[3] *International Center for Tropical Agriculture (CIAT), Tropical Forages Program, Kenya*

[4] *International Livestock Research Institute (ILRI), Kenya*

[5] *International Livestock Research Institute (ILRI), Uganda*

Pigs are sensitive to heat stress as they lack functioning sweat glands and have small lungs, which reduce their ability to disseminate heat by panting. Moreover, there is limited attention to impacts of climate change on pigs in Uganda by stakeholders despite the potential vulnerability of pigs to heat stress. The objectives of the study were to i) determine the heat-stress status in pigs, ii) analyse factors influencing heat-stress, and iii) explore the heat-stress adaptation options in Lira District, Uganda. Data on heat-stress indicators was collected from a survey of 104 households and measuring 259 pigs in Ojwina (Urban) and Barr (Rural) sub-counties of Lira district. Heat-stress indicators included skin temperature (ST=36.32°C±2°C), and rectal temperature (RT=39.06°C±0.83°C). Ordinary Least Squares linear regression analysis assessed the factors influencing each of the heat-stress indicators. Adaptation options were explored during four gender dis-aggregated focus group discussions with 15 male and 16 female participants in total. The preferred adaptation options were analysed using the average preference rating. According to the farmers, 51.6 % of the pigs were heat-stressed. The results showed that heat stress was influenced by the external temperature humidity index, pig management system, pig category, colour, heart girth, water quantity given, pig's body condition score and time of the day. The results showed that the most preferred adaptation options included constructing a high pig pen roof to facilitate easy air flow (Average Preference Rating = APR = 4.75); pouring water on the pigs (APR = 4.63); and allowing pigs to swim/wallow (APR=4.48). These heat stress adaptation options are suited to the local farm conditions in Lira and offer insight into appropriate techniques that could be applied elsewhere to improve livelihoods and food security. This study confirmed the importance of heat stress risk to pig production in Uganda, and more attention from stakeholders and policy makers is needed.

Keywords: Adaptation, climate change, heat stress, swine, Uganda

Contact Address: Paul Zaake, University of Nairobi, Institute of Climate Change and Adaptation, Nairobi, Kenya, e-mail: zaakepaul@gmail.com

The Role of Chicken Production Systems and Management Practices on Newcastle Disease Outbreaks in Kenya

Billy Okemer Ipara[1], David Jakinda Otieno[1], Rose Nyikal[1], Nabwile Stellah Makokha[2]

[1] *University of Nairobi, Dept. of Agricultural Economics, Kenya*

[2] *Kenya Agricultural and Livestock Research Organization (KALRO), Kenya*

Newcastle disease (ND) is the most prevalent chicken disease in many developing countries including Kenya. It accounts for over 80 % of chicken mortality rates, depriving farmers of their source of livelihood. It is the main challenge for farmers especially those rearing indigenous chicken under the free-range system. This is due to the lack of uniformity in practices, favouring the introduction and spread of the disease. This is worsened by the lack of information about how management practices contribute to the spread of ND. In Kenya, there exists limited information on how chicken production systems and management practices influence ND outbreaks. The current study aimed at assessing the role of chicken production systems and management practices on the frequency of ND outbreaks in Kenya. A Focus group discussion (FGD) consisting of chicken farmers, traders and veterinary officers was conducted to get insights on the chicken value chain and ND. A three-stage sampling procedure was used to select 192 and 140 chicken farmers in Kakamega and Machakos counties of Kenya. A poisson regression model (PRM) was used to estimate the effects of production systems, management practices as well as farmer attributes on the frequency of ND outbreaks. Descriptive results highlight the low access to institutional and support services like extension, trainings and credit. From the PRM results, practices like flock size, type of housing, housing composition, biosecurity and vaccination as well as farmer attributes like ND awareness and extension access had significant effects on the frequency of ND outbreaks. Based on the results, Counties should implement programs to recruit and deploy extension officers to facilitate delivery of information and extension. Credit service providers need to create affordable services and packages that target small-scale farmers. There is need to create more awareness among chicken farmers on aspects like disease detection, disease response and mitigation measures. Farmers should be sensitized on the need to adopt better feeding practices as well as proper housing for chicken so as to reduce the likelihood of birds coming into contact with disease spreading pathogens. Country governments should also develop relevant to facilitate efficient and effective vaccine delivery.

Keywords: Biosecurity, newcastle disease, production systems, vaccination

---

Contact Address: Billy Okemer Ipara, University of Nairobi, Dept. of Agricultural Economics, 29053, 00625 Nairobi, Kenya, e-mail: okemer96@gmail.com

Comparative Assessment of Age, Growth and Food Habit of the Black-Chinned Tilapia, *Sarotherodon melanotheron* (Rüppell, 1852), from Closed and Open Lagoons, Ghana

CEPHAS KWESI ZUH[1], SETH MENSAH ABOBI[2,1], BENJAMIN BETEY CAMPION[3]

[1] *University of Bremen, Fac. of Biology, Germany*

[2] *University for Development Studies, Fac. of Environment and Natural Resources, Ghana*

[3] *Kwame Nkrumah University of Science and Technology, Fac. of Renewable Natural Resources, Ghana*

The black-chinned tilapia *Sarotherodon melanotheron* is the most abundant fish species in the Nakwa (an open lagoon) and Brenu (a closed lagoon) in the Central Region of Ghana. Aspects of the life history characteristics and the ecology of the fish populations in both lagoons were studied to assess the bio-ecological status of this important resource. The size and weight of fish samples ranged between 3.9–11.5 cm total length and 0.960–27.299 g for Nakwa Lagoon and 5.6–12.8 cm total length and 3.160–29.810 g for the Brenu Lagoon. The estimated von Bertalanffy growth parameters were $L\infty$ = 12.04 cm; K =2.76 year^{-1} for the Nakwa Lagoon samples and $L\infty$ = 13.44 cm; K = 3.27 years^{-1} for Brenu Lagoon samples. Daily otolith incremental rate ranged from 0.01–0.03 mm per day and 0.01–0.02 mm per day for Nakwa and Brenu lagoons respectively. Stomach content analysis of the fish samples revealed that the species are planktivorous and the range of food varied between the lagoons. Green algae were the most prevalent food item in the stomachs of the fish samples from Nakwa with the frequency of 69 %, while diatoms (80.5 %) were most prevalent phytoplanktonic food item in for the fish in Brenu lagoon. The estimates of asymptotic length for the species in both lagoons are alarmingly close to known values of the species length at first sexual maturity and points to intensive fishing pressure. As a consequence, a more comprehensive sample-based survey is required in both lagoons to derive robust estimates of management reference points. The results of the stomach content analysis are beneficial to the construction of diet matrix for ecosystem models of the two systems.

Keywords: Age, food, Ghana, growth, lagoon, otoliths, tilapia

Contact Address: Seth Mensah Abobi, University for Development Studies, Fac. of Environment and Natural Resources
current address: University of Bremen, Fahrenheitstraße 6, 28359 Bremen, Germany, e-mail: abobi@uni-bremen.de

Potential of Essential Oils in Filling Gaps and Removing Traps for Sustainable Poultry Production

CHRISTABEL AGHOROWEI, AMAMIEGHAYE EDUM, RUTH TARIEBI S. OFONGO
Niger Delta University, Dept. of Animal Science, Nigeria

Essential oils extracted from plant materials are aromatic, volatile and oily liquids which can serve as alternatives to antibiotics. Essential oils have antimicrobial, anti-inflammatory, anti-oxidative and coccidiostatic properties. The essential oil of *Ocimum gratissimum* (lyn) a medicinal plant consumed as a spice in Nigeria was extracted to determine its in-vitro antibacterial activity against *Salmonella* enteritidis, *Salmonella* typhymurium and *Clostridia perfringens* isolated from poultry litter. The oil was extracted with the means of a Clevenger and placed in an amber bottle before antibacterial studies was carried out. The bacteria *S. enteritidis, S. typhymurium* and *C. perfringens* were isolated using bacteria specific agar after serial dilution of poultry litter collected from poultry pens. *Clostridium perfringens* was cultured on Reinforced *Clostridia* Agar while *Salmonella* were sub cultured from *Salmonella shigella* agar onto Mueller Hinton Agar to remove the effects of indicators and suppressive chemical agents in primary isolation media. The disc diffusion method was used to determine the zone of inhibition (ZOI – in mm) of *O. gratissimum* essential oil against the 3 bacteria. The oil was standardised to 80 %, 40 % and 20 % respectively with the aid of 2 % pharmaceutical grade laboratory soap.

The essential oil of *O. gratissimum* was elicited antibacterial property against all three bacteria tested. The ZOI against *S. typhymurium* was 40.00 mm (80 %), 35.00 mm (40 %) and 25.00 mm 20 %) respectively. Values recorded for *S. enteritidis* were 28.00 mm (80 %), 25.00 mm (40 %) and 23.00 mm (20 %). The essential oil cleared *C. perfringens* at all levels of treatment with the oil. A value of > 40.00 was recorded. From the results obtained it can be concluded that *O. gratissimum* essential oil has potential filling gaps and removing traps for sustainable poultry production in Nigeria.

Keywords: Antibiotics, *Clostridia perfringens*, essential oil, *Salmonella enteritidis*, *Salmonella typhymurium*

Contact Address: Ruth Tariebi S. Ofongo, Niger Delta University, Dept. of Animal Science, Wilberforce Island, P.M.B. 071, Yenagoa, Nigeria, e-mail: tariruth@live.de

Effects of Dietary Incorporation of Processed Guar (*Cyamopsis tetragonoloba*) Fortified with Enzyme on Broiler Performance

Mutaz Mahmoud, Elamin Khidir Elfadil, Asma Tameemeldar
University of Gezira, Poultry Production, Sudan

Using non-conventional feedstuffs in poultry diets to replace some of the common feed ingredients is a strategy suggested by poultry nutritionists to reduce the production cost and adding nutritional values. Guar meal could be included in poultry diets as a non-conventional protein source. A (2 × 3) factorial arrangement was used in a completely randomised design to study the effect of four levels of boiled guar seeds (0, 5, 10, and 15 %) and two levels of enzyme supplementation (0 and 500 mg kg $^{-1}$ on broiler performance. A total of 240 broilers were employed to 8 treatments and each treatment was replicated three times with 10 birds each. The experimental diet without guar and supplementation of enzyme was considered as the control treatment. Feed consumption, weight gain, and feed conversion ratio (FCR) were recorded. Two representative birds from each pen were selected and slaughtered. Relative weights of carcass, organs and cuts were measured. Blood samples were collected for determination of cholesterol, protein, calcium and phosphorus. Results showed that guar boiling processing reduced phytate and tannin content by 13 and 3 %, respectively. Birds fed on diets containing 5 or 15 % guar consumed lower feed when compared with those fed with diets without guar ($p \leq 0.01$). Weight gain and FCR were impaired when dietary inclusion of guar increased ($p \leq 0.01$). Enzyme supplementation improved the weight gain and FCR ($p \leq 0.01$). However, supplementation of enzyme had no significant effect ($p \geq 0.05$) on feed consumption. Dietary incorporation of guar by 0 and 5 % fortified with enzyme had the greatest influence on carcass weight. Thigh weight was not affected by experimental treatments, while breast weight was negatively affected ($p \geq 0.05$) by inclusion of guar. The weights of pancreas, liver, gizzard and proventriculus increased when dietary inclusion of guar increased. In conclusion, diets containing 5 % of guar and supplemented with enzyme recorded good performance results when compared with control treatment.

Keywords: Broiler, guar, performance, protein source

Contact Address: Mutaz Mahmoud, University of Gezira, Poultry Production, Elgamaa, Wad Medani, Sudan, e-mail: mutaz_babiker@yahoo.com

Broiler Chickens' Performance and Prebiotic-Potential of Wheat Offal and Palm Kernel Cake Supplemented with Xylanase

Oluwatomide Ariyo[1], Job Olutimehin Atteh[2], Oluseyi O. Oluwatosin[1], Esther Taiwo[1]

[1] *Federal University of Agriculture Abeokuta, Dept. of Animal Nutrition, Nigeria*

[2] *University of Ilorin, Dept. of Animal Production, Nigeria*

This study evaluated the prebiotic potential, performance and nutrient digestibility of broilers fed diets containing wheat offal (WO) and palm kernel cake (PKC) with or without xylanase supplementation. A 35-day feeding trial made up of a 3×2 factorial design was conducted using 288 1-day old Marshall broiler chickens. The birds were randomly allocated to six dietary groups. Each group was further divided into 6 replicates of 8 birds each. Birds were fed control (maize: 60 %), corn-wheat offal (30 % each) and corn-PKC (30 % each) based diets with or without xylanase supplementation. Nutrient digestibility trial was done on day 21 of the experiment using 2 birds per replicate. On day 35, 3 birds per replicate were sacrificed to determine the microbial profile assay. Their crops were aesthetically removed and placed in sterilized sample bottles (used to convey them to the laboratory for analysis). Data collected were subjected to Analysis of Variance with 5 % significance in a Factorial Design. Feed intake (FI) increased ($p < 0.05$) with wheat offal and PKC inclusion while weight gain (WG) and feed conversion ratio (FCR) were not affected. Xylanase supplementation had no effect ($p > 0.05$) on FI and WG but significantly improved the FCR (1.99). Xylanase supplementation did not influence ($p > 0.05$) crude protein (CP) and fat (CFa) retention but improved ($p < 0.05$) crude fibre (CF) digestibility (58.79 %). Birds fed control diet recorded a higher ($p < 0.05$) CP (77.83 %) and CFa (80.45 %) digestibility while birds fed test diets (WO and PKC) had higher CF digestibility. Total viable count (TVC) and feacal coliform count (FCC) was higher ($p < 0.05$) in the control groups as compared to groups fed WO and PKC. Xylanase supplementation and WO inclusion resulted into higher ($p < 0.05$) Lactobacillus count (1.46 cfu mL^{-1} and lower TVC and FCC. This study concluded that xylanase supplementation enhanced FCR, CF digestibility and prebiotic potential of broiler chickens fed wheat offal and palm kernel cake.

Keywords: Palm kernel cake, performance, prebiotic potential, wheat offal, xylanase

---

Contact Address: Oluwatomide Ariyo, Federal University of Agriculture Abeokuta, Dept. of Animal Nutrition, 110124 Abeokuta, Nigeria, e-mail: ariyooluwatomide@gmail.com

Response of Broilers to Dietary Inclusion of Non-Conventional Protein Sources Fortified with Enzyme Supplementation

Mutaz Mahmoud[1], Muataz Alfadol[2]

[1]*University of Gezira, Poultry Production, Sudan*

[2]*University of Bakhtalruda, Animal Production, Sudan*

Groundnut cake is used as main source of protein in poultry diets in many tropical countries. Due to competition between humans and livestock for groundnut, it is very important to find out alternative plant protein sources. The objectives of study were to assess the effect of using different non-conventional protein sources as partial replacement of groundnut cake (GNC), fortified with enzyme supplementation on broiler performance. A (4×3×2) factorial arrangement was used in a completely randomised design to study the effect of dietary inclusion of seed cakes (sesame (SSC), sunflower (SFC), cotton (CSC) and baobab seed (BSC)) in three inclusion levels replacing groundnut cake by (25, 33 and 50 %) and enzyme supplementation (with and without enzyme) to be compared with negative control (without enzyme) and positive control (with enzyme) on broiler performance. A total of 26 treatments were employed and each treatment was replicated three times with eight birds each. Feed consumption (FC), weight gain (WG), and feed conversion ratio (FCR) were recorded. Two representative birds from each experimental unit were selected and slaughtered. Final body and carcass weights were measured. Blood samples were collected for hematological profile and blood constituents. The results showed that, birds fed on 33 and 25 % BSC, control diet and 33 and 50 % SSC obtained the heaviest weights and carcasses ($p \leq 0.01$). Birds fed on control and BSC diets had the best FCR. Enzyme supplementation significantly improved WG and FCR ($p \leq 0.01$). However, the feed consumption was not affected by enzyme supplementation. CSC and SFC diets caused lower performance values when compared with control diet. Hemoglobin (Hb) and packed cell volume (PCV) were not affected by dietary treatments. The highest amounts of white blood cells (WBCs) were recorded with birds fed on 33 % SFC, 50 % SFC and BSC 25 %. Enzyme supplementation did not affect WBCs account. Blood cholesterol was not affected by experimental treatments. There were significant ($p \leq 0.01$) variations in blood total protein, due to protein source treatments. So, BSC and SSC up to 33 and 50 %, respectively could be included in broiler diets as partial substitution for GNC fortified with supplementation of enzymes.

Keywords: Broiler, non-conventional, performance, protein source

Contact Address: Mutaz Mahmoud, University of Gezira, Poultry Production, Elgamaa, Wad Medani, Sudan, e-mail: mutaz_babiker@yahoo.com

Sodium Diformate and Alkaloids in the Starter Diet of Young Broilers Improve their Performance against Negative and Positive Controls

CHRISTIAN LÜCKSTÄDT, STEVAN PETROVIĆ

ADDCON, Germany

Nutrition plays a crucial role in broiler production, particularly during the post-hatch period, since healthy gut development affects later performance. However, although poultry production often suffers from losses due to contamination with pathogenic bacteria, overuse of antibiotics, the development of resistant bacteria and their ill effects on the human population, have led to the prophylactic use of antibiotics being banned in animal farming in many countries. Although this ban is well deserved, the bacterial challenges experienced in the poultry industry mean that, it is still imperative to have some kind of tool to control bacterial infections and improve the performance of the farmed animal. Organic acids and plant extracts have proved especially effective in maintaining growth performance. Previous studies on the antimicrobial impact of these additives have placed less emphasis on the starter period. The objective of this experiment was therefore to evaluate the impact of dietary sodium diformate and alkaloids (traded as Formi® Alpha, ADDCON; hereafter referred to as NDF-A) in broiler starter diets until day 21, against both a negative and a positive control containing an antibiotic growth promoter (AGP; trimethoprim/sulfadiazine). 180 one-day old male broilers (Ross 308) were randomly allocated to one of three treatment diets with six replicates of 10 birds each in floor pens on a research farm in Iran. Experimental treatments were: negative control (NC); AGP; and 2 kg tonne^{-1} NDF-A. Broiler starter diets were provided as mash feed *ad libitum*. Body weight, feed intake and FCR were recorded after 21 days. Data were analysed and a confidence level of 95 % was defined. Growth performance results revealed a positive impact of the diformate-alkaloid additive. Dietary NDF-A improved body weight gain compared to both the NC and AGP diets (8.9 % and 2.6 %, respectively; $p < 0.01$). Feed intake of NDF-A differed only numerically (+6.1 % and +2.4 % against NC and AGP, respectively). FCR improved only numerically by NDF-A inclusion and varied between control (1.74) AGP (1.70) and NDF-A (1.69). Formi® Alpha improves growth performance in broilers during the crucial period of early growth, not only compared to a negative control but also compared to an AGP.

Keywords: Broiler performance, feed intake, poultry

Contact Address: Christian Lückstädt, ADDCON, Parsevalstrasse 6, 06749 Bitterfeld-Wolfen, Germany, e-mail: christian.lueckstaedt@addcon.com

Chelated Copper, Zinc and Manganese Improved Performance, Haematological and Serum Biochemical Indices of Layers (late-lay)

OLUWATOMIDE ARIYO[1], OLUSEYI O. OLUWATOSIN[1], OLUSESAN FAFIOLU[1], ABIODUN AYOOLA[2], OLUREMI OSINOWO[2], LUIS AZEVEDO[3]

[1] *Federal University of Agriculture Abeokuta, Dept. of Animal Nutrition, Nigeria*

[2] *Federal University of Agriculture Abeokuta, Centre of Excellence in Agricultural Development and Sustainable Environment, Nigeria*

[3] *Novus International Inc., Research and Development, United States of America*

The growth performance, haematological and serum biochemical indices of laying hens (late lay) fed diets supplemented with inorganic or chelated blends of copper, zinc and manganese were evaluated. A 77-day feeding trial was carried out using 540 (58 weeks old) Nera Black Hens. Birds were randomly allocated to five dietary treatment groups with 12 replicates of 9 birds each. The diets consisted of control (0, 0 and 0 mg kg^{-1} of Cu, Zn and Mn respectively), inorganic trace minerals (ITM) supplementation at 16, 64 and64 mg kg^{-1} of Cu, Zn and Mn, chelated trace mineral (CTM) supplementation at 16, 64 and 64 mg kg^{-1} of Cu, Zn and Mn (100 % CTM), CTM supplementation at 8, 32 and 32 mg kg^{-1} of Cu, Zn and Mn (50 % CTM) and CTM supplementation at 4, 16 and 16 mg kg^{-1} of Cu, Zn and Mn (25 % CTM) in that order. On day 77 of the experiment, 2.5 mL of blood was collected individually from 4 birds per replicate via brachial vein puncture and transferred to sample tubes containing EDTA for hematological analyses. Another 2.5 mL of blood was collected into sample tubes for serum biochemical analyses. Data collected were subjected to One-Way Analysis of Variance with 5 % significance in a Completely Randomised Design. Feed intake was increased ($p < 0.05$) with ITM supplementation while 50 % CTM and 100 % CTM reduced feed intake (117.29 and 116.18 g b^{-1} d^{-1} respectively). Kg feed per kg egg was better for diets supplemented with CTM. CTM supplementation at 50 and 100 % increased ($p < 0.05$) packed cell volume (PCV) while white blood cell count was similar ($p > 0.05$) across all treatment groups. Supplementation of CTM led to a higher ($p < 0.05$) total serum protein and albumin. Globulin, creatinine and ALT were not affected ($p > 0.05$) while 100 % CTM increased AST IU L^{-1}. ITM supplementation resulted in increased ($p < 0.05$) cholesterol (192.35 mg dL^{-1}) and LDL (140.47 mg dL^{-1}) levels in the sera while CTM supplementation increased ($p > 0.05$) HDL. This study revealed that CTM supplementation improved kg feed/kg egg, PCV, total serum protein and albumin of layers in late lay.

Keywords: Chelated trace minerals, haematology, inorganic salts, performance, serum biochemistry

---

Contact Address: Oluwatomide Ariyo, Federal University of Agriculture Abeokuta, Dept. of Animal Nutrition, 110124 Abeokuta, Nigeria, e-mail: ariyooluwatomide@gmail.com

A Comparative Study on the Performance of some White Leghorn Strains under Gezira State Conditions, Sudan

MOHAMED AHMED

University of Alneelain, Animal Production, Sudan

Data on the general performance of exotic egg type strains under Sudan condition is lacking. The main objective of this study is to compare the differences of some white leghorn strains in term of performance during growing, pre laying and laying period under local conditions. The present experiment was carried out in summer season during the period of first of May 2013 till the end of June 2014. Birds were caged randomly in an open-sided poultry house at University of Gezira. One hundred sixty two one-day-old female layer chicks from three white leghorn strains (Hyline, Hisex and Lohman) were used. The chicks were fed pre starter pellets from one day old up to the second week of age. The birds were provided with similar three types of rations, grower from 3 to 15 weeks of age and then pre layer ration from 16–21 weeks, finally layer ration was fed till the end of experiment. A completely randomised design (CRD) was used. Collected data were analysed using GLM procedure in SAS. Means were separated using Duncan's Multiple Range Test. Weekly body weight, body weight changes, feed intake, feed efficiency, livability percentage, hen day egg production, hen housed egg production and egg mass were reported. During growing period, Hyline strain had significantly ($P \leq 0.05$) showed the lower feed intake and improved feed efficiency versus Hisex and Lohman. During growing, pre-laying and the whole period, Lohman recorded significantly ($P \leq 0.05$) higher average livability% compared to Hyline. Nevertheless Hisex showed no significant ($P \geq 0.05$) difference versus both Hyline and Lohman. During laying period, no significant ($P \geq 0.05$) differences were observed in growth performance parameters among different strains. Moreover, no significant ($P \geq 0.05$) differences in average hen day egg production%, hen housed egg production% and average egg mass (g/hen/week) across different strains. It could be concluded that Hyline strain is superior in feed efficiency during growing period and egg weight while Hisex and Lohman strains have higher livability%.

Keywords: Egg production, layers, performance, white leghorn strains

---

Contact Address: Mohamed Ahmed, University of Alneelain, Animal Production, P.O. Box 12702, 11121 Khartoum, Sudan, e-mail: moha_555@hotmail.com

Assessment of the Efficacy of Bentonite-Montmorillonite Binder or Yeast to Alleviate Effect of Aflatoxin in Turkey Poults

ADEKOYEJO OYEGUNWA[1], EMMANUEL EWUOLA[2], ADEBISI AGBOOLA[2], EUSTACE IYAYI[2]

[1]*Tai Solarin University of Education, Agricultural Science, Nigeria*

[2]*University of Ibadan, Dept. of Animal Science, Nigeria*

In a 28-day feeding trial, the alleviation of the effect of aflatoxicosis in turkey was investigated by feeding 2 concentrations of yeast and bentonite-montmorillonite binders (BB) in diets contaminated with aflatoxin. One hundred and ninety two 21-day-old turkey poults were randomly allotted to six experimental diets in a completely randomised design (CRD) as follows: D1 (positive control without aflatoxin), D2 (negative control with 0.15 mg kg^{-1} of aflatoxin), D3 (negative control + 3 g kg^{-1} BB), D4 (negative control + 6g kg^{-1} BB), D5 (negative control + 1.5 g yeast kg^{-1} diet), D6 (negative control + 3.0 g yeast kg^{-1} diet). The experiment was conducted under standard experimental conditions and all animal management procedures were followed Addition of BB or yeast in the aflatoxin contaminated diets significantly improved the feed intake and body weight gain of turkey poults. Feed conversion ratio also showed significant improvement in turkeys fed positive control diets (D1), and diets that were treated with yeast and bentonite-montmorillonite binder in D3, D4, D5 and D6 when compared with the negative control diet (D2) which was not treated with any toxin binder. No significant difference ($p > 0.05$) was observed in the values of serum parameters in this study both for diets containing toxin binder and none. However, values of packed cell volume, haemoglobin and red blood cells were significantly improved with addition of yeast or BB. In conclusion, addition of yeast and bentonite-montmorillonite binder at the two levels in this study was able to bind the aflatoxin in the diet, thus making the aflatoxin in the diet ineffective.

Keywords: Aflatoxin, alleviation, bentonite-montmorillonite, turkey poults, yeast

Contact Address: Adekoyejo Oyegunwa, Tai Solarin University of Education, Agricultural Science, Ijagun, Ijebu Ode, Nigeria, e-mail: oyegunwaas@tasued.edu.ng

Haematological and Serum Biochemical Characteristics of Weaned Rabbits Fed Plantain Leaf and Concentrate

Glory Eddy Enyenihi[1], Udoh Akpan Inyang[1], Idorenyin Etuk[2]
[1] *University of Uyo, Dept. of Animal Science, Nigeria*
[2] *Federal University of Technology, Owerri, Animal Science and Technology, Nigeria*

An experiment was undertaken with 30 mixed bred rabbits (6–8 weeks), with initial weight of 0.7 ± 0.02 kg, to assess their haematological and serum biochemical response when fed plantain leaf and concentrate. The plantain leaf was offered at levels of 0 g, 25 g, 50 g, 75 g and 100 g in treatment 1 (control), 2, 3, 4 and 5 respectively, with six rabbits per treatment and for an eight week feeding trial. Control had concentrate alone. After the feeding trial, blood samples were collected from each rabbit through ear vein using a sterilized disposable syringe. The packed cell volume, haemoglobin, red blood cell, mean cell volume, mean cell haemoglobin, mean cell haemoglobin concentration, monocytes and Eosinophils were statistically similar ($p > 0.05$) amongst the five treatments, except total white blood cell, neutrophils, leukocyte counts and platelet counts which were significantly ($p < 0.05$) different. The alanine amino transferase, aspartate amino transferase, alkaline phosphatase activities and creatinine levels of rabbits fed plantain leaf were not significantly different ($p > 0.05$) from those fed the control treatment. The values for the haemotological and sera of the animals obtained in this study were within the normal range of healthy rabbits. The mean dry matter intake of rabbits were 55.91 g, 50.38 g, 72.53 g, 92.80 g and 113.10 g per day while their mean weight gain were 12.14 g, 4.11 g, 7.14 g, 6.25 g and 11.96 g per day for treatments 1, 2, 3, 4 and 5 respectively. This invariably suggests that feeding plantain leaf up to 100 g in rabbit diet per day will not illicit any deleterious effect on the blood profile of weaned rabbits and could go a long way in assisting to filling the gap in animal protein shortage.

Keywords: Electrolytes, haematology, plantain leaf, rabbits, serum biochemistry

Contact Address: Glory Eddy Enyenihi, University of Uyo, Dept. of Animal Science, 57 Nepa Line, Uyo, Nigeria, e-mail: gloryenyenihi@yahoo.com

Comparison between Concrete Floor and Deep Pit System for Production of Crossbred Pigs in Thailand

KORAWAN SRINGARM[1], CHAIWAT ARJIN[1], WOLFRAM SPREER[2], NATTAYA THONGDEE[1], WATCHARAPONG WATTANAKUL[3], NIRAPORN CHAIWANG[3], NUTTAWUT KRUTTHAI[3]

[1]*Chiang Mai University, Fac. of Agriculture, Dept. of Animal and Aquatic Science, Thailand*

[2]*University of Hohenheim, Dept. of Agricultural Engineering, Tropics and Subtropics Group, Germany*

[3]*Chiang Mai Rajabhat University, Fac. of Agricultural Technology, Thailand*

Formerly, in Thailand, pigs were kept by small holder farmers. Native Thai pigs were adapted to these conditions. Meanwhile, production patterns have changed and native Thai pig breeds have been improved by crossing with exotic breeds for a higher share of lean meat, resulting in more efficient production for the domestic market.

The objective of this study was to compare crossbred pigs (Thai native × Meishan × Duroc) kept on concrete floor to those kept in a deep pit system with respect to productive performance and microbial composition of the cecum content.

The experiment comprised 16 crossbred pigs, randomly assigned into two homogenous experimental groups kept on concrete floor (T1) and deep pit system (T2), with an initial weight of 15.6 kg and 15.1 kg, respectively. During the experiment, the weight of all pigs was recorded weekly and average daily gain (ADG) was calculated. Average dry matter feed intake (ADFI) was recorded for individual pigs and feed conversion ratio (FCR) was calculated. The cecal content was collected after the end of the experimental period of 140 days.

Data were analysed by ANOVA, showing that the productive performance of pigs in T1 was significantly (α=0.05) higher: ADG was 357.96 g and 310.70 g in T1 and T2, respectively. While no significant differences were found at FCR (T1: 7.73 (g g^{-1}), T2: 8.68 (g g^{-1})), ADFI was significantly higher in T1 (T1: 2798.25 g day^{-1}, T2: 2692.50 g day^{-1}. The population of *E. coli* (T1: 5.40 log CFU ml^{-1}, T2: 5.48 log CFU ml^{-1}) and *Lactobacillus* sp. (T1: 6.81 log CFU ml^{-1}, T2: 5.98 log CFU ml^{-1}) did not differ significantly, but the beneficial *Bifidobacterium* sp. population in the cecum of the concrete floor group (T1: 7.33 log CFU ml^{-1}) was significantly higher than in the deep pit group (6.51 log CFU ml^{-1}).

In order to exploit their full potential, new pig breeds require more energy-rich feed, improved hygiene and adapted keeping facilities. Even though the deep pit system is accepted among farmers for generation of good manure, this study shows that in terms of hygiene and productive performance of crossbred pigs, concrete floor systems are superior to deep pit systems.

Keywords: ADFI, ADG, *Bifidobacterium*, cecum, *E. coli*, *Lactobacillus*, microbial composition

Contact Address: Wolfram Spreer, University of Hohenheim, Dept. of Agricultural Engineering, Tropics and Subtropics Group, Garbenstrasse 9, 70593 Stuttgart, Germany, e-mail: wolfram.spreer@gmx.net

Evaluation of Prebiotic Activity of Conventional Fibre Feedstuffs in the Diets of Pigs

FEYISAYO OGUNJOBI[1], MAXWELL ADEYEMI[2], EMMANUEL OLUROPO AKINFALA[1]

[1] *Obafemi Awolowo University, Dept. of Animal Sciences, Nigeria*
[2] *Ondo State University of Science and Technology, Dept. of Animal Production and Health, Nigeria*

The use of antibiotics over the last few decades in the diets of pigs especially to enhance performance has gained relevance. The recent growing interest for increased antibiotic resistance has stimulated increasing concerns on alternatives to antibiotics. Sequel to this development, this study was carried out to evaluate the prebiotic activity of fibre feedstuffs. Twenty growing crossbred (Large White x Hampshire) pigs with a mean weight of 40 ± 0.50 kg were randomly allotted to four fibre feedstuffs (palm kernel cake (PKC), brewer's dried grain (BDG), wheat offal (WO) and rice bran (RB)) in a 14-day feeding trial. The experimental pigs were randomly allocated to five experimental diets in a completely randomised design with four animals per treatment. Diets T1, T2, T3, T4 and T5 contained 100 % maize (control), PKC, BDG, WO and RB respectively. The oligosaccharide contents of feedstuffs, the short chain fatty acids and the activities of bacteria in the guts of the animals were evaluated. Results showed significant differences ($p < 0.05$) in the arabinoxylan-oligosaccharide concentration among dietary treatments. The short chain fatty acid (SCFA) concentrations were significantly ($p < 0.05$) affected by dietary treatments both at the small and large intestines of pigs. Also, the SCFA concentrations produced in the large intestine were numerically ($p > 0.05$) higher than the small intestine. All the fibre feedstuffs promoted the growth of *Lactobacillus* only than maize in the small and large intestines. Findings from this study showed that PKC, BDG, WO and RB could elicit prebiotic activity in the gut of growing pigs.

Keywords: Fibre feedstuffs, short chain fatty acids, oligo saccharides

Contact Address: Maxwell Adeyemi, Ondo State University of Science and Technology, Dept. of Animal Production and Health, Km 6 Igbokoda Road Okitipupa, Okitipupa, Nigeria, e-mail: ade_maxwell@yahoo.com

Occurrence and Antibiotic Susceptibility Profile of Carbapenem-Resistant Enterobacteriaceae in Selected Aquaculture Ponds in Southwestern Nigeria

Elizabeth Omokoshi Joel[1], Abimbola Adekanmbi[2]

[1] *University of Ibadan, Dept. of Animal Science, Nigeria*

[2] *University of Ibadan, Dept. of Microbiology, Nigeria*

The indiscriminate use of antibiotics in aquaculture has contributed to the prevalence of antibiotic resistant bacteria in aquaculture ponds and this represents a major challenge to public health. The aim of this study was to determine the antibiotic susceptibility profile of carbapenem-resistant Enterobacteriaceae from selected aquaculture ponds in Awotan area of Ibadan, Oyo State. Water samples from fifteen (15) concrete and earthen aquaculture ponds were collected weekly for a period of two months. Isolation of bacteria was carried out using the standard pour plate technique on MacConkey agar supplemented with imipinem. Pink colonies presumptive of members of the Enterobacteriaceae were purified and characterised using conventional methods. The isolates were subjected to ten classes of antibiotics using the Kirby Bauer disc diffusion technique.

A total of forty four (44) carbapenem-resistant bacteria belonging to six genera namely: *Edwardsiella* (36.4 % n=16), *Salmonella* (2.3 % n=1), *Yersinia* (40.9 n=18), *Ewingella* (4.6 % n=2), *Shigella* (13.6 % n=6), and *Citrobacter* (2.3 % n=1) were obtained. 97.6 % of the isolates were resistant to imipinem, 88 % to amoxicillin-clavulanate while 83.7 % were resistant to cefpodoxime. All the isolates were resistant to ampicillin and cefotaxime with 11.6 % of them showing resistance to chloramphenicol and ciprofloxacin respectively. None of the carbapenem resistant bacteria in this study was positive for ESBL production.

The presence of carbapenem-resistant bacteria in aquaculture ponds in this study, suggests need for proper investigation, monitoring and management of aquaculture ponds from time to time to evaluate the emergence of bacterial resistance towards antibiotic. Also, the usage of antibiotic in aquaculture ponds should be well regulated.

Keywords: Antibiotic susceptibility profile, aquaculture ponds, carbapenem-resistance, Enterobacteriaceae

Contact Address: Elizabeth Omokoshi Joel, University of Ibadan, Dept. of Animal Science, Ibadan, Nigeria, e-mail: omobeth2006@yahoo.co.uk

Effect of Climate Change on Fish Migration and Fish Catches in Africa

EMEKA OGBONNA JOHN
Georg-August-Universität Göttingen, Dept. of Animal Sciences, Germany

The effect of climate change on water, natural resources, and on many sectors of African economies from agriculture, transportation and energy sector has been extensively researched. But the effect on marine aquaculture and fisheries ecosystem, a significant economic sector and a major source of food for large numbers of people in Africa, is yet to be deeply studied. This study evaluates the overall effect of climate change on fisheries catch and fish migration in Africa. Selected region in Africa were evaluated based on the fisheries catch and fish migration using past, current and future climate scenarios. The outcome of the research shows that climate change causes an increase in sea surface temperatures, with fish stocks migration toward cooler region away from equator leading to reduction in fish stock. It also affects the migratory patterns, birth/mortality rates and abundance of wild fish. Ocean warming in East Africa has already a disastrous effect on the coral reef. West African region such as Sierra Leone have experienced flooding of coastal zone due to rising sea levels, exacerbating the vulnerability of people living in the coastal zone. At the Global level, it is projected that climate change will decrease fish catches by 7.7 % and fish revenue at 10.4 % by 2050. This overall impact might mean about 26 % decrease in fish catch in West Africa region, and might be higher for parts of West Africa countries that are near to the equator, for instance Ghana fisheries will decrease by about 60 %, 56 % in Cote d'Ivoire, and 53 % in Nigeria. Roughly 12 million people are employed in the fisheries in Africa with ripple effects. For every fisherman's job, about 1.0 onshore job is created additionally in Mauritania, and 3.2 in Guinea. This result indicates that there are a huge multiplier effect of climate change impact on fisheries sector and will enable stakeholders to formulate policy intervention to alleviate poverty in the fisheries sector.

Keywords: Aquaculture, climate change, fisheries

Contact Address: Emeka Ogbonna John, Georg-August-Universität Göttingen, Dept. of Animal Sciences, Christophorusweg 12, 37075 Göttingen, Germany, e-mail: emekaogbonna.johm@stud.uni-goettingen.de

Comparative Analysis of Technical Efficiency in Integrated and Non-Integrated Aquaculture Farms in Rivers State, Nigeria

DUMKA BENSON

Federal College of Education (Technical), Omoku, Dept. of Agricultural Education, Nigeria

The study compared technical efficiency in integrated and non-integrated aquaculture farms in Rivers State, Nigeria. The integrated aquaculture farms produced feeds and fingerlings while the non-integrated aquaculture farms buy feeds and fingerlings from independent suppliers for fish production. Purposive sampling was used to select the 44 integrated aquaculture farms while multi-stage technique was used to sample the 126 non-integrated aquaculture farms. Data were analysed using descriptive statistics, annual depreciation and stochastic frontier model. The two production systems of integrated and non-integrated aquaculture production were efficient given the least efficiency of a farm as 50 %. The result of the study revealed that the integrated aquaculture farms had 62 % of the farms with indices of 0.96–1.00 as efficient while the technical efficiency of the farmers ranges from 88–98 % with a mean of 96 %. This indicates that the farmers need 4 % increases in productivity to achieve 100 % technical efficiency. The non-integrated aquaculture farms had 61 % of the farms with indices of 0.93–1.00 as efficient while the technical efficiency of the farmers ranges from 88–98 % with a mean of 93 %. This indicates that the farmers need 7 % increases in productivity to achieve 100 % technical efficiency. Based on the findings, the paper concluded that differential technical efficiency existed between integrated and non-integrated aquaculture farms in the study area. However, for this high rate of technical efficiency to be sustained in natural resource management; current technical and price information is needed by the aquaculture farmers which can only be disseminated by qualified and adequate extension workers.

Keywords: Aquaculture production, comparative analysis, stochastic frontier production model, sustainable resource management, technical efficiency

Contact Address: Dumka Benson, Federal College of Education (Technical), Omoku, Dept. of Agricultural Education, Omoku, Port Harcourt, Nigeria, e-mail: dumka.benson@yahoo.com

(Agro)forestry

Socio-economic aspects of tree-based systems

Forest Landscape Restoration in Indonesia: A Scaleable Smallholder Agroforestry Approach

Wolfgang Baum, Simeon Max

Fairventures, Germany

Large scale land use change from exploitation to regeneration must involve local people to ensure sustainability of operations. Environmental education and financial empowerment of the communities are crucial components. Fairventures has been operating in Central Kalimantan, Indonesia for the last 5 years socialising communities and individual smallholders for reforestation. We research cash crops with a clear market demand and test these with selected smallholder project participants. In this stage, we have a better understanding of promising tree species and a limited number of cash crops. We ensure smallholder ownership through their involvement from planting through management, harvest and sale. There are still plenty of unanswered questions that need to be adressed to strengthen the social, economic and environmental bases of our work.

Keywords: Agroforestry, business models, FLR, scale-up, smallholders

Contact Address: Wolfgang Baum, Fairventures, Hasenbergstraße 31, 70178 Stuttgart, Germany, e-mail: Wolfgang.Baum@fairventures.org

Forest Dependent People's Participation in Participatory Forest Management (PFM) Programmes in Kenya

ROBERT MBECHE[1], JOSIAH ATEKA[1], DONALD OGWENO[2], ANJA FASSE[3], ULRIKE GROTE[4]

[1]*Jomo Kenyatta University of Agriculture and Technology, Dept. of Agricultural and Resource Economics, Kenya*

[2]*Independent Forestry Consultant, Forestry, Kenya*

[3]*Hochschule Weihenstephan-Triesdorf University of Applied Sciences, Germany*

[4]*Leibniz Universität Hannover, Institute for Environmental Economics and World Trade, Germany*

Over the last two decades community participation in the management of state-owned forests has become a theme of policy and academic work in attempts to enhance sustainable forest management in many developing countries. In Kenya, the Forest Act 2005 entrenched Participatory Forest Management (PFM) through enabling the creation of Community Forest Associations (CFAs), authorised to co-manage the forests with Kenya Forest Service (KFS), the statutory forest agency to improve forest cover and their livelihoods. However, the degree to which these goals are met depends, in part, on accessibility and desirability of programmes targeting forest dependent people. Empirical research on factors motivating or dissuading forest dependent peoples' participation in CFA activities remains limited and a more complete understanding of these factors is critical to improving social equity outcomes. This paper investigates two issues that have received little research attention to date: the level of forest dependent people's participation at different stages of a participatory forest management (PFM) programme and factors that influence their level of participation. Data were collected from three CFAs in the Mt Elgon forest reserves, western Kenya through household surveys, key informant interviews, and focus group discussions. A participation index (PI) and binary logistic regression model were used to analyse the data. The results showed that slightly over a half of the respondents (51 %) were members of forest user groups (n=924). The results further revealed that the level of the forest users' PI was 40.7 %, 49 %, and 42.9 % at the planning, implementation, and monitoring stages, respectively suggesting low levels of participation. The logistic regression model showed that gender, household size, education level, land size, income from the forest, proximity to forest edge, household asset value and engagement in alternative livelihoods were found to be statistically significant ($p < 0.05$) predictors for the level of participation. Thus, our results suggest that the broader context of land tenure security and alternative livelihood strategy development may be critical in improving participation of forest dependent communities in CFA activities.

Keywords: Community forest associations, forest dependent people, Kenya, participation

Contact Address: Robert Mbeche, Jomo Kenyatta University of Agriculture and Technology, Dept. of Agricultural and Resource Economics, P.O Box 62000-00200 , Nairobi, Kenya, e-mail: rmbeche@jkuat.ac.ke

Inga and Camëntsá Ecosystem Services from Agroforestry Systems in the Sibundoy Valley, Colombia

Vandreé Julián Palacios Bucheli[1,2], Wolfgang Bokelmann[2]

[1] *Universidad de Ciencias Aplicadas y Ambientales, Colombia*
[2] *Humboldt University of Berlin, Faculty of Life Sciences, Germany*

Agroforestry systems play an essential role in the livelihoods and food security of the Inga and Camëntsá indigenous communities, as they supply provisioning, regulating and habitat, and cultural ecosystem services.. The agrosilvopastoral agroforestry system is of utmost importance for the development of the communities' culture. Nevertheless, the home garden is shrinking in size and gradually disappearing. The purpose of this research relates to economically valuing ecosystem services provided by agroforestry. Research for this study was conducted in five indigenous settlements of the San Francisco municipality. A census was carried out to determine the population (146 households), and a stratified random sample was used for the survey with a sample size of 67 households (45.8 %). Information required for the assessment is comprised of three data sets obtained from (i) cost-benefit analysis, (ii) contingent valuation, and (iii) travel cost analysis. Silvopastoral systems represent the highest economic value (965,800.81 € year^{-1}) despite its low value per hectare (4,601.24 € ha^{-1} year^{-1}). The low value per unit of land becomes relevant when it is aggregated to the total silvopastoral land area, which occupies the most extensive coverage (209.9 ha). Conversely, agrosilvopastoral systems have the highest value per hectare (7,607.41 € ha^{-1} year^{-1}), but the total value (518,825.05 € year^{-1}) position them as second in the list when aggregated to total land area (68.2 ha). Similarly, the agrisilvicultural, system has a high value per unit of land (5,722.54 € ha^{-1} year^{-1}), which is explained by the high value of output and input (2,239.7 € ha^{-1} year^{-1}) produced and demanded but it covers the smallest area (35.2 ha). The total economic contingent value adds up to 2,834.2 € ha^{-1} year^{-1}, which pertains to each hectare of agroforestry systems. Most of the families (88 %) were willing to protect home gardens. The total protected area adds up to around 71.7 ha, and the minimum monetary offset that households would be willing to receive per month is € 31.6, and the maximum is € 1,581.7. The economic valuation based on travel cost analysis shows a total value of around 648.4 € ha^{-1} year^{-1}.

Keywords: Agrisilvicultura, agrosilvopastoral, contingent valuation, indigenous communities, silvopastoral

Contact Address: Vandreé Julián Palacios Bucheli, Universidad de Ciencias Aplicadas y Ambientales, Cra 34 # 14 - 50, 520002 Pasto, Colombia, e-mail: vandreep@gmail.com

Landlords' perception of Trees in a Rural-Urban Continuum of Ghana

Jones Abrefa Danquah

University of Cape Coast, College of Humanities and Legal Studies, Dept. of Geography and Regional Planning, Ghana

Trees in built urban ecosystems are increasingly seen as a solution to reduce the impact of climate change on the wellbeing of urban dwellers. There is growing concern of Urban Heat Island (UHI) in the advent of climate change due to rising temperature and rapid development of grey infrastructure. This has become a major public health issue in densely urban areas across the globe. In Ghana, recurrent flooding, erratic rainfall patterns and heat waves particularly during the peak of the dry season have been recorded in major urbans areas. However, green infrastructure, particularly trees, in built ecosystems are seen as low cost intervention to address UHI, seasonal heatwaves and floods from stormwaters during the peak of the raining season in major urban centres in Ghana. For effective deployment of trees as green infrastructure in major cities it is imperative to understanding urban dwellers' perception of trees in built ecosystem. This will facilitate their maintenance and prevent vandalism. In this regard the study was conducted. I randomly sampled 600 landlords from three purposively selected major cities in the country. The cities were Kumasi and Sunyani from the middle belt and Cape Coast along the Atlantic coast of Ghana. Focus group discussion was conducted with 40 landlords to gain knowledge of landlords' understanding, attitude and perception of trees in their immediate surroundings. These results were used to develop a comprehensive interview schedule guide or questionnaire to interview 600 landlords, in order to solicit information on their perception. A five point Likert scale was employed from strongly agree (5) to strongly disagree (1) to assess their perception. The second part evaluated cultural and socioeconomic attributes and how they influence ones perception of trees. Data thus generated were analysed using multivariate analysis of variance (MANOVA). The results indicated that variables such as cultural tendencies, education, age, income, ethnicity and occupation positively enforce landlords' perception of trees. Benefits such as stormwaters reduction, air purification, medicinal, noise abatement, moderate microclimate and carbon sequestration were identify by landlords as key ecosystem services provided by urban trees. Medicinal properties of urban trees provided, a bases for vandalism.

Keywords: Ecosystem services, landlords, perception, trees, urban green space

Contact Address: Jones Abrefa Danquah, University of Cape Coast, College of Humanities and Legal Studies, Dept. of Geography and Regional Planning, Cape Coast, Ghana, e-mail: jones.danquah@ucc.edu.gh

Participatory Forest Management in West Usmabara-Tanzania. What Is the Community Perception on Success?

HUSSEIN LUSWAGA

Justus-Liebig University Giessen, Agricultural and Environmental Policy, Germany

The community participation in the participatory programs has been the mainstay approach in the management of forest in resource rich but less developed countries. The approach empowers the communities living around the forest to benefit and take part in the resource management. Tanzania is among the countries often cited as a success in the participatory program. Despite the increase in forest areas and villages under the participatory programme in Tanzania over the last two decades, considered as a success by policy makers', there is little insight into the communities' views regarding their participation in the forest activities. The communities are important stakeholders in the sustainability of forest management and their perceptions provide the realistic performance and success of the participatory program. Therefore, this study aimed at exploring the extent of the communities participation in the goals identified under the participatory programs in Tanzania. The survey was conducted in 159 households, randomly sampled from four villages bordering state and community forests, in West Usambara Mountains. Factor analysis was used from which three participation components, namely economic, decision making, and protection, were summarised. The participation components were then used to formulate the participation index which gauged overall participation intensity. Further, based on focal group discussion and pictorial presentation, the pebble distribution method was used to rank the community perception on the benefits and management activities related to the forest. The study observed significantly higher levels and intensity of participation in the villages around the state compared to those around the community forest. Training attendance, trust on institutions and location were significant predictors of participation intensity of the households. In conclusion, the perceived performance of the participatory programs in achieving the participation aims has only been moderate, with a better situation in the state-managed forest. Raising awareness and improving communication with villagers, fulfiling promises for the communities and provide more forest linked benefits are interventions which can improve the situation in West Usambara. This study implies the importance of considering the community views as important stakeholders in formulating participatory policy which consider community interest and the resource sustainability.

Keywords: Community participation, Lushoto, participatory forest management, Usambara Mountains

Contact Address: Hussein Luswaga, Justus-Liebig University Giessen, Agricultural and Environmental Policy, Giessen, Germany, e-mail: huslus@gmail.com

Forest Resources, Poverty and Inequality in Peruvian Amazon: The Role of Tenure Regimes and Remoteness

KARIN BEGAZO CURIE, LIESBET VRANKEN, KEWAN MERTENS
KU Leuven, Earth and Environmental Sciences, Belgium

Indigenous populations have historically been marginalised. Most indigenous people from Latin America live in villages with communal land ownership in remote areas. This population is highly dependent on forest resources by virtue of their traditions, culture, and forest knowledge but also due to remoteness and the presence of less degraded forests. Tenure regimes are considered to influence income and type of products extracted from forests. In Peru, according to law, all communal lands in forest areas are owned by indigenous population. Scarce literature analyses the relation among forests and livelihoods and most of previous studies could not assess to what extent the contribution of forests to livelihoods is mediated by either tenure regimes or remoteness. This research attempts to fill this gap in literature by focusing on the economic contribution of forest products to rural livelihoods and considering variations in tenure regimes (communal and private ownership) and remoteness. We use data from 400 household located in 50 villages in lowland forest areas in Loreto Region, in Peru. We use Ordinary Least Square and Fractional Response Models to assess total income and forest reliance, respectively. The Foster-Greer-Thorbecke (FGT) index of poverty and Gini coefficient were used to analyse differences in rural poverty and income inequality between villages with communal and private ownership. T-test was performed to evaluate forest income contribution on tenure regimes on the one hand and remoteness on the other. We found that forest income reduces inequality across households in villages in communal tenure. We show evidences that forests are an important income source for the poorest households and this contribution is more pronounced among households living in villages with communal tenure regime and in remote areas. Game meat and non-timber forest products are the most important subsistence income source for households living in remote villages and in villages with communal ownership. Our findings confirm the importance of forests for rural livelihoods in remote areas and in areas with communal land. This calls for policy interventions to develop alternative income sources and to avoid forests overexploitation which threaten the sustainability of forest resources.

Keywords: Forest dependency, forests resources, inequality, Peruvian Amazon, poverty, remoteness, sustainability, tenure regimes

---

Contact Address: Karin Begazo Curie, KU Leuven, Earth and Environmental Sciences, Leuven, Belgium, e-mail: kbegazo@lamolina.edu.pe

Forest Restoration Options on Degraded Hill Sites in the Pomacochas Basin (Amazonas, Northern Peru)

STEFAN HOHNWALD[1], HELGE WALENTOWSKI[2], MARTIN THREN[2], OSCAR A. GAMARRA TORRES[3], STEFFI HEINRICHS[4], HENRY HEINEN[2], ALEXANDER WIEGAND[2], ANA B. SABOGAL[5], STEFAN ZERBE[6]

[1]*HAWK- University of Applied Sciences and Arts Göttingen, Pedology, Geobotany, and Nature Conservation, Germany*

[2]*HAWK - University of Applied Sciences and Arts Göttingen, Resource Management, Germany*

[3]*University of Chachapoyas, INDES-CES- Instituto de Investigación para el Desarrollo Sustentable de Ceja de Selva, Peru*

[4]*Univ. of Goettingen, Silviculture and Forest Ecology of the Temperate Zones, Germany*

[5]*Pontifical Catholic Univ. of Peru (PUCP), Fac. of Liberal Arts and Humanities, Peru*

[6]*Free University of Bozen-Bolzano, Faculty of Science and Technology, Italy*

Although most of the northern Peruvian Andes are still dominated by diverse cloud and montane forests, agricultural patches and degraded forests are increasingly expanding, while natural habitats are suffering from insidious fragmentation. Subsequent soil degradation often leads to field abandonment, overused forests, or degraded pastures. However, ecological knowledge on the forest restoration potential, for instance on the dependencies of vegetation cover, soil conditions, exposition, and altitude, is scarce. Therefore, soil and vegetation patterns along nine transects within the upper Pomacochas Basin were investigated. Anthropogenic successional and disturbance grades, geological substrate, and altitude have the most important ecological impacts on vegetation and tree species composition (n=24 sites). Species respond to sandstone versus calcareous substrates, but also to depths of organic soil layers, and light conditions. The absence of organic layers under pastures contrasted with the accumulation of thick organic layers under forest cover. Vegetation composition at succession sites revealed certain starting points for restoration, like a herbal stage, bush stage, or secondary forest that will determine the length of regeneration paths. Pre-forest patches of *Alchornea* sp. and *Parathesis* sp. may act as habitat stepping stones for expeditiously restoring biocorridors of wildlife. Ecologically adapted reforestation, especially on less productive hill sides, can help to connect forests patches, including living fences as biocorridors. If well planned and managed, restoration measures may rebuild and maintain such unfragmented forests which would be of great benefit for the genetic interchange of forest plant and animal species like the flagship mammal species *Aotus miconax, Oreonax flavicauda, Tremarctos ornatus*, and *Puma concolour*, or for bird species like *Loddigesia mirabilis, Xenoglaux low-*

Contact Address: Stefan Hohnwald, HAWK- University of Applied Sciences and Arts Göttingen, Pedology, Geobotany, and Nature Conservation, Büsgenweg 1a, 37077 Göttingen, Germany, e-mail: stefan.hohnwald@hawk.de

eryi, and *Rupicola peruviana*. As the Pomacochas Basin is an important biodiversity hub in northern Peru, these afforested forest patches can help to secure the natural heritage. Thus, our findings can contribute to sustainable land use and conservation of a fragile ecoregion. The chances for strong implementation of nature conservation objectives and sustainability goals should be taken now, as the local population is interested in sustainable land use forms such as soft eco-tourism and agroforestry, while asking science for advice for an ecological sound development of the region.

Keywords: Andes, biocorridors, cloud forests, ecological sustainability, nature conservation, restoration, soils, trees

Identification of Gaps in the Community Forestry within the REDD+ Project in the Peruvian Amazon

Jeisson Rodriguez Valenzuela[1], Sven Wagner[2], Manh Hung Bui[3], Nicolas Jose Mesia Rojas[1], Sergio Parra González[4], Jeimy Katherin Feo Mahecha[5]

[1] *Technische Universität Dresden, Inst. of International Forestry and Forest Products, Germany*

[2] *Technische Universität Dresden, Institute of Silviculture and Forest Protection, Germany*

[3] *Vietnam National University of Forestry, Forest Inventory and Planning Dept., Vietnam*

[4] *University of the Llanos, Faculty of Agriculture, Department of Soil, Colombia*

[5] *Universidad de Ciencias Aplicadas y Ambientales U.D.C.A, Colombia*

Conservation concessions (CC) within the REDD+ project have been promoted as mechanisms to protect the primary forest from anthropogenic disturbances in the Peruvian Amazon. However, economic activities by human settlements located in buffer zones of those concessions, have affected the common forest resources given to these groups. Therefore, with a view to determining the effects of the forest utilisation on the communal land governed by the indigenous community of Gran Pajaten located in the buffer zone of the CC Montecristo (San Martin, Peru), different workshops (Focus group approach) were implemented with members of this community and environmental authorities. Outcomes highlighted two components that were discussed: 1) a social component, with respect to general information about economic activities of the people living in the buffer zone; 2) a forest component, related to forest management in the communal land. Results obtained confirmed that management plans implemented in the communal forest are based only on selective logging through inaccurate practices of motor-manual operations. However, they admitted that when the forest is logged, most of the time they do not care about the understory, focusing only on harvesting targeted timber species which can be cut, according to the agreement with the community. This agreement states: "It can be harvested up to two trees per person of the community, and only can be done with the purpose of self-consumption, for instance, fuelwood, construction of their houses, or improvements in their farms". Nevertheless, they cannot control activities by illegal loggers who extract the wood anonymously, which is a complex situation because those practices are carried out in remote areas, and sometimes is made by people of the same community or neighbouring communities. Furthermore, the information collected with these meetings and workshop with the community showed that although they promote some species through tree nurseries in their farms, they concentrate on a set of timber species with more economic interest or those that can represent some benefit to their agroforestry systems (AFS) with cocoa crops. Thus, the findings from this case study have confirmed poor practical knowledge by the community about sustainable forest management.

Keywords: Buffer zone, conservation concession, forest utilisation, illegal logging, indigenous community, Peruvian Amazon, REDD+

Contact Address: Jeisson Rodriguez Valenzuela, Technische Universität Dresden, Inst. of International Forestry and Forest Products, Pienner Straße 7, 01737 Tharandt, Germany, e-mail: jerovagro@gmail.com

Costs and Benefits of Sustainable Management of the Kyrgyz Walnut-Fruit Forests: Is the 2007-Logging Ban a Suitable Policy Instrument?

KLARA DZHAKYPBEKOVA[1], MYKTYBEK SABIROV[2], ZHAMILIA SULAIMANOVA[3], JYLDYZ SHIGAEVA[4], DAVLET MAMADJANOV[5], DIETRICH DARR[1]

[1]*Rhine-Waal University of Applied Sciences, Faculty of Life Sciences, Germany*
[2]*American University of Central Asia, Kyrgyzstan*
[3]*Weihenstephan-Triesdorf University of Applied Sciences, TUM Straubing, Germany*
[4]*University of Central Asia, Mountain Societies Research Institute, Kyrgyzstan*
[5]*National Academy of Sci. of the Kyrgyz Rep., Inst. of Walnut and Fruit Crops, Kyrgyzstan*

The walnut-fruit forests in Central Asia and in particular in Kyrgyzstan are a unique socio-ecological system and an important source of livelihood for the rural communities. Walnut-fruit forests have also been described as an important genetic resource and biodiversity hotspot, as well as providing essential environmental services. However, due to the political and socio-economic transformation after the collapse of the Soviet Union, the utilisation of these forests has increasingly become unsustainable, inducing significant degradation of these forests and a decline of their area. To protect these unique natural ecosystems a logging ban has been enforced as a National Law since 2007 which restricts the felling of trees in the walnut-fruit forests in Kyrgyzstan. However, this measure has been criticised as it is said to lead to the over-aging of walnut forests, which limits their natural regeneration potential; and deprive local communities of much-needed income opportunities from sustainable timber utilisation, which potentially aggravates rural poverty. Starting off a literature review on the effectiveness of logging bans in terms of their environmental and socio-economic outcomes, this study investigated the costs and benefits that would potentially be associated with sustainable forest management and timber harvesting operations in the walnut forests. It used a stand-based dynamic forest growth model developed from primary data from 19 forest plots, national forest thinning guidelines and inventory data. Data collected from a socio-economic survey (n=100), financial and Monte-Carlo analyses were used to determine the financial and economic profitability of sustainable forest management and timber harvesting operations for a period of 25 years. The results showed that 37 % of the surveyed forest plots were presented in an over-aged and stagnating condition. The dynamic forest growth model showed that the generated thinning plan allowed for the sustainable removal of 10,4 m^3 ha^{-1} timber in average every ten years depending on the on sanitary condition, age and density of a stand while ensuring the renewal of the over-aged forest plots. The economic evaluation indicated that only the forest plots with no additional planting activities had a positive NPV. Policy scenarios offering viable solutions were described, and relevant recommendations were provided.

Keywords: Central Asia, forest thinning, Kyrgyzstan, logging ban, sustainable forest use

Contact Address: Klara Dzhakypbekova, Rhine-Waal University of Applied Sciences, Faculty of Life Sciences, Marie-Curie-Str. 1, 47533 Kleve, Germany, e-mail: klara.dzhakypbekova@gmail.com

Complex Ecosystem Service Supply and Delivery in a Mountain Mosaic Landscape in Chiapas, Mexico

Alan Heinze[1], Thom Kuyper[2], Frans Bongers[1], Luis García Barrios[3], Neptalí Ramírez Marcial[4]

[1]*Wageningen University and Research, Forest Ecol. and Forest Manag., The Netherlands*
[2]*Wageningen University and Research, Soil Quality, The Netherlands*
[3]*El Colegio de la Frontera Sur, Depto. de Agricultura Sociedad y Ambiente, Mexico*
[4]*El Colegio de la Frontera Sur, Depto. Conservaciónde la Biodiversidad, Mexico*

Farming is the mainstay of a mountain community situated in the buffer zone of La Sepultura Biosphere Reserve in southern Mexico. Farmers depend on multiple ecosystem services, especially provisioning services like water, firewood, and ranching resources supplied by paddocks with a heterogeneous tree cover and terrain. Other local actors, especially conservation- and development-oriented stakeholders working in the reserve, are additionally interested in biodiversity, forest protection and carbon storage. Generally, they promote sustainable intensification of productive activities as part of a broader land-sparing approach. Support is upcoming for forest-based activities that seem successful in meeting both local livelihood aspirations and conservation goals.

We assessed ecosystem services through a social-ecological systems approach. To better understand ecosystem services supply, we conducted an integrated forest inventory in a selected rangeland area to compare ecosystem services among land cover types, viz. closed and open forests, agricultural lands and riparian corridors. We then focused on a forest-based provisioning service, pine resin, by using an ecosystem service mediating mechanism and factor framework.

Each land cover type provides a particular set of ecosystem services with synergies and trade-offs that farmers acknowledge. Ecosystem service supply and delivery do not match well, however. In the case of pine resin, actual production is variable and below potential production based on available resources. We identified different mechanisms, e.g. the management of paddocks, labour and technical skill, capital-labour relations, and an increasing appreciation of trees, that mediate the co-production of this ecosystem service and eventually how it contributes to farmer and community well-being. Additionally, there are contextual factors that influence these mechanisms, including access to tools, rules and values among group members, and support from outside institutions. With the resulting framework, we seek to contribute to landscape approaches, sustainable intensification and/or land use zoning, not just to safeguard biodiversity and enhance farmer production, but to support entire social-ecological systems for an adaptive provision of ecosystem services.

Keywords: Ecosystem services, Mexico, pine resin, production gap, social-ecological systems

Contact Address: Alan Heinze, Wageningen University and Research, Forest Ecology and Forest Management, Droevendaalsesteeg 3a, 6708 PB Wageningen, The Netherlands, e-mail: alan.heinzeyothers@wur.nl

Determinants of Forest Extraction among Rural Households in Mt Elgon, Kenya

PURITY YEGO[1], ROBERT MBECHE[1], JOSIAH ATEKA[1], ANJA FASSE[2], ULRIKE GROTE[3]

[1]*Jomo Kenyatta University of Agriculture and Technology, Dept. of Agricultural Resource Economics, Kenya*
[2]*Hochschule Weihenstephan-Triesdorf University of Applied Sciences, Germany*
[3]*Leibniz Universität Hannover, Institute for Environmental Economics and World Trade, Germany*

Forest extraction plays a vital role in the livelihoods of millions of poor rural households. Understanding the nature of forest dependence and determinants of forest extraction can prevent livelihood strains induced by forest degradation. This paper assesses the contribution, the nature, extend and determinants of forest extraction among rural households in Kenya. Data were collected from a random sample of 924 households in Bungoma and Trans-Nzoia counties in western Kenya. The level of forest extraction was analysed descriptively and measured in terms of the value of products extracted while a two-step Heckman model was used to assess the determinants of forest extraction. The results show that forest extraction is an important livelihood source for 48.9 % of the respondents with other key sources being agriculture (84.6%), formal and informal employment (40.4 %) and commercial business (9.5 %). The results further show that participation in forest extraction was generally higher for households with lower asset value, higher membership and headed by males. The main products extracted were firewood (61.1 %), food products such as wild fruits, vegetables and honey (31.8 %) and herbal medicine (4.2 %). The econometric results show that age of household head ($P = 0.048$), proximity to all weather roads ($P = 0.002$), and access to credit ($P = 0.007$) negatively influenced forest extraction decision while proximity to the market ($P = 0.000$), household size ($P = 0.073$) and membership in a forest user group ($P = 0.039$) had a positive influence. The level of extraction was negatively influenced by the age of the household head ($P = 0.000$), proximity to all weather roads ($P = 0.003$), membership in a farmer group ($P = 0.063$) and value of assets (P=0.068) while proximity to market ($P = 0.000$), household size ($P = 0.000$), membership in a forest user group ($P = 0.003$), value of shocks ($P = 0.012$) and engagement in farming ($P = 0.016$) positively influenced the extent of forest extraction. The study findings indicate that households in Mt Elgon rely heavily on subsistence products such as firewood and wild vegetables suggesting the need to promote alternative sources of energy such as energy saving stoves (*jikos*) and alternative livelihoods.

Keywords: Degradation, forest extraction, Heckman model, Kenya, livelihoods

Contact Address: Purity Yego, Jomo Kenyatta University of Agriculture and Technology, Dept. of Agricultural Resource Economics, 00100 Nairobi, Kenya, e-mail: yegopurity.c@gmail.com

Shocks and Coping Strategies of Forest-Dependent Households in Rural Kenya

Marlene Reimer[1], Robert Mbeche[2], Ulrike Grote[1]

[1]*Leibniz University Hannover, Institute for Environmental Economics and World Trade, Germany*

[2]*Jomo Kenyatta University of Agriculture and Technology, Dept. of Agricultural and Resource Economics, Kenya*

Households in rural areas of developing countries such as Kenya are often depending on natural resources that supplement consumption and income-generating activities. Especially forests play an important role in the lives of local communities as they help to sustain their livelihoods. But this dependence on forests induces many challenges. Deforestation, forest degradation and the outcomes of climate change are serious threats to forest-dependent households. In the case of Kenya, approximately 5,000 ha of forest are depleted every year. This development puts pressure onto local households by increasing the exposure to shocks, leading to irreversible asset losses that cause households to fall into poverty. This study therefore aims to assess the occurrence of shocks among these local communities, the welfare impacts as well as the households' coping capacities. Therefore, a descriptive analysis of the occurrence of shocks will be provided. Furthermore, the study includes an impact model which will assess the impact of shocks on households' welfare by using Propensity Score Matching. And lastly, an adaption model will be introduced to determine factors that influence the coping behaviour of rural households by using probit regressions. The analysis is based on cross-sectional data that was collected from 924 households in the region of Mt Elgon, Kenya, in November 2018. The survey addresses the effects of market-based incentives on forest conservation and development in rural areas of Kenya. Preliminary results suggest that almost every household had to face at least one shock during the last five years, the majority of them being agricultural. Especially droughts and floods but also large falls in sale prices for crops were most common. Furthermore, it could be seen that agricultural and social shocks were mainly counteracted with behaviour-based strategies while households who faced economic and health shocks often relied on asset-based strategies. But roughly 40 % of the shocks were not counteracted at all. This underlines the importance of further understanding of shocks and coping strategies. Until these risks are identified and eradicated, the vicious cycle of vulnerability will continue to diminish people's wellbeing.

Keywords: Coping strategies, forest-dependence, Kenya, shocks, vulnerability

Contact Address: Marlene Reimer, Leibniz University Hannover, Institute for Environmental Economics and World Trade, Schwester-Hildebranda-Straße 14, 49413 Dinklage, Germany, e-mail: marlene@reimer-dinklage.de

Maximizing Landscape Restoration to the Benefit of Smallholder Farmers in Forest Agro-Ecological Zones in Ghana

PATRICK OPOKU[1], DOROTHY ASARE AKOTO[2]

[1] *Deutsche Gesellschaft für Internationale Zusammenarbeit, Strengthening Advisory Capacities for Land Governance in Africa Programme, Ghana*

[2] *Forestry Research Institute of Ghana, Climate Change Department, Ghana*

Ghana is one of the countries in the world with high rates of deforestation. In fact, Ghana's deforestation rate, which stands at 2 % per annum, is higher than the average decline for Africa and West Africa, which are 0.62 % annual and 1.17 % annual, respectively. Now there is a strong emphasis on landscape restoration. Taking a cue from the cocoa sector in Ghana where the bulk of Ghana's enviable world record in cocoa production is planted by smallholder famers, it has been argued that plantation development and indeed landscape restoration can better be served by small-scale farmers as against large-scale developers. It is against these backgrounds that this study was conducted to analyse factors that affect the adoption and non-adoption of landscape restoration schemes by smallholder farmers. The study also identified best governance practices for enhancing landscape restoration. Data for the study was collected over a period of three months in two communities in the forest transitional zone of Ghana using mixed methods. The results showed that smallholder farmers genuinely care about tree planting, but are challenged by factors such as pest and diseases, wildfires, lack of capital and high transaction cost but low returns as factors affecting the adoption of plantation in Ghana. Moreover, respondents prioritised efficiency (efficient use of financial resources), equity (equal access to land by all persons) and effectiveness as the three most important governance issues to be considered in smallholder forest plantation development. Policy makers in Ghana should consider a mix of recommendations including giving farmers access to credits and markets; pest and disease control and good governance to trigger the needed participation in landscape restoration schemes by smallholder farmers in Ghana.

Keywords: Agro-ecological zone, landscape restoration, smallholders

Contact Address: Patrick Opoku, Deutsche Gesellschaft für Internationale Zusammenarbeit, Strengthening Advisory Capacities for Land Governance in Africa Programme, 7 Volta Street, P.O. Box KA 9698 Accra, Ghana, e-mail: opoku_patrick@yahoo.com

The Impact of ICRAF Tree Germplasm Distributions: *Calliandra calothyrsus* and *Gliricidia sepium* in Kenya

KAVENGI KITONGA[1], ALICE MUCHUGI[2], NELISSA JAMORA[3], MELINDA SMALE[4]

[1]*Crop Trust, Kenya*

[2]*World Agroforestry Centre (ICRAF), Genetic Resources Unit, Kenya*

[3]*Crop Trust, Germany*

[4]*Michigan State University, Agricultural, Food and Resources Economics, USA*

Tree-based production systems provide numerous ecosystem services that are important in sustaining the life of vast plant and animal populations. Although the optimisation of these services is contingent on sufficient tree diversity, drivers of change such as population pressure and related agricultural expansion have significantly contributed to tree diversity loss, with adverse consequences on ecosystem functionality. The realisation of this fact has led to significant efforts to conserve tree genetic diversity, in which the International Centre for Research in Agroforestry (ICRAF) plays a key role. This study investigates the impacts from the use of the two most popular species sourced from the ICRAF genebank, namely, *Calliandra calothyrsus* and *Gliricidia sepium*, among smallholder farmers. The study also examines the factors affecting agroforestry adoption, given the limited uptake of agroforestry interventions. A user survey was used to investigate the impacts of use, whereas Key Informant Discussions were employed to investigate constraints to adoption. Concerning constraints to adoption, we find that limited access to high quality germplasm, poor technical skills in producing high quality germplasm, and inadequate market incentives to produce high quality germplasm have indeed presented serious impediments to the scaling up of agroforestry interventions. We also find that fodder-specific tree aspects have contributed to differential success rates of adoption among the two fodder trees. Adoption of *Gliricidia sepium*, as compared to *Calliandra calothyrsus*, has been constrained by challenges associated with viability, palatability and non-immediate economic return. Concerning the impacts of use, we find that improved food security and incomes, increased milk production, and reduced vulnerability to drought were identified as the main benefits linked to the use of *Calliandra calothyrsus*. Improved food security, higher incomes and enhanced soil fertility were cited as the main use impacts associated with *Gliricidia sepium*. The findings demonstrate the important role of the genebank in conserving and distributing unique, high quality germplasm.

Keywords: Genebank, impacts, tree diversity

Contact Address: Kavengi Kitonga, Crop Trust, Nairobi, Kenya, e-mail: kavengik@gmail.com

Gendered and Age – Related Priorities for Food Tree Species to Address Food Security, Nutrition and Livelihoods: Participatory Research in Kitui County, Kenya

AGNES GACHUIRI, ANA MARIA PAEZ-VALENCIA, STEPHA MCMULLIN
World Agroforestry Centre (ICRAF), Kenya

Food trees have an important role to play in local food systems, they produce a variety of nutrient dense foods including fruits, leafy vegetables, nuts, seeds and edible oils. These foods can increase the nutritional quality of local diets, due to their micronutrients and seasonal availability. To better understand the importance and adoption of these trees in local food systems, gendered and age-related farmer preferences and priorities for species were investigated in Kitui County, Kenya. Participatory research was conducted in four villages, with 80 community representatives, 40 women and men each in eight Focus group discussions. Seasonality calendars were used for mapping of tree foods availability and pebble game was used for score -ranking preferred food trees and their functional uses. A total of 49 food trees species were listed, of which 65 % were exotics and could be used to fill food and nutrition gaps during food insecure months. In general, there was a high preference for exotic species, such as *Mangifera indica*, *Persea americana*, *Carica papaya* and *Citrus sinensis*. Knowledge on food tree species differed by gender and age, with older women listing the greatest number of species and indicating a higher preference for indigenous food trees compared to their male counterparts, citing the importance of these species for food, firewood and medicinal provisioning, however, with little market value. Older women preferred species such as *C. papaya*, *Vitex payos* and *Psidium guajava*, which are sold for income only in small quantities. In contrast, men prioritised food tree species potential economic value and other uses. Despite farmer preferences, challenges such as lack of availability of seedlings of improved varieties, prolonged droughts and scarcity of land. These constraints were gendered as well, with more younger women mentioning lack of knowledge about planting and management and socio-cultural restrictions e.g only having access to land when married; whereas younger men indicated the challenges of pests, limited markets,land scarcity and ownership. Understanding gendered and age-related priorities for food tree species, functional uses and planting in general, can support the development of preferred tree based agricultural interventions to address food security, nutrition and livelihoods

Keywords: Food security, food trees, gender, participatory research, priority setting

Contact Address: Agnes Gachuiri, World Agroforestry Centre (ICRAF), United Nations Avenue Gigiri, Nairobi, Kenya, e-mail: a.gachuiri@cgiar.org

Financial Profitability of *Gliricidia* Agroforestry Technologies in Dryland Areas of Dodoma Region, Tanzania

MARTHA SWAMILA[1], DAMAS PHILIP[2], ADAM MESHACK AKYOO[2], ANTHONY KIMARO[1], STEFAN SIEBER[3], TODD ROSENSTOCK[1]

[1] *World Agroforestry Centre (ICRAF), Tanzania Country Programme, Tanzania*

[2] *Sokoine University of Agriculture, Agricultural Economics and Agribussines, Tanzania*

[3] *Leibniz Centre for Agric. Landscape Res. (ZALF), PB 2, AG SUSLAND, Germany*

Optimal productivity of field crops in most dryland areas of sub-Saharan Africa including Central Tanzania is limited by declining soil fertility.The productivity of most crops, for example maize, in this regions is between 1–1.5 t ha^{-1} contributing to high food poverty levels of between 35.5 % and 51.4 % as well as deficiency of essential nutrients. As a way to deal with this problem, an extensive collaborative research undertaken by the World Agroforestry Centre (ICRAF) and local research institutions, led the development of *Gliricidia* agroforestry based soil fertility management technologies in dryland areas of Dodoma region as sustainable options to assist smallholder farmers with limited access to mineral fertilisers replenish their soils within a short time. While performance of gliricidia agroforestry technologies on crop productivity has been demonstrated, information on financial profitability of *Gliricidia* agroforestry technologies relative to conventional soil fertility management practices is lacking. Using farm budget models, this study was conducted to estimate financial profitability of gliricidia agroforestry based and conventional soil fertility management technologies (pigeon pea intercropping) in dryland areas of Dodoma region, Tanzania. Results show significant heterogeneity in profitability indicators with *Gliricidia*-maize-pigeon pea intercropping technology exhibiting the highest benefit-cost ratio (3.9) and fertilised maize monoculture exhibiting the lowest Benefit Cost Ratio (1.5). The returns to labour per person day was \$2.5 in *Gliricidia*, maize and pigeon intercropping, \$1.9 in maize-pigeon pea intercropping and \$ 0.2 in unfertilised maize monoculture. Therefore, it is concluded that, *Gliricidia* intercropping technology is more profitable than conventional soil fertility management technologies. The use of this technology could contribute to increased crop productivity and food security for poor resourced smallholder farmers in dryland areas of Dodoma region, Tanzania.

Keywords: Agroforestry, benefit-cost analysis, food security, land management, soil fertility

Contact Address: Martha Swamila, World Agroforestry Centre (ICRAF), Tanzania Country Programme, Dar Es Salaam, Tanzania, e-mail: marthaswamila@yahoo.com

The Role of Agroforestry in Sustainable Intensification of Cocoa Growing Systems Across West Africa: A Review

ABIGAIL TETTEY, RICHARD ASARE
International Institute of Tropical Agriculture (IITA), Ghana

Cocoa (*Theobroma cacao* L.) is one of the major smallholder cash crops in West Africa, whose productivity has stagnated at around 300-600 kg ha $^{-1}$ in the past decades. This has been attributed to poor soil fertility management, poor planting materials, varying degrees of pests and diseases incidence, and climatic pressures such as rises in atmospheric temperatures and rainfall variability. Improving productivity has cost West Africa about 2.3 million hectares of forest in the last three decades. Thus, the need for intensified production systems that improve productivity and at the same time promotes environmental integrity. Sustainable cocoa intensification involves a new vision of the cocoa economy whereby the crop is grown with the objective of increasing productivity while at the same time ensuring sustainability by protecting the environment. This work reviews the extent of cocoa agroforestry systems in Côte d'Ivoire, Ghana, Cameroon and Nigeria from 1980 to date. We considered 76 peer-reviewed articles related to shading and farm diversification in cocoa systems in the four countries. Literature search was done in search engines including Google Scholar, ResearchGate, ScienceDirect, Wiley Online Library and Springer Link. The nature of cocoa agroforestry strategies employed across these countries highlights on agronomic, economic and environmental benefits. Cocoa productivity continues to depend on increase in the production area with the use of agrochemicals, and the quest to reduce deforestation and conserve the environment have opposing goals. To enhance cocoa productivity sustainably, sector players are faced with trade-offs between yield increases in monocultures and environmental conservation in cocoa agroforest systems. Thus, ordinary intensification could decrease environmental resilience, making an adaptation of sustainable cocoa production practices vital. However, there is a need to consider understorey management and the composition of the agroforest structure. The disagreements on the technical characteristics of a cocoa agroforest and the lack of specific training for farmers is also of concern. We, therefore, suggest that cocoa agroforest system needs to take landscape and socio-economic issues into consideration and that extension programs should be tailored to encourage the adoption of this technique for sustainable cocoa production.

Keywords: Climate change, cocoa agroforests, cocoa productivity, deforestation, environmental integrity

---

Contact Address: Abigail Tettey, International Institute of Tropical Agriculture (IITA), Agostinho Neto Road Airport Residential Area, Accra, Ghana, e-mail: a.tettey@cgiar.org

Agroforestry Systems in Mozambique as Part of a South-South Cooperation Project

SIMONE P. FAVARO[1], CESAR MIRANDA[1], IVETE F. MALULEQUE[2], VANIA J. COSSA[2], FERNANDO M.A. FLUGENCIA[2], ARISTIDES C. MAMBA[2], ARNALDO M. JAMAL[2], LEONARDO D. PIMENTEL[3]

[1]*Empresa Brasileira de Pesquisa Agropecuária (EMBRAPA), Brazil*

[2]*Mozambican Institute for Agricultural Research, Mozambique*

[3]*Federal University of Viçosa, Brazil*

It is reported the elaboration process and field performance of a triangular-type international South-South technical cooperation partnership encompassing Brazil, Mozambique and donor agents from the northern hemisphere. The project was part of The Agricultural Innovation MKTPlaceplatform, coordinated by Embrapa, in partnership with Brazilian Federal University of Viçosa, MS Foundation, and the Mozambican Institute for Agricultural Research (IIAM). The framework of the evaluated technology was based on Agroforestry Systems envisaging the integration of both food and biomass for renewable fuels production as a driver to sustainable development of local family agriculture, by increasing availability of foodstuff and wood fuel sources for self-consumption and income. In such a system, it is expected improvements of soil properties and increasing awarenes towards conservation agriculture practices. The project was executed around the City of Nampula, Nampula Province, Northeast of Mozambique, in two sites: a local IIAM (on station) area, and an area belonging to a Small Farmers Association Mapwane, in Anchilo Community (on farm trials). Partners get together in the build-up of the proposal and all activities related to fieldwork. In the first two years of the project, main difficulties were legal aspects for a planned transference of Macauba palm (*Acrocomia aculeata*) genetic material from Brazil to Mozambique; low engagement of participating farmers, despite initial goodwill; trials with tree species with a long development cycle; lack of commitment from some initial partners; the instability of climate conditions. On the other hand, positive achievements of the project for both farmers and local technical team were the participative construction of the project, which resulted in a good exchange of experiences; good integration among the teams during fieldwork; female presence in the leadership and execution of the project; legacy of the implemented agroforestry systems and the building up of knowledge on conservation agriculture among the farmers. In conclusion, the cooperation was effective to strengthen ties between these nations of the South-South axis, effective exchange of experience among technical partners, as well as showing in the practice the innovative potential of agroforestry systems for the improvement of small farmer's livelihood.

Keywords: Biofuels, food production, small farm agriculture, sustainability

Contact Address: Simone P. Favaro, Empresa Brasileira de Pesquisa Agropecuária (EMBRAPA), Brasília, Brazil, e-mail: simone.favaro@embrapa.br

The (in)visible Components of Participatory Action Research

HEITOR MANCINI TEIXEIRA[1,5], MARGRIET GORIS[2,3], LEONARDO VAN DEN BERG[3,2], LUCAS CARVALHO GOMES[4,5], FELIX BIANCHI[4], MARIELOS PEÑA-CLAROS[1], IRENE MARIA CARDOSO[5]

[1] *Wageningen University and Research, Farming Systems Ecology (FSE), The Netherlands*
[2] *Wageningen University and Research, Forest and Nature Conservation Policy Group, The Netherlands*
[3] *Universidade Federal de Viçosa, Departamento de Economia, Brazil*
[4] *Wageningen University and Research, Farming Systems Ecology Group, The Netherlands*
[5] *Universidade Federal de Viçosa, Departamento de Solos, Brazil*

Scientists are increasingly recognising the importance of transdisciplinary collaboration as well as the incorporation of different worldviews in research. Such components are claimed to be crucial in scientific studies attempting to assess and design complex social ecological systems capable to provide multiple ecosystem services and conservation of natural resources. Despite the importance to engage local actors as co-creators in the research process using transdisciplinary and participatory approaches, few studies develop and assess effective methodologies for doing so. Research focus is often on data collection and interpretation, while social values and strategies to approach local actors and build a relationship of trust and collaboration are often overlooked. In this paper, we discuss participatory action research as an approach for engaging with local actors and to develop effective strategies for building more sustainable and resilient agri-food systems. Based on a case study in Brazil, we highlight six main components that are crucial for implementing participatory action research: (i) Collective definition of research questions; (ii) Participatory methodologies for building scientific knowledge; (iii) Sharing and discussing research results with local actors (iv); Integration between research and education; (v) Strengthening capacity building and interdisciplinary team work; and (vi) Social engagement with farmers. Our findings shows that building a team of researchers with different theoretical backgrounds and learning styles is important for effectively engaging with farmers and their organisations, and making research outcomes more relevant for society. The use of participatory methodologies is crucial not only to make this process possible, but also to generate valuable scientific data. However, engaging with farmers and different knowledge disciplines requires extra effort and time from researchers, which is not always valued or recognised by academia. Yet, participatory research processes can be facilitated by long-term local networks involving organisations such as universities, NGO's and farmer's cooperatives, associations and movements.

Keywords: Bottom-up, local actors, social network, social-ecological systems, transdisciplinary

Contact Address: Heitor Mancini Teixeira, Wageningen University and Research, Farming Systems Ecology (FSE), Wageningen, The Netherlands, e-mail: heitor.manciniteixeira@wur.nl

Agroforestry - a trendsetting and long-lasting technology

Oral Presentations

Posters

Reducing Forest Degradation: Energy Self-Sufficiency for Smallholder Farmers via Agroforestry Systems in Semi-Arid Tanzania

JOHANNES HAFNER[1], GÖTZ UCKERT[1], HARRY HOFFMANN[1], STEFAN SIEBER[1], ANTHONY KIMARO[2]

[1]*Leibniz Centre for Agric. Landscape Res. (ZALF), PB 2, AG SUSLAND, Germany*
[2]*World Agroforestry Centre (ICRAF), Tanzania Country Programme, Tanzania*

High dependency on firewood to meet households' cooking energy demand is one driver of forest degradation in rural Tanzania. In order to be energy-independent from external firewood sources small-scale farmers may take measures such as on-farm firewood supply as well as the use of energy-efficient cooking to balance domestic firewood consumption. We used the "controlled cooking test" protocol to determine the firewood consumption of three-stone-fire stoves and energy-efficient improved cooking stoves. We assessed the on-farm firewood production potential of *Gliricidia sepium*. In total we analysed 3 blocks with two intercropping treatments each with a plot size of 256 m^2: maize and *G. sepium* (treatment 1), and maize, pigeonpea and *G. sepium* (treatment 2). The *G. sepium* shrubs were planted in 2014 with a spacing of 4 x 4 m^2. For treatment 1, maize and *G. sepium* the wood biomass production was 36.7 kg (SD 12.3) in 2018, and 129.7 kg (SD 71.4) per 256 m^2 in 2019. For treatment 2, the wood biomass production was 19.7 kg (SD 1.6) in 2018, and 70.0 kg (SD 15.2) per 256 m^2 in 2019. The extrapolated data showed a firewood production potential of treatment 1 of 1432.7 kg, and of 767.8 kg for treatment 2 in 2018; in 2019 the potential area 5066.1 kg and 2731.8 kg per hectare.
The annual *G. sepium* firewood consumption of a five-head household to meet its cooking energy demand is 1298 kg with improved cooking stoves and 1815 kg with three-stone fire stoves. To determine the firewood self-sufficiency rate, we calculated the ratio of firewood production potential and consumption. In 2018, 88.3 % (treatment 1) and 47.3 % (treatment 2) with improved cooking stoves and 63.1 % (treatment 1) and 33.8 % (treatment 2) with three-stone-fire stoves of households' cooking energy demand could be covered by firewood from intercropped *G. sepium*. In 2019, the on-farm firewood production surpassed the demand: 312.2 % (treatment 1) and 168.4 % (treatment 2) with improved cooking stoves as well as 223.3 % (treatment 1) and 120.4 % (treatment 2) with three-stone-firewood of the households' cooking energy demand could be covered.

Keywords: Energy self-sufficiency, *Gliricidia sepium*, improved cooking stoves, on-farm firewood, semi-arid Tanzania, three-stone-fire stoves

---

Contact Address: Johannes Hafner, Humboldt-Universität zu Berlin, Albrecht Daniel Thaer-Institute of Agricultural and Horticultural Sciences (ADTI), Invalidenstraße 42, 10099 Berlin, Germany, e-mail: johanneshafner@gmx.net

Superfruit in the Niche - Underutilised Sea Buckthorn in Gilgit-Baltistan, Pakistan

MUHAMMAD ARSLAN NAWAZ[1], ASIF ALI KHAN[2], USMAN KHALID[3], ANDREAS BUERKERT[1], MARTIN WIEHLE[4]

[1]*University of Kassel, Organic Plant Production and Agroecosystems Research in the Tropics and Subtropics, Germany*

[2]*Muhammad Nawaz Shareef University of Agriculture, Plant Breeding and Genetics, Pakistan*

[3]*Information Technology University, School of Business and Management Sci., Pakistan*

[4]*University of Kassel, Tropenzentrum / ICDD / Organic Plant Production and Agroecosystems Research in the Tropics and Subtropics (OPATS), Germany*

Sea buckthorn is a medicinal plant with nutritious fruits, occurring throughout the temperate regions of the northern hemisphere. Given their nutritional properties considered as a "superfood", the berries have a large international market potential, particularly in China and Europe. Although sea buckthorn is widely spread in northern Pakistan, it is a neglected species there, as collection and drying of sea buckthorn berries is only performed at a rudimentary scale. Marketing is severely hampered by low raw product quality, varying price margins, and low local demand. This study provides comprehensive information about the current situation from collection to post harvest management of sea buckthorn fruits including the analysis of vitamin C under different sun and shade drying conditions. During 2017–2018 a total of 111 collectors and 17 commission agents were interviewed throughout Gilgit-Baltistan, Pakistan using semi-structured questionnaires. Sale's prices were low for the collectors (1.82 US$ kg^{-1}) leading to the fact that mostly poor households are involved in harvest and sale. Traditional sun drying, and storage conditions were inappropriate resulting in a decrease of chemical fruit quality: vitamin C concentration was only 50 % in sun dried as compared to shade dried fruits, which affects sale prices for most collectors (87 %). Only 16 % of the households were involved in value addition of sea buckthorn berries. Annual subsistence through by-product development was more rewarding (378 US$) than berry sale only (181 US$). Based on supply chain analyses, the non-coordination among actors and missing infrastructure pose major challenges affecting the efficiency of the local businesses at large. The study also showed an urgent need to set appropriate food quality standards, to increase communication among stakeholders, and to intensify training offers especially for collectors.

Keywords: Collection, drying, *Hippophae rhamnoides*, non-regulated price, post harvest handling, vitamin C

Contact Address: Martin Wiehle, University of Kassel, Tropenzentrum / ICDD / Organic Plant Production and Agroecosystems Research in the Tropics and Subtropics (OPATS), Steinstraße 19, 37213 Witzenhausen, Germany, e-mail: wiehle@uni-kassel.de

Promoting Agroforestry through Land and Benefit Sharing between Cocoa Mono-Croppers and Fruit Tree Producers (Cameroon)

Achille Jean Jaza Folefack[1], Dietrich Darr[2]

[1]*University of Dschang, Dept. of Agricultural Economics, Cameroon*

[2]*Rhine-Waal University of Applied Sciences, Faculty of Life Sciences, Germany*

Given higher yields compared to traditional cocoa production, a significant share of cocoa plantations in the Centre region of Cameroon have been converted to full-sun production systems in the past. However, compared to such systems the cultivation of cocoa under shade in agroforestry arrangements is associated with higher environmental benefits. Cocoa agroforestry also creates additional income from eco-certification programs such as Rainforest alliance or UTZ. While separated ownership of land and trees is common in this region, owners of cocoa plantations are hesitant to allow cultivation of fruit trees inside the cocoa farms as they are afraid of losing while sharing their land. This represents a significant limitation to the diffusion of cocoa agroforestry in Cameroon. Using the Coase Theorem approach, this study aims to compute the welfare gain and loss for cocoa mono-croppers and fruit producers as a function of the number of fruit trees planted on the shared land. From times series data collected in 15 villages between 2008 and 2019, we selected 215 cocoa farms, which integrated into their cocoa orchards the fruit tree species of safout (S; *Dacryodes edulis*), mango (M; *Irvingia gabonensis*) and/or ndjansang (N; *Ricinodendron heudelotii*). The results showed that, as the density of fruit trees increased, cocoa yield decreased and so did the cocoa mono-croppers' farm income, while fruit producers gained higher revenue. The socially efficient solution which satisfied both parties was achieved by growing 1111 cocoa plants ha^{-1} and 150 fruit trees ha^{-1}, resulting in a yield of 393 kg ha^{-1} for cocoa (C), 2580 kg ha^{-1} for S, 3200 kg ha^{-1} for M and 1175 kg ha^{-1} for N. In a C+S association for instance, cocoa croppers recorded revenue of 393,000 FCFA ha^{-1} (as compared to 600,000 FCFA ha^{-1} under mono-culture) versus 719,000 FCFA ha^{-1} for fruit tree producers. Revenues were increased in agroforestry systems composed of two or three fruit tree species. The paper discusses policy options to promote cocoa agroforestry while satisfying the interests of both cocoa mono-croppers and tree growers.

Keywords: Coase theorem, cocoa agroforestry, eco-agriculture, eco-certification, fruit trees, land sharing.

Contact Address: Achille Jean Jaza Folefack, University of Dschang, Dept. of Agricultural Economics, P.O. Box 222, Dschang, Cameroon, e-mail: ajazafol@yahoo.fr

Taking Stock of Carbon along a Land-Use Gradient in the Highlands of Java, Indonesia

JOEP VAN DER POEL[1], SARA ANDREA VACA SANCHEZ[2], AXEL MENTLER[2], KATHARINA KEIBLINGER[2], HATMA SURYATMOJO[3], KRISTIN BANYU RISANG HOBO[3], LIDIA SITUMORANG[3], NGADISIH NGADISIH[4], RIMA DARMAWANTI[4], REINHARD REBERNIG[5], HERMINE ZIMMERMANN[5], OLIVER GARDI[1], LINDSEY NORGROVE[1]

[1]*Bern University of Applied Sciences (BFH), School of Agricultural, Forest and Food Sciences (HAFL), Switzerland*

[2]*University of Natural Resources and Life Sciences Vienna (BOKU), Inst. of Soil Research, Dept. of Forest and Soil Sciences, Austria*

[3]*Universitas Gadjah Mada (UGM Yogyakarta), Fac. of Forestry, Dept. of Forest Resource Conservation, Indonesia*

[4]*Gadjah Mada University (UGM Yogyakarta), Fac. of Agricultural Technology, Dept. of Agricultural and Biosystem Engineering, Indonesia*

[5]*University of Natural Resources and Life Sciences Vienna (BOKU), Centre for Development Research (CDR), Austria*

Retaining or adding trees to cropping systems is widely assumed to increase carbon (C) stocks by building biomass, adding litter to soil, providing a stable microclimate, enhancing soil biological and physical properties and increasing temporal stability. Agroforestry may also provide landscape-level services such as watershed and biodiversity conservation and, on steep slopes in high rainfall areas, reduce the risk of soil erosion and landslides. Yet trees can compete with crops for space, nutrients and water and such trade-offs might be acute where population density is high, such as in Java Indonesia, where densities can exceed 900 km^{-2}.

We hypothesised that soil C stocks are higher in cropping systems where trees are included, and in abandoned tree regrowth. We selected two villages, Leksana and Penanggungan, in the Central Javan highlands (>1000 m a.s.l.). Here, maize, potatoes, vegetables and other crops are farmed on Nitisols, under a range of intensification scenarios, including or excluding trees. We compared cropped fields (i) with trees; (ii) without trees; (iii) without trees and with intensive pesticide and fertiliser use; and (iv) abandoned tree regrowth. On five fields of each type, we assessed tree vegetation (>2 cm diameter at breast height), soil C and bulk density (0–10 and 10–30 cm depths).

Trees on cropped fields were commonly two nitrogen-fixers, the native *Paraserianthes falcataria*, averaging 270 stems ha^{-1} and the exotic *Calliandra calothyrsus*, combined with both native and exotic fruit and timber trees. Mean aboveground C in trees was 6.3 Mg C ha^{-1}. In the abandoned regrowth, *P. falcataria* densities were lower

Contact Address: Joep Van der Poel, Bern University of Applied Sciences (BFH), School of Agricultural, Forest and Food Sciences (HAFL), Bern, Switzerland, e-mail: joep.vanderpoel@students.bfh.ch

(222 ha^{-1}) while densities of *C. calothyrsus* exceeded 11,000 stems ha^{-1} and above-ground C stocks were 60.4 Mg C ha^{-1}. Contrary to our hypothesis, soil C concentrations were high (5.1–7.9 % at 0–10 cm depth) and neither concentrations nor stocks were significantly different between any of the plot categories, ranging from 71–89 Mg C ha^{-1} to 30 cm depth. This indicates either that factors other than including trees have a stronger influence on soil C or more time is required to detect differences in such C-rich soils.

Keywords: Agroforestry, carbon stocks, Java, land-use gradient, space-for-time-substitution

Coffee Fruit Load Dynamics in Open and Shaded Systems Along an Altitudinal Gradient

ALEJANDRA SARMIENTO SOLER[1], PHILIPPE VAAST[2], MUNIR HOFFMANN[3], LAURENCE JASSOGNE[4], PIET VAN ASTEN[5,4], SOPHIE GRAEFE[6], MARIAN KOCH[1], REIMUND P. RÖTTER[1]

[1]*University of Goettingen, Dept. of Crop Sciences; Tropical Plant Production and Agricultural Systems Modelling (TROPAGS), Germany*

[2]*CIRAD & ICRAF, France*

[3]*Leibniz Centre for Agric. Landscape Res. (ZALF), Landscape Res. Synthesis, Germany*

[4]*International Institute of Tropical Agriculture (IITA), Uganda*

[5]*Olam International Ltd, Uganda*

[6]*University of Kassel, Organic Plant Production and Agroecosystems Research in the Tropics and Subtropics (OPATS), Germany*

It is well known that coffee yield is sensitive to climatic variations, in particular extended dry spells and extreme temperatures during the fruit development phase. Shading coffee is promoted as a management practice to reduce fruit drop by buffering extreme temperature and improving soil moisture. We investigated the effect of three coffee cultivation systems, i.e. Coffee-Open (CO), Coffee-Banana (CB) and Coffee-Shade tree (CT), along an altitude gradient on (a) microclimate (temperature, relative humidity) and soil water content, (b) fruit drop and branch die-back over two years of production (2015 and 2016). We monitored coffee fruit development on 810 coffee trees in 27 plots (30 coffee trees/plot), which were distributed homogeneously along an altitude gradient (1100–2100 m a.s.l.). Additionally, we recorded temperature and relative humidity in 18 of the 27 plots, and soil water content in 16 of the 27 plots. We found a lower temperature amplitude and lower maximum temperature in shaded systems, confirming the notion that shade improves microclimate for coffee grown beneath it. While, soil moisture did not differ across cultivation systems. CO had the largest fruit drop (56 ± 17 %), followed by CB (46 ± 10 %) and CT (36 ± 10 %) averaged for 2015 and 2016. In the mean CB experienced the largest fruit load per stem at harvest, as a result of intermediate fruit initiation and intermediate fruit drop along the production cycle. Moreover, differences in fruit load across systems varied along the altitude gradient. At low altitude, CT had lower fruit loads than CO and CB, while at high altitudes there were no differences in fruit load across systems. Our results indicated that 1) improved understanding of the cropping systems effect on coffee production cycle (fruit initiation and fruit drop) can help to develop shade management schemes which help to increase coffee yield, due to reduced fruit drop and branch die back. That 2) Effect of shade type and intensity in coffee production cycle varies along the altitudinal gradient, thus shade management strategies need to be adjusted to the prevailing environmental conditions.

Keywords: Altitude, climate change, *Coffea arabica*, fruit drop, fruit load dynamics, shade

Contact Address: Alejandra Sarmiento Soler, University of Goettingen, Dept. of Crop Sciences; Tropical Plant Production and Agricultural Systems Modelling (TROPAGS), Göttingen, Germany, e-mail: asarmie@gwdg.de

Reassessing the Role of Remnant Trees in Tropical Swidden Systems

ATA DAVATGAR, LINDSEY NORGROVE

Bern University of Applied Sciences, School of Agricultural, Forest and Food Sciences, Switzerland

In shifting cultivation (swidden), forest plots are cleared, cultivated for a short time then abandoned to fallow. In many parts of the tropics, plots are not clear-felled but certain trees are retained, providing services and also storing carbon. Carbon emission in such cyclical systems is less than that associated with permanent conversion practices as carbon is sequestered in the following fallow phase. The time averaged carbon of a swidden system is often estimated as half of the maximum carbon stocks at the end of the fallow phase and assumes negligible carbon stocks at the beginning of the crop phase. We conducted a systematic review to assess reasons why farmers retain remnant trees, where they do so, approximately what is the quantity of carbon retained in these trees and how does this alter the time-averaged carbon estimates of swidden cultivation based on clear-felling. Data from 103 publications in three agroecozones (humid, montane and dry forest) were included. We then combined our findings with data on the extent of swidden in ten countries in the tropics to estimate how much carbon in remnant trees is unaccounted for. The most commonly reported reasons for tree retention included future timber or fuelwood (n=34), use of fruits and leaves for food (n=24), use of other tree components for medical uses (n=22) and shade for other crops and workers on farm (n=21). Other reasons cited included soil fertility maintenance, seed flow under trees, supporting biodiversity, source of livestock forage, physical support for other crops such as yam vines in some areas. Some studies stated that due to the scarcity of tool such as chainsaw farmers left some hardwood species in field. We calculated that from ten countries alone, approximately 3 Pg carbon stored in remnant trees is unaccounted for so the global contribution of remnant trees in shifting cultivation to the "missing terrestrial carbon stock" would far exceed this.

Keywords: Remnant trees, shifting cultivation, time averaged carbon

Contact Address: Lindsey Norgrove, Bern University of Applied Sciences, School of Agricultural, Forest and Food Sciences, Länggasse 85, 3052 Zollikofen, Switzerland, e-mail: lindsey.norgrove@bfh.ch

Intercropping *Gliricidia sepium* Improves Crops Production and Agroecosystem Resilience in Kongwa District, Dodoma, Tanzania

ANTHONY KIMARO[1], JOHANNES HAFNER[2], GÖTZ UCKERT[3], STEFAN SIEBER[4]

[1] *World Agroforestry Centre (ICRAF), Tanzania Country Programme, Tanzania*

[2] *Humboldt-Universität zu Berlin, Albrecht Daniel Thaer-Institute of Agricultural and Horticultural Sciences (ADTI), Germany*

[3] *Leibniz Centre for Agric. Landscape Res. (ZALF), Research Area Land Use and Governance, Germany*

[4] *Leibniz Centre for Agric. Landscape Res. (ZALF), PB 2, AG SUSLAND, Germany*

Extensive grazing and the use of crop residues for cooking energy are major drivers of land degradation in semiarid zones. Poor soil health and unreliable precipitations in these areas also limit crop production and increase susceptibility to climate change. The impacts of climate change in agriculture are aggravated further by poor farming practices, including low use of inputs, especially mineral fertiliser. This study evaluated productivity and resilience effects of *Gliricidia sepium*-based intercropping systems. The split-plot design was adopted to test the effects of cropping system (sole maize, sole pigeonpea, maize-pigeonpea, maize-*G. sepium*, and maize-pigeonpea-*G. sepium*) and fertilisation (fertilised and unfertilised). Intercropping maize with *G. sepium* and pigeonpea improved crops (maize and pigeonpea) yield by up to 33 %, besides wood and fodder supply for improved livestock nutrition. Relative to maize monoculture, intercropping *G. sepium* with maize increase rainwater use efficiency up to 36 % (3.5 kg ha^{-1} mm^{-1}) and it was maximised (9.5 kg ha^{-1} mm^{-1}) during the drier season (2015), reflecting beneficial effects on soil moisture retention and resilience benefits of intercropping *G. sepium*. After 5 years wood produced in the maize-*G. sepium* treatment was 5.1 t ha^{-1}, which was enough to sustain a 5-member family for 3.1 years and 2.2 years when using improved cooking stoves and the 3-stone traditional firewood stoves, respectively. Utilizing this amount of wood offset about 1.58 tons of CO_2emissions. The economic benefits of integrating *G. sepium* compared to sole maize were 3- and 4- folds increase in return to labour per day worked and gross margin, respectively. Intercropping *G. sepium* improve crop production, build resilience of farming system and offsetting emissions

Keywords: Agroforestry, climate resilience, CSA, semiarid areas, soil moisture

---

Contact Address: Anthony Kimaro, World Agroforestry Centre (ICRAF), Tanzania Country Programme, P.O. Box 6226, Dar-es-Salaam, Tanzania, e-mail: a.kimaro@cgiar.org

Significance of Traditional Forest Management System in Biodiversity Conservation and Ecosystem Services Provision: Case Study of Osun-Osogbo Sacred Grove, Nigeria

Jonathan C. Onyekwelu[1], Bernd Stimm[2], Reinhard Mosandl[2], Lawal Amadu[1], Adewale O. Agbo-Adediran[1]

[1]*The Federal University of Technology, Dept. of Forestry and Wood Technology, Nigeria*

[2]*Technical University of Munich, Inst. of Silviculture, Center of Life and Food Sci. Weihenstephan, Germany*

Many Nigerian communities have age-long traditional systems of site protection before the advent of conventional forest management in the country. Nigeria has rich indigenous traditions of nature conservation. The role of traditional forest-related knowledge and forest management systems are increasingly being recognised as important for biodiversity conservation. In addition to biodiversity conservation, traditional natural resources management has supported livelihoods for many generations through provision of ecosystem services. This study investigated and compared the biodiversity status of sacred grove, primary and degraded forests with the view of assessing the importance of traditional forest management system in biodiversity conservation and ecosystem services provision. Osun-Osogbo sacred grove, two primary and two degraded forests were involved. Eight 20 m $\times$ 40 m sample plots were systematically laid in each site. In each plot, trees with dbh $\geq$ 10 cm were identified and their dbh measured. Regeneration was assessed within 5 m $\times$ 5 m quadrat. Factors contributing to preservation of sacred groves and provision of ecosystem services were investigated using semi-structured questionnaire. Generally, sacred groves harboured richer biodiversity and played significant role in biodiversity conservation compared to primary and degraded forests. Osun-Osogbo sacred grove had the highest species richness (61), diversity index (3.54), regeneration richness (66 spp.) and high species evenness (0.66). The frequency of endangered tree species in sacred groves was high. Except species evenness, there were significant differences ($p \leq 0.05$) between the diversity indices of the study sites. The better biodiversity indices of Osun-Osogbo grove compared to primary and degraded forests are indications of its importance in biodiversity conservation. The preservation of Osun-Osogbo grove was secured through belief and fear of deity, preservation of culture, place of worship, etc. while factors promoting biodiversity conservation in the grove include preservation of home of deity, tree felling being abomination, etc. Some ecosystem services provided by sacred grove include: healing, protection and peace, tourism, employment provision, spiritual worship, revenue generation. However, the rules and taboos used to preserve the grove are crumbling due to modernisation, advent of Christianity and Islam. This must be addressed if sacred groves are to continue to play their important role in biodiversity conservation.

Keywords: Biodiversity conservation, ecosystem services, Nigeria, primary and degraded forest, sacred groves, traditional forest management

Contact Address: Jonathan C. Onyekwelu, The Federal University of Technology, Dept. of Forestry and Wood Technology, P.M.B. 704, Akure, Nigeria, e-mail: onyekwelujc@yahoo.co.uk

Farmers' Use and Preferences of Trees in Bauchi State, Nigeria

ABUBAKAR YAHAYA TAMA[1], ANNA MANOUROVA[1], RAGHEB KAMAL MOHAMMAD[2], BOHDAN LOJKA[1]

[1] *Czech University of Life Sciences Prague, Fac. of Tropical AgriSciences, Czech Republic*
[2] *Salahaddin University Erbil, College of Agriculture, Iraq*

Trees play an important role to people of Bauchi State in many ways such as providing organic supplement to crop fields and most of the trees have multiple uses in form of productions such as medicine and fruits or services such as windbreaks and shades. Local farmers plant trees for the myriad of these uses which include: providing fuel wood, animal feeds, timber and stakes among other uses. The study aimed to identify the most important use of tree species by the farmers in Bauchi State, Nigeria. We also aimed to identify their preferences of tree species, to assess their abundance, identify agroforestry design and spatial distribution on farms. The fieldwork was conducted from June to September 2018 in both lowland and highland of Bauchi State. In total, 83 farmers were interviewed by semi-structured questionnaires while 52 tree species and their uses were identified in both agroecological zones. The results showed that farmers rely mostly on fruit and medicinal tree species such as *Mangifera indica*, *Citrus spp.*, *Anacardium occidentale*, *Psidium guajava*, *Azadichrata indica*, *Jatropha curcas*, *Moringa oleifera* and *Adansonia digitata*. Besides fruit and medicinal trees, service functions of trees such as fencing, shading, wind breaks play a crucial role. However, using trees for soil improvement was reported only in lowlands. Regarding the species preferences, *Adansonia digitata* has the highest priority among the farmers in the lowland, while *Parkia biglobosa* is the most preferred in the highland. *Moringa oleifera* was the most abundant species in both agroecological zones. In agroforestry systems, scattered trees were the most preferred in both lowland and highland while bush field was the most preferred location of farms in lowland and village field, the most preferred location in highland. To conclude, as no use of timber tree species was cited in both agroecological zones, we suggest farmers should incorporate timber species on their farms, especially *Pterocarpus erinaceous*, an indigenous tree that is known to grow fast and serve both as timber and medicinal species. Policy makers should introduce improved varieties of most preferred species. In the future, more studies have to be conducted on medicinal used trees in the study area as it was the second most mentioned use.

Keywords: Agroforestry, lowlands vs. highlands, multipurpose tree species, savannah, Sudan-Sahel zones

---

Contact Address: Bohdan Lojka, Czech University of Life Sciences Prague, Fac. of Tropical AgriSciences, Dept. of Crop Sciences and Agroforestry, Prague, Czech Republic, e-mail: lojka@ftz.czu.cz

Macaw Palm Mixed Cropping – What Are Suitable Cropping Options?

MELANIE HINDERER[1], CATHERINE MEYER[1], SÉRGIO MOTOIKE[2], GEORG CADISCH[1], THOMAS HILGER[1]

[1]*University of Hohenheim, Inst. of Agric. Sci. in the Tropics (Hans-Ruthenberg-Institute), Germany*

[2]*Federal Univerity of Viçosa, Department of Plant Science, Brazil*

Macaw palm (*Acrocomia sp.*) is an oil-bearing palm that gained a lot of interest in recent years due to its excellent bio-economic potential. Macaw palm has various traditional uses for food, animal feed, fibre and medicine. It is not competing with rainforest and fertile land and therefore a sustainable alternative to the African oil palm (*Elaeis guineensis*). The oils from the fruit pulp and kernel increased the importance of macaw palm because of their high nutritive value but also as source of biofuels. Macaw palm grows under a wide range of climatic conditions, from subtropical to tropical regions. It has a great potential for integration into silvopastoral and agrosilvocultural systems.

The aim of this study was to investigate (i) if macaw palm is suitable for integration into agroforestry and/or silvopastoral systems (ii) which spacing of the palm trees guarantees the best light conditions for such systems. Microclimate and photosynthetic active radiation (PAR) interception of *A. aculeata* were measured in three systems (alley cropped with coffee, a silvopastoral system and a sole stand) in the Zona de Mata region of Minas Gerais, Brazil.

Isotope discrimination was used as an indicator for water stress. The measured δ^{13}C values of coffee under shade trees or macaw palms were around the usual 28‰ for C3 plants, but under full sun, coffee's δ^{13}C ranged between 26 and 27‰, indicating slight water stress. Air temperature, relative humidity and soil temperature fluctuations were lower in the intercropped plots than in the full sun coffee plots. The maximum PAR increased with spacing from 540 to 1333 μmol m^{-2}s^{-1} in the 5 m by 4 m and 7 m by 4 m plot, respectively, allowing to grow other crops in between.

Macaw palm coffee intercropping and macaw palm silvopastoral systems can be considered a viable mitigation option for regions with climate change impact as it provides a prolonged harvesting window and a more diversified income for farmers. By integrating macaw palm lots into pastures, C4 grasses can establish between and shade areas for cattle are provided - a win-win situation for cattle raising without compromising macaw palm growth.

Keywords: Bioeconomy, climate change mitigation, surrogate for palm oil

---

Contact Address: Melanie Hinderer, University of Hohenheim, Inst. of Agric. Sci. in the Tropics (Hans-Ruthenberg-Institute), Auerbachstr. 16/2, 70794 Filderstadt, Germany, e-mail: mel.hinderer@gmail.com

Effects of Shading and Soil Moisture on *Brachiaria brizantha* Biomass in an Integrated Crop-Livestock-Forestry System

Sarah Glatzle[1], Marcus Giese[1], Sabine Stürz[1], Mariana Pereira[1], Roberto G. Almeida[2], Folkard Asch[1]

[1] *University of Hohenheim, Inst. of Agric. Sci. in the Tropics (Hans-Ruthenberg-Institute), Germany*

[2] *EMBRAPA Beef Cattle, Integrated Production Systems, Brazil*

Components of integrated production systems may compromise each other in competition for resources such as water and light. A smart and sustainable management of these resources is key to an overall successful system performance but often basic information are missing of how components and resources interact. This study analysed the seasonal dynamics of photosynthetically active radiation (PAR), soil moisture and grass biomass for integrated Crop-Livestock-Forestry (ICLF) demonstration plots in Campo Grande-MS, Brazil.

Data were collected at three ICLF plots consisting of grass pasture (*Brachiaria brizantha* BRS cv. Piatã) lined with east-west orientated strips of 20 m high Eucalyptus trees (*Eucalyptus urograndis*) in 22 m distance. PAR (AccuPAR CP-80), soil moisture (DELTA T FDR) and grass biomass (moving cages) were sampled in each plot in a line of five sampling points between tree rows to represent the shading gradient.

During rainy season (Dec–Feb) the PAR gradient was extremely high ranging from on average 1300 μmol m^{-2} s^{-1} (centre positions) to only 180 μmol m^{-2} s^{-1} near trees. In contrast due to lower inclination all points receive more or less the same amount of PAR (500 μmol m^{-2} s^{-1}) during dry season (Jun–Aug). For soil moisture we measured a clear gradient from on average 29–32 Vol % at centre positions to 20–25 Vol % near the tree lines. Dry season revealed same pattern with on average 3–5 Vol % lower values across the gradient. Biomass distribution showed a clear gradient as well with twice as much DM in centre positions for both seasons, while DM was about halved during dry season.

Although we recorded a very high variation of PAR across the gradient between trees during rainy season, soil water content appears dominant affecting grass biomass growth according to correlation analysis. This relation was not surprisingly tightened during dry season when soil moisture contents approach wilting points more often. However, differences in energy budgets across the gradient and seasons resulting from radiation inputs are tremendous and should be considered in follow up analysis and management strategies of ICLF systems in relation to tree distances and height.

Keywords: Brazil, grass biomass, PAR, soil moisture

Contact Address: Sarah Glatzle, University of Hohenheim, Inst. of Agric. Sci. in the Tropics (Hans-Ruthenberg-Institute), Stuttgart, Germany, e-mail: sarah.glatzle@uni-hohenheim.de

Site Adaptability of Mangrove Species in Myanmar

AUNG MYAT SAN, RALPH MITLÖHNER
Georg-August-Universität Göttingen, Burckhardt-Inst., Tropical Silviculture and Forest Ecology, Germany

Mangroves, halophytes, play an important role in the sustainability of coastal and marine ecosystems and provide protective and productive functions for the socio–economic development of the coastal communities. Myanmar, a tropical country in Southeast Asia, possesses about 2800 km of coastline along which mangroves occur in three different coastal regions. Due to anthropogenic influences and natural disasters, mangrove forests in Myanmar are being degraded at an alarming rate. Mangrove rehabilitation programs are being carried out in all the coastal regions and site-species matching is crucial for successful mangrove reforestation. Salinity is a limiting factor for the growth of mangroves and osmoregulation is one of the strategies mangrove species use to cope with saline site conditions. The study aimed to observe the interaction between the mangrove species and their growing sites in order to determine the most saline-tolerant species from respective intertidal zones. In each coastal region, two study sites were selected in distinct hydrological habitats; one on an island and one on the coast or delta area. Leaf and soil samples for 21 dominant true mangroves and associated species were collected. The osmotic potential of plants (midday and saturated) and soils were cryoscopically analysed. The difference in osmoregulation of individual species among different habitats and their relationship to their respective soils were determined. The osmotic potential of plants ranged from -1.2 to -7.1 MPa. The study revealed the applicational relevance of saturated leaf osmotic potential as an indicator of site conditions, due to its close proximity to soil osmotic potential. Optimally performing mangroves species from different habitats presented significant differences in osmotic potential. *Avicennia alba, Rhizophora mucronata* and *Rhizophora apiculata* from the lower intertidal zone; *Avicennia marina, Xylocarpus granatum* and *Bruguiera cylindrica* from the mid-intertidal zone; *Ceriops decandra* and *Excoecaria agallocha* from the high intertidal zone were the species with the lowest midday osmotic potential (the highest internal solute concentration) in their respective zones. These species can be proposed as saline-tolerant species for plantation establishment of mangrove restoration programs.

Keywords: Mangrove rehabilitation, osmotic potential, saline tolerance, site indicator, site-species matching

Contact Address: Aung Myat San, Georg-August-Universität Göttingen, Burckhardt-Inst., Tropical Silviculture and Forest Ecology, Büsgenweg 1, 37077 Göttingen, Germany, e-mail: aungmyatsan@gmail.com

Balanites aegyptiaca: A Multipurpose Tree Species for Forest Based Industry Development in Sudan

HANADI MOHAMED SHAWGI GAMAL[1], MAJDALDIN RAHAMTALLAH ABUALGASIM MOHAMMED[2]

[1] *University of Khartoum, Fac. of Forestry; Forest Products and Industries, Sudan*

[2] *Reseach Center, Sudan*

Sudan is endowed by a great diversity of tree species, nevertheless the utilisation of wood resources has traditionally concentrated to a few number of species. Most of the indigenous hardwood tree species in Sudan are used as charcoal, firewood and fuel wood due to the lack of information about their properties. Despite the richness of Sudan for most of the basic factors required to establish forest based industries, it still almost entirely depend on imports to satisfy its needs for industrial products such as pulp, paper, and fiberboard. There is an urgent need to evaluate the available local raw materials as potential sources for forest based industries. This would not only reduce imports, but it would also provide an economic incentive to the forestry and industrial sectors of Sudan.

The present study was carried out to investigate some wood properties of *Balanites aegyptiaca* and to assess its suitability for pulp, paper and flooring industries. This species is widely distributed and easily grown on large areas in Sudan. The wood materials were collected from four states in Sudan namely: Blue Nile state, Northern Kordofan state, Southern Kordofan state and White Nile state. Some anatomical, physical and mechanical properties were investigated. In anatomical investigations, fiber length, diameter, lumen diameter as well as double wall thickness were measured from which the Runkel ratio, slenderness ratio and coefficient of suppleness (or flexibility coefficient) were obtained. Concerning physical and mechanical investigations, wood basic density and hardness strength were determined respectively.

The results revealed that the anatomical, mechanical and physical wood properties of *Balanites aegyptiaca* may qualify for pulp, paper and flooring industries. The results of this study could enhance the establishment of such forest based industry in Sudan.

Keywords: *Balanites aegyptiaca*, development, forests based industries, Sudan

Contact Address: Hanadi Mohamed Shawgi Gamal, University of Khartoum, Fac. of Forestry; Forest Products and Industries, Alemarat Street 61, Khartoum, Sudan, e-mail: hanadishawgi1979@yahoo.com

Land-Use Intensification Gradient and Vegetation Diversity Loss Reduce Termite Abundance in Humid Lowland Forest of Cameroon

NIKOLA TEUTSCHEROVÁ[1], ANNA MANOUROVA[1], VÁCLLAV TEJNECKÝ[2], JAN ŠOBOTNÍK[3], DAVID SILLAM-DUSSES[4], JIŘI KINDL[3], BOHDAN LOJKA[1]

[1]*Czech University of Life Sciences Prague, Fac. of Tropical AgriSciences, Dept. of Crop Sciences and Agroforestry, Czech Republic*

[2]*Czech University of Life Sciences Prague, Fac. of Agrobiology, Food and Natural Resources, Dept. of Soil Science and Soil Protection, Czech Republic*

[3]*Czech University of Life Sciences Prague, Fac. of Forestry and Wood Science, Dept. of Forest Protection and Entomology, Czech Republic*

[4]*Paris 13 Univerisity, Laboratory of Experimental and Comparative Ethology, France*

Increasing pressure of human activities leads to a large-scale land-use change including deforestation, cropland expansion or agricultural intensification, which is reflected in the loss of soil quality. While forest conversion to cropland in slash-and-burn agriculture initially increases soil fertility by the input of ashes, the long-term soil degradation and biodiversity loss due to vegetation removal may have detrimental impacts on ecosystem services in the long-run. Apart from ash-originated nutrient input, agricultural activity as well as natural vegetation depend on nutrients originating from soil organic matter (SOM) decomposition, which is, particularly in tropics, mediated by termites. Therefore, termite species richness and abundance are considered suitable indicators of ecosystem functioning and nutrient cycling. In the present study, we evaluated the changes in vegetation structure, termite abundance and diversity as well as soil properties along the land-use intensification gradient in lowland humid forest of Cameroon. The four most common land-use systems included: primary forest (PF) which has been without disturbance for at least 70 years, secondary forest (SF) occasionally used for timber collection, cocoa agroforestry system (CA) and maize cropland (MC) established on area which was burnt 5 months before the study. All studied termite variables (abundance, genus richness, and diversity) decreased along land-use intensification gradient, and all were positively related to vegetation basal area, tree species diversity and richness, canopy cover and number of tree individuals. On the other hand, termite communities were negatively related to soil CEC, soil pH and Ca, which were all the highest in the MC when compared to other systems. Nevertheless, such amelioration of soil acidity and increase of base cations was likely a short-term result of slash-and-burn management, rather than a direct cause of termite abundance reduction. Therefore, the maintenance of diverse and abundant vegetation, either in its natural state or as diversified agroforestry systems, is the key mechanism to maintain functional termite communities, which can serve as ecosystem engineers and contribute to the long-term ecosystem productivity and soil health.

Keywords: Biodiversity, cocoa agroforestry, land-use system, soil fertility, vegetation structure

Contact Address: Bohdan Lojka, Czech University of Life Sciences Prague, Fac. of Tropical AgriSciences, Dept. of Crop Sciences and Agroforestry, Prague, Czech Republic, e-mail: lojka@ftz.czu.cz

Diversity of Natural Pollinators in Cocoa Agroforests in the Peruvian Amazon

Martina Snášelová, Jitka Straková, Nikola Teutscherová, Bohdan Lojka

Czech University of Life Sciences Prague, Fac. of Tropical AgriSciences, Czech Republic

Despite the importance of cocoa (*Theobroma cacao*, L.), fruit yield remain relatively low. A very low fruit set of cocoa relative to the numerous produced flowers has been connected to low pollination intensity. Abundance of cocoa pollinators (mostly Diptera: Ceratopogonidae) is influenced mainly by the availability of breeding material. One of the best breeding opportunities for pollinators represent ecosystems with heterogeneous shade cover and those where substrates of decomposing cocoa leaf litter and fruit husks are available in sufficient amount. Pollinator-friendly practices would lead to higher number of pollinators and consequently to a higher yield. The objective of this study was the evaluation of cocoa agroforests as a suitable habitat for pollinators in general and cocoa pollinators in particular and its subsequent potential to yield enhancement. The experiment was conducted in three different cocoa agroforestry systems in the Peruvian Amazon with different characteristics, including vegetation structure, canopy cover and soil cover. Insect trapping took place in parallel to monitoring the phenological patterns of the flowering and fruit set. Afterwards evaluation of captured insects was conducted. Chosen families of Hymenoptera and Dipetra order were determined to morphospecies and their abundance and diversity was calculated. The abundance of pollinators in general was the highest in the system with the highest number of species of shade trees and with the highest shade cover. There were no differences in insect species diversity and richness among systems. The abundance of Ceratopogonidae insects was very low in all systems and that is why they were excluded from our study. Other potential cocoa pollinators are small individuals from the Diptera order (namely the families Cecidomyiidae, Drosophilidae and Phoridae). Results indicated that agroforestry systems with adequate canopy shade cover and enough leaf litter could be considered as the best habitat for cocoa pollinator enhancement. However, an additional long-term survey needs to be done.

Keywords: Diptera order, flowering, *Theobroma cacao*, tree species diversity

---

Contact Address: Bohdan Lojka, Czech University of Life Sciences Prague, Fac. of Tropical Agri-Sciences, Dept. of Crop Sciences and Agroforestry, Prague, Czech Republic, e-mail: lojka@ftz.czu.cz

Domestication of Baobab (*Adansonia digitata* L.) in Kenya: Selecting Elite Mother Trees by Using a Multi-trait Approach

Katja Kehlenbeck[1], Monica A. Omondi[2], Justine Nyamweya[2], John Bosco Mukundi[2], Jens Gebauer[1], Fredah Karambu Rimberia[2]
[1]*Rhine-Waal University of Applied Sciences, Fac. of Life Sciences, Germany*
[2]*Jomo Kenyatta University of Agriculture and Technology, Dept. of Horticulture, Kenya*

Baobab (*Adansonia digitata* L.) with its highly nutritious fruit pulp is an important wild fruit tree in Kenya. The increasing demand for pulp from domestic and international markets puts pressure on wild baobab populations. Domestication and cultivation of baobab in agroforestry systems may result in better quality and higher quantities of baobab fruit pulp and simultaneously contribute to the conservation of this valuable plant resource. Within the Baofood project, this study aimed at developing a multi-trait approach for selecting elite baobab mother trees based on morphological characteristics and pulp chemical composition for future baobab domestication.
Two baobab populations were surveyed in Kenya and a total of 71 trees randomly selected. Ten fruits per tree were collected for assessing several quantitative and qualitative morphological traits. Pulp samples from each of the accessions were analysed for their soluble solids contents (°Brix) and total titratable acidity (TTA). Superior mother trees were identified by applying principal component analysis (PCA).
The studied 71 individual baobab trees were highly variable regarding morphological fruit characteristics. Mean fruit weight per accession ranged from 73–696 g with an overall median of 227 g. Total yield had a median of 54 kg per tree, ranging from 7 to 160 kg. Pulp weight also showed a high variability (median 38, range 11–136 g). Pulp proportion from the whole fruit ranged from 13 to 23 % with a median of 17 %. Median Brix was 11° (range 6–18°), while median TTA was 0.7 g equivalent malic acid/100 g fresh matter (range 0.4–1.2). A final PCA was performed using six traits with low levels of auto-correlations. Two elite mother trees were identified in each of the two research areas, with highest axis loadings for the traits 'yield' and 'Brix', while 'pulp proportion' was not considered as important by the PCA. After adding further fruit traits such as vitamin C content in future analyses, the finally selected trees should be used for vegetative propagation and tested on farms for their performance. Results of our study can contribute to better utilisation of baobab and to improved nutrition and livelihoods of rural communities in Kenya and beyond.

Keywords: Agroforestry, characterisation, conservation, fruit pulp, nutrient content

Contact Address: Katja Kehlenbeck, Rhine-Waal University of Applied Sciences, Fac. of Life Sciences, Marie-Curie Str. 1, 47533 Kleve, Germany, e-mail: katja.kehlenbeck@hochschule-rhein-waal.de

Inga edulis Legume Morphology and Genetic Structure in Wild and Cultivated Populations in Amazonian Peru

ALEXANDR ROLLO[1], MARIA MARGARIDA RIBEIRO[2], MARIE KALOUSOVÁ[1], BOHDAN LOJKA[1]

[1] *Czech Univerzity of Life Sciences Prague, Dept. of Crop Sciences and Agroforestry in the Tropics and Subtropics, Czech Republic*

[2] *Escola Superior Agrária do Instituto Politécnico de Castelo Barnco, Dept. of Natural Resources and Sustainable Development, Portugal*

Inga species (Fabaceae) are important components of Neotropical forests, as well as local food source. *Inga edulis* is one of the most used species in the Amazon region for fruits and shade tree. To the best of our knowledge little is known about *I. edulis* species' genetic diversity and structure in the wild and cultivated populations. In our study it was assessed the genetic diversity and structure using 259 trees sampled in five wild and 22 cultivated *I. edulis* populations in three different geographical regions (Selva Central, Ucvayali and Loreto) of Amazonian Peru. *Inga edulis* legume length was measured to highlight differences between wild and cultivated *I. edulis* trees. Using microsatellite markers it was determined the genetic diversity and structure of populations using analysis of molecular variance and Bayesian analysis. The average legume length in cultivated *I. edulis* trees (83 cm) was significantly larger then (39 cm) legume length average in wild trees. The Loreto region cultivated *I. edulis* trees had the highest legume length (148 cm) and lowest allelic richness. The expected genetic diversity and the average number of alleles was higher in the wild *I. edulis* compared to the cultivated *I. edulis* populations. A loss of genetic diversity was confirmed in the *I. edulis* cultivated populations. The species could have been simultaneously domesticated in multiple locations, usually with local origin. The original *I. edulis* Amazonian germplasm should be maintained, and cultivated population new germplasm influx from the wild populations could increase genetic diversity, provided that fruit yield will not be compromised.

Keywords: Agroforestry, Amazon basin, biodiversity conservation, domestication, edible fruit, Inga

---

Contact Address: Alexandr Rollo, Czech Univerzity of Life Sciences Prague, Dept. of Crop Sciences and Agroforestry in the Tropics and Subtropics, Kamycka 129, 16521 Prague, Czech Republic, e-mail: rollo@ftz.czu.cz

Genetic Diversity and Differentiation of *Olea europaea* subsp. *cuspidata* in the Hajar Mountains, Sultanate of Oman

NOURIN AHOSAN HABIB[1], MARKUS MÜLLER[2], OLIVER GAILING[2], KONSTANTIN V. KRUTOVSKY[2], ANDREAS BUERKERT[3], ANNE PATZELT[4], GHUDAINA AL ISSAI[4], MARTIN WIEHLE[5]

[1]*Khulna University, Forestry and Wood Technology Discipline, Bangladesh*

[2]*University of Goettingen, Inst. of Forest Genetics and Forest Tree Breeding, Germany*

[3]*University of Kassel, Organic Plant Production and Agroecosystems Research in the Tropics and Subtropics, Germany*

[4]*Diwan of Royal Court, Oman Botanic Garden, Oman*

[5]*University of Kassel, Tropenzentrum / ICDD / Organic Plant Production and Agroecosystems Research in the Tropics and Subtropics (OPATS), Germany*

Olea europaea subsp. *cuspidata* (Wall. & G.Don) Cif. (wild olive) is one of the six subspecies considered important for domestication purposes of olive as it bears favourable physiological and morphological traits. The distribution of wild olive stretches from South Africa over the Middle East and the Mediterranean region to China mainly at mid to high altitudes with reliable precipitation. Wild olive populations are considered isolated outposts of a formerly large and more or less connected *Olea europaea* population. Nowadays, wild olive populations are poorly connected, situated in fragile habitats, and hence more prone to additional stresses. These stresses include erratic weather extremes and anthropogenic influences such as illegal logging and livestock keeping activities that may result into further fragmentation of stands, shifts of demography, increased clonal growth, and lead to directional genetic change. Wild olive also grows in the Hajar Mountains, Sultanate of Oman, a habitat subjected to many anthropogenic disturbances. Therefore, the present condition of wild olive in the Hajar Mountains was studied to predict the future stability of this population. To this end a total of 362 individuals was inventoried and a subset of 188 individuals were genetically analysed at 12 microsatellite markers. Dendrological measures indicated disturbances mainly due to human activities. Height and crown area of individuals averaged 3.7 m (range: 0.2 to 12 m) and 14 m (range: 1 to 18 m), respectively. The stand conditional index was mainly 50 %. The identified genetic variation within the sampled areas was moderate (observed and expected heterozygosity: 0.58 to 0.55 and 0.64 to 0.62, respectively) with the absence of clones. Though significant genetic differentiation was found, there was no evidence of population sub-structure suggesting effective pollen and long-distance seed dispersal. Changes in gene flow patterns may be reflected in the next (seedling) generation. Further studies are necessary to detect any changes in the future and to develop sustainable conservation strategies for this species.

Keywords: EST-markers, genetic diversity, microsatellites, spatial genetic structure, wild olive

Contact Address: Martin Wiehle, University of Kassel, Tropenzentrum / ICDD / Organic Plant Production and Agroecosystems Research in the Tropics and Subtropics (OPATS), Steinstraße 19, 37213 Witzenhausen, Germany, e-mail: wiehle@uni-kassel.de

Multifunctionality in cultural landscapes

The Necessity and Challenges of Tropical Paludiculture

HANS JOOSTEN

University Greifswald, Inst. of Botany and Landscape Ecology; Partner in the Greifswald Mire Centre, Germany

The Paris Agreement has made the world simple. We have one common goal: to limit global temperature rise to clearly below 2°C. The physical consequence is, that we have to reduce global net CO_2 emissions by 2050 to 0 (zero). As, furthermore, these goals have to be reached "...in the context of sustainable development and efforts to eradicate poverty", the challenges are enormous.

Currently drained peatlands account for 5 % of all global anthropogenic GHG emissions, half of which coming from the tropics. Millions of hectares of drained peatlands will furthermore in future be flooded by the sea because of ongoing peat subsidence. The Paris Agreement and the Sustainable Development Goals imply that the use of drained peatlands is completely stopped and that worldwide until 2050 almost 20,000 km^2 of drained peatland will be rewetted annually. As the demand for biomass is growing, these peatlands must largely maintain their production function. Peatland agriculture and forestry must therefore rapidly advance the development of innovative, wet cultivation techniques, i.e. of paludicultures, which – per definition – minimise greenhouse gas emissions and stop subsidence.

The presentation will discuss the perspectives, options and major challenges of tropical paludicultures.

Keywords: Climate change, gas emisions, peatlands

Contact Address: Hans Joosten, University Greifswald, Inst. of Botany and Landscape Ecology; Partner in the Greifswald Mire Centre, Soldmannstraße 15, 17487 Greifswald, Germany, e-mail: joosten@uni-greifswald.de

Social Obstacles to the Management of Ecosystem Services in the Cubango-Okavango River Basin

STEPHANIE DOMPTAIL, MELLISSA TELLO

Justus-Liebig University Giessen, Inst. of Agricultural Policy and Market Research, Germany

The sustainable management of the Cubango-Okavango River Basin (CORB), southern Africa, is often framed at the transboundary level as a "conservation vs. exploitation" conflict opposing downstream to upstream nations. This masks the existence of more localised barriers to the sustainable management of ecosystem services (ESS) with regard to institutional and social challenges in the basin. Yet, social conflicts and challenges frame the capacity of local actors to engage in a transition towards sustainable resource management.

On the one hand, our contribution identifies key social challenges and conflicts related to the sustainable resource management in the basin. On the other, we identify the existing land use conflicts using the concept of ecosystem services and show which social conflicts they are attached to.

80 texts of interviews conducted from 2012 to 2013 with actors of 4 levels of governance (local, provincial, national and basin) were analysed with a qualitative and quantitative content analysis. Latent and revealed ecosystem service conflicts are identified using a diagnosis key for which the object of conflict was coded for, as well as the involved actors and the intensity of the conflict.

27 revealed and potential conflicts jeopardise the sustainable management of wildlife, crops and livestock, collected natural resources (e.g. wood), water and recreation. We found that social challenges are created by the emergence of new land uses and related ESS. As new parties come into play, the original land-users become marginalised. Yet, intra-community conflicts, such as gender inequality and the generational clash between youth and current land users, also burden the communities in their attempt to find sustainable land-use paths at the local scale.

Currently, smallholders are not actively shaping the debate over the relevance of using farming as a motor of rural development in the CORB. Beyond a technical support, means to facilitate the current political and social challenges to achieving sound resources use would make communities more sovereign to make their own choice towards a better livelihood.

Keywords: Conflict diagnosis, ecosystem services, land use science, social conflicts, social ecology

Contact Address: Stephanie Domptail, Justus-Liebig University Giessen, Inst. of Agricultural Policy and Market Research, Semckenbergstrasse 3, 35390 Gießen, Germany, e-mail: stephanie.domptail@agrar.uni-giessen.de

Hemp as a Multifunctional Crop to Promote Quilombola Communities of the Brazilian São Francisco Valley: A Case Study

DANILO CRISPIM MASSUELA, CINZIA PIATTI

University of Hohenheim, Institute of Social Sciences in Agriculture - Societal Transition and Agriculture, Germany

The Brazilian Sertão, a region chronically subject to famine and multiyear droughts, bear the worst distributive inequity and food security metrics in the country. Inside the region, the São Francisco river valley is the main agricultural landscape but also the biggest producer of narcotic cannabis in Brazil. The river lands have been the stage of violent and unequal forms of settlement, marginalizing traditional communities such as indigenous groups and the Quilombolas - a rural minority descending from slaves, historically subjected to food insecurity, racism and land confiscation; but it is also a prominent example as peasantry resistance on the fight for recognition and access to land, water and food. With the current political and economic crisis in Brazil, previous governmental programs and public policies aiming at rural development have been drastically reduced. In addition, the predatory interests of agribusiness in the region made smallholder farmers living in Quilombola communities suffer threats to their productive resources and hindrances to access constitutional rights such as land tenure, due to structural changes in governmental agencies. Often, community members join the traffic networking and crop narcotic cannabis for subsistence matters, facing violence and prosecution. But cannabis is also a multifunctional crop, drought tolerant, used for centuries to provide mankind nutrition, functional fibers and medicine. Recently, we have seen a relaxation of social stigma and consequent change in legislation, as a growing global trend towards the depenalisation and market regulation of light drugs, pushing the discussion towards the regulation of cannabis products for industrial (namely hemp), medicinal and recreational matters in Brazil. Such a trend could constitute a chance for legal markets creation, given hemps' rediscovered functions and features which respond to a growing consumer markets for ethical and sustainable products. From the Quilombola communities' perspective, this work aims at exploring the potential capacity of the landscapes in the São Francisco valley to provide industrial hemp products (seeds, leaves, fibers) to a growing global demand and potential local market in Brazil (food, cosmetics, industrial), countering social marginalization and environmental degradation in this tropical semi-arid region.

Keywords: Alternative food trends, cannabis, hemp, quilombola communities

---

Contact Address: Danilo Crispim Massuela, University of Hohenheim, Institute of Social Sciences in Agriculture - Societal Transition and Agriculture, Schloss Hohenheim 1 C (Museumsflügel), 70599 Stuttgart, Germany, e-mail: dcmassuela@protonmail.com

Ecosystem Services Valuation in Puerto Vallarta, Mexico for a Concurrent PES Scheme to Foster Adaptation to Climate Change

CARLOS MATIAS FIGUEROA[1], PILAR SALAZAR VARGAS[1], JOSÉ ALBERTO LARA PULIDO[2]

[1]*National Institute of Ecology and Climate Change, General Coordination of Adaptation to Climate Change, Mexico*

[2]*Ibero-American University, Transdisciplinary Center for Sustainability, Mexico*

Mexico introduced a nation-wide payment for ecosystem services (PES) scheme in 2003 as an economic incentive for forest owners to compensate them for conservation costs and expenditures incurred in while carrying out good land use practices. Due to budgetary constraints the Mexican government migrated from a purely public to public-private funding, called concurrent funds. To implement such scheme the National Forestry Commission (CONAFOR) requires information of the value of the ecosystem services and the land use opportunity cost. The National Institute of Ecology and Climate Change (INECC) developed such information for the watersheds that feed Puerto Vallarta in the state of Jalisco, Mexico. The location was chosen given its ecological and touristic importance and because it has been subject of the project "Conservation of Coastal Basins in the Context of Climate Change" to foster ecosystem-based adaptation (EbA) actions. The research intended to showcase the importance and value that upstream ecosystem services provide to tourism, the potential to implement EbA in the region and also to provide economic incentives to landowners to conserve forests and implement low-impact agricultural practices. To achieve these objectives a literary review was conducted to understand the environmental, social and economic dynamics in the area and to identify the priority ecosystem services, then the willingness to pay and accept (WTP and WTA) was estimated via a contingent valuation to consider the rural/urban and consumers/producers perspectives. The study focused on water provision, scenic beauty and carbon sequestration by the upstream forest in the region. During this contingent valuation, economic data regarding forestry and agriculture activities was also captured to calculate the opportunity cost to conserve the forest. The results showed that tourist and residents would be willing to pay $3,161 million pesos per year and farmers would be willing to accept $1,524.6 million pesos. This shows the great existing potential and feasibility to implement a concurrent PES scheme. A payment of $2,070 pesos per hectare would be a sufficient incentive to allow conserving 51.7 % of the forest areas of the watersheds. This research evidence the great potential to implement concurrent PES schemes in Mexico to conserve large forest areas while contributing to its Nationally Determined Contributions under the Paris Agreement.

Keywords: Concurrent funds, contingent valuation, ecosystem-based adaptation, opportunity cost, payment for ecosystem services

---

Contact Address: Carlos Matias Figueroa, National Institute of Ecology and Climate Change, General Coordination of Adaptation to Climate Change, Carretera Picacho-Ajusco 4219 Tlalpan, 14210 Mexico City, Mexico, e-mail: carlos.matias@inecc.gob.mx

Medicinal Plants as a Complementary Health Treatment by Small-Scale Farmers in South of Brazil

JULIA SEVERO DOS SANTOS[1], FRANCINE HENKE TESSMER[2], LAURA TELLES GOMES[1], ALINE NEUTZLING BRUM[1]

[1]*Federal University of Rio Grande, Nursing School, Brazil*
[2]*Fac. Anhanguera of Pelotas, Physiotherapy Dept., Brazil*

This study was carried out to understand the vulnerability situation of the small-scale farmers at Colônia Maciel, located in the South of Rio Grande do Sul, Brazil. Thus, to get presented results, 206 households were interviewed through semi structured questionnaires from September to December 2018. According to this survey, 65.4 % of population studied claims to have health problems. Among the major diseases: heart problems (31.5 %), auto immunes complications (27 %), respiratory limitations (11.7 %) and mental health disorders (10.8 %) were observed. The population surveyed was also asked about their different forms of health care, 71.2 % declare the use of local health service (UBS–Unidade Básica de Saúde). Further, 70.2 % make use of prescribed medication and 53.8 % make use of not prescribed medication to treat their illness. Still, was complementary form of health care, 79.8 % of population use medicinal plants to support formal treatment. Between the most used medicinal plants, stand out: marcela (*Achyrocline satureioides*) 15.9%; funcho (*Foeniculum vulgare*) 7%; camomila (*Matricaria chamomilla*) 6.5%; tansagem (*Plantago major*) 5.6%; boldo (*Peumus boldus*) 5.1%; cidreira (*Melissa officinalis*) 5.6%; and malva (*Malva sylvestris*) 4.2 %. These medicinal plants are mainly used for pain and stress alleviation. When asked about the origin of knowledge in medicinal plants, the studied population declared that information was transmitted by family over generations, about medicinal plants properties in this region. Thus, this study identified the need to rescue and intensify the popular and cultural knowledge about medicinal plants, as complementary health treatment.

Keywords: Health conditions, medicinal plants, small-scale farmers

Contact Address: Aline Neutzling Brum, Federal University of Rio Grande, Health Science, Rio Grande, Brazil, e-mail: neutzling@live.de

Ethnomedicine in the High Lands of Chiapas, Mexico

EDUARDO ALBERTO LARA REIMERS[1], YAMEN HOMAIDAN SHMEIT[1], ELOY FERNÁNDEZ CUSIMAMANI[1], JUAN MANUEL ZEPEDA DEL VALLE[2], DAVID JONATHAN LARA REIMERS[3]

[1]*Czech University of Life Sciences Prague, Fac. Trop. Agrisci., Dept. of Crop Sci. and Agroforestry, Czech Republic*

[2]*Chapingo Autonomous University, Regional Centers, Mexico*

[3]*Chapingo Autonomous University, Dept. of Forestry Engineering, Mexico*

Ethnomedicine is understood as a set of traditional knowledge, skills and practices based on theories, beliefs, and experiences of indigenous cultures. It is used as a health care source, as well as in the prevention, diagnosis, improvement or treatment of physical and mental illnesses around the world. Mexico has a long history in the use of ethnomedicine due to its biodiversity and multicultural (indigenous) groups, yet the use of medicinal plants is not only related to rural areas, economic status and cultural conditions.

The objective of this research was to record the medicinal plants used by 14 indigenous Tsotsil communities in the highlands of Chiapas, Mexico. The data was collected from 59 informants (61 % women, 39 % men) aged between 20 and 86 years, through a semi-structured questionnaire in the Tsotsil native language. A total of 59 species of medicinal plants belonging to 55 genera and 37 botanical families were reported. The Asteraceae family was the most cited, with 6 species (Uses Report = 51).

The highest cultural index (CI) was reported for *Matricaria chamomilla* (CI = 0.42), *Mentha sativa* (CI = 0.36) and *Ruta graveolens* (CI = 0.31). According to the informant Consensus factor (ICF) the main pathological categories treated were reproductive system diseases (ICF=0.80), respiratory infections (ICF = 0.75) and diseases of the digestive system (ICF= 0.70). The species reported and their diversity of uses provides the needs of the families with socioeconomic deficiencies.

Nowadays the use of medicinal plants is still high (visible) among the Tsotsil community but the diversity seems to be reduced in the recent years, and is gradually being replaced by modern medicine to heal general problems.

Keywords: *Asteraceae*, chiapas highlands, indigenous communities, medicinal plants, Tsotsil

Contact Address: Yamen Homaidan Shmeit, Czech University of Life Sciences Prague, Fac. Trop. Agrisci., Dept. of Crop Sci. and Agroforestry, Kamycka 933, Prague, Czech Republic, e-mail: yamen.shmait@gmail.com

Urban Ethnobotany: Medicinal Plants Used for the Treatment of Noncommunicable Diseases in Santa Marta, Colombia

Martina Tasev[1], Eduino Carbonó-Delahoz [2], Lukas Pawera[1], Cory Whitney[3], Naji Sulaiman[1], Vladimir Verner[1], Zbynek Polesny[1]

[1]*Czech University of Life Sciences Prague, Fac. of Tropical AgriSciences, Czech Republic*

[2]*Universidad del Magdalena, Herbario UTMC, Colombia*

[3]*University of Bonn, Inst. Crop Sci. and Res. Conserv. (INRES), Germany*

Colombian population is considerably threatened by noncommunicable diseases. However, as traditional medicine is still popular in Colombia, the local people do not hesitate to use medicinal plants for the treatment of a wide range of health disorders. This research aimed to inventory medicinal plants used for the treatment of obesity, diabetes, hypertension and heart diseases available within local herbal shops (herboristerías). Data were collected in the period from September 2016 till February 2017 in 7 herboristerías in the city of Santa Marta through semi-structured and open-ended interviews. First, the informants (herbalists in the sampled shops) was asked for the basic socio-economic information about the shop and the informant. Subsequently, the informants were asked to mention all plants prescribed for the treatment of each targetted disease and the plant parts used, mode of preparation and application. The study documented 53 folk taxa corresponding to 56 plant species, and belonging to 35 botanical families. The most frequently reported plant species were *Anacardium occidentale*, *Moringa oleifera* and *Bauhinia sp.* for the treatment of diabetes; *Marrubium vulgare* in obesity; *Salvia palifolia* for the treatment of hypertension; and *Phthirusa stelis* in heart diseases. Quantitative ethnobotanical indices were calculated to determine the most important species to treat each disease type and to compare the species variability among herboristerías. In general, there is a wide range of medicinal species. Across herboristerías 30 unique species were documented, meaning the Informant Consensus Factor was generally low. We also documented promising plant species which were not subjected to phytochemical research so far.

Keywords: Diabetes, heart diseases, hypertension, medicinal plants, obesity, South America

---

Contact Address: Zbynek Polesny, Czech University of Life Sciences Prague, Fac. of Tropical AgriSciences, Kamýcká 129, 16500 Praha - Suchdol, Czech Republic, e-mail: polesny@ftz.czu.cz

Ethnobotany of Wild Plants and Crop Wild Relatives in the Walnut-Fruit Forests of Kyrgyzstan

LINDA VLČKOVÁ[1], ADILET USUPBAEV[2], JYLDYZ SHIGAEVA[3], DIETRICH DARR[4], NAJI SULAIMAN[1], VLADIMIR VERNER[1], ZBYNEK POLESNY[1]

[1]*Czech University of Life Sciences Prague, Fac. of Tropical AgriSciences, Czech Republic*

[2]*National Academy of Sciences of the Kyrgyz Republic, Inst. of Biology and Soil Science, Kyrgyzstan*

[3]*University of Central Asia, Mountain Societies Research Institute, Kyrgyzstan*

[4]*Rhine-Waal University of Applied Sciences, Faculty of Life Sciences, Germany*

The walnut fruit-forests, situated in the mountains in the southwest of the Kyrgyz Republic, are of unique importance for global biodiversity and refuge of genetic resources such as walnut, apples, pears, plums, and several other species. Simultaneously, local people derive from these forests significant economic and ecological benefits. However, extensive grazing, the intensive collection of firewood, nuts, and fruits have caused significant damage to these ecosystems. The traditional botanical knowledge on useful plant species remains underdocumented and ethnobotanical research should contribute to preservation and transmission of this threatened knowledge on younger generations. This research was undertaken to assess the relative importance of various species for local populations as well as purpose and frequency of their use. Data collection was carried out via the free listing method and semi-structured interviews with 146 informants from five different study sites (Arslanbob, Toskool-Ata, Kyzyl-Unkur, Salam-Alik, Kara-Alma), between July and September 2018. A total number of 88 plant species from 37 botanical families were documented. The plant species reported provide food (39), medicine (51), fuelwood (28), material (14), and animal feed. The data collected included plant vernacular names, plant part used, particular uses, gathering sites and level of commercialisation. The most frequently documented families included Rosaceae, Asteraceae, Lamiaceae, and Polygonaceae. The ethnobotanical data were further analysed through quantitative indices, i.e. use reports, relative frequency of citation, use value, and cultural value. According to Use reports, the most representative species were *Juglans regia, Malus niedzwetzkyana, Malus sieversii, Prunus divaricata, Crataegus* spp., *Acer semenovii*, and *Acer turkestanica*. The results showed that walnuts, apples, and mushrooms were important sources of income for the communities studied.

Keywords: Biodiversity, forest, traditional knowledge, useful plants

Contact Address: Zbynek Polesny, Czech University of Life Sciences Prague, Fac. of Tropical Agri-Sciences, Kamýcká 129, 16500 Praha - Suchdol, Czech Republic, e-mail: polesny@ftz.czu.cz

Limitations and Impracticality of the Design-Based Sampling Method for Medicinal Plant Inventory in High Mountain Regions

Babu Dallakoti, Christine Fürst

Martin Luther University Halle-Wittenberg, Institute for Geosciences and Geography, Germany

Realizing the significant contribution of medicinal and aromatic plants in healthcare and cosmetic industries, the efforts have been put toward sustainable management of these natural products. Despite some attempts to address the issues of harvesting regime, socio-economic analyses and ethnobotanical surveys in relation to biodiversity conservation, the status quo from resource management perspective has not been changed significantly. As sustainable management requires reliable and scientific information on resource base, some design-based sampling methods adopted from forest or botanical inventories have been applied at different scales and time periods in different parts of the world to support their sustainability. Three resource inventories of different scales focusing on herbaceous medicinal plants from similar geographical areas of high mountain region of Nepal have been chosen for this studies to evaluate their validity from practical and scientific perspectives. Two of these inventories were carried out as a major part of a biodiversity conservation programs run by two development agencies in different districts. The third one was part of data collection for a research work by the principal author. All three inventories have basically selected the systematic sampling method in different variations. This paper presents the analyses of statistical rigorousness, reliability from practical view, uniformity and use of inventory data in resource management and research from all these three inventories. An exploratory method based on statistical analysis and designs in relation to cost-effectiveness has been applied for comparative analysis. This study concludes to conduct an extensive research on model-based sampling method to apply for medicinal plant inventory.

Keywords: Inventory, medicinal and aromatic plants, sustainable management

Contact Address: Babu Dallakoti, Martin-Luther University Halle-Wittenberg, Dept. Sustainable Landscape Development, Delitzscher Str. 159, 06116 Halle (Saale), Germany, e-mail: babu.dallakoti@geo.uni-halle.de

Determinants and Impact of Wild Fruit Collection on Food Security in Rural Zambia

RONJA SEEGERS

Leibniz University Hannover, Insitute for Environmental Economics and World Trade, Germany

Many sub-Saharan African countries are suffering from high rates of poverty, food insecurity and malnutrition. As one of these countries, Zambia shows a variety of wild fruit species that play an important role in food security. They are frequently used as supplements to staple food, particularly in times of poor harvest, are rich in vitamins and represent an easily accessible food source. The aim of this study was to 1) identify the determinants of wild fruit collection, and 2) examine the impact of wild fruit collection on food security.

For this purpose, a census of 215 households, focus group discussions and key informant interviews were conducted in Northern Zambia within the Food Security in Rural Zambia (FoSeZa) project. Both quantitative and qualitative analysis of data took place. Descriptive and regression analysis occurred by using Stata/IC 14.2.

The two most popular fruits were found to be *Uapaca kirkiana* and *Anisophyllea boehmii*. Fruits are mainly collected for own consumption, while sale of fruits is uncommon. Processing of fruits only takes place to a small extent. The quantity of wild fruits collected is positively and significantly affected by household size. Further factors such as size of land area, household's income and walking distance to gather wild fruits also influence the collected amount in dependency of the species.

As indicators for food security, the Food Consumption Score (FCS) and reduced Coping Strategy Index (rCSI) were calculated. Regression results reveal that the quantity of wild fruits collected has a positive and significant effect on food security based on the FCS, whereas the impact of collecting wild fruits on the rCSI is positive and insignificant.

Respondents highly evaluate wild fruits as nutrient source but high deforestation rates as well as increased population growth pose a threat to the availability of wild fruit plants. There is need to promote conservation management, cultivation practices and commercialization of wild fruits in order to ensure their long-term availability and maximize their impact on food security.

Keywords: Food security, malnutrition, sub-Sahara Africa, wild edible plants, wild fruits, Zambia

Contact Address: Ronja Seegers, Leibniz University Hannover, Insitute for Environmental Economics and World Trade, Elsa-Brändström-Straße 10, 30453 Hannover, Germany, e-mail: ronja.seegers@web.de

Determinants of Forest Dependent Households' Participation in PES Schemes in Kenya

Esther Waruingi[1], Robert Mbeche[1], Josiah Ateka[1], Anja Fasse[2], Ulrike Grote[3]

[1]*Jomo Kenyatta University of Agriculture and Technology, Dept. of Agricultural and Resource Economics, Kenya*

[2]*Hochschule Weihenstephan-Triesdorf University of Applied Sciences, Germany*

[3]*Leibniz Universität Hannover, Institute for Environmental Economics and World Trade, Germany*

In the face of increasing deforestation and forest degradation, Payment for Ecosystem Services (PES) schemes which pay land owners in cash or in kind – have been promoted as one of the mechanisms for enhancing forest conservation and livelihoods. The success or otherwise of these programmes is dependent in part on participation of forest dependent communities. This paper uses data from Plantation Establishment Livelihood Incentive Scheme (PELIS), a PES project under implementation in Mt Elgon, Kenya to examine factors that influence households' participation. Descriptive statistics defined participants while Tobit model estimated determinants of extent of participation in PELIS. The extent of participation was measured using the share of PELIS income out of total agricultural income. The results show that 52 % of the households participated in the programme. Of these, 90 % were male headed, had large household sizes, owned small land sizes, had lower asset value and low incomes illustrating a fair level of participation among the poor. Econometric results indicate that age ($p = 0.041$), attainment of tertiary education ($p = 0.074$) and livestock ownership ($p = 0.000$) negatively influence intensity of participation while household size ($p = 0.033$), membership to FUG ($p = 0.000$), membership to native community ($p = 0.026$) and distance to tarmac road ($p = 0.022$) positively influence intensity of participation. The results suggest a need for increased attention on institutional arrangements such as accountability in social groups and reduction of transaction costs to promote participation. Thus, policies and interventions to improve institutional arrangements are vital for attainment of greater participation by forest dependent households.

Keywords: Forest dependent people, Kenya, payment for ecosystem services, PELIS, Tobit model

---

Contact Address: Esther Waruingi, Jomo Kenyatta University of Agriculture and Technology, Dept. of Agricultural and Resource Economics, Nairobi, Kenya, e-mail: estherwangari16@gmail.com

Which Edible Insects Can Be Found in the Amoron'i Mania Region of Madagascar?

RAZAFINDRAKOTOMAMONJY ANDRIANANTENAINA,
RAVAOMANARIVO LALA HARIVELO

University of Antananarivo, Doctoral School Life Sciences and Environment, National Center for Applied Research for Rural Development (FOFIFA), Madagascar

In Madagascar, researchers have been studying the practice of insect consumption or entomophagy for only about a decade. In 2015, Randrianandrasana and Berenbaum mentioned that 53 species of edible insects were identified. According to a preliminary study in 2018, by a research team of the ProciNut project in the Amoron'i Mania region (central highlands), the local population ate insects especially during the rainy season when insects are abundant. That time of the year, food shortages can occur. However, aside from investigating which edible insects are known to the local population, the preliminary study could not determine which edible insects are naturally occurring. Therefore, an entomological inventory was carried out to check the presence of edible insects in the area. A standard method was used for sampling. Insects were collected along linear transects from each stratum of the forest (herbaceous, shrub, and tree level). Insects were actively caught using Japanese umbrella, filleting net, butterfly net, and Malaise trap. Where possible, the identification of captured insects was made in the field, using specific determinations or insect reference collection. As a result, we captured five orders and fifteen insect families in the study site. Coleoptera and Lepidoptera are the most common orders with seven and five families found, respectively. We identified two orders of edible insects: the Lepidopterans, wild silk worm, *Borocera cajanion* the shrub layer and in the Orthoptera order, two locust species, *Cyrtacanthacris tatarica* and *Gastrimargus africanus* on the herbaceous layer consisting mainly of grasses. These locust species are not pests of crops. The results acquired serve as basis for further research: The host plants from which insects were collected will be tested as food sources in feeding trials to assess the suitability of the collected species for farm based rearing. The number of generations per year that can be produced will determine the species potential for rearing.

Keywords: *Borocera cajanion*, *Cyrtacanthacris tatarica*, entomophagy, *Gastrimargus africanus* or *africanos*, insect rearing

Contact Address: Ravaomanarivo Lala Harivelo, University of Antananarivo, Doctoral School Life Sciences and Environment, Antananarivo, Madagascar, e-mail: lravaomanarivo@gmail.com

The Relationship between Land Sustainable Management, Stakeholder Pressure and Change in Ecosystem Services – Khorasan Province, Iran

NAGHMEH MOBARGHAEE[1], HOUMAN LIAGHATI[2], FAATEME MOHEBBI[2]

[1]*Shahid Beheshti University, Environmental Planning, Iran*

[2]*Shahid Beheshti University, Environmental Sciences Research Institute, Iran*

Measuring and anticipating the effects and changes of ecosystem services due to human activities, such as tourism, is a complicated issue in environmental assessments that can help to make the management plan suitable for the areas with remarkable natural value. The provision of environmental services map is a useful and common method used to provide information, although this method requires a lot of data. The purpose of this research is to investigate the trend of changes in ecosystem services in the "Darcastourism area in the northern Khorasan province with an emphasis on serving the storage and treatment of carbon and the quality of the habitat over a period of thirty years. In order to extract the changes in ecosystem services, land covers between the 1988 and 2018 was extracted in this area so that classes structure of ecosystem such as forests, rangeland and manmade and the course of its changes was be extracted within three decades. In order to estimate the value of recreation in the region, 384 questionnaires were completed among the local tourists. To do this research, ArcGis10.3 and Terrset's software was used. Finally to extract the changes in the ecosystem services such as carbon sequestrations and the quality of habitat, InVEST3.4.0 software was used. The results of this study indicate that the ecosystem services of carbon sequestration and habitat quality have declined over the past 30 years due to the expansion of human activities in the region. The value of each hectare for the tourism area was estimated. The inclination to pay for this region was estimated to be between 20 and 40 cents per ha. In addition, the results indicate that there are correlations between indigenous and the amount of entry fee, as well as being NGO member.

Keywords: Changes, ecosystem services, tourist, valuations, willingness to pay

Contact Address: Naghmeh Mobarghaee, Shahid Beheshti University, Environmental Planning, Velenjak, Tehran, Iran, e-mail: n_mobarghei@yahoo.com

Addressing Sustainability of Fresh Water Lake: The Ecosystem Valuation of Vellayani Lake in South India

Aswathy Vijayan[1], Elsamma Job[2], Archana Raghavan Sathyan[3]

[1]*Kerala Agricultural University, Dept. of Agricultural Economics, India*

[2]*Kerala Agricultural University (retired), India*

[3]*Kerala Agricultural University, Dept. of Agricultural Extension, India*

Fresh water lakes are sources of non saline water that can provide a manifold of direct and indirect ecosystem services to people. In a developing country like India, where availability of good quality water is crucial in determining the health security of people the presence of a freshwater lake benefits many stakeholders. A failure of information and lack of understanding on the economic value associated with any natural resource is the major reason for the degradation negative effects on sustainability. The Total Economic Value (TEV) encompasses the measure of the economic benefits of any environmental asset. The conflicting nature of interests by various stake holders is the major trap in addressing the sustainability issues. Vellayani lake, is the only fresh water lake in the Thiruvanathapuram district in Kerala state of India and is subjected to degradation due to urban pressures. So it is attempted to evaluate the TEV and prioritise the stakeholders based on the dependence on lake. The TEV is composed of provisioning service, cultural service, regulating and supporting functions. The provisioning ecosystem services are measured using market price and opportunity cost method were fishing, lotus collection, duck rearing, drinking water, irrigation water and bathing, which together contributed Rs.392.58 crore per year. The cultural services estimated using public pricing method included the services used by Centralised Sports Hostel for Canoeing and Kayaking and the annual boat race. The recreational and spiritual services value estimated using Travel Cost Method and the aesthetic value estimated using Hedonic Property Pricing accounted to Rs. 0.56 crore per year and Rs. 275.92 crore per year respectively. The Contingent Valuation Method revealed the value of regulating and supporting function as Rs 2.91 crore per year. Thus TEV of Vellayani lake estimated by the summation of economic value of various services was Rs 672.28 crore per year The value of ecosystem services expressed in the present study is an estimate of the benefits to the society and guides in prioritising the stakeholders. So these values can serve as communication tools for balanced decision making with respect to resource use and thus overcoming the traps in stakeholder management.

Keywords: Contingent valuation method, fresh water lake, hedonic property pricing method, sustainability, total economic value, travel cost method, Vellayani lake

Contact Address: Aswathy Vijayan, Kerala Agricultural University, Dept. of Agricultural Economics, Vellayani, 695522 Thiruvananthapuram, India, e-mail: aswathysuresh15@gmail.com

Assessment of Heavy Metal and Aromatic Hydrocarbons Concentration in Deposited Particulates by Plant Species in Ahvaz City

Bijan Khalilimoghadam, Attaollah Siadat, Ashkan Yousefi
Agricultural Sciences and Natural Resources University of Khuzestan, Dept. of Soil Science, Iran

Green space plants in contemporary cities are perfectly capable of deposition and retention of particulate matter and play a significant role in the purification of polluted urban air. The present paper is aimed at examining heavy metal concentration, aromatic hydrocarbons, and mineralogy of absorbed particulates by plant species in Ahvaz. To this end, samples were collected from different dominant leaf species from 10 locations in Ahvaz which had different land uses (industrial, residential, recreational, and high-traffic) in various time periods. The deposited particulate matter was measured in units of mass, area, and time. The total concentration of 29 elements (K, Na, Ba, Cu, Ni, Co, Ru, Mn, Zn, Pb, Sn, Li, Rb, Cr, B, Ga, Hf, Pd, Cd, Ti, Sr, P, Ag, U, Si, Mo, Fe, Al, and V) was measured after digestion with a mixture of perchloric and nitric acid at the ratio of 1:3 using ICP-Mass. Moreover, the concentration of aromatic hydrocarbons was measured using GC-MS. Changes in clay mineralogy of particulates and particle size distribution were studied using XRD and Master Sizer, respectively. The results of the present research revealed that steel industries in Ahvaz are one of the main sources of particulate matter, but due to the fact that these particles are primarily made of iron, their deposition is more likely to occur at distances close to this source. Particulate matter is generally composed of minerals such as calcite, silicate (quartz) and phyllosilicate. The results obtained from the decomposition of particulate matter indicated that the possible origin of elements in particulates is internal factors such as steel industries in Khuzestan province and urban traffic rather than external factors such as dust phenomenon. Carcinogenic polycyclic compounds are found everywhere in Ahvaz, the most important of which is benzo(a)pyrene, with concentrations ranging from 15 to 31 μg kg^{-1} of particulates. The findings of this study show that different plant species have a significant effect on the deposition of particulate matter and the purification of urban air. Therefore, the extension of green space using plant species with greater deposition can pave the way for reducing urban air pollution.

Keywords: Aromatic hydrocarbons, heavy metal, mineralogy, particulate

---

Contact Address: Bijan Khalilimoghadam, Agricultural Sciences and Natural Resources University of Khuzestan, Dept. of Soil Science, Mollasani, Ahvaz, Iran, e-mail: moghaddam623@yahoo.ie

Improving Seed Treatment Methods: A Key Factor to Reduce the Risk to Honey Bees and other Pollinators to Maintain Biodiversity

HAYDER ABDELGADER

Agricultural Research Corporation (ARC), Crop Protection Research Center, Sudan

Bees and other pollinators play a major role in maintaining the biodiversity in almost all environments. Pollinators support the reproduction of nearly 85 % of the world's flowering plants. However, these important species are endangered through the use of pesticides. In late April 2008, dust drift containing insecticide resulted in the largest bee poisoning in Germany for 30 years. The reason for these incidents was the contamination of flowering bee forage plants with dust particles abraded from maize seeds treated with the insecticide Clothianidin. Thus highly specialist techniques should be used when treating seeds with plant protection products to avoid such problem. The aim of this study was to improve the seed treatment methodology to reduce the drift generated from seeds by drilling and hence saving of bees and other pollinators as well as reducing the risk of people handling the treated seeds during the sowing activities and people located in the vicinity of the sowing site. The current study investigated the amount of drift generated from seeds of two varieties of cotton using two formulations of the neonicotinoid insecticide imadocloprid using the Heubach methods. The increase in percentages of drift generated with the use of Heubach Meter, through tested formulation of imadocloprid relative to the control treatment were found to be in the range of 336–378 % and 221–287% for the water dispersible powder formulation (WS) for Hamid and Barakat cotton varieties, respectively. For the Flowable Concentrate (FS) formulation the percentage increase in the drift over the control was ranging 82–95 % and 15–445 for Hamid and Barakat varieties respectively. The Heubach vaules were higher in case of WS formulations. They were ranging between 13.5–24.5 for Hamid variety and 23.3–25.4 for Barakat variety. The values for the FS formulation ranged between 7–8.8 and 2.64–14.7 for Hamid and Barakat, respectively. The pesticide residues measured were found to be more for WS formulation compared to FS formulation for both tested varieties. The results of the study indicated indicated that the Flowable concentrate formulation for seed treatment is better than the Water dispersible powder formulation and can play important role in improving seed dressing technology to save various pollinators

Keywords: Honey bees, insecticides, pollinators biodiversity, seed treatment

Contact Address: Hayder Abdelgader, Agricultural Research Corporation (ARC), Crop Protection Research Center, P. O. Box 126, Wad Medani, Sudan, e-mail: abdelgaderh@yahoo.com

Local Crop Diversity, Soil Properties and Altitude as Indicators of Carabidae Distribution in the Highland Coffee Farms

ANTHONY IJALA[1], SAMUEL KYAMANYWA[1], CHERUKUT SCOLA[1], CHRISTOPHER SEBATTA [1], JOHNNY MUGISHA[2], ERNST-AUGUST NUPPENAU[3], STEPHANIE DOMPTAIL[3], JENINAH KARUNGI[1]

[1]*Makerere University, College of Agriculture and Environmental Sciences, Dept. of Agricultural Production, Uganda*

[2]*Makerere University, School of Agricultural Sciences, Uganda*

[3]*Justus-Liebig University Giessen, Inst. of Agric. Policy and Market Res., Germany*

Carabids are sensitive to environmental changes, quantity of plant life and soil parameters, which can be influenced by biophysical conditions. In this study we carried out a two year research on the abundance and distribution of ground beetles (Carabidae) in coffee farmlands of Mt Elgon region of Uganda. The predictors under this study were altitude and cropping system and the response variables were abundance and distribution of Carabidae, abundance of semi-natural vegetation and selected soil parameters. Altitude was categorised as: low (1400–1499 m a.s.l), mid (1500–1679 m a.s.l) and high (1680–2100 m a.s.l); and cropping system as: Coffee monocrop; coffee+annual; coffee+banana; and coffee+banana+shade trees. For each altitudinal range, each cropping system was represented three times. The study was in two districts of the Mt. Elgon region making it a total of 72 study sites. The results from the generalised linear mixed models showed highly significant effects of altitude, cropping system, and the altitude*cropping system interaction on the abundance and distribution of the three Carabidae: *Anisodactylus* spp., *Chlanius* spp. and *Harpalus* spp. *Harpalus* spp. was significantly more abundant in coffee monocrop fields at lower altitude while *Chlanius* spp. was found more in coffee+banana+shade tree at mid altitude. Occurence of *Anisodactylus* spp. was higher in the coffee+annual system at high altitude. Bivariate correlations revealed a significant positive correlation between *Chlanius* spp. abundance and the abundance of semi-natural vegetation; and a negative correlation of *Chlanius* spp. and *Harpalus* spp. with the soil pH. These results showed differential preference of biophysical conditions of different Carabidae species indicating complexity in generating ecological management recommendations.

Keywords: Coffee farmlands, generalist predators, Mt. Elgon, semi natural vegetation, Uganda

Contact Address: Anthony Ijala, Makerere University, College of Agriculture and Environmental Sciences, Dept. of Agricultural Production, Kampala, Uganda, e-mail: tonyijala209@gmail.com

The Challenges of Smallholders and Small Entrepreneurs as Indirect Conservationists

LIZA MELINA MEZA FLORES

Las Americas' Fund, Projects Directorate, Peru

Peru is one of the richest countries in highly nutritious agrobiodiversity; however, the commercial varieties produced by research centres conquer the markets and consumers' preferences, contributing directly to the loss of indigenous genetic resources. The commercial varieties are more attractive over the native agrobiodiversity because of their higher crop productivity, bigger final products and shorter times to harvest. Moreover, even when highly nutritive and resistant to plagues, diseases and climate changes, the lack of economic incentives causes then the loss of local knowledge on traditional and sustainable management of the indigenous varieties and their consumption.

This situation is even more critical for smallholders in rural areas as it is too expensive to bring products to the niche markets for such special native products, as these markets are mainly in the capital cities of the regions. The lack of an efficient transport system and deficient conditions of roads, increases the inaccessibility to these markets. Moreover, these niche markets for native products, are also market for organic or ecological products. That means, the smallholders have to register their products as organic with an international certificate or in the PGS (participatory guarantee systems) to get premium prices. Both processes take time, require technical assistance, require social capital, and then the transaction costs increase.

There are several key actors, which should be involved in a strategy to increase nutrition based on native agrobiodiversity, such as the representatives of producers 'organisations, local governments, the Ministry of Environment, the Ministry of Agriculture, Ministry of Women's and Vulnerable Populations' Affairs, universities, ONGs and international funding organisations. But in my poster I will present that for sustainability it is more critical to include the niche markets, such as the Eco-fairs, ecological markets, producers' markets, and all the small entrepreneurs and smallholders, who are currently investing time and financial resources to promote special processed food with native products. These initiatives receive no support from the government, and their legal functioning is surrounded by bureaucratic procedures, high taxes, and no financial facilities, even though they are the only ones guaranteeing the sustainable conservation of indigenous agrobiodiversity and food security.

Keywords: Conservation, native agrobiodiversity, Peru, small entrepreneurs, smallholders, sustainability

Contact Address: Liza Melina Meza Flores, Las Americas' Fund, Projects Directorate, Av. Panamericana Sur 251, Lima 4 Lima, Peru, e-mail: lmeza@fondoamericas.org.pe

Analisis of Agreements Made by Public Prosecutor's Office of Goias in Meia Ponte River: Ecological Restoration of Legal Reserve Areas and Permanent Preservation Areas in Cerrado

LUCIANA RAMOS JORDÃO[1], SYBELLE BARREIRA[2]

[1]*State University of Goias, Law College, Brazil*

[2]*Federal University of Goias, Rural Development Department, Brazil*

Brazil is a country proud by its agriculture. Yet, it has major problems balancing agribusiness and attention to environmental law. Due to pressures made by farmers, the new president, Bolsonaro, elected in 2018, promised to reduce legal demands in order to avoid punishment for those who deforest areas protected by Brazilian Forest Code. Law has not changed yet, but many farmers fail to comply with regulations in the hope this will change soon. Public prosecutors work to make it possible for farmers to adjust their activities so that protected areas may remain covered with forest. To do so, Brazilian environmental law allows out-of-court agreements to avoid new lawsuits, to provide a faster response to those involved and to ease lengthy procedural system. Even though environment is considered a diffuse right, Public Prosecution Service try to reach some kind of settlement that make possible to preserve business activities and provide compensation to victims. However, considering the settlements signed in Goiás, one can notice that public prosecutors have difficulties enforcing these deals. In consequence, environmental problems take longer to be solved since the case ends up in judiciary system. There is no research that evaluates percentage of success obtained in these settlements. To do so, it's necessary to identify some agreements, considering a representative sample, and verify each case in order to determine if public prosecutors can reach a satisfying index that justify the efforts done and the time required to implement them. This, exactly, is the main research goal. The research aims to question effectiveness of environmental settlements by systematizing the information already collected by the Public Ministry. The research is based on the hypothesis that the noncompliance with the agreements occurs due to difficulties in the elaboration of adequate criteria for the activities of farmers and the peculiarities typical of Cerrado biome. Cerrado is a critical biome because any change done to it may aggravate climate changes in Brazil since the rains in Amazon Forest depend on the rivers located in Cerrado, the same happens to other Brazilian forests. Climate change consequences can reduce soil fertility, production and reduce food sovereignty.

Keywords: Cerrado, climate change, environmental law, rural development

Contact Address: Luciana Ramos Jordão, State University of Goias, Law College, Av. Perimetral Norte N. 4356 Casa 88b Cond. Alto Da Boa Vista, 74445500 Goiânia, Brazil, e-mail: lr.jordao@me.com

From Forest to Residential Housing? A Potential Victory over the "Private Profits, Public Impacts" Way of Proceeding

Victor D. Ávila-Akerberg[1], Tanya M. González-Martínez[1], Liza Melina Meza Flores[2], Marcela Nochebuena[3], Rubén Mayén[4], Florentino Morales-Solís[5]

[1] *Universidad Autónoma del Estado de México, Inst. for Agricultural and Rural Sciences, Mexico*

[2] *Las Americas' Fund, Projects Directorate, Peru*

[3] *Mexicans Against Corruption and Impunity Civil Association, NGO, Mexico*

[4] *Peoples United for the Sustainable Development of the Countryside and the City, NGO, Mexico*

[5] *Working for the Development of JIlotzingo, NGO, Mexico*

Mexico City Metropolitan Area (MCMA), one of the biggest worldwide, consumes significant environmental services, mainly provided by remaining forested areas at surrounding mountains, whose low lands have been taken by urban sprawling. In 2017 the local government allowed deforestation of 238 ha of well-preserved oak forest to favour the construction of "Diamond Forest", an urbanisation of 19985 new residences that will increase population density. The urban polygon is set on common land, partially overlapping with a natural protected area and local people have felt their traditional ways of life will be threatened, arguing forest destruction will negatively impact local nature and "its goods" (ecosystem services: 143 species, crop land, nutrient cycling, air quality, clean water provision), reducing life quality, and that local government will be obliged to provide services public-payed to the newcomers, with less natural resources. On top of it, governance is not clear and relevant stakeholders have been neglected: neither the environmental impact assessment, nor the public consultation have been taken in consideration for the approval of the project, and high public governmental figures along with a private real estate developer close to the former government appear to be involved in a non-transparent procedure. Interestingly, many different sectors of local society are overcoming the situation, from local urban neighbours next to the location, to assemblies of common-lands rural owners, have strongly requested scientific-based evidence and support from our academia to organise informed-public demonstrations and serious press investigations to raise awareness of the problem, bring a legal dispute, and to promote an actual sustainable practice for nature management. This is, therefore, a non-usual way to overcome a common situation of natural resources loss, self-organised and empowered by the civil society itself, who has built political capabilities based on legal rights and, very important, scientific-quality information about the potential environmental harms and ecosystem services loss, in order to defend their lands for the common good and sustainability. As a result of such actions, on January this year, a district judge ordered a temporary suspension of any building activities, which represents a precedent-setting victory, although partial, in an ongoing environmental conflict.

Keywords: Forest loss, natural resources governance, pro-environment actions, urbanisation

Contact Address: Tanya M. González-Martínez, Universidad Autónoma del Estado de México, Inst. for Agricultural and Rural Sciences, Campus El Cerrillo Piedras Blancas, 50090 Toluca, Mexico, e-mail: tanyamgm@gmail.com

Pro-Environmental Awareness of Nigerian Youths

LOVETH AIKOWE, JANA MAZANCOVA, MIROSLAVA BAVOROVA

Czech University of Life Sciences Prague, Fac. of Tropical AgriSciences, Czech Republic

Many studies on environmental knowledge have focused mostly on advanced regions, however developing countries need to step up their role in the fight against climate change which sets this study apart from other studies as it includes perspectives from developing countries. There are two opposing views on the relationship between environmental knowledge and pro-environmental behaviour (PEB). First, is the concept that environmental knowledge sets the basis for social behaviour that would lead to sustainable practice, however some cases have reported that having environmental education at university may not be enough to produce graduates who behave pro-environmentally in the society. Therefore, this study intends to address the environmental situation in Nigeria through the following research questions: What are the factors that determine pro-environmental behaviour and agricultural interest in Nigerian youths? What do Nigerian youths do (if they do anything e.g. recycling) in their daily lives to minimise their contribution to climate change? Does environmental knowledge translate to pro-environmental action among Nigerian youths? The methodology is based on a quantitative survey approach involving paper and online questionnaires targeting students sampled from two Nigerian universities focusing on agriculture and engineering. Data is coded then imputed for analysis using SPSS. For independent samples, comparisons between two groups of students (looking at gender and discipline) are carried out. To analyse student's knowledge, factor analysis is applied and generalised linear regression model is used to examine the main determinants of PEB. Theory of planned behaviour (TPB) is used to develop a model identifying/ explaining individual factors of PEB and pro-environmental awareness among Nigerian youths. Finally, sustainability knowledge is measured using the sustainability literacy test (SLT). Based on preliminary results, Nigerian students from agricultural university show higher awareness on climate change issue than students from other universities and higher PEB than other students. This study contributes to limited research on pro-environmental issues in Nigeria and can be replicated in other developing countries to bridge research gap in this field and regions.

Keywords: Agriculture, climate change, developing countries, pro-environmental behaviour, sustainable development, TBP

Contact Address: Loveth Aikowe, Czech Univeristy of Life Sciences Prague, Fac. of Tropical Agrisciences - Dept. of Economics and Development, Kamýcká 129, 16500 Prague - Suchdol, Czech Republic, e-mail: aikowel@ftz.czu.cz

Potential New Oil Crops for Edible and Non-Edible Purposes in Brazil

DÉBORA MADEIRA[1], KACILDA KUKI[1], SÉRGIO MOTOIKE[1], THOMAS HILGER[2], GEORG CADISCH[2]

[1]*Federal University of Vicosa, Department of Plant Science, Brazil*

[2]*University of Hohenheim, Inst. of Agric. Sci. in the Tropics (Hans-Ruthenberg-Institute), Germany*

In the current scenario of increasing demand for vegetable oils and fats, the diversification of oil-bearing species for commercial planting is a wise and necessary strategy. This action makes possible: (i) the reduction of the world´s dependency on the prevailing traditional oil crops; (ii) the sustainable development of new markets for non-food products, without affecting the food security of the global population, and (iii) the selection of oil crops adapted to the region of choice, reducing costs and environmental risks. The native Brazilian flora holds great biodiversity, with numerous oleaginous species. We present four marginal species, for their great potential for agricultural exploitation, rusticity, productivity and food and industrial aspects of their oils. *Acrocomia* sp. and *Syagrus coronata* are two palm species; the first one, with a broad geographic distribution, occupies mainly savannah and forest areas of the Southeast and Midwest of Brazil, while the latter populates especially the semi-arid regions of Southeast and Northeast Brazil. The fruits of *Acrocomia* sp. produce two types of oils: the pulp oil, rich in oleic acid and that of the kernel, rich in lauric acid. Estimates show that this palm can produce >5.0 t ha^{-1} of oil intended for food or non-food industry. The kernel of *S. coronata* contains oil composed of medium chain saturated fatty acids, and is suitable for human consumption because of its nutritional and organoleptic qualities. *Mabea fistulifera* is conspicuous in the Brazilian territory, colonizing degraded areas of Atlantic Forest, Cerrado and transition regions. The oil extracted from its seed, rich in polyunsaturated fatty acids, has applicability in the oil-chemical industry for the synthesis of polyols and polyurethanes. In the same way, the *Licania rigida* seed is an oil source for the paint, sealant and drying-oil industry, containing 70 % of lycanic acid. The oils of the four species can still be used in the pharmaceutical and renewable energy industries. Domestication is the first challenge for a sustainable and economically viable exploitation of these species, and the current *Acrocomia* sp. breeding programme can be used as a base model.

Keywords: Biodiversity, domestication, minor crops, potentials

---

Contact Address: Sérgio Motoike, Federal Univerity of Viçosa, Department of Plant Science, Rua Claudio Manoel470, 36576-298 Viçosa, Brazil, e-mail: sergiomotoike@gmail.com

Ethnobotanical Inventory of Medicinal Plants Used in Salamá District, Baja Verapaz, Guatemala

Barbora Fišerová, Eloy Fernández Cusimamani, Nayla Rodriguez
Czech University of Life Sciences Prague, Fac. of Tropical AgriSciences, Dept. of Crop Sciences and Agroforestry, Czech Republic

Medicinal plants for curing are important for local people in Guatemala, for whom its use is still common for medicinal purposes in their daily lives.
The study was focused on documentation of medicinal plants used in the traditional medicine in Salamá district, Baja Verapaz department of Guatemala. Ethnobotanical data were collected from 50 respondents (56 % women, 44 % men), through semi-open interview and by observations in the fields and speaking with local inhabitants which were randomly selected. The data were evaluated using the following indexes: use reports (UR), medicinal use value (MUV), frequency of citation (FC), relative frequency of citation (RFC) and fidelity level (FL).
A total of 44 species belonging to 41 genera and 26 botanical families were registered. The most dominant families were Lamiaceae (with 8 species; RU=12), Asteraceae (5; RU=11) and Rutaceae (3; RU=4). There were also found 21 categories of various diseases treated with medicinal plants, the largest representation was gastrointestinal disorders (76 %) and breathing system problems (47 %). According to the quantitative data evaluated, *Matricaria chamomilla* was the species of greatest use and importance (UR= 5; RFC= 0.5; MUV= 0.12; FC=26). The most used parts of the plants for the treatment of different diseases were leaves (55 %) and the used preparation mode was decoction (73 %).
Majority of people were using plants for curing themselves for over 10 years, mainly because of family traditions. People mostly buy those plants on the local market because they are easily available.

Keywords: Ethnobotany, gastrointestinal disorders, Lamiaceae, *Matricaria chamomilla*, medicinal plants, traditional medicine

---

Contact Address: Nayla Rodriguez, Czech University of Life Sciences Prague, Fac. of Tropical AgriSciences, Dept. of Crop Sciences and Agroforestry, Prague 6 - Suchdol, Czech Republic, e-mail: rodriguez@ftz.czu.cz

Constraints of Agricultural Development in the Context of Environmental Conservation for Protected Areas, Vietnam

NHUNG NGUYEN THI TRANG[1], CUONG TRAN HUU[1], LEBAILLY PHILIPPE[2]

[1] *Vietnam National University of Agriculture, Marketing, Vietnam*

[2] *University of Liegè, Agro-Bio Tech, Belgium*

The study was conducted in 2018 in Xuan Thuy national park (XTNP), a large wetland ecosystem located in northern Vietnam. The survey identifies existing farming practices and constraints of agricultural development in the context of protecting ecosystem for the 15,100 ha protected area. We used in-depth interviews for 12 staffs of local authorities and semi-structured survey for 234 farmers living adjacent the park. Farmers have been applied diverse culture practices (mono cropping, intensive aquaculture and integrated agro-forestry) but there were excessive and imbalance use of fertilisers compared with local standards. Moreover, farms have been managed with wide range of pesticides, antibiotics and various agrochemicals without carefully record keeping and less concern on environment protection. Even though farmers aim at increasing quantity of products but yields were still lower than those in other areas in Vietnam and gradually decrease recently. We have found various social-economic-environmental-institutional reasons behind sustainable agriculture prospect. Almost of farmers relied on purchased inputs while external input costs have been increasing in particularly hired labour. This leads to the fact that conservation was rank at least important while profitability was most incentive as reported by prevailing cultivators. Most of farm managers lacked of marketing activities and faced price squeeze when selling farm products. Moreover, disease outbreak, un-controlling of exotic snails and water conflicts were environmental problems preventing farmers to tackle alone. Agricultural development is a difficult task and need supports from many authorities (XTNP managers, communal authorities, irrigation branch) but their staffs lacked of capacity and connection with locals in conservation programs. Public sectors have been providing general environmental news and meetings (waste collecting or cleaning rivers) but rarely information and training on environmental friendly farming. Private input dealers played essential role in introducing new techniques but finally their purpose was selling companies' inputs. Thus, it urges on us actions for heightening awareness of farmers, strengthening capacities of staffs in accordance with more communication with farmers. In addition, improving farm productivity simultaneously minimising impairments to the water body should be addressed. Economic sustainability could also be improved through strengthening marketing strategies and forming farmers' market groups.

Keywords: Constraints, environment protection, protected areas

Contact Address: Nhung Nguyen Thi Trang, Vietnam National University of Agriculture, Marketing, Trau Qui Town Gia Lam District, 100000 Ha Noi, Vietnam, e-mail: thuytrangnhung@gmail.com

Nutrition, food quality and food processing

Food supply chain innovations to enhance food security

Oral Presentations

Posters

Prospect of Organic Agriculture in Achieving Food Security in the Least Developed Countries

NIRAJ PRAKASH JOSHI[1], LUNI PIYA[2]

[1] *Hiroshima University, Taoyaka Program, Japan*

[2] *Hiroshima University, Graduate School for International Development and Cooperation, Japan*

Organic agriculture (OA) is praised unequivocally for its environmental and health benefit. But OA, is questioned about its potential impact on global food security due to its low yield reported in many systematic yield studies in developed countries, more specifically USA, Canada and Europe. There is very limited study on the yield of OA and or its socio-economic impact of OA in Least Developed Countries (LDCs). LDCs have a high prospect of conversion to OA because of a very low level of external inputs such as synthetic chemical fertilisers and pesticides use in their agriculture system. At the same time, a haphazard use of these external inputs are having serious consequences on human health as well as the environment. This paper aims to analyse prospects of OA in achieving food security goal of LDCs. FAOSTAT data on food balance and literature are the main source of data. Yield ratios of crops show higher yield from OA compared to conventional agriculture in LDCs. Food supply is growing at higher rate in LDCs than Northern America between 1961 and 2011. However, growth rate in nutrient supply in LDCs is meager. Hence, current food supply in the region is not able to supply the standard nutrient requirement. Cereal is the single most important food item contributing to nutrient supply in LDCs indicating lack of nutrient diversity. Under such context OA would increase food supply of all food categories. Consequently, LDCs can achieve the standard nutrient requirement and help in dealing with persistent undernourishment in LDCs. However, concerted effort is necessary for wider acceptability of OA in LDCs.

Keywords: Food production, food supply, North America, nutrient supply, yield ratio

Contact Address: Niraj Prakash Joshi, Hiroshima University, Taoyaka Program, 1-5-1 Kagamiyama, 739-8529 Higashi Hiroshima, Japan, e-mail: nirajpj@hiroshima-u.ac.jp

How Can Urban Agriculture Contribute More to the Well-Being of Farmers in a Disfavoured Environment?

Erik Engel[1], Nicole Paganini[2], Tinashe Kanosvamhira[3], Luisa Chicamisse Mutisse[4], Abongile Mfaku[3], Ivo Cumbana[4], Anja Schelchen[2], Anja Kuehn[2]

[1]*Frankenfoerder Forschungsgesellschaft, Germany*

[2]*Humboldt University Berlin, Centre for Rural Development, Germany*

[3]*University of the Western Cape, Geography and Environmental Studies, South Africa*

[4]*Eduardo Mondlane Univerrsity, Mozambique*

Urban agriculture (UA) can have multiple benefits, from feeding people to creating green oasis in sealed environments. The action-research presented here seeks to contribute to increasing the benefits of urban agriculture for food and nutrition security (FNS) and income for disfavoured populations in Cape Town and Maputo. In both cities UA plays a negligible role for FNS and to the food systems as a whole. History and structural differences have shaped very different urban landscapes:

Nearly 40‘000 persons are depending on urban agriculture in Maputo. Over 15’000 farmers (most of them organised in associations) and backyard gardeners sell mainly cabbage and lettuce on local markets. The majority of sales happens via intermediaries. The lack of crop rotation increases the pressure of pests and diseases, which is countered by (often uncontrolled use of) pesticides. Most farmers have a fairly diverse diet. Like other low-income groups of Mozambique, they are nonetheless moderately food insecure.

The importance of urban agriculture in Cape Town rests mainly on its contribution to more diverse diets of the farmers as well as in the social networks the farmers are establishing. Despite an “urban agricultural policy”', farmers receive little support from government. NGOs are the main source for inputs and technical knowledge. There are few local markets in the townships, and township vegetable mainly furnish (predominantly white, inner-city) markets with the help of small sales enterprises usually founded by NGOs. In that way, farmers receive support and income, but are, simultaneously, highly dependent on “their” NGO. Self-organisation so far has been weak. Social processes in the townships - just as the entire food system of the city- are still dominated by dynamics originating in the apartheid and colonial heritage. Activities in and around urban agriculture can contribute to lowering social barriers and to food diversity of involved households. It cannot be a supplement for serious social and economic policies necessary in the disadvantaged neighbourhoods.

UFISAMO research covered value chains, organisational structures, consumption habits and knowledge transfer. An overview of results, good practices and recommendations to strengthen the respective roles of UA will be presented.

Keywords: Cape town, food and nutrition security, food systems, Maputo, urban agriculture

---

Contact Address: Erik Engel, Frankenfoerder Forschungsgesellschaft, Invalidenstrasse 42, 10115 Berlin, Germany, e-mail: engel@frankenfoerder-fg.de

Nutritional Status of Adolescent Girls across the Rural-Urban Interface of Bengaluru

Ashwini B C[1], Umesh K B[1], Veerabhadrappa Bellundagi[1], Stephan von Cramon-Taubadel[2], Anjali Purashotham[2]

[1]*UAS GKVK Bengaluru, Dept. of Agricultural Economics, India*
[2]*University of Goettingen, Dept. of Agricultural Economics and Rural Development, Germany*

Nutrition is essential for growth and development of human beings and the nutritional status is the state of individual's health determined by what we eat. A proficient nutritional status of a country's population is responsible for its socio-economic development. Increased employment opportunities among men and women have a direct influence on the lifestyle changes, diet and consumption patterns which in turn influence health and nutritional status, especially among women and adolescent girls. Adolescence is the growth stage between infancy and adulthood of a person and is broadly considered as the age ranging from 10 to 19 years. Although adolescents are the most valuable human resources, their health has been neglected for many years because they were considered to be less vulnerable to disease than young children or the very old.
Among the adolescents, girls are the worst sufferers of malnutrition especially in rural areas of Asian countries. Regrettably, assessment of the nutritional status of adolescent girls has been the least explored area of research in India. Carried out in the framework of the Indo-German collaborative research project 'The Rural-Urban Interface of Bengaluru - A space of Transition in Agriculture, Economics and Society', subproject 'Food Insecurity at different stages of Urbanisation', funded by the Department of Biotechnology (DBT) of the Government of India, the present study focused on the nutritional status of adolescent girls in the rural-urban interface of Bengaluru, India. Primary data was collected along a north and south transect crossing through Bengaluru and covered 467 respondents (adolescent girls). Anthropometric measurements and multinomial logistic regression were used to analyse the magnitude and factors influencing their nutritional status across the city's rural-urban interface. Factors such as age, education, wealth index and location dummy had a significant influence on the nutritional status of adolescent girls. In particular, urbanisation had a direct impact on the nutritional status of adolescent girls: when moving along the gradient from rural to urban of both south and north transect, the magnitude of overweight increased to the tune of 33%. Urban adolescent girls were relatively more prone to overweight because of change in lifestyle and consumption of unhealthy food items. The study revealed that rural girls were suffering from undernutrition and urban girls were suffering from overnutrition, i.e. clearly demonstrating the double burden of malnutrition.

Keywords: Adolescent girls, anthropometric measurement, multinomial logistic regression, nutritional status, urbanisation

Contact Address: Ashwini B C, UAS GKVK Bengaluru, Dept. of Agricultural Economics, Bengaluru, India, e-mail: ashwinismile813@gmail.com

Nutritional Status of Women and Children in the Rural-Urban Interface of Bangalore

K. Geetha[1], D. Vijayalakshmi[1], Shilpa Yatnatti[1], Christoph Dittrich[2]

[1] *University of Agricultural Sciences, Food Science and Nutrition, India*
[2] *University of Goettingen, Inst. of Geography: Human Geography, Germany*

Dual burden malnutrition continues to be a major public health concern in India. Determination of forms of malnutrition clusters is important for the development of appropriate interventions. The study investigated nutritional status of women and children in the rural-urban interface of Bengaluru. A total of 300 middle income families spread across rural, transition and urban regions of north and south transects of Bengaluru was selected by purposive random sampling. Nutritional status of women respondents and children in the family were assessed by comparing the somatic measurements (height, weight, mid upper arm circumference) and their indices (Body Mass Index, Waist Hip Ratio) with standards. According to Body Mass Index the majority (63 %) of women were overweight or obese across the rural-urban gradient. Though there was increase in per cent overweight and obese women from rural to urban it was statistically non-significant. Similar trend was also observed for Waist Hip Ratio classification of women. When body weight is considered, among children 52.1 per cent were below 90 per cent of IAP (Indian Academy of Paediatrics) standard in rural, followed by 51.6 per cent in transition and 40 per cent in urban. Interestingly in urban 35.3 per cent of children were above 100 per cent standard. The per cent children with height meeting above 100 per cent of standard was almost equal in all the gradients. It is noteworthy that children falling below 90 per cent of standards were more in transition (22.6 %), followed by rural (16.7 %) and urban (15.3 %). MUAC an indicator of muscle mass was less than 90 per cent IAP standards in majority of the children irrespective of gradients. In all, the anthropometric measurements compared with standards indicated maximum percent of rural children to meet less than 90 per cent standards compared to transition and urban. Maximum per cent of underweight children were in rural (25.5 %), whereas over weight were more in urban (14.3 %). These findings determine pattern of malnutrition influenced by growing urbanisation in Bengaluru on surrounding rural localities with respect to nutritional status of women and children.

Keywords: BMI, children, somatic measurements, women

Contact Address: Shilpa Yatnatti, University of Agricultural Sciences, Food Science and Nutrition, Uas Gkvk, 560065 Bengaluru, India, e-mail: shilpayatnatti@gmail.com

Crop Diversification, Household Decision-Making, and Dietary Diversity: A Panel Data Analysis of Ethiopian Rural Households

Orkhan Sariyev, Tim K. Loos, Ling Yee Khor
University of Hohenheim, Inst. of Agric. Sci. in the Tropics (Hans-Ruthenberg-Institute), Germany

Sub-Saharan Africa is home to most of the ultra-poor population. Many of whom experience alarmingly high nutritional deficiencies. Considering that most of the ultra-poor are subsistence farmers, the importance of production diversity for better nutrition seems obvious. Both on-farm diversification and dietary diversity are determined by different demographic, socioeconomic and institutional factors. In this regard, we employ panel data analyses of 363 Ethiopian smallholders surveyed in 2014 and 2016. The objectives are (i) to identify the main determinants of crop diversity captured by the Simpsons Index of Diversity and simple crop count, (ii) to estimate the effect of crop diversification on household dietary diversity, and (iii) to determine the effect of women's participation in decision-making on both crop and dietary diversity.
With regard to the first objective of the study, we employ the quasi-maximum likelihood estimation method for a fractional response with correlated random effects and Poisson fixed-effects models for SID and crop count, respectively. A Poisson regression model is employed to capture the effects of crop diversity indicators on dietary diversity. Furthermore, regarding the third objective, we capture women's participation in decision-making processes regarding the type of crop produced and food purchased and identify their association with the respective outcomes in the estimations.
We find a positive effect of farm size and adoption of crop rotation and/or intercropping practices on crop diversity, while an increase in non-farm income contributes negatively. Female participation in decision-making regarding crop production positively contributes to the number of crops produced. Results also show that, for smallholders with no or very little non-farm income, crop diversity is important to achieve better dietary diversity irrespective of the distance to the market. On the other hand, households with some non-farm income are observed to have no significant gain from further crop diversification. Moreover, female participation in decision-making regarding food purchases results in higher dietary diversity. Our analyses contribute to the growing literature on the linkage between crop diversification and dietary diversity and shows a significant positive association between women's decision-making and both aspects of household livelihood.

Keywords: Crop diversification, decision-making, dietary diversity, Ethiopia

Contact Address: Orkhan Sariyev, University of Hohenheim, Inst. of Agric. Sci. in the Tropics (Hans-Ruthenberg-Institute), Wollgrasweg 43, 70599 Stuttgart, Germany, e-mail: o.sariyev@uni-hohenheim.de

The Food Security Standard: It Fills a Gap but has some Traps

TINA BEUCHELT[1], RAFAËL SCHNEIDER[2], LILIANA GAMBA[3]

[1]*University of Bonn - Center for Development Research (ZEF), Ecology and Natural Resources Management, Germany*

[2]*Deutsche Welthungerhilfe e.V., Policy and External Relations, Germany*

[3]*WWF, Germany*

Sustainability standards address risks and problems in international agricultural value chains. Despite certifying agricultural goods that are produced in food insecure countries, most standards do not particularly address food security. A consortium of the University of Bonn and the NGOs Welthungerhilfe and WWF therefore developed the "Food Security Standard" (FSS) which includes a set of criteria and indicators that can be integrated into any sustainability standard. The FSS is tested in five pilot countries with different crops on plantations and in smallholder settings in Asia, Latin-America and Africa. The accompanying research identifies means to easily measure food security at local and national level. It further analyses the applicability of the FSS, its legitimacy and reliability of results. Over 25 key persons were interviewed, more than 80 farmers and workers and six multi-stakeholder workshops conducted in five pilot countries and Germany.

Results show that the FSS can be integrated into a normal audit of a sustainability standard and does not require significant additional efforts. With structured interviews and clear criteria and indicators, the FSS is able to reliably capture the food security situation of workers, farmers and communities and identifies gaps. While plantations are expected to be able to comply with the FSS requirements, only better-off smallholder farmers are probably able to meet all criteria, though still might need some external support. Plantations, smallholder organisations or their representatives indicated their willingness to implement the FSS but not without market demand and price premiums.

Severely food insecure farmers are not expected to be able to comply with the FSS. To avoid their exclusion from (sometimes lucrative) markets, an option is to set up a "Food Security Sensitive Management", keeping the idea of the FSS. This alternative approach should include all actors along the value chain to uptake and proof their responsibility for food security when buying sustainability certified products from hungry smallholders. The question of how much the private sector can contribute to increase smallholders' food security, how to avoid "green washing"' on social issues and how compliance with the FSS could progressively be achieved, needs further investigation and stakeholder discussions.

Keywords: Certification, plantation, smallholder, sustainability standard

Contact Address: Tina Beuchelt, University of Bonn - Center for Development Research (ZEF), Ecology and Natural Resources Management, Genscherallee 3, 53113 Bonn, Germany, e-mail: beuchelt@uni-bonn.de

Perceptions and Practices of Rural Women on Cultivation and Consumption of Vegetables and Fruits: An Intervention for Food-Related Health Problems in Amhara Region, Ethiopia

SIMEGN ALAMIREW[1], BERNHARD FREYER[1], JIM BINGEN[2]

[1]*University of Natural Resources and Life Sciences Vienna (BOKU), Div. of Organic Farming, Austria*

[2]*Michigan State University, Dept. of Community Sustainability, United States of America*

A diet that is deficient in nutrients can lead to health problems. Rural women and children are the most vulnerable social groups in terms of food security and balanced diets. Many studies in Ethiopia are witnesses that women and children are mostly affected by malnutrition, undernutrition, non-communicable diseases and other food related health problems. Also certain cultural eating practices and knowledge deficits can lead to food related health problems.

The aim of this study is therefore to understand the rural women's attitudes in cultivating vegetables and fruits for their own consumption, as one approach to secure healthy diets. A survey with 320 purposely-selected women in eight rural *Kebeles* (Amhara region) was done and data analysed with descriptive statistics. The findings of this study indicate that there is neither any eating habit nor little trend of cultivation of vegetables and fruits; and there is no or little awareness about the significance of vegetable and fruits for a healthy diet. Moreover, women perceived that fruits and vegetables are only for unhealthy/ill and weak individuals, which is a widespread conviction in their communities and also women accept such kind of cultural belief. Furthermore vegetables and fruits are understood as luxury foods which are not very important to their health. Economic and educational status of women and availability of fruits and vegetables also determine their perceptions and habits of consumption of vegetables and fruits. In cases, women produce vegetables and fruits, they mostly sell to secure their income. Hence appropriate food practice programs such as providing intensive training for rural women on how to cultivate vegetables and fruits as well as habituating diverse feeding practices, including vegetables and fruits, and awareness raising about the relevance of vegetables and fruits for their health, are highly recommended.

Keywords: Balanced diet, diverse foods, perception and practice, unhealthy food, vegetables and fruits

---

Contact Address: Bernhard Freyer, University of Natural Resources and Life Sciences (BOKU), Div. of Organic Farming, Gregor Mendel Straße 33, 1180 Wien, Austria, e-mail: Bernhard.Freyer@boku.ac.at

Marketing Channels for Underutilised Indigenous Fruit Tree Products: The Case of Baobab Pulp on Seed in Kenya

GEORGE KAIMBA[1,2], MUENDO MUTUKU KAVOI[1], DAGMAR MITHÖFER[2]

[1]*Jomo Kenyatta University of Agriculture and Technology, Dept. of Agricultural and Resource Economics, Kenya*

[2]*Rhine-Waal University of Applied Sciences, Faculty of Life Sciences, Germany*

Underutilised indigenous fruit trees have become part of the income generating opportunities for local communities living in arid and semi-arid lands of developing countries. The baobab tree is a good example of an underutilised indigenous fruit tree in Kenya that provides fruits and vegetables. In the past decade, the tree has generated much attention owing to its enormous nutritive attributes. Despite the high nutrient and polyphenol content, coupled with acceptance as food ingredient by the European Union (EC 2008) and the US Food and Drug Administration (FDA 2009), the tree and its products remain underutilised in Kenya. This is due to limited empirical research on the tree, especially the marketing structure of unprocessed baobab fruit and pulp in Kenya. This study therefore analyses the determinants of collectors' participation in different marketing channels using 270 baobab collectors in three counties in Kenya. A multinomial logit regression is employed in the empirical analysis. The results show that baobab collectors participate in the domestic markets through three main channels, namely: assemblers, rural wholesalers and urban buyers. The majority of the collectors traded via the assemblers and rural wholesalers. Price and non-price factors classified under a) human capital factors such as gender, age, number of children, marital status, number of trees, collection point, selling experience, other incomes and number of buyers known; b) transactional factors which include price of baobab, wage rate, transport cost, packaging costs distance to market, product form and price awareness in other markets; and c) institutional factors such as access to credit, which all significantly influence collectors' marketing channel decision. The main policy recommendations include improving road networks to reduce on transport costs; capacity building to sensitize and promote awareness of the importance baobab of products in order to increase its demand domestically; improve institutional services in order to promote baobab collectors' access to credit, marketing information and training.

Keywords: Kenya, Baobab pulp, malnutrition, market development, marketing channels, poverty reduction

---

Contact Address: George Kaimba, Rhine-Waal University of Applied Sciences, Faculty of Life Sciences, Marie-Curie-Str.1, 47533 Kleve, Germany, e-mail: kinyuakaimba@jkuat.ac.ke

Status and Prospect of Edible Insects in Myanmar

AYE AYE MYINT

Yezin Agricultural University, Department of Animal Science, Burma

People in different regions of Myanmar consume a large number of insect species. The consumption of insects by rural and urban residents will improve the food security. However insect consumption in urban areas is for snacks especially at night as they are a preferred food with beer and other beverages. Most edible insects traditionally have been wild harvested. Some species are available only seasonally while others are collected throughout the year. The time of insect harvesting from wild habitats depends on specific species behaviour and life cycles. While crickets (*Acheta domesticus*) are one of the most prefer insects in Myanmar, cricket farming is not well developed yet. Small-scale cricket farming requires relatively little investment. But the main constraints are lack of knowledge of farming techniques and inexperience in marketing. This study aims to help local farmers to adopt the simple method of the cricket farming practices to enhance food security, food safety and livelihood opportunity for women in selected areas of Myanmar. House cricket *A. domesticus* can be reared using a simple method with e.g. carrot, pumpkin and Chinese cabbage as feeds. Total life cycle duration is about 52 days at 27–30 °C. This method can reduce the cost of feeds and farmers will be able to get a quick financial return on investment. Silk worm *Bombyx mori* is one of the potential edible insect in some areas of Myanmar. Silk worms have been reared by the silk industry. Small-scale producers not only support the regional silk industry but also consume the silk worm pupae, an important by-product at the household level. Training is needed to provide farmers, extension agents, and households with knowledge and skills on insect farming, marketing, nutrition and product development. The results of this study can support the government to design policies to promote the insect sector in Myanmar and to raise awareness among consumers on nutrition and health benefits of eating insects.

Keywords: Cricket, consumption, marketing, production, silkworm

Contact Address: Aye Aye Myint, Yezin Agricultural University, Department of Animal Science, Yezin, 15013 Zayarthiri, Burma, e-mail: ayeayemyint2006@gmail.com

Crop Diversification and Participation of Farm Households in Different Marketing Chains of Finger Millet and Maize Crops vis-à-vis Food Security Status in Rural-Urban Interface of Bengaluru

VEERABHADRAPPA BELLUNDAGI[1], UMESH K B[1], ASHWINI B C[1], PRASANNA KUMAR P S[1], STEPHAN VON CRAMON-TAUBADEL[2]

[1] *UAS GKVK Bengaluru, Dept. of Agricultural Economics, India*

[2] *University of Goettingen, Dept. of Agricultural Economics and Rural Development, Germany*

The linkage between agriculture and nutrition is complex and often debated in India's policy discourse. The enigma of being the fastest growing economy and yet the largest home of under- and mal-nourished population takes away the sheen from India's economic achievements. Diversification of agriculture in favour of competitive and high-value enterprises is considered an important strategy to augment farm income, generate employment, alleviate poverty and conserve soil and water resources. In the recent past, agricultural diversification occurred largely through crop substitution. Within the Indo-German research project '*The rural-urban interface of Bengaluru - a space of transition in agriculture, economics and society*', sub project '*Food insecurity at different stages of urbanisation*' funded by the Department of Biotechnology, Government of India, we examined crop diversity, use of marketing chains and food security status of farm households (HHs). Primary data was collected from 659 HHs situated along a south and 616 HHs located along a north transect across the rural urban interface of Bengaluru. Crop diversification was analysed using Herfindahl index (HI), with a HI close to zero representing complete crop diversification and a value close to one representing complete specialisation.

A larger area was cultivated along the south (212.58 ha) than along the north transect (178.28 ha). However, crop diversity was higher along the north (0.46) than the south transect (0.49), and highest specialization was found the rural-urban transition zone. Participation of HHs in marketing chains of finger millet and maize revealed that finger millet is mainly grown for family consumption rather than for market sale. Producers realized a higher millet price on farmers' markets than through other marketing channels, and food security status was also higher in HHs using this channel. In case of maize, HHs selling to the Agricultural Produce Marketing Committee realised higher prices than HHs selling on other markets. Overall, farm HHs along the south transect were more food secure than those on the north transect. The volume of marketed surplus was higher for maize than for the staple crop finger millet, but there is a larger scope for finger millet production in Karnataka as it is rainfed.

Keywords: Crop diversification, finger millet, food security, Herfindahl index, maize, marketing channels

Contact Address: Veerabhadrappa Bellundagi, UAS GKVK Bengaluru, Dept. of Agricultural Economics, 560065 Bengaluru, India, e-mail: veeru.b4619@gmail.com

Food Transition Across Rural-Urban Gradient of Bengaluru, Karnataka

K. Geetha[1], Vijayalakshmi D[1], Shilpa Yatnatti[1], Christoph Dittrich[2]

[1]*University of Agricultural Sciences, Food Science and Nutrition, India*
[2]*University of Goettingen, Inst. of Geography: Human Geography, Germany*

Urbanisation and economic growth leading to life style changes are major factors responsible for food and nutrition transitions, a shift from indigenous traditional food to diversified global foods. The present study was undertaken to explore food transitions across rural-urban gradients of the South Indian mega city of Bengaluru covering a northern and a southern transect. From the geocoded localities 300 middle income households comprising of rural (n=100), transition (n=100) and urban area (n=100) were selected by purposive random sampling technique. Information pertaining to food habits, dietary diversity and food intake was elucidated through standardised questionnaire. The findings revealed, that quite all of urban respondents (99.0 %) had three meals per day, but in rural environment it was only 67%. Interestingly, majority of rural families were non-vegetarians (94.0 %) against transition (82.0 %) and urban (85.0 %). Statistically meal pattern and food habits were significant across the rural-urban gradient. Differences in inclusion of high value foods such as fruits, vegetables, processed foods, meat and eggs contributed to slight changes in household dietary diversity score from rural to urban. It was observed that consumption of cereals, oils and fats, and sugars, which are predominant sources of energy, was more than the recommended dietary allowances in all three study areas in both the transects. However, highest intake of cereals was observed in the most rural parts of both transects. Intake of oils, fats and sugars was comparatively higher in urban families. Interestingly, an increasing trend in consumption of milk and milk products was observed from rural to urban gradient, Its adequacy ranged from 57.57% (rural) to 77.92% (urban) in the north transect. However, in the southern transect it ranged from 53.90% to 70.07% among rural and urban, respectively. Consumption of vegetables was higher among all the study areas compared to roots, tubers and green leafy vegetables. Fruit consumption was comparatively higher among females in south transect compared to the north. These transitional changes in food consumption in the rural-urban food system were attributed to availability, accessibility and diversity in food across the rural-urban gradient which may impact on nutritional and health status of families.

Keywords: Dietary diversity, food intake, rural-urban gradient

Contact Address: Shilpa Yatnatti, University of Agricultural Sciences, Food Science and Nutrition, Uas Gkvk, 560065 Bengaluru, India, e-mail: shilpayatnatti@gmail.com

Physicochemical Quality and Microbial Safety of Cow Milk Sold in Ruzizi Lowland, South-Kivu, Democratic Republic of Congo

BWIHANGANE BIRINDWA, CIZA AZINE, JACKSON ISHARA, SIMON BAENYI

Evangelical University in Africa, Animal Science, DR Congo

Cow milk is one of the most preferred foods in Democratic Republic of Congo and in South-Kivu in particular due its nutritional quality. However, it IS easily contaminated by bacteria. Milk provides relatively fast income for small producers. However, milk quality depends on many environmental factors. Nutritional quality and microbial safety of cow milk sold in Ruzizi lowland in South-Kivu was investigated. In this regard, a total of 120 samples were collected, there of 90 and 30 in Buvaku and Ruzizi lowland, respectively. For both areas of study samples were collected at each level of the value chain including farm, wholesalers and retailers. The results showed that the type of milking used in Ruzizi lowland is totally manual. The majority of breeders (65 %) prepare teats before milking. In Ruzizi lowland all farmers maintain livestock buildings using insecticides and 55 % of them use fungicides as disinfectants. Fever is the most common clinical sign in farms in the Ruzizi plain (55.0 %), followed by diarrhea (30.0 %) and mastitis (15.0 %). To prevent the various diseases in the herd, a large part of the farmers use drugs (75.0 %) for a good health of the animals whereas others substitute the medicines by medicinal plants (25.0 %) that they administer in the ration. In addition, 92.5 % of the wholesalers are women while 7.5 % are men and they are all married. Most sellers face some difficulties including lack of equipment (45.0 %) and finished product (32.5 %). All sellers pack milk in containers and do not practice any type of preservation. Furthermore, the pH of the milk varied throughout the value chain with an average of 6.2 at the farm level, 5.4 at the wholesale level, and 5.0 at the retail level. Total titratable acidity did not vary significantly throughout the chain, with an average 16.8°D. All collected samples were contaminated but the level of TAMF (Total Aerobic Mesophilic Flora) contamination varied between each level of the value chain. All samples contained total and faecal coliforms due to the lack of hygiene during milking; however, no *Salmonella* was found in milk.

Keywords: Bukavu, cow milk, microbial safety, physicochemical quality, Ruzizi lowland

---

Contact Address: Bwihangane Birindwa, Evangelical University in Africa, Animal Science, Q. Ndendere Panzi, 3323 Bukavu Town, DR Congo, e-mail: adjibir@yahoo.fr

Potential of Edible Insects to Combat Hidden Hunger in Sub-Saharan Africa

Nils Nölle[1], Alexandra Baisch[1], Yisa Njowe Kieran Bong[2], Elizabeth Kusia[2], Subramanian Sevgan[2], Hans Konrad Biesalski[1]

[1] *University of Hohenheim, Inst. for Biological Chemistry and Nutrition / Food Security Center, Germany*

[2] *International Centre of Insect Physiology and Ecology (ICIPE), Kenya*

Edible insects are gaining attention as an alternative food source to fish and livestock. However research has mainly focused on the protein content of edible insects, while their micronutrient content and thus their potential to combat hidden hunger, is far less studied. Therefore, edible insects, commonly consumed in Kenya and Uganda, two countries, which show a high risk to suffer from hidden hunger, were collected. Samples included long horned grasshoppers (*Ruspolia differens*), crickets (*Gryllus bimaculatus*), as well as several kinds of caterpillars (*Gonimbrasia zambesina, Cirina forda*). If available, samples were collected fresh and processed, e.g. boiled or fried, to account for possible losses during processing. Samples were analysed for selected (pro)vitamins (riboflavin, carotenoids) via HPLC and for dietary minerals (e.g. iron and zinc) measured either by ICP-MS or ICP-OES. To ensure consumer safety, levels of heavy metals were also analysed, using either ICP-MS or atomic absorption spectrometry.

All analysed samples showed high contents of riboflavin (1.18–3.11 mg/100 g dried sample material). Grasshoppers analysed with their wings on contained remarkably higher amounts of lutein, zeaxanthin and beta-cryptoxanthin compared to their plucked counterparts. Beta-carotene was contained within all samples, ranging from 1.82 to 49.70 mg/100 g dried sample material contained in crickets. All analysed samples contained substantial amounts of iron (33.05–1078.57 $mg\,kg^{-1}$ dried sample material) and zinc (49.91–131.88 $mg\,kg^{-1}$ dried sample material), and, surprisingly, also calcium (223.30–1458.99 $mg\,kg^{-1}$ dried sample material). All samples showed negligible levels of cadmium and mercury, but some had elevated levels of lead (0.10–3.15 $mg\,kg^{-1}$ dried sample material).

Therefore insects, consumed in Kenya and Uganda show great potential for delivering high levels of micronutrients in particular those related to hidden hunger.

Keywords: Alternative food sources, insects, malnutrition, micronutrients, nutrition

Contact Address: Nils Nölle, University of Hohenheim, Inst. for Biological Chemistry and Nutrition / Food Security Center, Garbenstraße 30, 70599 Stuttgart, Germany, e-mail: noelle@uni-hohenheim.de

Participatory Value Chain Development to Enable Farmer Groups to Add Value to Pineapples in Uganda

MICHELLE MAUSBACH[1], MARGARETA LELEA[1], OLIVER HENSEL[2], BRIGITTE KAUFMANN[1]

[1] *German Institute for Tropical and Subtropical Agriculture (DITSL), Germany*

[2] *University of Kassel, Agricultural and Biosystems Engineering, Germany*

In Uganda, pineapples are one of the most important horticultural cash crops supporting the livelihoods of numerous small-scale farmers. However, due to their high moisture content of over 80 % and their perishable nature, pineapples are prone to spoilage and wastage. Although fruit processing is promoted for its potential to avoid post-harvest losses and add value to the local pineapple supply chain, particularly small-scale processing faces numerous constraints under the prevailing low-external input conditions.

Within the frame of a transdisciplinary research project that seeks to reduce losses and add value in small-scale East African food chains, this study aims at co-developing feasible small-scale pineapple syrup production with improved product quality and expanding marketing possibilities together with pineapple value chain stakeholders in Uganda.

Qualitative data was obtained during fieldwork in Masaka District in Uganda between April - July 2018 by applying a participatory and actor-oriented approach using a series of group meetings (n=27 with 2–15 participants each) conducted with two farmer groups that produce syrup, complemented with semi-structured interviews (n=20) and participant observations (n=4).

The first part of the results reveals motivations, benefits and challenges of pineapple syrup production from the perspective of the farmers and established processors. The second part of the results describes the participatory value chain development process and focusses on quality criteria and the actions performed to achieve a high-quality syrup by the farmer groups. Finally, the results show how farmer groups co-developed a context-specific assessment procedure to maintain and improve syrup quality that reflected product quality standards demanded by the market. This led to an iterative review of their business objectives to promote value addition.

The study highlights the importance of applying collaborative learning approaches within participatory value chain development for stakeholder empowerment to co-design innovations that they can put into action.

Keywords: Co-design innovations, collaborative learning, pineapple, participatory methods, Uganda

---

Contact Address: Margareta Lelea, German Institute for Tropical and Subtropical Agriculture (DITSL), Steinstraße 19, 37213 Witzenhausen, Germany, e-mail: m.a.lelea@ditsl.org

Residues of Melamine Persist in Meat of Broiler Fed In-Feed Larvacide after Mandatory Withdrawal Period

Oluwafolakemi Adeleye[1], Oluseyi O. Oluwatosin[2], Adebayo V. Jegede[3], Olusesan Fafiolu[3], Adeoye J. Oyewusi[3], Olumuyiwa Tunde Sigbeku[4]

[1]*Africa Centre of Excellence in Agricultural Development and Sustainable Environment, Livestock Science and Sustainable Environment, Nigeria*

[2]*World Bank Africa Centre of Excellence in Agricultural Development and Sustainable Environment (CEADESE), FUNAAB, Nigeria, Livestock Science and Sustainable Evironment Programme, Nigeria*

[3]*Federal University of Agriculture, Abeokuta, Dept. of Animal Nutrition, Nigeria*

[4]*National Agency for Food and Drug Administrtion and Control, Animal Feed and Premix Division Veterinary Medicine and Allied Products Directorate, Nigeria*

Larvacides are commonly used in poultry feed to alter the moulting stage of houseflies (*Musca domestica*) which hatch on droppings, hence reducing the population of flies and smell nuisance in poultry houses. Cyromazine, an active ingredient in larvacides, has melamine as a metabolite which became a public health concern after the death of 9 infants and hospitalisation of 294,000 others after taking melamine tainted infant formula. However, cyromazine is widely used to reduce smell from poultry litter and to increase nitrogen content which usually translates to higher weight gains in broilers without the knowledge of its residual effect on tissues of animals. In this study, residues of in-feed larvacide in broiler tissues was investigated.

All protocols used in this study were approved by the Animal Care and Use Review Committee guidelines of Centre of Excellence in Agricultural Development and Sustainable Environment, Federal University of Agriculture, Abeokuta, Nigeria. One-hundred and sixty day old Arbor Acre broilers of approximately 40 g body weight were used in the study. Four diets were formulated to contain cyromazine at 0, 0.25, 0.50 and 0.75 g kg^{-1} and they were assigned to 4 dietary groups consisting of 4 replicates per treatment of 10 birds each in a completely randomised design for 42-days feeding trial. A bird per replicate (4 birds $treatment^{-1}$) was sacrificed to harvest tissue for residue determination at week 7, 8, 9 and 10 to establish a 7, 14, 21 and 28 days withdrawal period. Results indicated that cyromazine residue in meat (thigh and drumstick) were higher ($p < 0.05$) in the treated groups than the control group, which also contained residue of melamine. In conclusion, residues of cyromazine and melamine were left in the tissues of broiler chickens even up to 28 days withdrawal period, which is higher than the maximum allowable limit by WHO.

Keywords: Cyromazine, larvacide, melamine, residue, chicken meat

Contact Address: Oluwafolakemi Adeleye, Africa Centre of Excellence in Agricultural Development and Sustainable Environment, Livestock Science and Sustainable Environment, No 4 Faseun Avenue Ebenezer Obantoko, Abeokuta, Nigeria, e-mail: folake2016@yahoo.com

The Impact of Irrigation on Households' Food Security in Rural Ethiopia: An Application of Difference-in-Difference Technique

MEAZASHWORK GETIE ERUCKNEH[1], BAMLAKU ALAMIREW ALEMU[2]

[1]*Nile Insurance, Ethiopia*
[2]*Addis Ababa University, Development Studies, Ethiopia*

In Ethiopia, irrigation development for rural households emerged as a development strategy to fight against the recurrent drought since a long time back. This is believed to augment the otherwise meagre incomes households earn from rain-fed agriculture. However, studies evaluating the contribution of such interventions are scanty in the country limiting our knowledge and the possible strategies to be developed for policy uptake. The main objective of this study is, therefore, to evaluate the impact of irrigation on the food/nutrition security level of households using Food Consumption Score. Food Consumption score indicates not only the type of food households consume but also the quality of the foods. This suggests that it helps also measure nutrition aspect of the food households consume. For this reason, the outcome variable for food security is measured using this uni-dimensional measure; Food Consumption Score. Data emerged from two waves of panel data collected by the Ethiopian Central Statistics Agency and World Bank. Data analysis was carried out using descriptive statistics and econometric analyses involving pooled regression (for robustness check). To account for the problem of unobserved heterogeneity, we used Difference-in-Difference model with fixed effects regression technique. The result shows that, after controlling for the effects of time-invariant unobserved factors, participating in irrigation has a positive and significant impact on food score of rural households. Therefore, all involved actors in rural development should put in place a concerted effort to realise the role irrigation could potentially play for rural development in general and food security of households in Rural Ethiopia in particular.

Keywords: Difference-in-Difference, food security, households, impact evaluation, irrigation

Contact Address: Bamlaku Alamirew Alemu, Addis Ababa University, Development Studies, Addis Ababa, Ethiopia, e-mail: bamlakalamirew@gmail.com

Impact Evaluation of a Cooling Truck on the Camel Milk Value Chain in Isiolo, Kenya

MELINA GERHARD[1], PASCALE WAELTI[1], GENEVIEVE ATIENO OWUOR[2], DAVIS IKIROR[2], SARA IMBACH[3]

[1]*Bern University of Applied Sciences; School of Agricultural, Forest and Food Sciences, International Agriculture, Switzerland*

[2]*Vétérinaires Sans Frontières (VSF) Suisse, Kenya*

[3]*Vétérinaires Sans Frontières (VSF) Suisse, Switzerland*

Providing livelihood in times of drought, the camel is a well-adapted animal to dry areas in the Horn of Africa and its milk plays an important role in the diet of pastoralist people. Some of them have moved into urban centres like the Eastleigh neighbourhood in Nairobi, where camel milk is mainly supplied by Anolei camel milk cooperative from Isiolo. When the milk was transported to Eastleigh with public buses, several inconveniences for the Anolei cooperative members as well as the Eastleigh milk dealers arose. Among others, milk got lost or spoiled and it was brought to Nairobi on unsteady times. As a consequence, the cooperative acquired a cooling truck in October 2017, partly subsidised by Vétérinaires Sans Frontières (VSF) Suisse. Since then, the truck is bringing a daily amount of approximately 4000 l of milk from Isiolo to Eastleigh.

This study assessed the impact of the cooling truck on the camel milk value chain from Isiolo to Nairobi. A value chain analysis determined benefits, downsides and social changes for the situation with and without cooling truck. A cost-benefit analysis established the economic viability of the truck for the cooperative income and a partial budget analysis gave insight into the changes for individual members.

According to Anolei Cooperative members, the benefits of the truck outweigh the constraints. Less spoilage and loss of milk, a reliable time schedule, convenient pick-up spots and more free time are only few of the advantages of the truck. On a purely economic level, the truck costs outweigh the benefits for the cooperative, that is to say that net present value and internal rate of return are both negative while benefit-cost ratio is below 1. Moreover, impacts on the individual members' household budget are slightly negative. The main reason for the adverse economic outcome is the unexploited potential of the truck. Its capacity could be utilised better if Anolei Cooperative was transporting milk from non-members or other cooperatives. Another way to boost total additional benefits is to increase proceeds by elevating transportation fees. These alternatives are discussed in the poster.

Keywords: Arid lands, camel milk, cooling truck, value chain

Contact Address: Melina Gerhard, Bern University of Applied Sciences; School of Agricultural, Forest and Food Sciences, International Agriculture, Kuehrainweg 3, 4805 Brittnau, Switzerland, e-mail: melina.gerhard@gmx.net

Sustainability Hot Spot Analysis of Insect Supply Chains for Food and Feed

DARYA HIRSCH, MARTIN HAMER, ISABELLE HIRSCH

Bonn-Rhein-Sieg University of Applied Sciences, International Centre for Sustainable Development (IZNE), Germany

In western countries, using insects as food and feed is still at a very early stage in comparison to many countries in Global South, where insects are an integral part of people's nutrition. Globally, experts predict a market growth for insect-based products and insects itself as food within the next years, based on higher demand through an increasing world population and the need of alternative protein resources for human nutrition. Nevertheless, insect rearing is new everywhere and the larger scale production systems are mostly found in western countries. This raises the question if insect production can remain sustainable in future, in terms of social acceptability and economic viability as well as resource effectiveness and efficiency. One of the possible approaches to measure the sustainability of value and supply chains holistically, but in a simple manner, is the Sustainability Hot Spot Analysis (SHSA), developed by the Wuppertal Institute (Germany). We applied this approach for the first time on insects for food and feed. We supplemented the existing SHSA method with economic categories, since insect supply chains are hardly researched due to rare practises of insect farming. So far, economic aspects have not been considered in the current literature on SHSA. Beside this, ethical and ethnical aspects need to be taken into account in the social categories. Ethnical aspects contain e.g. justifying the killing of insects in different religions, whereas the ethical aspects refer to welfare issues such as animal husbandry, herd management as well as transport. Our preliminary analysis showed that within the phase of insect rearing, hot spots were identified for energy consumption caused by e.g. fodder production as well as in disease and pest management. In addition, further hot spots were identified in the processing phase also concerning energy consumption due to e.g. grinding, dehydrating, mechanical drying as well as freeze-drying. The next planned step is to put the focus of the SHSA on selected edible insect species, like field crickets (*Gryllus bimaculatus*), bamboo worm (*Omphisa fuscidentalis*), palm weevil (*Rhynchophorus ferrugineus*) as well as black soldier fly (for feed) (*Hermetia illucens*) and conduct additional stakeholder analyses.

Keywords: Economy, environmental and social impact, insect farming and processing, LCA, resource efficiency

Contact Address: Isabelle Hirsch, Bonn-Rhein-Sieg University of Applied Sciences, International Centre for Sustainable Development (IZNE), Grantham-Allee 20, 53757 Sankt Augustin, Germany, e-mail: isabelle.hirsch@h-brs.de

Food Security of Dairy and Non-Dairy Farming Households in Northwest Cameroon

RAJA RAM MEHTA, EVA SCHLECHT

University of Kassel / University of Goettingen, Animal Husbandry in the Tropics and Subtropics, Germany

Food insecurity and lack of nutritional diversity remain a challenging problem in many central African countries. This study was carried out within the BMBF-funded UrbanFood^{Plus} project and focused on the city of Bamenda in Northwest Cameroon. There, due to market failure and political crisis, the conditions for small-scale urban dairy farming greatly deteriorated since 2016. By comparing urban households (HHs) having dairy cattle and selling milk on the informal market (group I) to urban HHs having dairy cattle but selling no more milk since 2016 (group II) and to urban HHs that sold all dairy cattle after the market failure (group III), it was tested whether owning dairy cows as such, and also selling milk, has an impact on the family's food diversity and food security. Altogether 152 HHs comprising 1,147 individuals were surveyed between March and June 2018. Three sets of questionnaires, namely a 7-day food diary, the standard household food insecurity access scale (HFIAS) questionnaire and a general questionnaire on HH characteristics were used to assess the diversity and frequency of food intake, food insecurity status and demographic and socio-economic HH condition. Based on the acquired information, food consumption score (FCS), a measure of food diversity, and HFIAS, an indicator for food security, were calculated, and related to HH characteristics. FCS revealed that group I had the greatest food diversity, followed by group III, and group II. Likewise, group I was the most food secure according to HFIAS, but group III was less food secure than group II. Both production diversity score (PDS) and number of income sources showed a significant positive correlation with FCS and a negative correlation with HFIAS. According to ordinary least squares (OLS) regression, an HH head with secondary or higher-level education had a significantly negative impact on both the indicators. Diversification of income sources through training on agriculture-related skills, encouragement for dairy farming along with development of dairy infrastructure and diversification of farm activities seem to be crucial to improve the food security status of agriculture-based HHs in Bamenda.

Keywords: Dairy cattle, food diversity, food security, indicators, Northwest Cameroon

Contact Address: Raja Ram Mehta, University of Kassel / University of Goettingen, Animal Husbandry in the Tropics and Subtropics, Göttingen, Germany, e-mail: manava.evol@gmail.com

A Customized Assessment Tool to Differentiate Safety and Hygiene Practices in Emerging Dairy Chains

JAMES LEDO, KASPER HETTINGA, PIETERNEL LUNING

Wageningen University and Research, Food Quality and Design Group, The Netherlands

In emerging dairy chains, inconsistent product quality and safety issues are prevalent. The need for continuous improvement in practices to meet rapidly growing demands for dairy products quality and safety is imperative. However, existing assessment tools are often based on advanced food safety management systems and standards, which present an inaccurate picture of practice performance in emerging dairy chains. This study presents the development of a customised tool to assess and differentiate levels of safety and hygiene practices in emerging dairy chains. The tool consists of indicators and corresponding grids to assess and differentiate the levels of safety and hygiene practices at the farm, during transportation, milk bulking, and retail points crucial for microbiological and chemical (i.e. aflatoxin) safety. The assessment tool was piloted in Tanzania, as an example of an emerging dairy chain, using interviews, farm visits and audio-visual assisted observations. Thirty-eight smallholder farmers and three large-scale farms were interviewed and their farm practices observed. Overall, the customized assessment tool was able to differentiate accurately safety and hygiene practices of the farmers. The obtained profiles of farmers' safety and hygiene practices provided a basis for stepwise improvement of practices of individual farmers in emerging dairy chains. Moreover, the profiles provided a starting point for development of training programs customised to the knowledge and skills needs of groups of farmers with similar profiles. The tool is useful for a comprehensive baseline and post-intervention assessment to give evidence of the impact on farmers' safety and hygiene practices achieved after an intervention.

Keywords: Dairy chain actors, milk, milk safety, on-farm practices

Contact Address: James Ledo, Wageningen University and Research, Food Quality and Design Group, Bornse Weilanden 9, 6708 WG Wageningen, The Netherlands, e-mail: james.ledo@wur.nl

Developing Processing Technologies of Edible Insects with Innovative Approaches to Enhance Rural Nutrition in Madagascar

TOLOJANAHARY CHRISTIAN RATOMPOARISON, JEAN ROGER EMILE RASOARAHONA, FELAMBOAHANGY RASOARAHONA

University of Antananarivo, College of Agricultural Sciences, Food Science and Technology Department, Madagascar

The richness of flora and fauna biodiversity with more than 80 % endemicity is an asset for rural nutrition in Madagascar. The wild resources collected by communities (plants, mushrooms, insect ...) are little or not transformed for lack of adequate technology or the quantity harvested does not allow it. This study values the knowledge of communities that have consumed insects for centuries by developing small and medium scale edible insect processing technologies. Two main species of wild insects consumed are collected in the wild and preserved by lyophilisation before analysis. The results of micronutrients and macronutrients analysis are compared with those of individuals of the same species reared under controlled conditions. Production (breeding) and processing (from harvesting to slaughtering to long-term preservation) trials were conducted with the effective participation of selected farmers in the Malagasy highlands. Insects were dried and powdered, and the meal thus obtained was incorporated to daily consumed products at different rates (10 %, 15 % and 20 %) to assess the acceptability and preference of the community. The technologies developed and validated by local stakeholders are appropriate to the peasant reality in terms of equipment availability and investment cost. The energy balance was controlled through the use of hybrid solar-wood combustions energy. Results of nutritional values of processed products and their acceptability are presented and discussed. In the median to long term, the innovative processing technologies presented can have a positive impact on the nutrition of a population with chronic deficiency of protein and oligo-elements and have the potential to become an additional source of income for rural households with little means (landless). The national policy is still to be developed with the actors concerned to further promote the edible insect sector. The results of this study can contribute to this development.

Keywords: Acceptability, insect meal, micronutrients, participative research, processing, protein

Contact Address: Tolojanahary Christian Ratompoarison, University of Antananarivo, College of Agricultural Sciences, Food Science and Technology Department, Campus Ambohitsaina, 175 Antananarivo, Madagascar, e-mail: christianratompoarison@gmail.com

Complementing Sustainable Resource Management with Land Use Intensification to Overcome Poverty in Ethiopia

KARIN GAESING

University of Duisburg-Essen, Inst. for Development and Peace (INEF), Germany

The research project "Ways out of extreme poverty, vulnerability and food insecurity" conducted by the Institute for Development and Peace of Duisburg University with funds from the German Federal Ministry for Economic Development and Cooperation (BMZ) aims to develop recommendations for development cooperation regarding how to reach extremely poor, vulnerable and food insecure people and to effectively improve their lives. The research approach is based on identifying good practice projects and assessing their outreach approaches, project activities, lessons learnt and above all their impact with regard to the reduction of food insecurity, vulnerability and poverty.

One of these good practice projects is situated in Ethiopia. Around Debre Tabor in the highlands of Northern Ethiopia, people have been food insecure and extremely poor, eking their living from 0.5 ha of land per family. They grew cereals and pulses and left the land bare for the rest of the year. The implementation of the LANN+ (Linking Agriculture and Natural Resource Management for Nutrition Security) approach of Welthungerhilfe enabled them to harvest three times a year on their same 0.5 ha with a diversification of crops including fruit and vegetable and a significant increase of productivity. They apply small scale irrigation as well as improved agronomic practices such as the efficient use of rain water and sustainable natural resource management. Nowadays, they not only manage to feed their families, but they are also able to send their children to school, improve their housing, participate in savings and credit groups and start petty trading and other income generating activities.

The research team found the Debre Tabor project to be one of the good practice examples to sustainably overcome hunger, malnutrition and poverty. Research results from Debre Tabor and other projects point to the conclusion that sustainable land use change and natural resource management on smallholder farms provide a safe and lasting way out of poverty and food insecurity.

Keywords: Land use intensification, natural resource management, northern Ethiopia

Contact Address: Karin Gaesing, University of Duisburg-Essen, Inst. for Development and Peace (INEF), Duisburg, Germany, e-mail: kgaesing@inef.uni-due.de

Enhancing the Sustainability of Insect Rearing Systems for Nutrition through a Multidimensional Potential Assessment Framework

INGO WAGLER

Center for Development Research, University of Bonn, Ecology and Natural Resources Management, Germany

The utilisation of insects for benefits in resource efficiency and nutrition is gaining momentum. In several countries, both, in the global North and South, insect rearing systems are being developed and introduced at a rapid pace. At the same time, expressions of concern about underestimated risks and reports of failures are also increasing. As yet, there is no standard or systematic approach to assess the actual potential insects represent for nutritional goals in a given context. The research project introduces a multidimensional approach to inform context-specific adaptations and thereby enhance the sustainability of the rearing systems and the corresponding value chains.
Our investigation is based on a three-dimensional framework, developed to allow a comparative evaluation of the primary social, ecological and entomological variables related to insect rearing. The approach was applied to a case study in Sandrandahi Commune in the Amoron'i Mania Region in the central highlands of Madagascar. Studies comprised a primary data collection phase and an ongoing secondary phase. The primary phase was conducted during an interdisciplinary research mission in October 2018 and two follow-up visits in 2019. The main research methods employed were gender and age disaggregated thematic focus group discussions, key stakeholder interviews, transect walks and insect inventories. Secondary data featured demographic, seasonal/climatic and biological data on different insect species.
We concluded that the potential for insect rearing in the case study area is significant. Needs and acceptance are high in the social dimension, as is a widespread demand for resource efficient nutrient sources, especially proteins in the ecological dimension. In the entomological dimension, a major obstacle identified was the fact that no single species meets joint criteria of being widely accepted as food, well adapted to the ecological conditions, especially climate and known for its suitability for domestication. Results also suggested that significant constraints in the availability of labour and investment, in suitable and sustainable feed sources and the inefficiency of market supply chains will have to be addressed by a rearing system design.

Keywords: Insect farming, Madagascar, nutrition, social-ecological systems

Contact Address: Ingo Wagler, Center for Development Research, University of Bonn, Ecology and Natural Resources Management, Genscherallee 3, 53113 Bonn, Germany, e-mail: wagleri@uni-bonn.de

Comparative Evaluation of the Composition, Digestibility and Functionality of Chemically Modified Protein Isolates from Soya Bean and some Under-Utilised Legumes

OLUWATOYIN ALETOR[1], FUNMILOLA ISHAYA[2], VALENTINE ALETOR[3]

[1]*The Federal University of Technology, Akure, Chemistry, Nigeria*

[2]*Federal Polythenic, Ado-Ekiti, Dept. of Science Technology, Nigeria*

[3]*University of Africa, Nutrition and Biochemistry, Nigeria*

Native proteins, particularly those from indigenous under-utilised legumes have limited application especially in industrial food systems largely because of information dearth on their quality and functionality. Because of the rising cost of soya bean, there is renewed interest in evaluating the potentials of alternatives. Consequently, two (2) under-utilised legumes, namely: Pigeon pea (PP) (*Cajanus cajan*) and African yam bean (AYB) (*Sphenosstylis sternocarpa*) were processed into their protein isolates using alkaline (NaOH) solubilisation and acid (HCl) precipitation at their various isoelectric pH. The protein isolates were modified using acetic anhydride. The protein isolates and their chemically modified forms were thereafter analysed with respect to their proximate composition, metabolisable energy (ME), *invitro* multi-enzyme protein digestibility (IVPD) and functional properties. The findings were compared with the more conventional soya bean protein isolate (SI). On the average, the PP and AYB had 92.54 and 90.13 g/100, SI had 85.8 g/100g crude protein while the modified under-utilised legumes had 92.4 and 90.5 g/100g, respectively. Ash was higher in protein isolate of PP (3.6 g/100g), AYB (3.3 g/100g) and low in SI (1.0 g/100g) while the modified form had 1.9 and 1.8 g/100g, respectively. However, ME was highest in the Soya isolate (512.4 Kcal/100g) than those of PP and AYB which ranged between 369.01 and 380.07 Kcal/100g. The protein isolates of PP and AYB were more digestible and ranged between 91.0 and 98.8 % when compared to 85.9 % in Soya bean. The Soya bean isolate had better water holding capacity (WHC), oil holding capacity (OHC) and foaming stability than the PP and AYB Isolates. The foaming capacity, emulsion capacity and emulsion stability of the modified PP and AYB protein isolates were generally higher than those of soya bean isolates. Modification generally improved protein functionality when compared with the unmodified isolates. Given the higher *in vitro* protein digestibility and other functional attributes of pigeon pea and African Yam Bean than Soya isolates, it was concluded that these under-utilised legumes seeds could serve as useful alternatives for the much more expensive soya bean.

Keywords: Chemical modification, protein isolates, soya bean, underutilised legumes, pigeon pea, African yam bean

Contact Address: Oluwatoyin Aletor, The Federal University of Technology, Akure, Chemistry, PMB 704, Akure, Nigeria, e-mail: toyinaletor@yahoo.com

Antimicrobial Activity of *Lactobacilli* and *Bifidobacteria* Isolates against Pathogenic Bacteria in Dairy Product

HASSAN ABDEL MONEEM IBRAHIM EL DEMERDASH
Arish University, Dept. of Food and Dairy Sciences and Technology, Egypt

Food-borne illnesses are still a major concern for consumers, the food industry and food safety authorities. Meanwhile, consumers have been questioning the safety of synthetic preservatives of food.

Four lactic acid bacteria were isolated from a traditional Egyptian fermented dairy product. The isolates were identified using genus-specific PCR technique as *Lactobacillus plantarum* HM1, *Lactobacillus rhamnosus* HM1, *Bifidobacterium bifidum* Bb11 and B12.These isolates showed different antibacterial agents against pathogens in their culture cell filtrates. The protective effects of isolated lactic acid bacteria against pathogenicity induced by *Staphylococcus aureus*, *Salmonella, Listeria* monocytogenes and *E. coli* O157:H7. The antimicrobial agents were bacteriocin and organic acids. HPLC analysis indicated that the isolates produced lactic and acetic acids. Propionic acid was produced only by bifidobacteria isolates. The inhibition activity of these substances has been reported to be strain-dependent. Inhibition zones diameters were affected by the addition of Proteinase K and / or neutralisation of cell filtrates. All isolates were observed to behave a good antagonistic activity against the tested indicator strains with differences in size of inhibition zone (mm), meanwhile *Listeria* monocytogenes and *E. coli* O157:H7 were not inhibited by the extract of *Lactobacillus rhamnosus* HM1and *Bifidobacterium bifidum* Bb11 respectively.

Supernatant and cell pellets of isolates were used as biopreservation system in yoghurt and Karish soft cheese. The sensory and microbiological analyses were followed during the storage at 4°C for 21 days. Data showed that no apparent or detectable of pathogenic bacteria growth to the end of the storage period up to 21days. Sensory evolution showed that, use of the combination between supernatant and cell pellets at concentration of 1:1, enhanced the antimicrobial activity and improved the organoleptic properties and the viability of starter culture of the resulted dairy products. These results indicate that isolated lactic bacteria can be used as control agent for food contamination by pathogenic bacteria as bio-preservation system.

We can conclude that, the application of antimicrobial compounds-producing protective cultures may provide an additional parameter of processing in order to improve the safety and ensure food keeping quality.

Keywords: Antimicrobial activity, biopreservation, isolation, pathogens

Contact Address: Hassan Abdel Moneem Ibrahim El Demerdash, Arish University, Dept. of Food and Dairy Sciences and Technology, Arish, Egypt, e-mail: eldemerdashhassan@gmail.com

Nutrition security

Women's Age and Eating Time Determine Impact of Nutrition Education on Dietary Diversity Scores of Women and Children in Uganda and Kenya

Irmgard Jordan[1], Bárbara Fróes[1], M. Gracia Glas[1], Anna Röhlig[1], Samwel Mbugua[2], Julius Twinamasiko[3], Sahrah Fischer[4], Lydiah M. Waswa[2], Johnny Mugisha[5], Jeninah Karungi[5], Elizabeth Kamau[2], Thomas Hilger[4], Ernst-August Nuppenau[6]

[1]*Justus-Liebig University Giessen, Center for International Development and Environmental Research, Germany*

[2]*Egerton University, Department of Human Nutrition, Kenya*

[3]*Makerere University, Dept. of Agribusiness and Natural Resource Economics, Uganda*

[4]*University of Hohenheim, Inst. of Agric. Sci. in the Tropics (Hans-Ruthenberg-Institute), Germany*

[5]*Makerere University, School of Agricultural Sciences, Uganda*

[6]*Justus-Liebig University Giessen, Inst. of Agric. Policy and Market Res., Germany*

Poor dietary diversity lead to malnutrition, especially for vulnerable individuals such as mothers and infants. The objective of this study was to assess the impact of nutrition education (NE) on time-use and dietary diversity of women and children in Kapchorwa District, Uganda, and Teso South Sub-County, Kenya.

In June 2016 and 2017 cross-sectional agriculture-nutrition surveys were conducted, targeting approx. 830 farm households with children aged below five years. Women Dietary Diversity (WDDS) and Child Dietary Diversity Scores (CDDS) were calculated based on data from 24h-dietary-recalls. Time use was assessed based on 24h-physical-activity-recalls. NE was conducted and partly linked with agriculture extension (AGNE). Groups with no intervention were included in analysis for comparison. The results showed that a gendered division of labour, and beyond the time spent on farming, women were engaged in child care and domestic activities, had therefore a longer workday and significantly less free time than men. The control group had a significant decrease in ΔWDDS in the year of the intervention which was a drought year. The age of the mother explained partially the variance in ΔWDDS but differently in both countries.

In Uganda children of the NE group showed a significant increase in CDDS (M diff=0.31 point). Differences in Δeating time of mothers was statistically significant related to CDDS ($p = 0.045$).

Differently from the results of Uganda, the moderation analysis of Kenya confirmed that participation in the intervention resulted in a slight increase in women's dietary diversity scores. In terms of development, according to Bonferroni post hoc test, the

Contact Address: Irmgard Jordan, Justus-Liebig University Giessen, Center for International Development and Environmental Research, Senckenbergstr. 3, 35390 Gießen, Germany, e-mail: Irmgard.Jordan@ernaehrung.uni-giessen.de

changes on scores were statistically significant higher in the AGNE (M diff=0.580 points, $p = 0.013$) and NE groups (M diff=0.613 points, $p = 0.007$) when compared to the control group. The eating time of mothers in Kenya was positively statistically significant associated with CDDS after the intervention in all three groups. This indicates that the more time mothers spent eating/feeding their children, infants' diets became more diverse.
Attention should be given to gender-time-use in agriculture and nutrition intervention to enhance impact of nutrition education and agriculture interventions.

Keywords: Agriculture innovations, gender, Kenya, nutrition eucation, time use, Uganda

The Potential of Sustainable Antimicrobial Additives for Food Packaging from Native Plants in Benin

BARBARA GÖTZ[1], MATHIAS HOUNSOU[2], SYLVAIN DABADÉ[2], THOMAS HAVELT[3], MICHAELA SCHMITZ[3], DJIDJOHO JOSEPH HOUNHOUIGAN[3], JUDITH KREYENSCHMIDT[1]

[1]*University of Bonn, Inst. of Animal Sciences, Germany*

[2]*University of Abomey-Calavi, School of Nutrition, Food Science and Technology, Benin*

[3]*University of Applied Sciences Bonn-Rhein-Sieg, Germany*

In sub-Saharan Africa, food losses occur mostly in the early steps of the supply chains. The main reasons are damages due to handling during harvest, transport, and accelerated spoilage due to environmental and hygienic conditions. Packaging solutions can help to decrease losses, but packaging waste is an increasing issue in developing countries, and fossil resources are limited. Therefore, the implementation of active biobased and biodegradable packaging solutions is of great interest. Active packaging solutions can increase shelf life and food safety of perishable foods by inhibiting the growth of bacteria. A sustainable approach is the usage of phytochemicals from local plants as additives for active packaging. Consequently, extracts from 16 plants from Benin were screened for their antimicrobial activity against spoilage bacteria of perishable foods, foodborne pathogens and relevant fungi. The antimicrobial activity of ethanolic extracts was determined using the agar diffusion assay and identifying the minimal inhibitory concentration. Also, the potential of synergetic effects of combinations of extracts was studied using the same methods. The active phytochemicals were identified by different chemical analyses. As another important factor in food spoilage, also the analysis of the antioxidative potential was performed. Half of the studied plants showed activity against spoilage and pathogenic bacteria. For example, *Staphylococcus aureus* was inhibited by eight ethanolic plant extracts (n=16) with an inhibition zone of 14-17 mm (well=8mm). Additionally, fungal growth was inhibited. These results could be confirmed by the chemical analyses of the phytochemicals and the antioxidative activity. Positive synergetic antimicrobial effects of combinations of plants give further opportunities in packaging design. The results of the antimicrobial screening showed that native plants from Benin have potential as an additive against pathogens and spoilage organisms of perishable foods. Integrating such antimicrobial additives, in biobased and biodegradable materials, offers advanced opportunities in sustainable food packaging solutions. Besides, the prolongation of shelf life of a few days can deliver an important contribution to the reduction of food losses.

Keywords: Active packaging, antimicrobial, biobased, food loss, sustainability, Western Africa

Contact Address: Barbara Götz, University of Bonn, Inst. of Animal Sciences, Katzenburgweg 7-9, 53115 Bonn, Germany, e-mail: bgoetz@uni-bonn.de

Consumer Awareness and Attitude towards Baobab Products in Kenya

COLLINS KIRUI KIPROTICH[1], KAVOI MUTUKU MUENDO[1], DAGMAR MITHÖFER[2]

[1] *Jomo Kenyatta University of Agriculture and Technology, Kenya*

[2] *Rhine-Waal University of Applied Sciences, Faculty of Life Sciences, Germany*

Baobab (*Adansonia digitata* L.) shows high market potential in arid and semi-arid areas of Kenya. It is widely traded by few actors, but its products occupy a small market share. The benefits of baobab products are yet to be communicated to consumers. This study aimed to examine awareness and attitudes of rural and urban consumers in Kenya towards baobab products. Data on socio-economic characteristics, baobab product awareness, and attitude towards baobab products was collected from 353 consumers in rural and urban markets in Kenya. A pretested questionnaire was administered to each consumer through personal interviews. Descriptives were used to analyse awareness level on baobab products. Zero-truncated Poisson regression model was used to asses factors influencing awareness of baobab products. Descriptive and exploratory factor analysis was used to assess the attitudes of the local consumers towards baobab products. The results showed a mean awareness score of 5.92 and 6.45 for urban consumers and rural consumers respectively, this is a clear indication of low awareness level compared to the maximum awareness score of 21. The model results revealed that age ($p < 0.05$), gender ($p < 0.01$) and group membership ($p < 0.01$) influenced consumer awareness positively, while income level (0.05) had a negative influence. Descriptive showed that considerable percentage of consumers expressed positive attitudes, with (95 %), (77 %) expressing positive attitude on baobab pulp being food for all, and health benefits accrued from consumption of pulp respectively. Exploratory factor analysis produced five factors that explained 63.18 % of the total explained variance. The first factor was 'health, nutritive value and culture' which accounted for 17.01 % of the total explained variance. Other factors were 'product availability and freshness' (12.50 %), production process and affordability (11.56 %), taste and income value (11.55 %), age and social class (10.56 %). Group membership is essential, it provides links to access a variety of information that is vital in raising baobab products awareness level. Finally, improvement of product labelling, certification and freshness, awareness creation through formal and informal education will positively shape consumers attitudes.

Keywords: African baobab, attitudes, awareness, utilisation

Contact Address: Dagmar Mithöfer, Rhine-Waal University of Applied Sciences, Faculty of Life Sciences, Marie-Curie-Str. 1, 47533 Kleve, Germany, e-mail: Dagmar.Mithoefer@hochschule-rhein-waal.de

Effect of Sieve Particle Size on Quality Attributes of Peeled and Unpeeled Orange Fleshed Sweet Potato Composite Flours

Solomon Kofi Chikpah[1], Joseph Kudadam Korese[2], Oliver Hensel[1], Barbara Sturm[1]

[1] *University of Kassel, Agricultural and Biosystems Engineering, Germany*

[2] *University for Development Studies, Agricultural Mechanization and Irrigation Technology, Ghana*

Orange fleshed sweet potato (OFSP) is an important source of vitamin A and other health promoting nutrients. With the need to preserve these nutrients, processing of OFSP into high quality flour is seen as a timely strategy. Substituting wheat flour with OFSP flour in processed food products could reduce foreign exchange outlays, create new markets for producers, and result in especially increased vitamin A consumption among consumers provided there is adequate retention of ß-carotene during processing. The objective of this study was to assess the effect of sieve particle size on the physicochemical, functional, and nutritional properties of peeled and unpeeled OFSP composite flours. Peeled and unpeeled OFSP slices were pretreated by soaking in sodium metabisulphite solution (0.5 % w/v), dried in hot air dryer at 60 °C, milled separately and sieved using 250 μm or 500 μm mesh particle size. Each OFSP flour was used to prepare blends of OFSP and wheat flours and evaluated for quality using standard procedures. The sieve particle size had no significant ($p > 0.05$) effect on moisture content, water activity and CIE colour attributes of OFSP flours. However, the peeled OFSP composite flours had higher L*, b* and C* values than unpeeled OFSP flours. The sieve particle size significantly ($p < 0.05$) affected water absorption capacity (WAC), swelling capacity (SC) and water solubility (WS). The OFSP flours sieved with 500 μm mesh particle size had significantly ($p < 0.05$) higher WAC and SC but lower WS as compared to 250 μm sieve. The peeled OFSP flours had significantly ($p < 0.05$) higher WS but lower WAC and SC than the unpeeled OFSP flours. Generally, OFSP flours had lower pasting temperature and peak time than 100 % wheat flour. The final trough, setback viscosity and pasting temperature did not vary ($p > 0.05$) between the corrective unpeeled and peeled OFPS flours. The sieve particle size did not significantly ($p > 0.05$) affect the nutritional composition of flours. With exception of Vitamin C, unpeeled OFSP whole flour had the highest phytonutrients composition and antioxidant activity while wheat flour obtained the least values. Both unpeeled and peeled OFSP flours sieved with 250 μm or 500 μm have high potentials as ingredients for novel food developments.

Keywords: Composite flour, functional properties, orange fleshed sweet potato, phytonutrients, sieve particle size, total antioxidant activity

Contact Address: Solomon Kofi Chikpah, University of Kassel, Agricultural and Biosystems Engineering, Nordbahnhofstrasse 1a, 37213 Witzenhausen, Germany, e-mail: schikpah@uds.edu.gh

Characterisation of Macauba (*Acrocomia aculeata*) Kernel Proteins and Evaluation of their Solubility Properties

SÉRGIO HENRIQUE TOLEDO E SILVA[1,2], LIDIANE BATAGLIA SILVA[2], STEPHANIE BADER-MITTERMAIER[2], GABRIELE DOER[2], PETER EISNER[1,2]

[1] *Technical University of Munich (TUM), School of Life Sciences Weihenstephan, Germany*

[2] *Fraunhofer Institute for Process Engineering and Packaging IVV, Germany*

Macauba palm (*Acrocomia aculeata*), a perennial and fructiferous palm tree native to tropical and subtropical Americas and the extensive Macauba palm tree reservations in Brazil, is considered as a promising alternative for vegetable oil production. Its productivity is estimated to be from 2,500 to 5,000 kg of oil ha^{-1} y^{-1} and the oil from Macauba kernel is economically attractive due to the high content of short chain fatty acids. After de-oiling, the kernel meal offers a protein concentration between 30–50 %, thus highly suitable for the production of vegetable proteins. Hence, characterising the Macauba kernel proteins is imperative in order to develop an extraction process that promotes the sustainable use and the complete economical exploitation of this yet underexploited wild plant species.

The present work aims at the characterisation of the proteins from de-oiled Macauba kernel meal and the evaluation of their solubility properties under different extraction conditions. The findings showed that the globulins (salt soluble proteins) accounts for approximately 60 % of the total extractable proteins of Macauba kernel. Electrophoretic pattern performed by sodium dodecyl sulfate-polyacrylamide gel electrophoresis demonstrated that the Macauba globulins are mainly composed by the 7S (48.5 kDa) and the 11S (43.6 kDa) globulins. The isoelectric point of the 7S and 11S globulins, determined by two dimensional electrophoresis, ranged from 4.1–5.9 and from 3.3–6.8, respectively.

Minimum solubility of Macauba kernel proteins was observed at pH 4.0 with 10.9 % of protein recovery (PR). At low ionic strength 0.1 mol L^{-1} of NaCl), the maximum PR at pH 6.0–9.0 ranged from 25.6 to 31.2 %, whereas at higher salt concentration (0.50–0.75 mol L^{-1} of NaCl), PR increased to 64.8–73.1 % at same pH range. Protein extraction yields depend strongly on the properties of the different protein fractions present in the raw material, which can be separated by changing the conditions of the milieu. With the results herein described, it will be possible to design a process for extraction of the majority of the proteins available in the Macauba kernel meal. Furthermore, the present study also contributes to leverage the sustainable use of Macauba fruits, a renewable resource for diversifying feedstock and ensuring biodiversity.

Keywords: *Acrocomia aculeata*, globulins, Macauba, protein extraction, protein solubility, sodium dodecyl sulfate-polyacrylamide gel

Contact Address: Sérgio Henrique Toledo E Silva, Technical University of Munich (TUM), School of Life Sciences Weihenstephan, Weihenstephan, Germany, e-mail: sergio.toledo@ivv.fraunhofer.de

Urban Agroecology for Health and Wellbeing

MAI ABBAS, ALEX FRANKLIN, CHIARA TORNAGHI

Coventry University, Centre for Agroecology, Water and Reilience, United Kingdom

The number of people who are living in urban areas is constantly increasing. The historical rise of urbanisation alongside industrialisation has produced a rift between humans and nature and a rupture in soil nutrient cycling between town and country; these are some manifestations of what in literature is known as the metabolic rift, a concept originated in Karl Marx work. This rift detaches people from their own land, knowledge, and food. Accordingly, few citizens in urban areas have direct experience with growing food and the number of people who are lacking the knowledge to prepare their own food from basic ingredients is increasing. Conditions of urbanisation largely subject people to dependency on processed, non-seasonal/fresh food. This attribute of urbanisation is negatively reflected on the population's health and wellbeing. Maintaining the health and wellbeing of a city's population can be considered the drivers for food system change. Urban agroecology can be an approach to heal this rift and maintain a state of health and wellbeing among citizens. It is emerging as a specific type of urban agriculture that includes the ethical, social, cultural, political and productive approaches typical of peasant agroecology. The goal of urban political agroecological practices is to change the often un-resourceful, day-to-day functioning of urban life particularly in relation to soil care, socio-economically just food systems. The focus in this research will be on the power of these practices to create a change and help people to adjust and respond to various stimulations in their life. This helps them to reconnect and redefine their relationship with food and having a sense of wellbeing. The target group of the research is pregnant women or/and with children up to two years old (a stage in life that induces reflections on the link between food and health). A series of action learning workshops (kitchen waste, growing food and cooking healthy meals) will be carried out. The aim of this paper is to report on a review of literature and provide insight into the relationship between agroecology, food knowledge and wellbeing. It is expected that this research will provide an evidence base for policy intervention to promote community-based urban agroecology and to focus on transformative life moments for building awareness.

Keywords: Food knowledge, nutrition, urban agroecology, wellbeing

---

Contact Address: Mai Abbas, Coventry University, Centre for Agroecology, Water and Reilience, Ryton Gardens, Wolston Lane, CV8 3LG Coventry, United Kingdom, e-mail: ad1212@coventry.ac.uk

Food Sovereignty in Central America: A Comparison of Conventional and Regenerative Practices in Rural Communities

RICARDO ROMERO-PEREZGROVAS[1], VICTOR ARBOLEDA[2], HAROL REVELO[3], JUDITH HECK[4], CHARLES FRENCH[5]

[1]*Sustainable Harvest International, Field Programs, Mexico*

[2]*EARTH University, Agronomy, Ecuador*

[3]*EARTH University, Agronomy, Costa Rica*

[4]*Deutsche Gesellschaft für Internationale Zusammenarbeit, PROCAMBIO, Honduras*

[5]*University of New Hampshire, UNH Extension, United States of America*

Rural communities in Central America have been facing a series of intricate problems regarding crop production since the 1970s. On one hand, the agro-industrial complex dedicated to export crops (bananas, sugar cane, papaya, cacao, coffee, oranges, etc.) have continued to accumulate the most fertile and suitable lands. Large private landowners, as well as private companies, have relied on the key principles of conventional farming, namely: monocropping, high external input use, mechanisation, improved seed varieties and/or hybrids and cheap local labour. Public policies, marketing and the perception of peasants transitioning being transformed into agricultural workers have all served to promote the conventional crop production system as a panacea and the only possible way to achieve modernity and progress.

All these factors have contributed to a copycat effect in rural communities, where peasants try to imitate and apply the principles of conventional agriculture. With low capital and in marginal lands (steep slopes, swamps, isolated, etc.) mechanisation has been almost impossible to achieve, but the use of improved seeds and synthetic inputs has been extended through government subsidies programs and international cooperation agencies using the agro-industrial complex as their supplier.

High synthetic chemical use has created an environmental and health crisis as well as debt and economic dependency for peasants' households. Monocropping and the preference for cash crops has negatively affected the diets of rural households.

This study compared the production diversity and its effects on dietary conditions of rural households that have adopted the principles of conventional agriculture with those households that have kept regenerative agriculture principles (crop diversity and associations, local seeds and crops, organic fertilisation, etc.).

The study was based on direct measurements of crop diversity in plots and food diversity in kitchens, as well as semi-structured surveys in a stratified sample of 540 households distributed among 25 rural communities of three countries (Belize, Honduras and Panama).

The results concluded that households practicing regenerative agriculture have a more diverse diet, more protein consumption, and less dependency on external purchases when compared to households practicing conventional agriculture.

Keywords: Central America, food sovereignty, nutrition, regenerative agriculture

Contact Address: Charles French, University of New Hampshire, UNH Extension, Durham, United States of America, e-mail: charlie.french@unh.edu

Pathways Leading to Diversified Diets - A Retrospective Analysis of a Participatory Nutrition-Sensitive Agricultural Project

Julia Boedecker[1], Carl Lachat[2], Patrick Van Damme[3], Francis Odhiambo Odour[1], Gina Kennedy[4], Céline Termote[1]

[1]*Bioversity International, Healthy Diets from Sustainable Food Systems, Kenya*
[2]*Ghent University, Dep. of Food Technology, Safety and Health, Belgium*
[3]*Ghent University, Dept. of Plant Production - Lab. for Tropical Agronomy, Belgium*
[4]*Bioversity International, Italy*

A previous impact study that assessed the effectiveness of participatory farm diversification and nutrition education on dietary diversity in Western Kenya showed an increased dietary diversity of young children. In this present study we would like to identify the pathways, and the behaviours explaining those pathways, that led to the increased dietary diversity.
While existing frameworks summarise pathways from agriculture to nutrition outcomes, they do not explain the behaviour along the pathways. This is unfortunate as such insights could guide future nutrition-sensitive agricultural interventions. Data collection and analysis will thus be inspired by a widely-used framework related to nutrition-sensitive agriculture, the Pathways from Agriculture to Nutrition and a model for behaviour change, the COM-B model.
The study design consists of a qualitative cross-sectional study applying 10 focus group discussions and 5 key informant interviews. The focus group discussions are conducted with community members who have actively participated in the participatory farm diversification and nutrition education while the key informant interviews involve local authorities who have been accompanying the communities in this.
For data analysis we chose to apply a process that is inspired by the constant comparison analysis. Based on the collected information own frameworks of agriculture-nutrition linkages will be developed and the behaviour and motivations along these frameworks will be identified. The triangulation will aim to maximise the understanding of the pathways leading to more diverse diets. Gender disaggregation during data collection and analysis will enable the interpretation regarding gender equality and women empowerment aspects.
The study is on-going and results will be presented.

Keywords: Agricultural-nutrition pathways, community-based participatory approach, dietary diversity, nutrition-sensitive agriculture

Contact Address: Julia Boedecker, Bioversity International, Healthy Diets from Sustainable Food Systems, c/o ICRAF, P.O. Box 30677, 00100 Nairobi, Kenya, e-mail: j.boedecker@cgiar.org

Enablers, Difficulties and Hindrances of Nutrition Recommendations Application in Karamoja, Rural Uganda

TINA KOCH[1], GUDRUN STALLKAMP[2], M. GRACIA GLAS[1], IRMGARD JORDAN[1]

[1]*Justus-Liebig-University Giessen, Center for International Development and Environmental Research, Germany*

[2]*Welthungerhilfe e.V., Head Office Kampala, Uganda*

In developing countries, nutrition education is often implemented as part of multi-sectoral programmes. Different interventions included within the larger project framework targeting at diverse underlying causes of malnutrition. Still, putting gained knowledge from nutrition education into practice often remains a challenge.
Aim of this study was to identify enablers, difficulties and hindrances of nutrition recommendations application, given during nutrition education, within a Welthungerhilfe project.
A quantitative study was carried out with women (n= 62) in two respective Welthungerhilfe project villages in Karamoja, rural Uganda. Five nutrition sessions were offered to participants between February and March 2018. Three weeks after the last nutrition session, women were interviewed (April 2018), using standardised and open questions. Indicators, influencing each nutrition recommendations application were identified and quantified.
Main enablers for change in dietary behaviour were affordability of food (financial means generally given), food availability, knowledge gain from nutrition education and temporary financial means. At the same time, affordability of edibles (financial means generally not given) was identified being a common difficulty and hindrance alike. Special enablers were taste and food preparation. A positive attitude towards alcohol was another hindrance identified.
Nutrition education embedded within a multi-sectoral programme provided a good platform for provisioning specific nutrition recommendations. Understanding women's perceptions on nutrition recommendations application can shed light on enabling and hindering factors for putting gained knowledge into practice. Identified factors within this study, such as food availability or knowledge gain from nutrition education should be considered as immediate or short term factors. Nevertheless, they can help in finding specific adaptations for nutrition education as well as for other project components within the multi-sectoral programme.

Keywords: Enablers, hindrances, nutrition education, nutrition recommendations, rural Uganda

Contact Address: Tina Koch, Justus-Liebig-University Giessen, Center for International Development and Environmental Research, Giessen, Germany, e-mail: tina.koch@ernaehrung.uni-giessen.de

Dietary Intervention for Management of Obesity among Rural Farm Women

K. Geetha, Geetha Yankanchi, Netravati Hiremath
University of Agricultural Sciences, Food Science and Nutrition, India

Obesity and related co-morbidities are major health problems throughout the world. In India, the increased level of obesity is primarily associated with the transformation from rural to urban lifestyle. Dietary approaches for weight loss have gained widespread interest and considerable importance. Dietary fibre being one of the essential components of the healthy diet has gained the important physiological and metabolic functions in the management of obesity. Hence, the present study was undertaken to develop high fibre food mix and to assess its efficacy on obese subjects. The high fibre food mix was developed by using region specific foods with high fibre content. The efficacy of food mix was assessed through dietary intervention. Intervention was carried out on obese rural farm women for a period of 120 days by providing 1/3 rd daily requirement of protein and energy. Impact of food mix was evaluated by assessing the somatic and biochemical parameters at pre and post dietary intervention. The developed mix contained protein (15.80±0.32g), fat (2.6±0.12g), energy (320 kcal), carbohydrate (60.75g) and dietary fibre (29.5±0.91g) per 100 g of the mix. The significant reduction was observed in weight (65.34±10.97 kg to 63.23±10.60 kg), body mass index (27.84±4.14 to 26.95±4.06) and hip circumference (105.23±9.96 cm to 104.93±9.90 cm) from pre to post intervention ($P \leq 0.05$) period. Similarly, significant reduction was also observed in cholesterol, low density lipoprotein and fasting blood sugar from pre to post test. The food based dietary intervention with high protein and dietary fibre showed desirable reduction in body weight and biochemical parameters. Hence, the inclusion of high fibre food mix in daily diet helps in management of body weight and lipid profile.

Keywords: Body mass index and lipid profile, dietary intervention, obesity

Contact Address: Geetha Yankanchi, University of Agricultural Sciences, Food Science and Nutrition, Room No. 126 and 127 North Block Uas Gkvk, 560065 Bengaluru, India, e-mail: geethanutrition@gmail.com

Households Food Consumption and Nutrient Deficiencies Trends in the Democratic Republic of Congo

PATRICE MIRINDI[1], JOHN MBURU[1], JOHN ULIMWENGU[2], WIM MARIVOET[3]

[1] *University of Nairobi, Dept. of Agricultural Economics, Kenya*

[2] *International Food Policy Research Institute (IFPRI), Africa Region, United States*

[3] *International Food Policy Research Institute (IFPRI), African Region, Senegal*

All the countries members of the United Nation are committed to achieving SDG 2. Africa has seen the least progress in terms of improvement towards malnutrition prevalence. Particularly in DRC, around 4.5 million people are chronic food insecure, the country has been ranked 176th out of 189 countries figuring in the in the Human Development Index. Due to lack of data, it is hard to have indicators that give a current state of the food security and nutrition yet the situation is extremely alarming. Different studies have been conducted to give an overview of the nutritional status and economic changes in the country. Moreover, studies have shown that DRC faces a higher variability in time and space in its economy that should lead to nutrition transition. Unfortunately, there is no information on nutrition deficiencies and household behaviour or choice for diet in order to understand their motivation for the selection of food. Therefore, the purpose of this study was to assess the consumption trends of selected food groups and nutrient deficiencies. In order to achieve this, we analysed the food budget share, the food purchased composition and mapped the nutrient deficiencies as well as their trends between 2005 and 2012 for urban and rural areas of the 26 provinces of DRC. The study used secondary data from the National Household Surveys collected between 2005 and 2012. Our results suggested that households spend ¾ of their budget on food. We were able to classify DRC in five different groups having almost similar food intake. Moreover, about 60 per cent of the budget allocated to the food is spent on cereals, root and tubers as well as meat and fish. We also found that there is a deficiency in zinc, calcium, iron and vitamin B12 in almost all the provinces for both rural and urban areas. For protein, calories, folate, and vitamin A the consumption is quite acceptable. However, the trends seem to be negative in almost all the provinces. Therefore, propoor policies, multi-stakeholder parternships as well as nutrition education is needed to alleviate food insecurty and malnutrition in DRC.

Keywords: Democratic Republic of Congo, food consumption trend, food groups, nutrient deficiency trend, nutrition

Contact Address: Patrice Mirindi, University of Nairobi, Dept. of Agricultural Economics, Anp Hostel Cavs University of Nairobi, Nairobi, Kenya, e-mail: patricemirindi@gmail.com

Perceptions of Rural and Urban Population of Insect Consumption: Does Entomophagy Have a Future in Madagascar?

NARILALA RANDRIANARISON, HERIMIHAMINA ANDRIAMAZAORO
University of Antananarivo, Procinut Project, Madagascar

According to FAO (2013), the constant increase of the world population, the scarcity of natural resources, the strong pressure on the environment, are all constraints guiding reflections and discussions of researchers around the world on the possibility of considering insect consumption as a sustainable alternative to cope with a growing demand for animal protein. This is also true for the case of Madagascar where poverty and food insecurity dominate, and the impacts of agriculture and livestock farming on the environment are increasingly felt. Historically, entomophagy is already a secular practice. Insects have already been included in the traditional diet of Malagasy population. This communication aims to explain the perceptions of rural and urban population of insect consumption. The reasons for the consumption or not of insects were studied by considering various categories of population.
This communication lean on fieldworks conducted in the southern and central Madagascar highlands (Amoron'i Mania and Analamanga). The data were collected through surveys using a semi-structured questionnaire among various population groups, including 80 farmers, 20 civil servants, 20 workers in private sector, 20 in free-trade zones, 20 in informal sector. The results showed that entomophagy is a practice that interests more rural than urban population. Insects are more popular in rural areas where some people prefer to eat them instead of meat. People who do not consume them are rare; the reason is the allergy problems that can occur. In urban areas, psychological barriers affect negatively the development of entomophagy. For people who have never eaten insects, the main reason is the 'disgust' that comes from eating insects. The origins of disgust are rooted in the culture and education, which have a major effect on dietary habits. Some people have mentioned the danger that can be caused by consuming insect, especially the allergy case, and therefore refuse to consume it. Others just do not want to taste. The physical form of the insect (appearance, size), is at the origin of this feeling. However, these category of people agree to eat only if the insects are processed (e.g. into flour) and used as ingredients for other foods.

Keywords: Eating habits, entomophagy, insects, Madagascan highlands

Contact Address: Narilala Randrianarison, University of Antananarivo, Procinut Project, Antananarivo, Madagascar, e-mail: narilalar@yahoo.fr

Marketing and Consumption of Edible Insects in the Urban Center of Yangon, Myanmar

MYINT THU THU AUNG, JOCHEN DÜRR
University of Bonn, Center for Development Research (ZEF), Germany

In Myanmar, insect-eating is a traditional habit in rural, food-scarce areas. Nowadays, insects have become popular also in big cities. Edible insects vary from region to region and include crickets, bamboo caterpillars, water beetles, honey bees, giant water bugs and big ants. Despite the fact that entomophagy is widespread in Myanmar, research on its market potential is scarce. This study explores the potential and constraints of edible insect consumption in the urban centre of Yangon by using a qualitative data collection technique through face-to-face in-depth interviews with vendors and consumers at five market places. Results revealed that crickets are the most popular edible insect. Crickets are collected from nearby Bago region, and from the more distant States of Karen, Mon and Shan. There are two cricket seasons when they are sold fresh or fried in downtown markets and also offered as a special snack in bars. In recent years, cricket prices have increased which hindered development of demand. Five types of consumers and seven types of non-consumers could be categorised, depending on their reasons to buy or not edible insects. Reasons for consumption comprise good taste, high nutritious value, and natural and healthy food considerations. People who do not eat insects mentioned hygiene factors, cultural and religious beliefs, anxiety of the appearance of insects, health problems, suspicion of the use of insecticides for gathering, unaffordable high prices, and doubts about the freshness of the product. To overcome the main constraints, differentiated strategies would be necessary according to the non-consumer group. Hygiene standards and regulations could help convince non-consumers who fear that insects are a non-fresh and impure product. Awareness-building on collecting methods would be important for non-consumers who believe that crickets are gathered by using insecticides when they are actually caught by means of light traps. Promotion of value-added products may persuade non-consumers by taste. Finally, the problem of limited supply and high prices could be solved by increased production of edible insects through systematic rearing.

Keywords: Consumer groups, crickets, entomophagy, insect rearing

Contact Address: Myint Thu Thu Aung, University of Bonn, Center for Development Research (ZEF), Genscherallee 3, 53113 Bonn, Germany, e-mail: myintthuthuaung@gmail.com

The Prevalence of Anaemia among Children in Northern Uganda: Urban-Rural Comparison

JUDITH WOLLMERSHÄUSER, HANS KONRAD BIESALSKI
University of Hohenheim, Inst. of Biological Chemistry and Nutrition, Germany

Affecting 2.36 billion people worldwide, anaemia is a relevant, global public health problem with far-reaching consequences for human health as well as the social and economic development of a country. More than half of all cases are caused by iron deficiency. According to the World Health Organisation, 56 % Ugandan children aged 6 to 59 months are anaemic having a hemoglobin level below 11.0 g dl $^{-1}$ The aim of the study was to assess the prevalence of anaemia in Northern Uganda among the vulnerable group of children aged 6 to 59 months, focusing on the residence as an influence factor of the multifactorial aetiology of anaemia in developing countries.
A total of 273 children, treated in a health centre in the north of Uganda, were part of the investigation. In addition to the measurement of the haemoglobin level and anthropometric data, a sickle cell test was conducted as well as an interview of the attendant person concerning several determinants of anaemia.
The study detected a prevalence of anaemia in children of 90.8 %. Classified by severity, 42.3 % were affected by severe, 38.2 % by moderate and 10.3 % by mild anaemia. The mean haemoglobin concentration was 7.45 g dl $^{-1}$. Furthermore, the urban-rural comparison showed significantly higher hemoglobin values among children from urban areas, but not a significantly lower prevalence of anaemia. A rural residence was significantly associated with a higher severity of anaemia.
Despite the lower prevalences in comparable previous studies, anaemia remains one of the most important health challenges among children under five years of age, especially in northern Uganda. In future a comprehensive analysis and the identification of the main risk factors of anaemia in this region could be beneficial to improve already existing interventions for the reduction and prevention of anaemia in a more efficient way and furthermore to adress the predominantly rural population.

Keywords: Anaemia, influencing factors, residence, rural, Uganda, urban

Contact Address: Judith Wollmershäuser, University of Hohenheim, Inst. of Biological Chemistry and Nutrition, 70599 Stuttgart, Germany, e-mail: judith.wollmershaeuser@uni-hohenheim.de

Constraints of Home Gardens to Impact Nutritional Anemia: A Case Study of Eco-Sustainable Garden Empowering Mbororo Women in Cameroon

NKAANAH FOMEFRET INES INGRID[1], PRIDE ANYA EBILE[2]

[1] *University of Bamenda, Fac. of Health Sciences, Cameroon*

[2] *University of Hohenheim, Dept. of Crop Science, Germany*

Malnutrition is a severe public health as it leads to increase mortality and morbidity. The cheapest intervention used to improve nutritional intake and outcome has been food-based approaches like home gardening, bio-fortification, and keeping of small animals. The project, "Eco-sustainable Gardens Empowering Mbororo Minority Women" is a home garden project in the North West region of Cameroon designed using the concepts of nutrition-sensitive agriculture. Though these home gardens grow vegetables rich in micronutrients such as iron and vitamin A, they most often do not often translate into the micronutrients of the beneficiaries.

Iron deficiency anemia has several consequences on the immune and mental function, as well as general wellbeing of every individual. Women, in particular, have increased iron demands due to menstruation, pregnancy, and lactation. This study aimed at assessing if the home garden project "Eco-sustainable Gardens Empowering Mbororo Minority Women" had an effect on anemia amongst the beneficiaries.

A case-control study design was used in the study to assess nutritional anemia amongst mbororo women with similar socio-economic characteristics. Hemoglobin levels of Mbororo women in the home garden project community ongoing for two years versus those without the home garden were measured (using a Urit 12 hemoglobinometer) to depict iron deficiency anemia and values compared.

There was no significant difference between hemoglobin levels of the Mbororo women with the home gardens and those without. Both communities were anemic. Although home gardens address the problem of food insecurity, other sources of micronutrients such as animal sources and supplements should be considered in these communities.

Keywords: Iron deficiency, malnutrition, nutritional anemia

Contact Address: Nkaanah Fomefret Ines Ingrid, University of Bamenda, Fac. of Health Sciences, Eloumden Simbock, Yaounde, Cameroon, e-mail: nkaanahinesingrid@gmail.com

Nutritional Potential of Indigenous Vegetables Growing in Understory Coffee Agroforests of Yayu, Southwestern Ethiopia

HABTAMU S. ARAGAW[1], DANIEL CALLO-CONCHA[2], NOHR DONATUS[3]
[1]*Ethiopian Institute of Agricultural Research (EIAR), Ethiopia*
[2]*University of Bonn, Center for Development Research (ZEF), Germany*
[3]*University of Hohenheim, Inst. of Biological Chemistry and Nutrition, Germany*

Ethiopia is confronted with the paradox of hosting hundreds of edible plants showing a high level of food and nutritional insecurity. Ethiopians' diets heavily depend on staples and often lack protein and micronutrients. Hence, a large section of the population, particularly children and women, are malnourished with respect to one or several nutrients. In southwestern Ethiopia, indigenous vegetables growing within agroforestry systems have the potential to fulfill the nutritional demands of local people. Therefore, this study assessed the nutritional potential of indigenous vegetables existing in understory coffee agroforests of Yayu, SW Ethiopia. An extensive ethnobotanical household survey (n=300) was done to document edible plants. Biochemical analyses, i.e., proximate food composition, vitamin and mineral content determination of selected vegetables, were conducted using standard analytical methods. The results showed four vegetable species namely; *Amaranthus graecizans*, *Hypolepis sparsisora, Portulaca oleracea* and *Solanum nigrum* having higher protein, fat, provitamin A, calcium, magnesium and iron contents compared to regularly cultivated crops. However, the vegetables showed relatively low calorific content. Crude protein content ranged from 15.62 to 19.26 g/100 g edible portion (EP) in four of the determined vegetables. Calcium content varied largely between 585–785 mg/100 g but was notably high for *A. graecizans* (2065 mg/100 g EP). The iron content varied from 24.14 g/100 g in *S. nigrum* to 91.29 g/100 g in *A. graecizans*. Provitamins A were high in *A. graecizans* (75 μg retinol activity equivalent (RAE) /100 g EP) and *S. nigrum* (62.5 μg RAE/100 g EP). Calorific value was relatively high in *A. graecizans* (142 kcal/100 g) and *H. sparsisora* (143 kcal/100 g dry EP). In conclusion, underutilised indigenous vegetables are potential sources of dietary nutrients locally needed. Particularly notable as providers of protein, energy, minerals, and provitamins A are *A. graecizans, P. oleracea* and *S. nigrum*.

Keywords: Diet diversity, indigenous vegetables, nutritional potential, Yayu biosphere reserve

Contact Address: Daniel Callo-Concha, University of Bonn, Center for Development Research (ZEF), Genscheralle 3, 53113 Bonn, Germany, e-mail: d.callo-concha@uni-bonn.de

Are Vegetables Foods or Just a Sauce?

JILL MARIE FERRY[1], ANNET ITARU[2], LYDIAH M. WASWA[2], M. GRACIA GLAS[1], DAISY ALUM[3], MARGARET KABAHENDA[3], IRMGARD JORDAN[1]

[1]*Justus-Liebig-University Gießen, Nutritional Science, Germany*

[2]*Egerton University, Kenya*

[3]*Makerere University, Uganda*

An adequate consumption of fruits and vegetables contributes to a lower risk for non-communicable diseases. Fruits and vegetables are not only characterised by a high micronutrient content but also an important source of bioactive plant components with health benefits. This study focused on perception and attitudes towards vegetables to better understand willingness to plant and consume vegetables at household level.

In Feb. 2019, 24 focus group discussions were held in Kapchorwa District, Uganda, and Teso South sub-County, Kenya. Women, male and youth farmers were invited to discuss separately about legumes, vegetable and fruit availability in the upcoming 5 months. Further topics were food preparation techniques, taste, perception and attitudes towards the foods.

The food lists included between 12-25 different green leafy vegetables (GLV) which were either available at home only, both at home and on the market or at the market only. The list was shortest among the youth groups, and about the same length among women and men. Most groups respondents, given a choice they rather purchase fruits instead of vegetables. Only 2-3 GLV species were regularly consumed and considered eligible to be served to guests. Tomatoes and onions were just used as spices to most GLV dishes. Mixing of GLV to change taste and to cover shortages was rarely done. The portion sizes were often 'too little' compared to the recommended amounts for vegetables and fruits because they are 'just the sauce' to the staple.

The discussion confirmed that GLV are perceived as sauce and not considered as important component for a healthy diet. Gaps in the provision of vegetables are usually not closed by purchases on the market although GLV can be found there. The value of mixed dish composition for example enhancing bioavailability of plant based iron by consuming vitamin C rich fruits next to the vegetables is not known. Agriculture extension and nutrition counselling need to join efforts to raise awareness about the nutrient value of vegetables to sustainable raise consumption levels.

Keywords: Healthy diet, Kenya, sauce, Uganda, vegetables

Contact Address: Irmgard Jordan, Justus-Liebig University Giessen, Center for International Development and Environmental Research, Senckenbergstr. 3, 35390 Gießen, Germany, e-mail: Irmgard.Jordan@ernaehrung.uni-giessen.de

The Effect of Lactic Acid Fermentation on Cassava Leaves

Ziba Barati, Sebastian Awiszus, Sajid Latif, Joachim Müller
University of Hohenheim, Inst. of Agric. Sci. in the Tropics (Hans-Ruthenberg-Institute), Germany

Cassava leaves are mostly considered a neglected part of the cassava plant. In some countries like Nigeria or Congo, cassava leaves are traditionally consumed with starchy dishes. Cassava leaves are a rich source of protein, vitamins, carotenoids and minerals. However, cassava leaves also contain a considerable amount of cyanogenic glucosides, which should be removed before human consumption. This study examined the effect of lactic acid fermentation on cassava leaves. The cassava leaves bought from the local market in Stuttgart were used. The cassava leaves were fermented naturally or with starter culture from the lactic acid bacteria *Bifidobacterium*, *Lactobacillus acidophilus* and *Streptococcus thermophilus* for 21 days at 37°C. The characteristics of cassava leaves and fermented cassava leaves were measured by pH, electrical conductivity, colour, dry matter, ash, crude fiber, crude protein, glucose, total phenolic content and cyanide content according to standard methods. It was found that the pH, electrical conductivity, crude fiber, crude protein, glucose and cyanide content was significantly ($p < 0.05$) decreased in fermented cassava leaves. On the other hand, the ash content and total phenolic content was significantly ($p < 0.05$) increased in fermented cassava leaves. Furthermore, the colour parameters were significantly ($p < 0.05$) influenced by fermentation and became darker. However, there was no significant differences between naturally fermented cassava leaves and fermented cassava leaves with a starter. The results show that lactic acid fermentation can be used as a method to detoxify the cassava leaves and to increase and retain the valuable nutrients of the cassava leaves for human consumption.

Keywords: Cassava leaves, cyanogenic glucosides, detoxification, fermentation, human foods

Contact Address: Ziba Barati, University of Hohenheim, Inst. of Agric. Sci. in the Tropics (Hans-Ruthenberg-Institute), Garbenstr.9, 70599 Stuttgart, Germany, e-mail: barati@uni-hohenheim.de

Comparative Nutritional Analysis of Paste Developed from *Tricosanthes Cucumerina* Linn and *Lycopersicon Esculentum* L. Mill. in Nigeria

OLAIDE RUTH ADERIBIGBE[1], FOLASADE OLABIMPE ADEBOYEJO[2], MARY OLUWAKEMI AFOLABI[2], JOSEPH KUDADAM KORESE[3], SOLOMON KOFI CHIKPAH[4]

[1]*National Horticultural Research Institute, Product Development, Nigeria*

[2]*University of Ibadan, Food Tecnology, Nigeria*

[3]*University for Development Studies, Agricultural Mechanization and Irrigation Technology, Ghana*

[4]*University of Kassel, Agricultural and Biosystems Engineering, Germany*

Tricosanthes cucumerina Linn. is one of the indigenous underutilised vegetables found in South-Western Nigeria and is used in rural areas as a substitute for vine tomato (*Lycopersicon esculentum* L. Mill.) due to its sweet tasting, aromatic, and deep red endocarp pulp. It is known as snake gourd, viper gourd, snake tomato or long tomato in many countries. Little attention is being paid to this crop despite its rich nutrient content and ease of cultivation. This study explored the potential of the crop in development of tomato paste; its beta-carotene, lycopene, vitamin C and mineral content was compared to that of vine tomato paste. The fully ripe fruits of the two vegetables were harvested from a local farm in Ibadan, Nigeria. The pulp of the snake tomato was extracted, concentrated, poured into a sterilized glass jar and corked. As for the vine tomato, wholesome fruits (roma variety) were washed, milled, concentrated, poured into sterilized glass jars and corked. All laboratory analyses were done using standard procedures. Paste from snake tomato had significantly ($p < 0.05$) higher content of β-carotene, vitamin C and lycopene (2.02. 0.30 and 1.34 mg g^{-1}, respectively) than paste from vine tomato (1.03, 0.14 and 0.42 mg g^{-1}, respectively). Likewise, mineral elements such as Fe, Zn, Ca and K were significantly ($p < 0.05$) higher in paste from snake tomato (3.44, 0.86, 13.38, and 320.25 mg/100g, respectively) than paste from vine tomato (2.69, 0.24, 12.01 and 179.6 mg/100g, respectively). However, sodium was significantly ($p < 0.05$) higher in vine tomato paste than snake tomato paste (59.2 mg/100g vs 6.35 mg/100g). In conclusion, the paste from snake tomato pulp has higher bioactive compounds and minerals than vine tomato paste. These essential nutrients contained in snake tomato make the fruit and its paste of nutritional and health importance. Thus, it can serve as a good substitute or complementary raw material for tomato paste industry.

Keywords: Bioactive component, mineral, tomato paste, underutilised plant

Contact Address: Olaide Ruth Aderibigbe, National Horticultural Research Institute, Product Development, 1 Jericho G.r.a, 20002 Ibadan, Nigeria, e-mail: connecttolaide@yahoo.com

Composite Flour Blends: Effect of Particle Size of Peeled and Unpeeled Orange Fleshed Sweet Potato Flours on Quality Characteristics of Cookies

SOLOMON KOFI CHIKPAH[1], OLAIDE RUTH ADERIBIGBE[2], BARBARA STURM[1], ALBERT ESPER[3], OLIVER HENSEL[1], JOSEPH KUDADAM KORESE[4]

[1] *University of Kassel, Agricultural and Biosystems Engineering, Germany*

[2] *National Horticultural Research Institute, Product Development, Nigeria*

[3] *Innotech, Innotech Stuttgart, Germany*

[4] *University for Development Studies, Agricultural Mechanization and Irrigation Technology, Ghana*

Orange fleshed sweet potato (OFSP) has great potentials to combat malnutrition; especially vitamin A deficiency due to its high beta-carotene and other essential nutrients. The OFSP flour has the potentials to replace wheat flour in bakery foods production. This will reduce wheat flour importation and increase intake of essential nutrients in sub-Saharan Africa countries. The quality attributes and consumer acceptability of baked products may be affected by flour particle size and percentage of different flours in composite flour formulations.

Thus, this study investigated the effect of sieve particle size on the physical, nutritional, antioxidant and sensory properties of peeled and unpeeled OFSP composite flour cookies. Peeled and unpeeled OFSP flours sieved through 250 μm or 500 μm mesh particle size were produced. OFSP and wheat composite flours were made and used to produce cookies baked at 170°C for 11 minutes. The quality attributes and consumer acceptability tests were conducted in accordance with international standard methods. The 100 % wheat flour cookie was used as control. The evaluations showed sieve particle size had no significant ($p > 0.05$) influence on the baking loss, thickness, diameter, spread ratio, water activity (aw) and CIE colour attributes (L*, a*, b*, chroma and hue angle) of cookies. However, both 100 % peeled and unpeeled OFSP and their composite flours cookies had significantly ($p < 0.05$) lower baking loss, high diameter and spread ratio than the control cookies. Similarly, proximate, mineral and bioactive compounds as well total antioxidant activity of cookies were not significantly ($p > 0.05$) affected by sieve particle size. The unpeeled OFSP flours cookies had significantly ($p < 0.05$) lower crude protein but higher ß-carotene, total phenols, flavonoids and antioxidant activity. The sensory assessment however showed that sieve particle size had significant ($p < 0.05$) influence on overall acceptability of cookies. The colour, aroma, crispiness and taste were not significantly affected by sieve particle size. Cookies prepared from OFSP flour sieved with 250 μm mesh particle size performed better in terms of overall acceptability. Generally, 10–50 % peeled and 10–20 % unpeeled OFSP composite flours sieved through 250 μm mesh particle size and control cookies had the best overall acceptability.

Keywords: Bioactive compounds, composite flour, cookies, orange fleshed sweet potato, physical properties, sensory properties

Contact Address: Joseph Kudadam Korese, University for Development Studies, Agricultural Mechanization and Irrigation Technology, Fac. of Agriculture, POB TL 1882, Nyankpala campus, Tamale, Ghana, e-mail: jkorese@uds.edu.gh

Chemical Composition of Baobab (*Adansonia digitata*) Stem Bark and Fractionation of its Ethanolic Extract Using Column Chromatography

Kayode Adebisi Arowora, Ojochenemi Ejeh Yakubu, Ayobami Jahdahunsi Kukoyi

Federal University, Wukari, Biochemistry, Nigeria

The African Baobab tree (*Adansonia digitata*) and its related species belong to the family Malvacca. The tree is a traditional food plant in Africa, but is little-known elsewhere. Baobab has been suggested to have the potential to improve nutrition and health benefits of phytochemical substances which have been related to the action of the compounds such as anti-oxidants and inhibitors of lipid peroxidation. In view of aforementioned, this study investigated, the proximate, phytochemical and mineral composition as well as fractionation of phytochemicals in stem bark of baobab which was carried out using column chromatography. The stem bark samples were collected from Wukari Local Government Area, Taraba state in Nigeria, air-dried and pulverised. Proximate analysis of the pulverised samples were determined by using the standard methods of AOAC. Findings showed that using the column chromatographic fractionation, the following ranges were observed: Total antioxidant capacity (735.29–5001.00 mg ml^{-1}), total phenolic content (4.29–1415 mg ml^{-1}), total flavonoid content (1.11–22.16 mg ml^{-1}), metal chelating activity (300.00–8476.92 mg ml^{-1}) and beta carotene content (14.81–129.63 mg ml^{-1}). Proximate analysis showed high content of soluble carbohydrates (73 %). With the aid of high performance liquid chromatography (HPLC), varying levels of phytochemicals were detected with the highest level being observed for phenols to be 152.4 mg/100g, while the least was determined to be 6.54 mg/100g for flavonoids. Other phytochemicals observed were tannins, alkaloids, glycosides, terpenes and steroids. The mineral composition in baobab stem bark was carried out using atomic absorption spectrophotometer, which showed the presence of phosphorus (185.56 mg/100g), copper (12.33 mg/100g), zinc (10.43 mg/100g), iron (8.47 mg/100g), calcium (312,47 mg/100g), magnesium (128.62 mg/100g), sulphur (0.085 mg/100g), sodium (145.35 mg/100g), manganese (81.36 mg/100g), potassium (85.34 mg/100g), and chromium (0.86 ug/100g). This study revealed the presence of phytochemicals that could be used in chemotherapy of some human diseases, especially the presence of cardiac glycosides which can be beneficial to the heart. The proximate composition can be exploited for its nutritional benefits in food supplements, nutraceuticals and functional food production.

Keywords: Baobab, proximate, minerals, phytochemicals, stem bark

Contact Address: Kayode Adebisi Arowora, Federal University, Wukari, Biochemistry, 200 Katsina Ala Road P.M.B. 1020 , Wukari, Nigeria, e-mail: aroworak2002@gmail.com

Optimisation of Drying Conditions for Cassava Foam Powder Production and Properties of Cassava Foam Powder

OLUWATOYIN AYETIGBO[1], SAJID LATIF[1], ADEBAYO ABASS[2], JOACHIM MÜLLER[1]

[1] *University of Hohenheim, Inst. of Agric. Sci. in the Tropics (Hans-Ruthenberg-Institute), Germany*

[2] *International Institute of Tropical Agriculture (IITA), Tanzania*

The objective of the study was to optimise the drying conditions for cassava foam into foam powder, which is a novel form of cassava product and to characterise the physicochemical properties of the foam powder. White-flesh cassava foam was produced by whipping the cassava pulp with 14.97 % glycerol monostearate (GMS) colloid and 0.51 % sodium carboxy-methylcellulose (NaCMC) for 2.07 min, and yellow-flesh cassava foam was produced by whipping the cassava pulp with 14.29 % GMS colloid and 0.6 % NaCMC for 2 min. Both foams were dried to powders at 50 °C, 65°C and 80°C in drying pans of 6 mm, 8 mm and 10 mm thickness. Using response surface method, the optimal conditions for drying was identified based on criteria of maximum first and second falling rate diffusivities and minimum time to dry to 10 % moisture content. The optimum drying conditions for white cassava foam was at 80 °C in 10 mm thick pans, while optimal drying conditions for yellow cassava foam was at 80°C in 8 mm thick pans. For drying of yellow cassava foam, for instance, the actual and predicted values for first falling rate diffusivity, second falling rate diffusivity and time required to dry to 10 % moisture content were 2.7×10^{-9} m^2 s^{-1} and 2.2×10^{-9} m^2 s^{-1}, 1.2×10^{-8} m^2 s^{-1} and 0.8×10^{-8} m^2 s^{-1}, and 5.3 h and 6.3 h, respectively. The dried white and yellow cassava foam powders had significantly lower total cyanogenic glucosides content, total carotenoids content, apparent amylose and swelling power, but significantly higher water solubility, water absorption capacity, oil absorption capacity and least concentration for gelation compared to that of dried non-foamed cassava pulp powder. Scanning electron micrographs revealed interactions between starch granules and the colloids that keep the white and yellow foams stable during drying.

Keywords: Cassava foam powder, diffusivity, falling rate, oil absorption capacity, response surface method, Stabiliser

Contact Address: Oluwatoyin Ayetigbo, University of Hohenheim, Inst. of Agricultural Engineering, Tropics and Subtropics, Garbenstr. 9, 70599 Stuttgart, Germany, e-mail: ayetigbo_oluwatoyin@uni-hohenheim.de

Effect of Lactation Season on Awassi Sheep Milk Quality and Fatty Acid Profiles

MOUNTAHA YOUSSEF[1], MUHI EL-DINE HILALI[2], AAD POULIN[3], KHALED HUSHAYMI [4], SAAB ABI SAAB[5]

[1]*Holy Spirit University of Kaslik (USEK), Faculty of Agricultural and Food Sciences, Lebanon*

[2]*International Center for Agricultural Research in the Dry Areas (ICARDA), Sustainable Intensification and Resilience of Production Systems Program (SIRPS), Jordan*

[3]*Notre Dame University (NDU), Louaize, Lebanon*

[4]*Lebanese Agricultural Research Institute (LARI), Lebanon*

[5]*Faculty of Agricultural and Food Sciences, Holy Spirit University of Kaslik (USEK), Lebanon*

Awassi sheep milk is the main milk used to produce traditional dairy products in the Middle East. Sheep management during lactation is affecting milk quality including fatty acid profiles. Generally, farmers graze their animals and offer a little supplementation during spring. Later on, during late spring and summer, concentrates are provided, mainly barley grains. The offered supplementation can be characterised as an energy rich unbalanced diet. Changes in milk composition due to the progress of lactation and traditional feeding regime can affect yields of traditional dairy products, such as white cheese, whereas changes in fatty acid profiles can affect the quality and nutritional value of ghee, an important product in the Middle East.

Data were collected at control days (30, 40, 70, 100, 120, 130 days after lambing) from eight lactating Awassi ewes at Terbol research station, Lebanon. The ewes were put under conventional feeding regime. Milk production was measured, and milk samples were obtained. Milk composition and milk fatty acid profiles were analyzed.

Average milk production during the study period was 568 g day^{-1}. Milk production decreased significantly by 15 % with the progress of milking season. Milk protein and total solids increased by 19 %, whereas the fat content increased by 60 % ($p < 0.01$). These results allow the suggestion to producers to produce traditional yogurt at the beginning of the milking season and to produce cheese at late spring, since cheese yield will be higher at that stage. The milk fat quality was affected by the progression of the milking season and the change in feeding regime, which utilises more stubbles and concentrates. Saturated fatty acids showed an increase of 8 % towards the end of the milking season ($p < 0.01$) with a remarkable increase in C14:0 and C16:0, 23 and 27 % respectively, whereas C4:0 to C12:0 decreased ($p < 0.01$). Monounsaturated and polyunsaturated fatty acids were decreasing by 20 and 18 %, respectively, ($p < 0.01$). Moreover, the omega-3 fatty acids decreased by 75 %, whereas omega-6 fatty acids increased by 13 % ($p < 0.05$). Therefore, ghee produced in spring has a better nutritional value as it will be richer in bioactive compounds such as CLA and omega-3 fatty acids.

Keywords: Awassi sheep, fatty acid profile, lactation stage, milk composition

Contact Address: Muhi El-Dine Hilali, International Center for Agricultural Research in the Dry Areas (ICARDA), Sustainable Intensification and Resilience of Production Systems Program (SIRPS), P.O. Box 950764, 11195 Amman, Jordan, e-mail: m.hilali@cgiar.org

The Significance of Net Covers in Postharvest Quality Management: Increasing and Preserving Health Promoting Carotenoids in Vegetable Amaranth

ELISHA OTIENO GOGO

Pwani University, Crop Sciences, Kenya

Vegetable amaranth are not only important sources of nutrients, but also rich in health promoting compounds, such as carotenoids. However, in the open field, the crop suffers from insect pest and diseases, and unfavourable weather conditions that severely affect its quality, including carotenoids. The present study evaluated the effect of net covers on carotenoids in vegetable amaranth (*Amaranthus cruentus* L. cv. Olevolosi). The experiment was laid in a randomised complete block design with three replications. The crop was grown at Egerton University, Kenya for 8 weeks either under net covers (white, 0.9 mm pore size) or in the open (control). Microclimate (soil moisture, temperature, relative humidity and PAR) was monitored throughout the production period. After harvest, the leaves from each treatment were kept at room temperature conditions (temperature; 20 ± 3 °C, relative humidity 65 ± 5 %) for ten days. Data were collected at harvest and during storage, at 2-days interval. Carotenoids (lutein, lycopene and ß-carotene) as well as chlorophyll a and b were extracted using acetone-hexane and analysed using UV-VIS spectrophotometer. Net covers resulted in increased lutein (20 %), lycopene (15 %), ß-carotene (26 %) and chlorophyll b (40 %) content compared with the control. However, chlorophyll a content was not affected by net covers. Net covers also led to reduced loss of carotenoids and chlorophylls content during storage. The findings demonstrate the potential of using net covers in improving nutritional quality of vegetable amaranth.

Keywords: Agricultural nets, *Amaranthus* spp., antioxidative compounds, postharvest loss reduction, protected culture

Contact Address: Elisha Otieno Gogo, Pwani University, Crop Sciences, P.O. Box 195, 80108 Kilifi, Kenya, e-mail: e.gogo@pu.ac.ke

New technologies for sustainable agriculture and food processing

Oral Presentations

Posters

Effect of Pretreatment and Integration of Superabsorbent Polymers (SAP) as Desiccants on Quality Parameters of Frafra Potato (*Solenostemon rotundifolius*) during Solar Drying

Joseph Kudadam Korese[1], Matthew Atongbiik Achaglinkame[1], Olaide Ruth Aderibigbe[2], Oliver Hensel[3], Albert Esper[4], Barbara Sturm[3]

[1] *University for Development Studies, Agricultural Mechanization and Irrigation Technology, Ghana*

[2] *National Horticultural Research Institute, Product Development, Nigeria*

[3] *University of Kassel, Agricultural and Biosystems Engineering, Germany*

[4] *Innotech, Innotech Stuttgart, Germany*

Frafra potato (*Solenostemon rotundifolius*) is one of the underutilised and highly perishable food crops in Ghana. Being rich in starch and various bioactive compounds, it offers a dual-functionality. The study was undertaken to obtain *Solenostemon* flour that can be utilised as an ingredient for food processing in Ghana and Africa at large. Specifically, the aim of this study is to evaluate the effect of various pretreatments viz. blanching, sodium metabisulphite (5 % w/v) and control on quality attributes of *Solenostemon* flour during forced convection solar drying. Additionally, as solar drying is only satisfactory so long as the sun is shining, this study also investigated the effect of incorporation of SAP as a desiccant on continuous drying of *Solenostemon* at night-time. Towards these objectives, fresh *Solenostemon* (6 mm slice thickness) was subjected to each pretreatment method and then solar dried under two test conditions: without SAP and with incorporation of SAP at sunset till 6:00 am the following day. During the drying process, ambient weather conditions as well as parameters of the drying air and moisture loss were monitored. Additionally, the final moisture content, water activity and some nutritional attributes for each pretreatment and test condition were evaluated and compared. Results showed that the control was averagely superior except for iron and vitamin C, and β-carotene content of the flour where blanching and SMBS were higher, respectively. Incorporation of SAP at night-time reduced the drying time by 11 %, 12 % and 19 % respectively for blanching, sodium metabisulphite and control as compared to drying without SAP, with a significant effect (P–706; 0.05) on the water activity, iron, vitamin C, total phenols of the dried product except on colour (P–707; 0.05). Pretreatment with integration of SAP had a synergistic effect (P–706; 0.05) on most of parameters measured. From this study, it can be hypothesised, pretreatment of *Solenostemon* tubers may not be necessary before solar drying except when iron, vitamin C and β-carotene are of key interest. Also, SAP as a desiccant could successfully be incorporated in solar dryers for air dehumidification in order to ensure continuous night drying of products.

Keywords: Frafra potato, pretreatment, quality attributes, solar drying, super absorbent polymer

Contact Address: Joseph Kudadam Korese, University for Development Studies, Agricultural Mechanization and Irrigation Technology, Fac. of Agriculture, POB TL 1882, Nyankpala campus, Tamale, Ghana, e-mail: jkorese@uds.edu.gh

Evaluation of Efficiency of two Local Improved Stove Models in Honduras and Panama

Bert Kohlmann[1], Ricardo Romero-Perezgrovas[2], Mildred Linkimer[1], Maria Fernanda Romero Rubiano[1], Victor Arboleda[3], Sam Bentson[4]

[1]*EARTH University, Costa Rica*
[2]*Sustainable Harvest International, Field Programs, Mexico*
[3]*EARTH University, Agronomy, Ecuador*
[4]*Aprovecho Research Center, United States of America*

Around the world almost two billion people depend on biomass as their fuel source for cooking. The intense use of biomass puts pressure on the biomass sources of the world including forests. In Central America rural households depend almost entirely on wood as its cooking fuel. An inefficient wood combustion on traditional stoves emits smoke containing Particulate Matter (PM), CO and CO_2. Continued exposure to the smoke contributes to respiratory diseases. Women and children spend the most significant amount of time in the kitchen, and are therefore the most vulnerable people for contracting these diseases. Due to these two reasons a) environmental pressure, b) health hazards, an effort to develop improved stoves with better combustion that consume less wood and emits less smoke have been a constant in the region since the 1980´s. This study evaluated two improved stove models (DAMAK and Mani). For the laboratory evaluation the Water Boiling Test (WBT) was used with a Portable Emissions Measurement System (PEMS), allowing measurements of CO, CO_2 and PM. Wood consumption was also evaluated using a volume method. The results then were compared to the ones obtained by other two commonly used improved stoves models in the region, the Patsari and the Perfection. The performance results of the DAMAK model were 63 % less wood consumption and 50 % less CO emissions, but with 9.4 % more PM emissions when compared with the Patsari. There were not statistically significant differences in CO_2 emissions. These laboratory tests were also compared with data obtained on field evaluations. Using an Indoor Air Pollutant Meter and a wood volume method, a total sample of 174 stoves were evaluated on stoves used by final users in their own kitchens, 92 improved models (DAMAK and Mani) and 82 traditional stoves in rural communities of Panama and Honduras. The three measured variables (CO, PM and wood consumption) had a statically significant difference between the two groups. Improved stoves emitted less CO and PM and consumed less wood. Local improved stove models have a history of better acceptance and adoption by final users when compared to models imported from other regions.

Keywords: Biomass, Central America, improved stoves

Contact Address: Bert Kohlmann, EARTH University, San José, Costa Rica, e-mail: bkohlman@earth.ac.cr

Toward Automated Biodiversity Research on the Tropical Ecosystem Using Artificial Intelligence

Abir Bhattacharyya[1], Lennart Jansen[2], Matthias Kleinke[2], Ronny Hartanto[1]

[1] *Rhine-Waal University of Applied Sciences, Fac. of Technology and Bionics, Germany*

[2] *Rhine-Waal University of Applied Sciences, Fac. of Life Sciences, Germany*

Biodiversity plays an important role in the tropical ecosystem. Many forests have been planted as they promise. These conversions have a bad impact on the overall ecosystem as many of the biodiversities on the original forest have gone during the process. How much species still exists on the plantation or certain fields? This is a major issue in the order to determine the biodiversity of the plantation or field. The biodiversity can be measured using different methods, such as counting the number of individuals - or even families - in a given area. Due to the size of the field, an automated process could be a great help to produce those metrics. This paper presents algorithms for species classification using machine learning. The algorithms show a positive result, where the precision of 61 %, could be improved after the system was trained using a sparse dataset. In the preliminary stage, the systems were trained using few species only using convolutional neural networks to check the feasibility and challenges. Furthermore, more species may be included in the training sets and the algorithm may be used to detect the species in real time. For the purpose of getting the biodiversity index, having an algorithm which could be the number of different species on the data sets would be enough. The system does not need to tell what the name of the particular species in detail. The final result is not only available for the tropical ecosystem but also any ecosystem where the biodiversity index of the plants needs to be evaluated.

Keywords: Artificial intelligence, biodiversity, tropical ecosystem

Contact Address: Matthias Kleinke, Rhine-Waal University of Applied Sciences, Fac. of Life Sciences, Marie-Curie-Str. 1, 47533 Kleve, Germany, e-mail: matthias.kleinke@hochschule-rhein-waal.de

Development of a Decision-Making Tool for Solar Energy Implementation along the Rice Value Chain

Ana Salvatierra-Rojas, Florian Männer, Victor Torres-Toledo, Joachim Müller

University of Hohenheim, Inst. of Agric. Sci. in the Tropics (Hans-Ruthenberg-Institute), Germany

Energy supply in agri-food value chains is challenging in several African countries, where the supply is weak and relies mostly on fossil fuels. Nowadays, there are different renewable energy solutions which could cover the energy demand at different levels of the agri-food chains. Solar photovoltaic (PV) systems and solar thermal devices are the leading solar technologies used for applications like solar PV pumps for irrigation systems, solar dryers, solar cooling systems, solar mills, and solar cooking, among others. Moreover, researchers are continuously working on the development of different energy sources or on the optimisation of the existing solutions to improve the efficiency of various machines on several value chains. Nevertheless, the implementation of solar energy systems throughout Africa is limited, mainly due to the lack of information on new research related to solar energy applications in agriculture, the energy demand of various machinery utilised along the value chain and lack of performance evaluation of a solar photovoltaic (PV) system. It is for this reason that farmers, producers, stakeholders, policy makers, governmental and non-government agencies require rapid analytical tools to support the decision-making process for advising proper implementation. Therefore, a decision-making tool was developed using a graphical interface designed in MATLAB®/Simulink®. Taking as a baseline example the rice value chain in Benin, it was assessed two potential scenarios. The first scenario the partial substitution of fossil fuels by simulating the energy supply of a hybrid design (PV systems plus diesel generator-system) and the second scenario is the total replacement of fossil fuel by simulating the energy supply of solar PV systems. This tool evaluates the optimum PV system taking into consideration many factors, such as local weather data, baseline information of the energy consumption of different machines operated in the rice value chain, machine capacity, and the economic data in terms of the payback period. At the same time, this tool offers information on the potential mitigation of greenhouse emissions, due to the reduction of fossil fuel consumption by the integration of solar PV systems in the value chain.

Keywords: Decision-making tool, hybrid PV, diesel systems, rice value chain, solar energy

---

Contact Address: Ana Salvatierra-Rojas, University of Hohenheim, Inst. of Agric. Sci. in the Tropics (Hans-Ruthenberg-Institute), Garbenstrasse 9, 70599 Stuttgart, Germany, e-mail: Ana.SalvatierraRojas@uni-hohenheim.de

On-Farm Wastewater Treatment Using Biochar from Local Agroresidues Promotes Safer Irrigation Water for Food Production in Developing Countries

Korbinian Kaetzl[1], Manfred Lübken[2], Gülkader Uzun[2], Tito Gehring[2], Edith Nettmann[2], Kathrin Stenchly[1], Marc Wichern[2]

[1] *Universität Kassel, Grassland Science and Renewable Plant Resources, Germany*

[2] *Ruhr-Universität Bochum, Institute of Urban Water Management and Environmental Engineering, Germany*

In this study, the suitability of an anaerobic biofilter (AnBF) as an efficient and low-cost wastewater treatment for safer irrigation water production for sub-Saharan Africa was investigated. To determine the influence of different ubiquitous available materials on the treatment efficiency of the AnBF, rice husks and their pyrolysed equivalent, rice husk biochar, were used as filtration media and compared with sand as a common reference material. Raw sewage from a municipal full-scale wastewater treatment plant pretreated with an anaerobic filter (AF) was used in this experiment. The filters were operated at 22°C room temperature with a hydraulic loading rate of 0.05 $m\,h^{-1}$ for 400 days. The mean organic loading rate (OLR) of the AF was 194 ± 74 and 63 ± 16 gCOD $m^{-3}\,d^{-1}$ for the AnBF. Fecal indicator bacteria (FIB) (up to 3.9 log10-units), bacteriophages (up to 2.7 log10-units), chemical oxygen demand (COD) (up to 94 %) and turbidity (up to 97 %) could be significantly reduced. Additionally, the essential plant nutrients nitrogen and phosphorous were not significantly affected by the water treatment. Overall, the performance of the biochar filters was significantly better than or equal to the sand and rice husk filters. By using the treated wastewater for irrigating lettuce plants in a pot experiment, the contamination with FIB was more than 2.5 log-units lower (for most of the plants below the detection limit of 5.6 MPN per gram fresh weight) than for plants irrigated with raw wastewater. Respective soil samples were minimally contaminated and nearly in the same range as that of tap water.

Keywords: Agricultural residues, anaerobic wastewater filtration, biochar, fecal indicator bacteria, low-cost biofiltration, rice husk, water reuse

---

Contact Address: Korbinian Kaetzl, Universität Kassel, Grassland Science and Renewable Plant Resources, Steinstrasse 19, 37213 Witzenhausen, Germany, e-mail: kaetzl@uni-kassel.de

Can Farmer-to-Farmer Mechanisation Enterprises Contribute to Filling the Technology Adoption Gap in Zambia?

Diana Jaime[1], Johannes Mössinger[1], Manuel Narjes[2], Thomas Berger[1]

[1] *University of Hohenheim, Inst. of Agric. Sci. in the Tropics (Hans-Ruthenberg-Institute), Germany*

[2] *University of Hohenheim, Inst. for Farm Management, Germany*

Zambia is a relatively land-abundant country; 58 % of its territory has a medium-high potential for agricultural production, but less than 14 % is currently under cultivation. This inefficient use of land resources can be explained, among others, by the low adoption of improved technology, inappropriate use of inputs and vulnerability to weather conditions.

In northern Ghana, mechanisation has contributed to cropland expansion during the last 15 years as a growing share of medium and large scale farmers have acquired agricultural machinery. These technology adopters have increased the supply of Mechanisation Hire Services (MHS) improving tractors accessibility to those farmers who are unable to buy their own. Similarly, medium scale farmers started offering MHS in Zambia since 2011, yet at a lower rate of technology adoption when compared to northern Ghana.

The objective of this study is to assess the financial viability and sustainability of farmer-to-farmer MHS enterprises in Zambia as a means to reduce the technology adoption gap among smallholders. To achieve this goal, 30 interviews were conducted with Tractor Service Providers (TSPs) across the country to collect data related to farming activities and machinery services. The information of one representative TSP was used, to parameterise a single agent mathematical programming model using the agent-based software package MPMAS. The TSP agent considered a 15-year-period planning horizon corresponding to a tractor's estimated lifetime, including crop production, farm, and off-farm income, labour, and machinery management. Results showed that MHS provision increases the TSP's average discounted cash surplus by 47 % suggesting that it is a profitable activity with high potential to promote mechanisation. However, the supply chain has to overcome some deficiencies like a demand with low purchasing power, high transaction costs and expensive maintenance and breakdown repairs that discourage the expansion of services. Mechanical training, mechanisms that reduce logistic costs and education on business management can contribute to reducing bottlenecks in the supply of MHS. The public sector should focus on improving the agribusiness environment and access to land and credit for smallholders to provide the demand side with enough liquidity to guarantee the sustainability of the business.

Keywords: Mathematical programming, mechanisation enterprises, sustainability

Contact Address: Diana Jaime, University of Hohenheim, Inst. of Agric. Sci. in the Tropics (Hans-Ruthenberg-Institute), Alte Dorfstrasse 41, 70599 Stuttgart, Germany, e-mail: dianax.jaime@uni-hohenheim.de

Renewable Energy Development in Southeast Asia: A Review on Trade-Offs with Environment and Nature Conservation

SANTI PRATIWI, NATALY JUERGES

University of Goettingen, Dept. Forest and Nature Conservation Policy, Germany

Southeast Asian countries are rich in renewable energy resources, including hydro, wind, solar, geothermal and bio-energy (biomass, bio-fuel, and bio-gas). Blessed with this enormous potential, countries try to fulfil the energy security and climate change mitigation targets by promoting renewable energy development. This paper reviewed several studies showing that there are numerous trade-offs between renewable energy development and environment and nature conservation. The most frequently reported trade-off is coming from hydro power development in Cambodia, Lao PDR, Vietnam, Myanmar, and Malaysia. The construction of dams producing hydroelectric power results in land use change and habitat loss, effecting biodiversity and often water quality. Another important trade-off is caused by bio-energy, for example by the expansion of oil palm plantations which are reported as the main reason of deforestation and wildlife extinction in Indonesia. Trade-offs further include the possibility to emit greenhouse gasses and air pollution which contrasts with the climate change mitigation action. Another trade-off occurs in geothermal power development, which decreases soil and water quality, resulting in landslide and soil erosion in Indonesia. There were no studies found about trade-offs related to wind power, likely because there is not so much wind power development yet in Southeast Asia.
Reflecting on the existing impacts, it is clear that renewable energy is not always sustainable. Limited financial resources make advanced technology and infrastructure unaffordable, resulting in poor quality of equipment and facilities, damaging the environment. These trade-offs also have further implications for social and economic aspects, especially for livelihood near the project areas. Further scientific research and evaluation will be needed to better understand impacts and find ways of balancing the trade-offs in an optimal way.
Lack of appropriate policies and regulation considering the environmental impact of the project can also trigger environmental damage. Government commitment and intervention will be vital to drive the industry in a sustainable manner and cause less negative impacts to environment and nature. A good policy formulation with strong implementation in the field can achieve sustainable renewable energy development without harming the environment.

Keywords: Environment, nature conservation, renewable energy, Southeast Asia, trade-offs

Contact Address: Santi Pratiwi, University of Goettingen, Dept. Forest and Nature Conservation Policy, Albrecht-Thaer-Weg, 22F / 11, 37075 Göttingen, Germany, e-mail: spratiw@gwdg.de

Mechanical Properties of Native Tree Species for Soil Bioengineering in Northeastern Mexico

REBECA ZAVALA[1], ISRAEL CANTU SILVA[1], HUMBERTO GONZÁLEZ RODRÍGUEZ[1], TETSUYA KUBOTA[2]

[1]*Autonomous University of Nuevo Leon, Faculty of Forest Sciences, Mexico*

[2]*Kyushu University, Faculty of Agriculture, Japan*

In recent years, the effect of soil bioengineering has played a very important role on slope stability. However, our area of study is constantly under the influence of small-scale earthquakes and extreme events of heavy rainfall which cause potentially unstable conditions on the slopes. The mechanical properties of the root systems tensile strength (Ts) and modulus of elasticity (Eroot) of four native species were analysed for a potential use as soil bioengineering elements. We investigated if tensile strength (N/mm^2) and modulus of elasticity of roots (N/mm^2) was different between studied species: *Cercis canadensis, Celtis laevigata, Quercus rysophylla* and *Ligustrum lucidum*. The species considered were selected based on their native characteristics and widespread existence on the slopes. Regarding tree forest species, the tests were conducted with the Universal Testing Machine Shimadzu type SLFL-100KN. The relationships among root diameter, tensile strength (Ts), and modulus of elasticity (Eroot) was negative and could be fitted with a power regression equation, showing highly significant values $p < 0.01$. *Celtis laevigata* showed the maximum value of tensile strength (Ts) 28.11 N/mm^2 while the minimum value of tensile strength was observed in *Ligustrum lucidum* 5.27 N/mm^2. For the variable modulus of elasticity (Eroot) *Celtis laevigat*a showed the maximum value of 90.01 N/mm^2 while the minimum value of modulus of elasticity was observed in *Ligustrum lucidum* 29.16 N/mm^2. Results of mechanical proprieties showed the following ascending order: *Ligustrum lucidum* < *Quercus rysophylla* < *Cercis canadensis* < *Celtis laevigata*. Likewise, *Celtis laevigata* showed the highest tensile strength and modulus of elasticity of all investigated species.

Keywords: Modulus of elasticity, root, tensile strength

Contact Address: Rebeca Zavala, Autonomous University of Nuevo Leon, Faculty of Forest Sciences, Corregidora ·530 Sur, Linares, Mexico, e-mail: rebecagzg@gmail.com

Utilisation of Traditional Processed and Preserved Cowpea Leaves in the Coastal Region of Kenya

JOSHUA OWADE[1], GEORGE ABONG'[1], MICHAEL OKOTH[1], AGNES MWANG'OMBE[2]

[1] *University of Nairobi, Dept. of Food Science, Nutrition and Technology, Kenya*

[2] *University of Nairobi, Dept. of Plant Science and Crop Protection, Kenya*

Cowpea leaf is one of the African indigenous vegetables utilised as food in sub-Saharan Africa. Whereas the seasons of glut have massive spoilage and postharvest losses of the vegetable, seasons of drought experience scarcity and less utilisation of cowpea leaf. Through customized techniques, traditional communities have sought to improve its availability and utilisation. However, the efficiency of these techniques in improving the utilisation of this vegetable in and out of season is yet to be established. The current study sought to establish the traditional food preservation and processing techniques and their efficiency in improving utilisation of the vegetable. The study used a cross-sectional design with random sampling of 205 households in Taita Taveta County which ranks among the top producers of the vegetable in Kenya. Results showed that of the households that produced cowpea leaves, about three quarters (73.7 %) utilised cowpea leaves as a priority vegetable compared to 34.6 % and 19.0 % who favoured kales and cabbages. Boiled, sun-dried and blanched cowpea leaves were the most utilised forms of the vegetable by 81.5 %, 44.9 % and 16.5 % of the households. The traditional preservation methods practised in the households were sun-drying (77.5 %), blanching (27.3 %) and a combination of the two (54.1 %). Drought and low production quantities constrained the utilisation of the vegetable in 83.4 % and 51.2 % of the households. During drought, 42.9 % of the households utilised dried vegetables which would keep for up to one year. Source of the vegetable and the person who determines the food to be bought in the household were significantly ($p < 0.05$) associated with the utilisation of dehydrated vegetables during scarcity periods. Households whose production of the vegetable was severely challenged by access to seed, weeds, massive postharvest losses and seed scarcity had odds ratio of 7.2, 0.3, 0.3 and 0.2, respectively, of drying cowpea leaves for later use. In conclusion, drying of cowpea leaves enhances the utilisation of the vegetable in the area. The up-scaling of the technique can be used to increase availability of the vegetable to improve its utilisation.

Keywords: Cowpea leaves, preservation, processing, indigenous vegetable, utilisation

Contact Address: Joshua Owade, University of Nairobi, Dept. of Food Science, Nutrition and Technology, Nairobi, Kenya, e-mail: owadehjm@gmail.com

Freeze-Thaw Pre-Treatment Optimisation for Cassava Tubers to Improve Peeling Efficiency

ZIBA BARATI, SAJID LATIF, SEBASTIAN ROMULI, JOACHIM MÜLLER
University of Hohenheim, Inst. of Agric. Sci. in the Tropics (Hans-Ruthenberg-Institute), Germany

Cassava peeling is a very important operation step in cassava processing. Different shapes and sizes of cassava tubers are the major challenges in cassava peeling. In this study, the effect of freeze-thaw pre-treatment (FTP) on the peeling process of cassava tubers was investigated. A prototype cassava peeling machine was used with five rotating cylindrical abrasive brushes. Frozen cassava tubers purchased from a local market in Stuttgart were used for this study. The length and weight of the cassava tubers varied from 200–280 mm and 500–900 g, respectively. The operational parameters were rotational speed of brushes (550–1150 rpm), peeling time (1–5 min), thaw temperature (50–90°C) and time of thawing (0–120 s). Response surface methodology using central composite design was applied to optimise FTP to improve the peeling process of cassava tubers. Peeled surface area and peel loss were measured as the responses. Additionally, the quality of peeled cassava tubers after freeze-thaw pre-treatment in terms of starch content and structure of the starch granules was investigated. After 30 runs in the central composite design, multivariate correlation was established through reduced cubic model with R^2 of 0.85 and 0.99 for peeled surface area and peel loss, respectively. Results reveal that peeled surface area and peel loss were significantly influenced by rotational speed of the brushes, peeling time and thawing time ($p < 0.05$). Under optimal peeling conditions, rotational speed of 1000 rpm, peeling time of 3 min, thaw temperature of 60°C and thawing time of 90 s, the peeled surface area and peel loss was around 94.9 % and 21.7 %, respectively. It was found that the freeze-thaw pre-treatment had no negative effect on the quality of cassava tubers. The results show that the application of FTP by optimising the ratio of peel loss and peeled surface area can improve the peeling process of cassava tubers.

Keywords: Cassava root peeling, central composite design, freeze-thaw treatment, peel loss, response surface method

Contact Address: Ziba Barati, University of Hohenheim, Inst. of Agric. Sci. in the Tropics (Hans-Ruthenberg-Institute), Garbenstr.9, 70599 Stuttgart, Germany, e-mail: barati@uni-hohenheim.de

A Systems Approach to Biodigester Evaluation: Analysing Bioresources to Asses Farm Integration in the Arsi Zone, Ethiopia

Mia Schoeber[1], Bernhard Freyer[1], Gerold Rahmann[2]

[1]*University of Natural Resources and Life Sciences (BOKU), Dept. of Sustainable Agricultural Systems, Austria*

[2]*Thuenen Institute of Organic Farming, Germany*

Smallholder farms in the Arsi zone, Ethiopia, have an average size of 3.3 ha, are situated at an altitude of around 2200 m with annual rainfall between 800–1300 mm and temperatures between 10–25 °C. Agriculture in this region has a high potential, but is limited by water logging, surface runoff, high soil erosion and decreasing soil fertility. Traditionally, livestock manure on smallholder farms is primarily used as a fuel and the remainder as fertiliser on fields. Competition between these two applications can be eliminated by biodigesters, in which manure is transformed to biogas utilised for light and cooking, and bioslurry, a nutritious organic fertiliser. In this study we analyze, how far biodigesters as part of the farming system can cover farm own energy demand and increase nutrients and organic matter supply in the form of bioslurry/compost to the crops to increase farm income and the sustainment of soil and future crop productivity. Bioresource management of 47 farms have been analyzed, and mainly qualitative data collected to identify opportunities and constrains of biodigester use.

Results show that bioslurry partly contributes to cover fertiliser demand, as bioslurry production is limited by biodigester size, labour and manure due to relatively short stable periods. The quantity of biogas produced is also able to partly cover household energy demand. Manure is still - to some extent - used as a fuel, since the biogas stove does not support cooking the locally favoured “injera”.

Since the share of animal manure is limited for bioslurry production, and already available cooking facilities for injera is not integrated, recirculation of nutrients and organic matter stays insufficient. Recommendations include forage legumes and shrubs, and grasses into the farm system, which provide quality feed, increasing milk and cow dung production, and carbon and nitrogen stored in below-ground biomass. Improved bioslurry storage facilities and provision of a stove supporting the cooking of injera will increase biodigester efficiency and contribute to an overall increase of farm income.

Keywords: Arsi zone, Ethiopia, bioresources, household biodigester, integrated smallholder farm

Contact Address: Mia Schoeber, University of Natural Resources and Life Sciences (BOKU), Dept. of Sustainable Agricultural Systems, Vienna, Austria, e-mail: miaschoeber@hotmail.com

Development of a Photovoltaic Driven Ventilation System to Modified Traditional Ethiopian Gombisa for On-Cobs-Maize Drying and Storage

CHEMEDA ABEDETA GARBABA[1], OLIVER HENSEL[2]

[1] *Jimma University, Horticulture and Plant Sciences , Ethiopia*

[2] *University of Kassel, Agricultural and Biosystems Engineering, Germany*

Unsafe moisture content at loading and the climatically uncontrolled nature of traditional storage structure (gombisa) together with ventilation dependent on wind alone, results in mycoflora growth and development on maize in the system. Therefore, this study aimed to develop and test a photovoltaic driven ventilation system fitted to a gombisa for natural air in-bin drying of on-cobs-maize and increased shelf life of the stored product. A modified gombisa was constructed from locally available materials in Germany. An appropriate fan type and size, humidistat set at 70 % and two 20 Wp photovoltaic panels were utilised for ventilation purpose, fan control and to power the fan, respectively. In total 1.76 m^3 of on-cobs-maizes with an average moisture content of 0.22 on d.b. ($kg\ kg^{-1}$) were used for the study. Data was collected on solar irradiance, photovoltaic voltage, current, inlet duct air velocity and temperature and relative humidity inside the storage system. Similarly, moisture content of on-cobs-maize, ambient temperature and relative humidity data were also collected for both experiments. The result for the temperature and relative humidity trends revealed higher variability and fluctuation for ambient compared with inside the modified gombisa. Ventilation of on-cobs-maize for 10–12 days resulted in a reduction of moisture content (d.b.) to almost 0.14 ($kg\ kg^{-1}$) which generally is considered safe for mould growth conditions. A computational fluid dynamics simulation result revealed the uniformity of the drying of on-cobs-maize using the ventilation system fitted to the modified gombisa. Secondary data of solar irradiance obtained from Jimma area, Ethiopia compared to the current experiment show higher energy availability, demonstrating high potential to apply ventilation and drying system to the region. Storing maize inside modified gombisa played a role in protecting the stored product from outside weather conditions. Also, monitored temperature, relative humidity and energy output showed the system was able to bring the product to safe moisture content for storage without mould development. This promising research result needs to be tested and validated in tropical regions of the world.

Keywords: Maize in cobs, modified gombisa, relative humidity, temperature, ventilation system

Contact Address: Chemeda Abedeta Garbaba, Jimma University, Horticulture and Plant Sciences , Jimma, 000 Jimma, Ethiopia, e-mail: chemedaa@yahoo.com

Pyro-Gasification of Cassava Peels for Concurrent Production of Syngas and Biochar

THIERRY GODJO

University of Sciences, Technologies, Engineering and Mathematics, Doctoral School of Sciences, Technologies, Engineering and Mathematics, Benin

In Benin, cassava is an important component of the agricultural economy. It represents 8.3 % of the agricultural gross domestic product. However, the difficulties to energy access and the waste management from cassava processing are obstacles for its development. In order to fill these gaps, a pyro-gasification process using an up flow syngas reactor has been developed. The aim of this work is to provide an alternative solution represented by use of cassava waste peelings as feedstock in a pyro-gasifier, working with low values of equivalence ratio (ER), in order to obtain syngas and biochar. The tests carried out have shown that the developed pyro-gasifier can be able to recover syngas energy of approximately 82 % of the mass of treated peelings and produces about 18 % of charcoal. In addition, from an energy point of view, the recovered syngas correspond to 74 % of the energy of the treated peelings, while the obtained charcoal (18 % of raw material) corresponds to 26 % of total energy. The pyro-gasifier's flow rate is 10 kg h^{-1}. The flame temperature was recorded as 255°C. As for performance compared to gari, the hourly capacity of gari cooked with the energy produced is 14 kg against 5 kg obtained with the conventional process. The average specific consumption of fire wood, expressed as weight of wood per kilogramme of gari produced, is 2.40 according to the conventional process and 0.00 for the pyro-gasifier. Finally, results show that energetic value of syngas from cassava waste peelings is suitable for the production of heat for cooking gari while the biochar can be used as fuel and carbon sinks.

Keywords: Benin, Biochar, cassava peels, energy recovery, *Manihot esculenta*, pyro-gasification, syngas

Contact Address: Thierry Godjo, University of Sciences, Technologies, Engineering and Mathematics, Doctoral School of Sciences, Technologies, Engineering and Mathematics, B.P. 2282 Goho, 01 Abomey, Benin, e-mail: thierrygodjo@hotmail.com

Biogas Energy Potential in Syria: Prospects and Challenges

GHAITH HASAN, HYNEK ROUBIK, JANA MAZANCOVA, JAN BANOUT
Czech University of Life Sciences Prague, Fac. Tropical AgriSciences, Czech Republic

Syria suffers from a catastrophic civil war and a strong embargo that have negative consequences for all sectors and affects the country's exploitation of underground energy resources. Current low quantities of energy production and still increasing demand for reliable energy consumption open potentials for alternative sources of energy. Organic waste from different agricultural, industrial and domestic production has been increased in the last decades. However, methods of organic waste management are often ineffective in terms of health, environment and economic sustainability. Hence, it has become a necessary goal to seek effective technology that can convert organic waste into an energy source. The experience of producing biogas from agricultural and animal waste in developing countries is seen as one of the appropriate ways for generating renewable environmentally friendly energy. The main objective of this research is to investigate the feasibility of biogas production in Syria, its prospects and challenges. A full profile about the biogas potential resources and biogas plants history in this Mediterranean country is provided. A comparison between the situation in Syria and its neighbour countries is shown in order to provide a glimpse about the biogas situation in the Middle East area. The study found that although there is a real crisis in the securing of energy resources in Syria, biogas production technology has not been widely deployed yet due to the economic, technical, social, and other causes and difficulties. The study found that the basic components of the application of biogas production technology in Syria are available through the presence of suitable quantities of organic waste and the moderate climate in the region.

Keywords: Anaerobic digestion, biogas technology, developing countries, organic waste, renewable energy

Contact Address: Ghaith Hasan, Czech University of Life Sciences Prague, Fac. Tropical AgriSciences, Kamýcká 129, 16521 Prague, Czech Republic, e-mail: hasan@ftz.czu.cz

Change in Quality Properties of Anna Apple During Cold Storage

MOHAMED ABDEL FATTAH ABDEL MAKSOUD, AYMAN AMER EISSA, EHAB EL SAEIDY, SAID FATHI EL SISI

Minoufia University, Dept. of Agricultural Engineering, Egypt

Cooling is one of the main techniques used to preserve the postharvest quality of horticultural products. A uniform distribution of humidity, prevents the formation of hot spots within or between bins, and maybe thereby helps to reduce water loss from the fruit therefore reflect of its quality. So, relative humidity control is essential during cold storage. Colour and firmness are important quality parameters which determine consumer acceptability of fruits like apple. The aim of this work was studying the effect of relative humidity (90 %, 95 % and a not fixed range of 55–70 %), position of apple inside pallets and cooling time in a cold storage at 10°C on quality properties of Egyptian Anna apple (firmness, colour, total soluble solids and reducing sugar). The obtained results showed that, the maximum firmness with high significance for all treatments was 35.2±0.158 N and 23.95±0.158 after 10 and 40 days, respectively, at 95 % RH for the middle of pallet A in the cooling room. The maximum L* value was 69.8±0.42 at middle of pallet A at 95 % of RH after 10 days cold storage, while after 40 days it reached 62.9±0.42.The lowest mean value of parameter a* was observed at bottom of pallet A (-2.833±0.177) at 90 % of RH after 10 days cold storage. The maximum total soluble solids (%) recorded were 10.969±0.058, 10.668±0.058 and 11.005±0.058 for apples stored 40 days for bottom and meiddle of pallet A and top of pallet B without control of RH (range 55–70 %). The maximum reducing sugar values (%) of 13.667±0.07, 13.371±0.078 and 13.667±0.078 were observed in apples stored 40 days at 90 %, 95 % and without control of RH at the top of pallet B and it was concluded that the optimum quality properties of stored apples were achieved at 95 % RH and in the middle position of pallet A after 40 days of cooling.

Keywords: Cold room, colour, firmness, post harvest quality, reducing sugar, relative humidity, total soluble solid

Contact Address: Mohamed Abdel Fattah Abdel Maksoud, Minoufia University, Dept. of Agricultural Engineering, 32116 Shibin El-Kom, Egypt, e-mail: mohamed.abdelmaksoud@agr.menofia.edu.eg

Greenhouse Solar Dryer with Biomass Furnace for Coffee Farmer Communities

ROMUALDO MARTINEZ, ROBELYN DAQUILA
Philippine Center for Postharvest Development and Mechanization, Philippines

A greenhouse solar dryer with biomass furnace was developed to enable coffee farmer communities in the Philippines to dry harvested coffee even during prolonged rainy periods and with minimal contamination from dust, insects, animals and other sources. The dryer had parabolic-shaped, greenhouse-like structure, 7.2 meters long, 5.1 meters wide and 2.7 meters high and used twin-wall transparent polycarbonate sheets. It was installed on 0.1 m thick concrete pavement. Inside the dryer were two drying tables 6 meters long and 1 metre wide that could accommodate 180 kg of harvested coffee cherries or de-pulped coffee. Outside one end of the dryer was a biomass furnace with two heating ducts that passed across the dryer below the drying tables and connected to two chimneys outside the opposite end. It provided the needed heat during prolonged rainy periods. On one end of the dryer were three exhaust fans that simultaneously drew in ambient through vents on the opposite end of the dryer and drew moist air out. The fans were powered by a deep-cycle battery charged by solar cell panels. Dryers were installed in three coffee communities. Results of drying tests showed that on the average, the drying time of coffee cherries was reduced from 14 days to 7 days as compared against direct sun drying. Likewise, drying of de-pulped coffee was reduced from 7 days to 4 days as compared to direct sun drying. With the use of the biomass furnace, drying was unhampered by prolonged rainy periods. Occurrences of molds on coffee were prevented and product quality remained high because drying delays were minimised. Because of the dryer, coffee farmers were able to sell their dried coffee at a higher price.

Keywords: Biomass, coffee, dryer, energy, solar

Contact Address: Romualdo Martinez, Philippine Center for Postharvest Development and Mechanization, CLSU Compound, 3120 Munoz, Philippines, e-mail: romualdomartinez@yahoo.com

Improved Design of Automated Cassava Peeling Machine for Large Scale Industries

OLAWALE JOHN OLUKUNLE[1], OLUWATOYIN OLUKUNLE[2]

[1]*The Federal University of Technology, Akure, Dept. of Agricultural and Environmental Engineering, Nigeria*

[2]*The Federal University of Technology, Akure, Dept. of Biotechnology, Nigeria*

A review on cassava peeling machines in Nigeria was conducted recently with a view to present objective appraisal of existing peeling machines and hence identify knowledge and technology gap in cassava peeling mechanisation. The processing of cassava tubers into various industrial or human uses require many unit operations which have been effectively mechanised with the exception of peeling. Cassava peeling mechanisation has therefore been identified as a global challenge for design engineers particularly in countries where cassava and its products form staple food for the people. The features, prospects and limitations of cassava peeling machines in Nigeria presented are diverse and multidimensional. A more efficient cassava peeling machine would be required to reduce tuber loss and improve peeling efficiency through proper design of the peeling system. In order to develop a peeling machine that will satisfactorily peel the tubers with near zero tuber loss and high output capacity, analysis of the peeling pattern in automated peelers was considered to produce an improved model. The limitations identified in the review were used as basis for the design of the improved model of the automated cassava peeling machine. Special features in the improved design include modification of the knife edge peeling tool, incorporation of brush peeling tool as peeling guide and optimisation of various machine and operational parameters. Results show reduction in tuber losses of up to a range of 2 to 5 % at machine speed of 200 to 450 rpm. The peeling efficiency increased appreciably between 85 % and 95 % and machine capacity increased between 750 and 1200 kg h^{-1}. Again, this design was found useful for industrial purposes in order to meet the growing need for primary and secondary products from cassava. The machine was less labour intensive, has low repair and maintenance costs, high output capacity, low tuber loss and the machine requires moderate technical skill to operate. One outstanding merit of this improved design is its ability to peel small size tubers and the machine could be adapted to peel other smaller tubers and fruits such as cocoyam, potato and plantain.

Keywords: Cassava peeling machine, peeling efficiency, tuber loss

Contact Address: Olawale John Olukunle, The Federal University of Technology, Akure, Dept. of Agricultural and Environmental Engineering, Praise Close, Peace Avenue, FUTA South Gate, Akure, Nigeria, e-mail: wale_olukunle@yahoo.com

(Socio)economics

Agricultural transition and rural employment

Oral Presentations

Posters

Cardamom Production and the 'Good Life': Agricultural and Social Change in East Nepal

MARIE-LUISE MATTHYS

University of Bern, Interdisciplinary Centre for Gender Studies, Switzerland

What is a "good life"? In countless writings, Eastern and Western thinkers alike have dwelled on this fundamental philosophical question. Through implementing development policies, governments aim at making their citizens' lives better, thereby implicitly defining what a good life is. In everyday practices, citizens enact their own modes of life, each person holding an individual perspective of what makes one's life good. This research is about the interplay of philosophy, agricultural policy, and rural people's practical understandings of the "good life" in the context of agricultural development.

About a decade ago, farmers in East Nepal started taking up commercial cardamom production, an agricultural development strategy fostered by the Nepalese government. Today, cardamom is the major income source in the region, and everyday life is constantly changing. Is this development fostering "good lives", according to the rural population? An innovative combination of qualitative methods including participatory photography is used to trace answers to this question, involving 58 respondents of different classes, genders, and castes.

Preliminary findings suggest that for the majority of respondents a "good life" is a life with few hardships, peaceful family relations, a good future for their children, substantive individual freedoms, as well as "sukha", i.e. happiness or joy. Respondents argue that, in general, cardamom contributes to this kind of good life because its production requires less physical effort and yields more income than previous crops. Notably, the respondents view an increase in income not as an end in itself – instead, they value money as a useful means to foster the above dimensions of the "good life". However, the benefits of cardamom production are not equally distributed: limited availability of land and irrigation force numerous small-scale farmers to take up causal labour on wealthier people's farms, potentially exacerbating existing inequalities. In addition, respondents are worried about the spread of plant diseases jeopardising sustainability, which points to the farmers' dependency on the new cash crop. Most strikingly, wealthy and poor farmers alike do not want their children to continue agricultural production: eventually, a "good life" is the life of an employee residing in town .

Keywords: Agricultural change, commercialisation, good life, inequality, intersectionality, participatory photography

Contact Address: Marie-Luise Matthys, University of Bern, Interdisciplinary Centre for Gender Studies, Mittelstrasse 43, 3012 Bern, Switzerland, e-mail: marie-luise.hertkorn@izfg.unibe.ch

Determinants of the Adoption of Sustainable Intensification in Southern African Farming Systems: A Meta-Analysis

QIAN GUO, EMMANUEL OLATUNBOSUN BENJAMIN
Technische Universität München (TUM), Dept. of Production and Resource Economics, Germany

Climate change and environmental degradation are major threats to sustainable agricultural development in sub-Saharan Africa, especially, Southern Africa. Thus, the concept of sustainable intensification (SI), i.e. getting more output from less input using certain practices that are environmental friendly, has become an important topic among researchers and policy makers in the region in the last three decades. A comprehensive review of literatures on SI in the region identify nine relevant drivers of adoption of SI among (smallholder) farmers in Southern Africa. These drivers include (i) age, (ii) size of arable land, (iii) education, (iv) extension, (v) gender, (vi) household size, (vii) income, (viii) membership in farming organisation and (ix) access to credit. We present the results of a meta-analysis of 21 papers on the impact of these determinants on SI adoption among (smallholder) farmers in Southern African Development Community (SADC) using random-effects estimation techniques for the true effect size. While our result suggests that variables such as extension, education, age and household size may influence the adoption of SI in SADC, factors such as access to credit and women smallholder farmers are also of great importance. The challenges credit constraint poses to farmer´s ability (especially women) to adopt sustainable practices can be perceived as a major hindrance to rural agricultural development in Sub-Saharan Africa. One the other hand, credit constraint has also been observed to promote sustainable agricultural practices, among smallholder farmers, if monetary incentives and compensations are adequate. Decision-makers should therefore concentrate efforts on these factors in promoting SI across the SADC.

Keywords: Adoption, climate change, effect size, meta-analysis, random-effects model, smallholders, Southern Africa Development Community (SADC), sustainable intensification

Contact Address: Emmanuel Olatunbosun Benjamin, Technische Universität München (TUM), Dept. of Production and Resource Economics, Alte Akademie 14, 85354 Freising, Germany, e-mail: emmanuel.benjamin@tum.de

Influence of Conservation Policies and Deforestation on Forest and Agricultural Income in Ecuadorian Tropical Forests

Tatiana Ojeda Luna[1], Eliza Zhunusova[2], Sven Günter[2], Matthias Dieter[2]

[1] *University of Goettingen, Graduate School Forestry and Agricultural Sciences, Germany*
[2] *Thünen Institute of International Forestry and Forest Economics, Germany*

Natural resources contribute with more than 50 % to rural incomes at the Ecuadorian tropical forest frontiers. Due to high deforestation Ecuador implemented two major conservation policies: (i) Protected areas (PAs), a command-control policy; and, (ii) Socio Bosque programme (PSB), an incentive-based mechanism. PAs and PSB protect forest lands by restringing the access for a long-term perspective. Given the dependency of rural farmers to natural resources, they might be influenced by processes that limit the access or availability of these resources. Studies on household income in Ecuador are still scarce and do not include conservation policies and deforestation into the analysis. Our research incorporates two regions located in the tropical lowland forest frontiers: (i) the Central Amazon (CEA), with low deforestation and high forest cover, where PAs were implemented; and, (ii) the Northwestern Coast (NOC), with high deforestation and few forest remnants, where PSB was established. Based on socio-economic data from 1300 household surveys applied from August 2016 to August 2017, we performed econometric analysis to determine how conservation policies and deforestation influence forest and agricultural income; we included households' characteristics, natural assets and access to markets as control variables. In the NOC our results showed that deforestation has a positive relation with forest income; while PAs limit forest income and facilitate agricultural income. When analysing the effect of PAs on indigenous landholders, we observed that indigenous can still have a high forest income even under the presence of a PAs; this is due to the big forest lands owned by them. In the CEA, PSB did not show significant effect, but deforestation showed a positive influence on both income sources. Forest loss can bring immediate cash revenues; but, in a long term perspective deforestation leads to resources depletion which might affect rural income. Our results allow identifying households that can be impacted by conservation strategies and deforestation, which is useful for policy decisions and for the success of in-situ conservation.

Keywords: Incentive-based conservation, livelihood strategies, protected areas, rural income, Socio Bosque

---

Contact Address: Tatiana Ojeda Luna, University of Goettingen, Graduate School Forestry and Agricultural Sciences, Schönningstedter 112, 21465 Reinbek, Germany, e-mail: t.ojedaluna@stud.uni-goettingen.de

Rural-Urban Transformation Processes in India and West Africa

Andreas Buerkert[1], Hanna Karg[2], Chickadibburahalli T. Subbarayappa[3], Dimba Chowdappa Hamumanthappa[4], Z. Mudalagiriyappa[3], Ellen Hoffmann[1], Eva Schlecht[5], Tobias Plieninger[6], Stephan von Cramon-Taubadel[7], K.B. Umesh[8]

[1]*University of Kassel, Organic Plant Production and Agroecosystems Research in the Tropics and Subtropics, Germany*

[2]*University of Freiburg, Dept. of Environmental Social Sciences and Geography; Physical Geography, Germany*

[3]*University of Agricultural Sciences, Dept. of Soil Sci. and Agricultural Chemistry, India*

[4]*University of Agricultural Sciences, AICRP on Agroforestry, India*

[5]*University of Kassel / University of Goettingen, Animal Husbandry in the Tropics and Subtropics, Germany*

[6]*University of Kassel / University of Goettingen, Social-Ecological Interactions in Agricultural Systems, Germany*

[7]*University of Goettingen, Dept. of Agricultural Economics and Rural Development, Germany*

[8]*University of Agricultural Sciences, Department of Agricultural Economics, India*

In recent decades global population growth and rural-urban migration have led to rapidly increasing transformation of landuse systems with concomitant changes in agricultural intensities and enhanced reliance on the multiple services that agro-ecosystems are expected to provide. This is particularly visible in the dynamic interface of megacities in Asia and Africa. By 2050 the global share of urban population has been estimated to reach 68 % compared with 55 % today, the majority of which is accounted for in Asia while Africa shows the highest urban growth rate. The corresponding loss of agricultural areas around cities, which are of particularly high productivity, has been estimated at 2 % globally, with 60 % of it occurring in Asia. Serious threats to agricultural production, urban food security, natural habitats, biodiversity, and air and water resources are the consequence. Based on a comparative analysis of the status quo in the rural-urban interface of Bengaluru in South India and of Ouagadougou and Tamale in West Africa we use a social-ecological systems framework to quantify spatial changes in landuse patterns, wetland structure and flows of resources that affect the sustainability of the three cities. Key bottlenecks of future development are water cycling and use, strategic planning in the use of rural and urban spaces in a multifunctional landscape and maintenance of soil productivity.

Keywords: Inda, matter flow, multi-functional landscapes, resource governance, rural-urban interface, urbanisation, West Africa

Contact Address: Andreas Buerkert, University of Kassel, Organic Plant Production and Agroecosystems Research in the Tropics and Subtropics, Steinstraße 19, 37213 Witzenhausen, Germany, e-mail: buerkert@uni-kassel.de

Reaching the Ultra-Poor through Livelihood Asset Transfer: Impacts on Rural Income Diversification and Migration in Ethiopia

TADIWOS ZEWDIE, JAMES WARNER, GETAW TADESSE

International Food Policy Research Institute (IFPRI), Africa, Ethiopia

Social protection programs in Africa and South Asia are relatively successful at reaching moderately poor households but ultra-poor tend to inadequately served or bypassed. In view of this, Ethiopia has piloted innovative livelihood asset transfer programme between 2016 to 2018 to enhance the rate of graduation from the national Productive Safety Net Program (PSNP). It provides a direct transfer of a livelihood investment grant together with human resource development training. Similar programs are already being replicated in several other countries. Little is known whether the programme has the potential to help the ultra-poor move onto a sustainable trajectory out of poverty. Using a household panel data and difference- in- difference estimation approach, we examined the impact of the livelihood transfer programme on economic improvement of ultra-poor households. The results show that the livelihood transfer has a positive and statistically significant effect in increasing household income diversification and per capita cash income; and reduced rural migration. Further analysis by gender indicates that female-headed households gain higher per capita income than male-headed households. However, the latter have gained more diversified income (or new investments from the transfer resource) than the former. We also find evidence that per capita income gains and rural migration reduction from programme participation is statistically significant across income groups implying that the programme is indeed pro-poor. Thus, it can help to enhance the rate of graduation from social protection supports. Overall, the pilot livelihood asset transfer programme impacts have proven to be potentially scalable in the country and elsewhere.

Keywords: Ethiopia, income diversification, livelihood asset transfer, migration, ultra-poor

---

Contact Address: Tadiwos Zewdie, International Food Policy Research Institute (IFPRI), Africa, Addis Ababa, 27886/1000 Addis Ababa, Ethiopia, e-mail: t.zewdie@cgiar.org

Self-Organised Natural Resource Management in Burkina Faso

JANA HEROLD, KARIN GAESING

University of Duisburg-Essen, Institute for Development and Peace, Germany

The research project "Ways out of extreme poverty, vulnerability and food insecurity" conducted by the Institute for Development and Peace at the University of Duisburg-Essen with funds from the German Federal Ministry for Economic Development and Cooperation (BMZ) aims to develop recommendations for development cooperation regarding how to reach extremely poor, vulnerable and food insecure people and to effectively improve their lives. The research approach is based on identifying good practice projects and assessing their outreach approaches, project activities, lessons learnt and above all their impact with regard to the reduction of food insecurity, vulnerability and poverty. Among the research methods are focus group discussions, wealth rankings and intensive interviews with project staff and beneficiaries.

One of the good practice projects researched by the INEF team is an initiative by the two NGOs Initiative Développement Durable (IDD) and Terra-Verde in Burkina Faso. The two NGOs put great emphasis on informing and sensitising smallholder farmers for the need for sustainable resource management and on building their self-help capacity to plan, organise and implement these activities themselves. With the assistance of the project, farmers mainly build stone bunds along contour lines for better rainwater infiltration in their fields. This activity alone allows farmers to significantly increase their yields. In addition, it turns barren land into cultivable fields. The research team found the project to be one of the good practice examples to sustainably overcome hunger, malnutrition and poverty. Research results point to the conclusion that farmers are willing and able to actively engage in sustainable natural resource management if they are properly informed about the consequences of continuing to maintain the status as compared to the benefits of applying stone bunds and other soil and water conservation measures.

Keywords: Burkina Faso, natural resource management, soil fertility, stone bunds

Contact Address: Jana Herold, University of Duisburg-Essen, Institute for Development and Peace, 47048 Duisburg, Germany, e-mail: jana.herold@inef.uni-due.de

Challenges Facing the Quality Management along Organic Agriculture Value Chains in Southwestern Nigeria

OLUBUNMI AYOBAMI DUDUYEMI
Obafemi Awolowo University, Dept. of Animal Science, Nigeria

Organic agriculture in Nigeria is still at its infancy, representing a production system that enhances and sustains the health of soils, ecosystems and people. It relies on ecological processes, biodiversity and cycles adapted to local conditions, rather than the use of inputs with adverse effects. This combines tradition, innovation and science to benefit the shared environment and promote fair relationships and a good quality of life based on principles of Ecology, Health, Fairness and Care.

The development vision by Nigeria's government conceptualises a transformation in organic agriculture that would ensure food security, the right to sustainable development for all and adaptation to the climate change challenge. The transformation agenda is meant to wean Nigeria off food imports by boosting domestic food production. This entails reforms in the input supply regime, a targeted region specific increase in the output of priority commodities, post-harvest systems development, a strong orientation towards agri-business and promoting value-addition in the product chain. Despite these policy interventions, the organic agriculture sector is still largely faced with a lot of quality management issues across its value chain segments relating to market access and orientation, available resources and physical infrastructures.

This multiplicity of intertwined factors and forces cannot be resolved by the monolithic conventional agriculture approach to increasing yields, rather, a holistic system that provides a whole range of products and services that are required to sustain the livelihoods of the population, is resilient and supports the ecosystem functions that underpins organic agriculture production, including a living soil, and in the context of sustainable development and poverty eradication, empowers poor farmers in order to secure their food security and equitable returns on their investments.

Therefore, overcoming these challenges will result in an affordable and appropriate approach for increasing local food security and resilience to drought as well as regenerating the productive capacity of degraded lands. Hence, the rising global demand for organic agriculture products based on their superior health and environmental characteristics has created new market opportunities for farmers in Southwestern Nigeria to leverage on.

Keywords: Nigeria, organic agriculture

Contact Address: Olubunmi Ayobami Duduyemi, Obafemi Awolowo University, Dept. of Animal Science, 220005 Ile Ife, Nigeria, e-mail: bunmid2000@yahoo.com

Providing Smallholder Farmers with Weather-Related Information to Build their Resilience to Climate Variability: A Qualitative Exploration

Rosaine Nerice Yegbemey[1], Ghislain B. D. Aihounton[2], Aline M. Aloukoutou[3]

[1]*University of Parakou, Department of Rural Economy and Sociology, Benin*

[2]*University of Parakou, Laboratoire d'Analyse et de Recherches sur les Dynamiques Economiques et Sociales (LARDES), Benin*

[3]*Bureau de Recherche et de Développement en Agriculture (BReDA- ONG), Benin*

This paper is aimed at exploring possible impact pathways of a hypothetical intervention which consists in providing smallholder farmers with weather-related information. The study used a qualitative approach based on focus group discussions with smallholder farmers and agricultural extension officers to build a Theory of Change (ToC) of the intervention of interest. The field work was conducted in three villages of the municipal area of Bembèrèkè in North Benin (West Africa). The results suggested that providing smallholder farmers with weather-related information has the potential to help them in taking informed production decisions. Possible decisions include the right timing of key activities such as planting or sowing, application of inputs (e.g. fertilisers, herbicides and insecticides), etc. Through informed production decisions, smallholder farmers can better allocate their production resources and record higher yield and income. The ultimate impact is that smallholder farmers will have better life—livelihoods. There is several weather-related information that can be potentially useful for smallholder farmers. There are also several dissemination channels, some based on the social network of the community and others on Information, Communication and Technology (ICT) tools that can be used. Each channel has some strengths and weakness and the best approach would probably depend on the settings of the intervention area. Though the expected impact is straight forward, a number of assumptions that needs to be in place for an impactful intervention. For instance, weather-related information needs to be accurate, available in a timely manner and easy to use (read and interpret). Furthermore, the described possible impact pathways need to be tested rigorously through policy-oriented research.

Keywords: Benin, impact pathways, theory of change, weather-related information

---

Contact Address: Rosaine Nerice Yegbemey, University of Parakou, Department of Rural Economy and Sociology, Parakou, Benin, e-mail: ynerice@gmail.com

Power-Relations through Institutional Bricolage in the Rural Farmers' Decision to Transit towards Agroecology in Peru

MARCO FROLAIN DE LA CRUZ SÁNCHEZ[1], JOOST DESSEIN[2]

[1]*KU Leuven Public Governance Institute, Belgium*

[2]*Ghent University, Dept. of Agricultural Economics, Belgium*

Empirical evidence increasingly suggests that while the uptake of agroecology brings significant benefits to ecosystems, health, and communities, yet achieving a transition towards agroecology can still be complicated. In this article we contribute to this literature by offering a power-sensitive analysis of the transition towards agroecology. We build on insights from critical institutionalism and introduce a complementary analytical view for examining the institutional arena in which actors' interactions influence individual decisions to adopt agroecological practices. We label this bundle of interactions as 'chaotic relational patterns'. Our empirical data is drawn from a study of a program in rural villages in Peru, designed to encourage farmers' to take up agroecology. Our case study demonstrates that the local dynamics influenced agricultural transition and that their evolution exposed more complex mechanisms of power-play in farmers' interactions. We identify two power-play mechanisms that influenced farmers' decisions to adopt agroecological practices: copying and learning - which entails learning from peer farmers with more resources before adopting an agroecological production model, and the manageability of the risks - which was adopted as a strategy for managing the multiple risks of adopting agroecology, and aligning their economic and social interests with those of the agroecology programs.

Keywords: Agriculture transition, agroecology, critical institutionalism, power relations

Contact Address: Marco Frolain De La Cruz Sánchez, KU Leuven Public Governance Institute, Huikstraat 25, 2000 Antwerp, Belgium, e-mail: m.delacruzs@pucp.pe

A Mixed-Method Systematic Review to Assess Household Characteristics and Livelihoods in Rural Zimbabwe

GRACE MUDOMBI-RUSINAMHODZI[1], ANDREAS THIEL[1], GEORGE OWUOR[2]
[1]*University of Kassel, International Agricultural Policy and Environmental Governance, Germany*
[2]*Egerton University, Agricultural Economics and Agribusiness Management, Kenya*

Livelihood activities that are dependent on natural resources form a continuum of people-forest interactions. They often constitute multifaceted livelihood strategies consisting of various intertwined components that strive to balance livelihood activities and forest resources conservation. Livelihood strategies are linked to social relations, habits, norms, customs and beliefs, and code of forest practice. Hence, households' decisions on forest resources use are influenced by a more complex set of factors with implications on rural livelihoods and the level of motivation for conservation of forest resources. Productive bricolage thus offers an approach to examine role of non-timber forest products (NTFPs) in rural livelihoods, everyday decisions on natural resource use and forest resource changes in an integrated manner. Productive bricolage gives insight into the drivers of changes in forest resources and the constraints and opportunities that determine people's choices regarding forest resource use. This perspective would improve understanding of how NTFPs production interacts with other livelihood strategies of producers, how important NTFPs production is to their rural households and whether NTFPs production has the potential to incentivize forest resources conservation. This study uses a mixed-method systematic approach to provide a meta-synthesis of the socio-economic demographic characteristics of households, and the role of NTFPs in rural livelihoods through building insights from previous studies on various communities in Zimbabwe in order to understand how use of forest resources can effectively contribute to both livelihood enhancement and forest resources conservation. Findings from a review of 29 primary studies and 10 grey literature publications show that livelihood activities include dryland crop production, gardening, livestock production, formal employment, casual labor remittances, fishing, selling of crafts, selling of NTFPs, shoe repairing/tailoring, brick molding and beer brewing. Crop and livestock production are directly dependent on forest resources and incomes from selling crafts and NTFPs. Land cover is directly affected by agricultural expansion as well as agricultural management options. Households that are *de jure* female-headed, large in household size, have no remittances, have small land holdings and fewer livestock are more dependent on forest resources. The socio-economic contributions of forest resources to rural livelihoods range between direct use values (income, energy, shelter, food, medicine, crafts) to indirect use values (regulating services). Though aggregate contribution of NTFPs to rural households is significant, it varies amongst different social actors and groupings.

Keywords: Conservation, forest resources, livelihood activities, meta-synthesis, productive bricolage

Contact Address: Grace Mudombi-Rusinamhodzi, University of Kassel, International Agricultural Policy and Environmental Governance, Steinstrasse 19, 37213 Witzenhausen, Germany, e-mail: mudombigrace@yahoo.com

Public Images of Dairy Production Systems among Urban Citizens in Bogotá, Colombia

WIEBKE WELLBROCK[1], STEFAN BURKART[2], KAREN ENCISO[2], ANDREA KNIERIM[3]

[1] *Independent Researcher, Germany*
[2] *International Center for Tropical Agriculture (CIAT), Trop. Forages Program, Colombia*
[3] *University of Hohenheim, Inst. of Social Sciences in Agriculture, Germany*

In Colombia, the consumption of fresh milk and milk products is steadily increasing. This rise demands an intensification and modernisation of the current dairy production system. To do so, intensification of dairy production is envisioned, posing a threat to the livelihoods of smallholders which make up 81 % of dairy producers in Colombia. The aim of this study is to investigate the public image of dairy production systems among urban citizens in the city of Bogotá in order to find out what type of dairy production system is culturally sensitive and socially acceptable in Colombia. Designed as an explorative case study, 30 participants were interviewed using a triangulated research method including a word association task, a drawing exercise and a semi-structured interview. The results show that participants are aware of two distinct types of dairy production systems in the country- industrial production systems and smallholder production systems, but they show more affinity for smallholders. This is because participants appreciated values such as naturalness and care for the animals and associated these values with smallholders. At the same time, participants were concerned about health, hygiene and food safety standards of milk and milk products coming from smallholders. Industrial production systems were associated with modern technologies as well as high hygiene and food safety standards. In order to ensure culturally sensitive and socially acceptable dairy production systems in Colombia, it is necessary to enhance the development of smallholder dairy farms and to provide education and support for better health, hygiene and food safety standards on these farms.

Keywords: Colombia, dairy production systems, public images, urban citizens

Contact Address: Wiebke Wellbrock, Independent Researcher, Weidengarten 7, 53129 Bonn, Germany, e-mail: wwellbrock@gmail.com

Socio-Economic Contribution of Indigenous Goats to Smallholders in Low Input Crop-Livestock Production System in Malawi

WILSON KAUMBATA[1], LIVENESS BANDA[1], GÁBOR MÉSZÁROS[2], TIMOTHY GONDWE[1], M. JENNIFER WOODWARD-GREENE[3], BENJAMIN ROSEN[3], CURTIS VAN TASSELL[3], JOHANN SÖLKNER[2], MARIA WURZINGER[2]

[1]*Lilongwe University of Agriculture and Natural Resources, Animal Science, Malawi*

[2]*University of Natural Resources and Life Sciences, Vienna (BOKU), Dept. of Sustainable Agricultural Systems, Austria*

[3]*United States Department of Agriculture / Agricultural Research Service (USDA/ARS), United States of America*

Indigenous goats are important in resource-poor communities because they provide tangible benefits such as cash revenue from animal sales, meat for home consumption, manure, skins, and fiber. They are also a source of intangible benefits, e.g. savings, insurance, and for socio-cultural purposes. Despite the valuable contributions of goats to the livelihoods of millions of resource-poor farmers in developing countries, goats have been overlooked in national agriculture development strategies and prejudice towards other livestock species like cattle still exist. This study was conducted to determine socio-economic contributions of indigenous goats to smallholders in low input crop-livestock production system where community-based goat breeding programs (goat CBBPs) are being implemented. The aim was to provide government and other development agencies with the information which can form the basis for policy, technical and financial support for the goat CBBPs. Data was collected through a 12-month flock and household (137 households) monitoring study between August 2017 to July 2018. Data was analysed using enterprise budgeting and cost-return analysis.

The results showed that indigenous goat enterprises in smallholder farms are profitable and economically viable. The mean annual gross margins per flock and per goat was €101.00 and €17.60, respectively. The average return on capital invested was 24.6 %, exceeding the prevailing average commercial deposit rate (8 %) by several folds. Goats accounted for 61.2 % of the total livestock household income representing the biggest contributor, while cattle, pigs and chickens contributed 17.6 %, 15.5 % and 4.1 %, respectively. Sale of live goats constituted the major (79.2 %) proportion of the total offtake rate, suggesting that goats are primarily kept for generation of cash revenues. Inclusion of intangible benefits of goats significantly increased the mean annual gross margins and the return on capital by 60.3 %, reflecting the importance of socio-economic roles goats play in providing current and future economic stability to rural households' economy. Hence, programs like goat CBBPs are meant to harness the potentials of indigenous goats to optimise their contributions towards reduction of rural poverty and hunger. Therefore, financing and supporting scaling up of such programs is a meaningful direct investment into the development of rural economy.

Keywords: Breeding program, community-based, gross margins, rural poverty, tangible and intangible benefits

Contact Address: Wilson Kaumbata, Lilongwe University of Agriculture and Natural Resources, Animal Science, P. O. Box 219 Lilongwe Malawi, Lilongwe, Malawi, e-mail: wilk18@yahoo.com

Calling for Mechanisation: Farmers' Willingness to Pay for Small-Scale Maize Shelling Machines in Tanzania

BEKELE HUNDIE KOTU[1], GUNDULA FISCHER[2], ADEBAYO ABASS[2], IRMGARD HOESCHLE-ZELEDON[3], MATEETE BEKUNDA[2]

[1]*International Institute of Tropical Agriculture (IITA), Ghana*
[2]*International Institute of Tropical Agriculture (IITA), Tanzania*
[3]*International Institute of Tropical Agriculture (IITA), Nigeria*

While mechanisation is an important complement, and in some cases a necessary condition, to agricultural intensification, humans are the main power source of agricultural production in sub-Saharan Africa. This is partly because of the failure of the supply driven mechanisation policies promoted in SSA until 1980s which gave marginal attention to the demand side of mechanisation. We conducted a study in three districts of central Tanzania with the aim of assessing farmers' willingness to pay (WTP) for two type of small-scale (diesel-powered and electric-powered) maize shelling machines and identifying factors affecting their WTP. We collected data from randomly selected 400 farmers constituting about equal number of both gender categories. We considered three mechanisation approaches namely: 1) the rental service model (RSM), 2) the group ownership model (GOM), 3) the private ownership model (POM). We used interval estimation econometric model to analyse the data. Our results show that mechanisation of maize shelling can be promoted in the study areas following the three business models. More than 98 % of the farmers can fit to the RSM at existing market prices. The GOM is feasible for about 64 % of the farmers considering the diesel machine while it can work for about 90 % of the farmers considering the electric machine. The mean group sizes suggested by the farmers are ten and eight people for the diesel machine and the electric machine, respectively. Mechanisation of maize shelling is also possible through the POM for about 8 % (diesel machine) and 64 % (electric machine) of the farmers. The factors that positively influence farmers WTP include experience in using machines to shell maize, livestock wealth, cost of hired labour, volume of maize production, and off-farm income during off-harvest season whereas the factors that negatively influence farmers WTP include gender of respondents (being female), active female labour in the household, and age of respondents.

Keywords: Group ownership, maize shelling machine, private ownership, rental service, Tanzania, WTP

Contact Address: Bekele Hundie Kotu, International Institute of Tropical Agriculture (IITA), Sagnarigu, TL6 Tamale, Ghana, e-mail: b.kotu@cgiar.org

What makes Small Bamboo and Rattan Handicraft Enterprises Successful? A Case Study from Chuong My District, Hanoi Capital, Vietnam

VAN HIEP TRAN[1], DIETRICH DARR[2], JÜRGEN PRETZSCH[1]

[1]*Technische Universität Dresden, Inst. of International Forestry and Forest Products: Tropical Forestry, Germany*

[2]*Rhine-Waal University of Applied Sciences, Faculty of Life Sciences, Germany*

With a long-standing history, Vietnam is reputed for the production of handicraft bamboo and rattan products such as kitchen utensils, home decor objectives, storage bins, trays, baskets, and gifts. These products are not only commercialised but also represent Vietnamese cultural and historical values. Small bamboo and rattan handicraft enterprises (SBRHEs) in Vietnam cooperate with households in the handicraft villages to produce these handicraft products, so SBRHEs creates employment opportunities and contributes to rural poverty reduction. However, most SBRHEs currently face significant constraints such as the use of outdated technology, lack of product innovation, shortage of capital, unstable labour resource, and poor market access. In addition, the enterprises face stiff competition from international bamboo and rattan manufacturers from China and other countries. This puts SBRHEs at high risk of disappearance and may lead to unemployment and increasing poverty among handicraft households. On the other hand, a few SBRHEs from the study region have grown successfully and well established themselves in international markets over the last years. Yet, a comprehensive analysis of key factors determining the upgrading of SBRHEs is lacking to date. This paper demonstrates preliminary results from a case study undertaken to close this gap.

In-depth interviews have been conducted with representatives from six small bamboo-rattan handicraft enterprises, five bamboo traders, and twelve handicraft households in Chuong My district, Hanoi, to further complement qualitative insights, and group discussions and expert interviews have been made to validate the findings.

This paper shows that SBRHEs highly depend on handicraft households for producing handicraft bamboo and rattan products, and participating in trade fairs plays an important role in accessing the market. Moreover, we argue that the entrepreneur's education and work experience also play a significant role in explaining SBRHEs upgrading. Furthermore, the role of the government and institution support in SBRHEs upgrading are delineated.

Keywords: enterprise upgrading, handicraft, bamboo, rattan, non-timber forest products, commercialisation, rural development, small enterprises

---

Contact Address: Van Hiep Tran, Technische Universität Dresden, Inst. of International Forestry and Forest Products: Tropical Forestry, Hochschulstr.20, 01069 Dresden, Germany, e-mail: tranvanhiep@humg.edu.vn

The Wellbeing of Smallholder Coffee Farmers in Mount Elgon Region- A Qualitative and Quantitative Analysis of a Rural Community in Eastern Uganda

ANNA LINA BARTL[1], ISABEL BARTL[2]

[1] *Georg-August-Universität Göttingen, Agricultural Economics and Rural Development, Germany*

[2] *University of Applied Sciences and Art (HAWK), Dept. of Social Work and Health, Germany*

For many smallholder farmers in the Mount Elgon region of Uganda, Arabica coffee cultivation is the major activity for earning their living. It is well-known that smallholder coffee farmers often do not live under conditions that surpass existential needs. On the contrary, different levels of wellbeing and perceptions of smallholder coffee farmers are not well examined. In this study, we classify different levels of wellbeing and investigate how the own wellbeing is perceived.

We use quantitative data of 431 smallholder coffee farmers in the Mount Elgon region to identify different levels of wellbeing using the material, the social, and the personal wellbeing dimensions. In addition, 10 qualitative interviews with the households' mother help to understand correlations between household wellbeing, composition, challenges, and possibilities with regard to health care, and water supply.

Results indicate three levels of wellbeing and an impact of household composition on the levels of wellbeing. Furthermore, we found that the sense of personal and social wellbeing strongly correlates with the households' economic situation. Besides, available healthcare provisions are linked with very high costs; one consultation of a doctor nearby may constitute costs exceeding an eleventh of an average annual household income from coffee selling. Even so, most of the farmers do not perceive life quality influencing factors like the lack of health insurance, health care nearby, water supply, or the lack of infrastructure as one of their major deficiencies. In contrast, deficiencies that impact the constitution for farm management activities are perceived as more constraining.

For all life quality influencing factors, better educational opportunities are mentioned from the mothers' perspective to provide the best solution for better levels of wellbeing for future generations.

Keywords: *Coffea arabica*, Mount Elgon, smallholder coffee farmers, Uganda, wellbeing

---

Contact Address: Anna Lina Bartl, Georg-August-Universität Göttingen, Agricultural Economics and Rural Development, Göttingen, Germany, e-mail: anna.bartl@uni-goettingen.de

Community Irrigation Sharing Arrangement and Smallholders' Market Led Agriculture - A Case of Odisha State, India

Surajit Haldar[1], Smita Susovita[2], Debdutt Behura[2]

[1]*Justus Liebig University Giessen, Institute for Agricultural Policy and Market Research, Germany*

[2]*Odisha University of Agriculture and Technology, Bhubaneswar, Dept. of Agribusiness Management, India*

Community irrigation sharing arrangement (CISA) is known for its manifold socio-cultural-economic benefits irrespective of its place of formation and scale of operation. This study attempted to answer how such a sharing arrangement has shifted smallholders' land use decision especially when the market is favourable. A mixed method of survey was adopted to collect primary data from 60 farmers of 20 CISA from 12 villages of two adjoining districts, viz. Cuttack and Jagatsinghpur, Odisha State, India during November 2018 March 2019. Focus group discussions were conducted to collect information on land use and land cover change (LULCC), irrigation provision and market accessibility by CISA farmers. Based on these three parameters, surveyed farmers were post classified into four different groups namely highly diversified irrigated agriculture for direct market supply (HDIAM), moderately diversified limited irrigated agriculture for direct market supply (MDLIAM), moderately diversified limited irrigated agriculture for indirect and limited market supply (MDLIAIM) and least diversified mostly irrigated agriculture for direct market supply (LDMIAM). HDIAM and MDLIAM farmers were diversifying their land use mostly to off-season vegetables that fetch a remunerative price. For MDLIAIM and MDLIAIM farmers, sugarcane occupied more than 50% of cultivable land. Their market access was limited by a distant secondary market and poor transportation and communication facility.

The farmers were personally interviewed to elicit information on crop choice and irrigation accessibility before and after CISA formation. By using difference-in-difference approach it was observed that the previous rice-based cropping pattern, i.e. rice (rainy season)-green gram (winter season)-fallow (summer season) changed to the vegetable based cropping pattern, i.e. rice (rainy)-vegetables (winter)-vegetables (summer) for HDIAM and MDLIAM farmers. MDLIAIM farmers though shifted to vegetables and sold at farm gate price to traders but also specialized in sugarcane cultivation taking the existing facility of contract farming with the sugar mill. For LDMIAM, rainy season rice area was reduced to half of the gross cropped area and replaced by sugarcane after CISA formation. Thus, an irrigation provision through CISA enhanced farm economics while the market determined LULCC.

Keywords: Community!irrigation sharing arrangement, difference-indifference, market led agriculture

Contact Address: Surajit Haldar, Justus Liebig University Giessen, Institute for Agricultural Policy and Market Research, Senckenbergstrasse 3, 35390 Giessen, Germany, e-mail: surajit.haldar@agrar.uni-giessen.de

The Spatial and Land-Use Dimension of Food Systems in Transition: The Case of Northern Vietnam

VINCENT LINDERHOF[1], MARIEKE MEESKE[2], VASCO DIOGO[1]

[1] *Wageningen University and Research, Wageningen Economic Research, The Netherlands*
[2] *PAX, The Netherlands*

In Vietnam, production choices on crop cultivation, fertiliser use, pesticide use and water use affect the landscape and the availability of food. Choices on pesticide use also affect production and the safety of food and the health of people in the region. Earlier studies in Vietnam have shown that pesticides end up in surface water which is then used for drinking water, irrigation or in fish tanks. In fact, pesticide use also affect landscape characteristics. Hence we explore the interactions between food production, food consumption and the landscape. The research question was how do landscape elements, incorporating environmental aspects affecting health, affect food consumption. We estimated the relationship between several household level characteristics, as well as spatial features of the locations where these households reside, on the consumption value and caloric value of six food groups: cereals (including rice), fruit and vegetables, meat, fish and seafood, eggs and milk, and food away from home. With the Vietnam Household Living Standard Survey (VHLSS) for 2013/14, we analysed the diets for different groups of consumers. Spatial information on the share of surface water and the share of urban area was measured at district level and added to the VHLSS data. Households in districts with higher shares of surface water were estimated to have a higher share of fish consumption and a lower share of meat consumption. From an environmental and health perspective, when surface water would be contaminated, households in water-rich areas thus would have a higher probability of being affected by pesticides due to higher fish consumption. In districts with higher shares of urban areas, households consumed higher shares of fruit and vegetables as well as fish. These households consumed lower shares of meat and cereals. From an environmental and health perspective, households in these areas are more prone to health hazards related to higher levels of pollution from water though higher levels of fish consumption and through consumption of irrigation fruit and vegetable.

Keywords: Environmental pollution, food consumption, landscape indicators, northern Vietnam

Contact Address: Vincent Linderhof, Wageningen University and Research, Wageningen Economic Research, Prinses Beatrixlaan 582, 2595 BM The Hague, The Netherlands, e-mail: vincent.linderhof@wur.nl

Factors Affecting Meat Consumption in Sulaymaniyah City of Iraqi Kurdistan

Niga Abdalla

Czech University of Life Sciences Prague, Fac. of Tropical Agrisciences - Dept. of Economics and Development, Czech Republic

Global average meat consumption increased considerably during the last fifty years; the biggest share of the increase was coming from emerging countries. Iraqi Kurdistan is part of one of the emerging countries, the area faced an economic crisis between 2014 and 2018 that could make a considerable change in the level of food consumption.

The main aims of the study were to determine the effect of demographic characteristics (gender, age and education), household characteristics (household size and income), and some aspects of consumer behaviour on meat consumption, as well as change in the patterns of meat consumption during the crisis in Sulaymaniyah city of Iraqi Kurdistan.

This study was conducted using quantitative exploratory questionnaire survey. A total number of 233 questionnaires was filled through quota sampling. Data analysis programme SPSS was used for statistical data analysis, i.e. multiple linear regression model and Chi-squared test. The result shows that among the most important factors that have a statistically significant relation to meat consumption are; household income, gender, preferences for the taste of a specific type of meat and fat content in the meat. While characteristics such as age, education, household size, buying meat for the household and concerns about animal welfare did not have a significant effect on meat consumption in the area.

The study can fill a gap in the field of consumer behaviours and be used as a source for future research since there are limited sources in that field. Also, the meat producers and sellers can take advantage of the findings from the study to understand consumer preferences and satisfaction and use more appropriate marketing strategies.

Keywords: Consumer behaviour, economic crisis, household income, meat production, red and white meat

Contact Address: Miroslava Bavorova, Czech Univeristy of Life Sciences Prague, Fac. of Tropical AgriSciences - Dept. of Economics and Development, Kamýcká 129, 16521 Prague - Suchdol, Czech Republic, e-mail: bavorova@ftz.czu.cz

Learning and extension

Resilience and Vulnerability of Agroecology Peasant Schools in Post-Conflict Colombia

Giovanna Chavez Miguel
Leibniz Centre for Agric. Landscape Res. (ZALF), Land Use and Governance, Germany

Agroecology peasant schools emerge often in face of insufficient state support in rural areas in Colombia, as grass-root initiatives that utilise agroecological education as the main instrument of social transformation for promoting alternative approaches to sustainable territorial development. The diverse actors involved in the school activities engage in a constant network of interactions and knowledge exchange, which enables them to build a pool of collective skills to help reconstitute the shattered social capital base and mitigate the hazards left behind by conflict. Peasant schools in Colombia have not been broadly studied and there remains a need to systematize the experiences for abstracting the lessons learned in each school. This research aims to argue that peasant schools can play an important role in promoting sustainable practices in resource management, and in strengthening organisational aspects at farm level that harmonise efforts and increase inter-agency collaboration. Analysing through a lens of resilience and vulnerability, this study investigates the factors that strengthen or weaken these organisations and analyses the current volatile policy setting that might determine whether they prevail or disappear in a complex post-conflict context. For these purposes, a three-month field work was carried out in different regions of the Colombian Andes by visiting multiple initiatives working in rural education with an agroecological focus. This study applied an ethnographic mixed methodology in form of semi-structured interviews and focus group discussions with different target groups. Preliminary results include the development of a typology outlining the main characteristics, functionality and structure of each school and the a description of the historical and geographical contexts in which they emerge and operate. Further, the study derives attributes and indicators in a participatory manner to assess the resilience and vulnerability of schools in the current economic, sociopolitical and environmental settings. The first policy analysis reveals that the policies instated after the peace agreements continue on favouring agroindustrial development over small farms, putting pressure on peasant groups in terms of land access, security, autonomy and reduced state support.

Keywords: Agroecology, campesino, Colombia, peasant, post-conflict, rural education, school

Contact Address: Giovanna Chavez Miguel, Leibniz Centre for Agric. Landscape Res. (ZALF), Land Use and Governance, Müncheberg, Germany, e-mail: giovanna.chavez.m@gmail.com

Silent Comics for Agricultural Extension of Rural Communities in SW Madagascar

KATHRIN STENCHLY[1], TOBIAS FELDT[2], DAVID WEISS[3], JESSICA ANDRIAMPARANY[4], ANDREAS BUERKERT[1]

[1] *University of Kassel, Organic Plant Production and Agroecosystems Research in the Tropics and Subtropics, Germany*

[2] *University of Kassel / Georg-August-Universität Göttingen, Animal Husbandry in the Tropics and Subtropics, Germany*

[3] *Science & Art, Painting & Graphic Design, Germany*

[4] *University of Kassel, Centre for International Rural Development, Germany*

The distribution of silent comic illustrations can facilitate the communication and transfer of scientific recommendations about sustainable land management (SLM) to local communities in countries where many people are illiterate. However, since there are cross-cultural differences in "visual languages", visualisation styles needs to be carefully selected as well as locals' comprehension of the illustrated recommendation evaluated systematically. Three agricultural recommendations were chosen for comic-style illustrations, distributed to six communities in the Mahafaly region of southwestern Madagascar and evaluated using a three-step, interdependent approach. The silent comics illustrated (i) composting of manure and its application to improve soil fertility; (ii) cautious utilisation of succulent silver thicket as supplementary forage; and (iii) sustainable harvesting practice of wild yam. Results revealed that general understandability strongly depended on the community that was surveyed and on the environmental subject that was illustrated. We found a strong relationship between the general understandability of comics and the divergence that exist in communities' socio-economic structure. Education level was an important factor that explained a better understanding of respondents for the comic that illustrated compost production, but not for comics that illustrated sustainable usage of silver thicket and wild yam harvest. Willingness to follow the recommended practice was impaired when respondents valued no change to the improved technique compared to the common one. Effects of respondents' socio-economic characteristics on the implementation of the recommended practice could not be clarified within this study due to the small subset of data. Based on the evaluation of recurring comments made by respondents and interviewers, we conclude that comics can be a useful communication tool to increase locals' awareness and comprehension for SLM practices. This, however, requires that drawing details used to facilitate farmers' ability to adopt a point-of-view inside the comic story are used thoughtfully as they might interfere with the central message.

Keywords: Agriculture, environmental protection, graphic novel, Mahafaly Plateau, respondents' understanding, SLM practices, visual communication

---

Contact Address: Kathrin Stenchly, University of Kassel, Organic Plant Production and Agroecosystems Research in the Tropics and Subtropics
current address: University of Kassel, Competence Centre for Climate Change and Adaptation (CliMA), 37213 Witzenhausen, Germany, e-mail: stenchly@uni-kassel.de

Exploring Smallholder Farm Decisions through Typologies and Serious Gaming

Mirja Michalscheck[1], Jeroen C.J. Groot[1], Irmgard Hoeschle-Zeledon[2], Gundula Fischer[3], Bekele Hundie Kotu[4], Pablo Tittonell[5]

[1] *Wageningen University and Research, Farming Systems Ecology Group, The Netherlands*
[2] *International Institute of Tropical Agriculture (IITA), Nigeria*
[3] *International Institute of Tropical Agriculture (IITA), Africa RISING, Tanzania*
[4] *International Institute of Tropical Agriculture (IITA), Ghana*
[5] *National Institute for Agricultural Technology (INTA), Natural Resources and Environment Program, Argentina*

While scientists and businesses keep developing ingenious technical 'solutions' for agricultural challenges in development contexts, technology adoption by smallholder farmers remains low. It seems as if scientists and farmers hold a different understanding of the problem, of the solution or about the enabling environment for technology adoption. In order to better understand how farm-level decisions come about, this study explores dynamics of land allocation decisions among smallholder farm households in Northern Ghana, using typologies and serious gaming. We employed a statistical farm and a participatory farmer typology, which revealed differences in farm resource endowment and in individual farm-related interests and constraints respectively. We then developed a closed, experimental serious game to determine the power-backing of individual interests as well as negotiation dynamics in order to explain how divergent individual interests may translate into household-level decisions. The serious game simulated the process of a household-level negotiation on land allocation between a male household head, a wife and the eldest son of a hypothetical medium resource endowed farm household. Contrary to the general local cultural narrative, we found that the wife and the son had a significant influence on the household-level decision outcome. The household-level outcome was, furthermore, more profitable as well as agro-biologically and nutritionally more diverse and productive as compared to the household heads' individual suggestion. Our finding highlights that the integration of diverse perspectives led to a robust 'solution', supported by all stakeholders. Scientist and businesses may hence improve the identification of problems and solutions by means of participatory and gender transformative approaches. Concerning the negotiation process, power was found to be exerted, withheld or overruled and it seemed as if, even when power was withheld or overruled, it still had an importance in time and across decision-domains — a feature of complex rather than linear systems. One has to consider that non-adoption of a technology might be the best choice for a farmer, given his or her interests, priorities and understanding of whole-systems consequences. In conclusion, scientists and businesses may provide better farmer support when acknowledging household-level decision-making dynamics as the core enabling environment for any proposed change.

Keywords: Land allocation, northern Ghana, participatory typology, power dynamics

Contact Address: Mirja Michalscheck, Wageningen University and Research, Farming Systems Ecology Group, P.O. Box 430, 6700 AK Wageningen, The Netherlands, e-mail: mirja.michalscheck@wur.nl

Challenges and Opportunities for the Implementation of an Innovative Solar Milk Cooling System in Kenya

FARAH MRABET[1], JULIET KARIUKI[2], VICTOR TORRES-TOLEDO[1], JOACHIM MÜLLER[1]

[1] *University of Hohenheim, Inst. of Agric. Sci. in the Tropics (Hans-Ruthenberg-Institute), Germany*

[2] *University of Hohenheim, Social and Institutional Change in Agricultural Development, Germany*

While the need to preserve evening milk by smallholder dairy farmers is well understood, much less has been documented on the adoption dynamics in small scale systems for addressing these demands.

The solar milk cooling system (SMCs) represents an innovative solution to cool milk on farm level. It is possible to use the SMCs to cool up to 40 liters of evening milk notably in off grid contexts. To enhance understanding of SMCs adoption, this study explores the introduction and implementation of SMCs in Western-Kenya. A mixed methods approach was adopted in which quantitative data (milk volumes, number of farmers, payments) and qualitative data (surveys, semi-structured interviews, focus groups) were collected. Two systems were installed in March 2018 to collect the evening milk which was previously not being collected due to lack of storage facilities. The study also implemented a willingness to pay study and the results used to develop a payment plan signed by both groups to pay for the system through monthly installments. Additionally, a one year monitoring study was conducted in order to capture the reality and dynamics around the system. The results show that the implementation of the technology in Kenya and its success depends largely on non-technological factors: market insecurities (the price difference between the informal and the formal market); institutional challenges (payments modes, governance structures); seasonality of milk production; and the social relationships between the different actors. These results also reflect the finding that both of the study regions were unsuccessful in reaching the targets that were discussed in the agreements. While cooling is indeed needed in these regions, unless the above mentioned challenges are addressed, unstable and inconsistent adoption will undermine the overall potential benefits occurring from the sale of cooled evening milk. This study presents an example of the importance of a multi-disciplinary approach to understand the local context where an innovation is tested.

Keywords: Adoption, cooling, innovation, milk, small-scale farms, technology

Contact Address: Farah Mrabet, University of Hohenheim, Inst. of Agric. Sci. in the Tropics (Hans-Ruthenberg-Institute), Garbenstr. 9, 70599 Stuttgart, Germany, e-mail: farah.mrabet@uni-hohenheim.de

Leveraging Collaborative Learning to Build Capacity for Entrepreneurship together with Rural Women's Groups in Nigeria

KAYFE NWOSU[1], MARGARETA LELEA[2], SIMON OYEGBILE[3], BRIGITTE KAUFMANN[2]

[1]*University of Göttingen and University of Kassel, Germany*

[2]*German Institute for Tropical and Subtropical Agriculture (DITSL), Germany*

[3]*National Horticultural Research Institute, Nigeria*

In UPGRADE Plus, a project that aims to co-develop processing of underutilised species together with women's groups in West Africa, entrepreneurial capacity building is approached through the application of collaborative learning within a transdisciplinary research design. This epistemological orientation recognises participating women as custodians of knowledge specific to the contextual system within which they operate and thus uses methods that create action opportunities for them to decide what is feasible and what is not. Within the frame of this broader project, this study analyses the different levels of engagement of women's groups selected to participate in this transdisciplinary research. The interactions with the women were guided by the collaborative learning (CL) framework which follows a series of iterative phases of team building, dialogue, discovery and application. The phases of dialogue and discovery were facilitated with six steps which first brought forward existing knowledge on entrepreneurship within the women's groups and in later stages allowed for learning exchanges with other functioning businesses within the region leading towards the development of their own business plans. The study employed a qualitative approach to analyse themes that emerged from 25 group meetings and 12 individual interviews with four rural women's groups in south-western Nigeria from December 2018 until April 2019. Results show how groups respond differently to the same CL process, with some groups advancing more quickly than others. Although the groups share similarities in culture, occupation, religion and age distribution, it is their differences in intrinsic motivation, group cohesiveness, previous experiences and individual and group foresight that determined the pace of the group, that is, their ability to learn together. Capacity building goals could only be achieved with women's groups who were willing to go beyond transactional relationships. We found that the time and energy required for active engagement in the CL process revealed groups that were ready for longer term collaboration. Critically, the sense of ownership that is built through the active engagement is essential for projects that aim to sustainably improve rural livelihoods through entrepreneurial activities.

Keywords: Business, learning, transdisciplinary research, women

Contact Address: Kayfe Nwosu, University of Göttingen and University of Kassel, Goßlerstraße 13, 37073 Göttingen, Germany, e-mail: kayferosaire@gmail.com

Conditions and Potentials for Education for Sustainable Development in Primary Schools in North-East Madagascar

JANNA NIENS, SUSANNE BÖGEHOLZ

Georg-August-University Goettingen, Didactics of Biology, Germany

Madagascar is a biodiversity hotspot with many endemic species. This makes the country's natural habitats especially worth protecting. However, the growing population, rising demand for natural resources and destructive agricultural practices result in a burden on the resting primary forest. In order to meet the demands of present and future generations while also protecting the biodiversity, sensitizing the local population for biodiversity-related values and sustainable land use is indispensable and can be realised through education.

In Madagascar, where most children only graduate from primary school, the integration of Sustainable Development (SD) related content is especially relevant for primary level. Obstacles in the Malagasy education system are manifold: Insufficient financial support causes bad conditions of school buildings and lack of school equipment. Also, many primary school teachers did not attend professional pedagogical training and are therefore highly underqualified and the official school curriculum is described as outdated as it hardly contains locally relevant topics.

The project "Competencies for rural Madagascar" aims at assessing the leverage points of Education for Sustainable Development (ESD) in Malagasy primary schools. It is part of the transdisciplinary joint project Diversity Turn that investigates sustainable land use in vanilla farming in the SAVA-region in northeast Madagascar. The project has three foci:

1. Conditions of primary education in the SAVA-region
2. Curricular prerequisites for ESD in Madagascar
3. Prerequisites of primary school teachers in the SAVA region regarding ESD issues

Findings from a quantitative study about the school structure in N = ca. 120 schools based on the international studies TIMSS (2011/2015), PIRLS (2011) and PASEC (2014) as well as preliminary results of a quantitative teacher study (N= 302) about SD-related knowledge will be presented and discussed. Within the scope of transdisciplinary research, the potential use of the results by local stakeholders to promote ESD in primary education will be pointed out.

Keywords: Education for sustainable development, Madagascar, primary education

---

Contact Address: Janna Niens, Georg-August-University Goettingen, Didactics of Biology, Waldweg 26, 37073 Göttingen, Germany, e-mail: janna.niens@uni-goettingen.de

Water Decline and Sustainable Livelihood: A Narrative of Adaptation Options in Southern Lake Chad Basin

OLUSOLA ADEOYE

Foundation for Environmental Mediation, Right and Livelihood Research, Nigeria

Humanity is faced with manifold changes in the global environment. These have consequence on the capacity of the earth system to sustain life. The effect the changes produce carries the capacity to make the world faces a future of very limited vital natural resources. With severe impacts of climate change, Lake Chad remains one of the world's social-ecological systems that provide essential resources to support the livelihoods, health and development of human societies. Since the historic drought of the 1970s, Ninety percent of the water of the lake has dried up making over 46 million of its riparian population to face limited access to drinking and irrigation water, fisheries production and transportation. Through the lenses of post-development theory of international relations and complex interdependence theory, the initial study investigated the inter-play between global environmental change (climate change) and threat to human security in the Nigerian section of Lake Chad. For the presentation, this paper will highlight the link between the globally driven climate variability, violent conflicts and water resource scarcity and the direct implications on the escalation of poverty, food security and income generation. Specifically, the paper will offer an alternative narrative on adaptation strategies among family farmers and artisanal fishers in three selected communities. The overall study becomes important because family farming and fisheries constitute the majority of the world's small and medium enterprises. In the selected communities, they are considered key to poverty alleviation and food security. Identifying the local adaptation strategies in these communities is important in order for policymaker to make informed decision on sustainable water management for food production. To obtain data for the study mixed methods including extensive fieldwork during which field observations and interviews were conducted. Coupled with review of recent literature, data obtained were analysed to identify various adaptation strategies among the farmers and the fishers.

Keywords: Climate change, human security, lake Chad, resource scarcity

Contact Address: Olusola Adeoye, Foundation for Environmental Mediation, Right and Livelihood Research, 41 Oluwalogbon Street Ketu, Lagos, Nigeria, e-mail: anchorpeace@gmail.com

Modelling Adoption and Welfare Impacts of Agricultural Upgrading Strategies (UPS) among Rural Smallholders in Tanzania

CLAUDIO PAUL NGASSA, ANJA FASSE

Weihenstephan-Triesdorf University of Applied Sciences,, Environmental Policy and Resource Economics, Germany

The uptake of agricultural technologies among the rural smallholders is considered as an important way of improving household's welfare in developing countries. Unlike much of the literature which focused on examining adoptions of agricultural innovations at either crop-level or at a definite stage of the value chain and associated impacts, this study examines rural households' adoption of a bundle of upgrading strategies (UPS) along the local traditional AVCs and evaluates the associated welfare impacts in terms of household's income and food security covering access, availability, utilisation and stability. We use panel data of 820 rural households, collected in 2014 and 2016, from four treatment and two control villages located in Chamwino and Kilosa districts in Tanzania. Out of 820 households in our sample size, 486 are treated households while 334 are non-treated households. In modelling UPS adoption and average treatment effects (ATT and ATU), we adapt a panel data endogenous switching regression (ESR) model to circumvent selection bias emanating from both time varying and time invariant observed and unobserved heterogeneity. As a robustness check, a difference-in-difference (DID) model is estimated. Results show that, adopting upgrading strategies enhance households' incomes and households' food security status in all four pillars significantly. We suggest that efforts aimed at raising households' income and food security status in the rural areas, should focus on promoting adoption of upgrading strategies on several components of the traditional AVCs such as natural resources, crop production, processing, waste management and bioenergy, markets, and consumption for improved rural households welfare.

Keywords: Agricultural value chain, panel data endogenous switching regression model, Tanzania, upgrading strategies, welfare impacts

---

Contact Address: Claudio Paul Ngassa, Weihenstephan-Triesdorf University of Applied Sciences,, Environmental Policy and Resource Economics, TUM Campus, 94315 Straubing, Germany, e-mail: claudio.ngassa@hswt.de

Factors Influencing the Adoption of Organic Production Systems in Vegetable Farming

MAGDALENA WANAWAN, VIRGINIA TAPAT

Regional Field Office, Cordillera Administrative Region (DA-RFO, CAR), Philippines

Conventional production of vegetables is predominantly practised in Benguet Province, the cradle of semi-temperate vegetables in northern Philippines. With the growing interest in healthy food consumption, some farmers embraced organic vegetable farming. This study aimed to determine the influential factors in producing organically grown vegetables considering the nutritional and economic importance of this crop. It also endeavoured to identify prevailing problems and concerns related to organic vegetable production. Vegetable growers from six municipalities who opted to shift to organic farming served as respondents for the study. An interview schedule was used to collect quantitative data while informal group discussions, key informant interviews, and farm reconnaissance yielded qualitative data. Qualitative analysis showed that the goals and motivations of farmers in adopting organic vegetable production systems are economic benefits, ecological integrity, health benefits, and favourable conditions. Regression analysis revealed that the income from organically grown vegetables had the highest influence in the utilisation of organic production systems. Certified organic vegetables can command steady premium price while the value of conventional vegetables is highly fluctuating. Other predictor variables are : diversity of vegetables grown, credit assistance and number of support services espousing organic farming. The support services mostly accessed by the respondents include educational visits, trainings and demonstration farms. The farmers' discernments on the five technology attributes posited by Rogers (1983), namely compatibility, complexity , trialability, observability and relative advantage were also considered. Among these attributes, observability surfaced to have a significant influence in the adoption of organic vegetable production systems and techniques. Meanwhile, one of the main problems in organic farming is the migration of pests to the organically-managed fields when the surrounding conventional farms are sprayed with pesticides. Another concern is the rehabilitation period of the soil which takes 3–5 years and the yields are affected while the field is adjusting. The findings suggest enabling environment policies and support interventions considering the influencing factors and problems in growing vegetables using organic production options.

Keywords: socio economic factors, support services, technology attributes, organic production systems

---

Contact Address: Magdalena Wanawan, Regional Field Office, Cordillera Administrative Region (DA-RFO, CAR), Research Division, Sto. Tomas Road, 2600 Baguio City, Philippines, e-mail: magdalwan@yahoo.com

Farmers' Adaptation and Coping of Climate Variability-Induced Shocks in Ethiopia: Disentangling Household-Specific Determinants of ex-ante and ex-post Strategy Choices

HABTAMU DEMILEW YISMAW[1], CHRISTIAN TROOST[1], MOTI JALETA[2], THOMAS BERGER[1]

[1] *University of Hohenheim, Inst. of Agric. Sci. in the Tropics (Hans-Ruthenberg-Institute), Germany*

[2] *International Maize and Wheat Improvment Center (CIMMYT), Ethiopia*

Climate variability-induced shocks and their profound impact in reducing agricultural production, inflating agricultural input and output prices, and deteriorating smallholder farm households' welfare have been posing formidable policy challenges in Ethiopia for several decades. Smallholders' welfare, on the other hand, depends on their choice of strategies to deal with these shocks before and after their occurrence, which are in turn highly farmer- specific. This paper examines household-specific determinants of farmers' choice of ex-ante and ex-post adaptation strategies to climate-induced shocks in Ethiopia. Baseline data collected by CIMMYT from 898 households in the major maize growing hotspots in Ethiopia is used for analysis. Logistic principal component analysis (LPCA) is applied to identify dominant strategies while multivariate probit analysis (MVP) is applied to model farmers interdependent adaptation decisions.

Findings show that farmers choice of adaptation and coping strategies are highly idiosyncratic and heterogeneous. Gender, education, market information, farming knowledge and experience, participation in rural institutions, social networks, resource endowments and farmers' expectation are the major drivers of farmers choices of ex-ante adaptation as well as ex-post coping strategies. Results from both LPCA and MVP show that most of the strategies farmers choose are complementary strategies. This implies that there is no single best strategy that works for all farmers, instead farmers use multiple adaptation and coping strategies tailored for their socio-economic settings. Findings also show complementarity of strategies is stronger in ex-ante strategies than ex-post strategies.

Based on this we suggest that analysis of robust climate adaptation interventions should focus on the household level. In addition, knowledge creation on innovative adaptation options such as application of shock resistant varieties is a key area of intervention to improve farmers resilience to climate variability-induced shocks. Moreover, strong aftershock household asset restoration and development schemes should help farmers get back in track by preserving their productive farm assets.

Furthermore, this paper identifies farmers dominant strategies by applying dimensionality reduction of binary data using LPCA. Results of LPCA were compared with classical linear PCA and exponential family PCA and LPCA was found to be more appropriate in dimensionality reduction of binary data.

Keywords: Adaptation, coping, Ethiopia, logistic PCA, multivariate probit

Contact Address: Habtamu Demilew Yismaw, University of Hohenheim, Inst. of Agric. Sci. in the Tropics (Hans-Ruthenberg-Institute), 70593 Stuttgart, Germany, e-mail: h.demilewyismaw@uni-hohenheim.de

Uptake of Integrated Soil Fertility Management Techniques in Maize Mixed Farming Systems of East and Southern Africa: Case of Malawi's Rift Valley Escarpments

POWELL MPONELA[1], GRACE VILLAMOR[1], LULSEGED TAMENE[2], SIEGLINDE SNAPP[3], QUANG BAO LE[4], CHRISTIAN BORGEMEISTER[1]

[1]*University of Bonn, Center for Development Research (ZEF), Ecology and Natural Resources Management, Germany*

[2]*International Center for Tropical Agriculture (CIAT), Agroecosystems and Sustainable Landscapes Research Area, Malawi*

[3]*Michigan State University, Plant Soil and Microbial Sciences, United States of America*

[4]*International Center for Agricultural Research in the Dry Areas (ICARDA), Jordan*

Addressing the dual challenges of increasing food production and reducing environmental degradation requires first an an understanding the farmers intentions and actions. As farmers strive to earn a living on their land units, they perceive the environmental resources and adapt to changes by either taking autonomous proactive actions or imitate others resulting in heterogeneous farming households and farmlands. Given several approaches that have been rolled out since 1980s - including sustainable land management, climate smart agriculture and integrated soil fertility management - farmers allocate resources to technologies that either maximise soil fertility improvement or minimise risk of soil degradation. Among the 30 % of total farming households sampled from 5 villages in rural Malawi, inorganic fertilisers were used by 90 %, legumes by 72 % and organic manures by 57 %. Empirical evidence from the double hurdle model shows that none of the household and farm factors consistently influenced the two decisions (adopt and intensify) for all the three technologies. Positive effects from increasing labour supply are offset by the negative effects associated with increasing dependency ratios. However, women empowerment could increase the probability of legume cropping but with negative effects on organic manure application. It is evident also that there could be spill-over effects in that if a driver is only important for a particular soil fertility technology, it could have indirect effects on associated technologies. On the basis that all factors are dynamically changing with different magnitudes and confidence intervals, sustainable intensification could be achieved by harnessing the positive effects while concurrently reducing the negative influences.

Keywords: Household decision, integrated soil fertility management, intensification, rift valley escarpments, smallholder farmers

Contact Address: Powell Mponela, University of Bonn, Center for Development Research (ZEF) - Ecology and Natural Resources Management, Genscherallee 3, 53113 Bonn, Germany, e-mail: powell.mponela@uni-bonn.de

Forage Technology Adoption Studies in Bovine Livestock Production Systems: a Field Open to Innovation

KAREN ENCISO, MANUEL DIAZ, NATALIA TRIANA-ANGEL, STEFAN BURKART
International Center for Tropical Agriculture (CIAT), Colombia

Sustainable intensification through improved forages has proven to be one of the most promising strategies for increasing productivity in tropical livestock systems and mitigating environmental impacts. However, reported adoption rates remain low for many developing countries. Different studies have been commissioned to analyze, through various methodological approaches, factors affecting adoption decisions. This research provides a literature review in order to i) identify the variables included in forage adoption studies and their possible impact on the adoption decision, and ii) analyse the different methodologies used, highlighting their strengths and weaknesses. Through an exhaustive search, a total of 27 studies (published between 2000 and 2019 in Scopus and Science Direct databases) were analyzed. Results show that the topic has been approached mainly from a micro perspective (focused on the primary producer) based on the theories of utility maximisation (TUM) and Planned Behaviour (TPB). Among the explanatory variables included most frequently in TMU studies are age, educational level, herd size, resource allocation, technical assistance and membership in an association. The effect of these variables on technology adoption has been inconsistent and ambiguous: while some studies report significant and positive effects, others find the opposite. Studies applying TPB were more consistent, mostly correlating the intentions of the producers with social attitudes and norms and to a lesser extent with the control of perceived behaviour. However, in most of the reviewed studies, focus was put on evaluating the intention to adopt and not on the actual adoption. In parallel, other, still incipient, types of studies emerged, that question the suitability of predicting adoption from a micro perspective, highlighting instead macro approaches from an innovation system perspective. According to the results obtained from this literature research, it is recommended to consider the use of a mixed-methods approach allowing a deeper understanding of adoption and dissemination processes *in situ*. Two aggregation approaches are then proposed in a complex and dynamic system, which include not only the perception of the adopter but also the surrounding social structure, the historical context in which the adopter lives and the process of diffusion of innovations.

Keywords: Cattle production, dairy production, diffusion of innovations, Latin America, mixed-methods approach, sustainable intensification

Contact Address: Stefan Burkart, International Center for Tropical Agriculture (CIAT), Trop. Forages Program, Km 17 Recta Cali-Palmira, Cali, Colombia, e-mail: s.burkart@cgiar.org

Unlocking Barriers to Adoption and Scaling of Climate Smart Cocoa Practices in Ghana

MUSTAPHA ALASAN DALAA[1], RICH KOFI KOFITUO[1], ISSAC ALVIN AMOAH[2], LAURENCE JASSOGNE[3], RICHARD ASARE[1]

[1]*International Institute of Tropical Agriculture (IITA), Ghana*

[2]*University of Ghana, Dept. of Agricultural Extension, Ghana*

[3]*Opus Insights, Uganda*

Cocoa production in West Africa has been adversely affected by climate change at varying degrees. The cocoa farming areas in Ghana vary according to severity in impact of climate change and has been delineated into three climatic impact zones namely; Transform, cope and adjust zones. For years, cocoa industry technical experts have recommended Good Agricultural Practices (GAPs) without considering the different farmer typologies across the climate impact zones. The Consultative Group for international Agricultural Research (CGIAR) through the International Institute of Tropical Agriculture (IITA) in Ghana recently documented and aligned climate smart cocoa (CSC) practices across the three impact zones to help farmers mitigate against the effect of Climate change. The aim of this study was to identify farmer typologies in the different climate impact zones and how this affect adoption of CSC recommendations.

Data was collected using semi-structured questionnaire from 270 cocoa farming household on socio-economic characteristics and intensity of CSC implementation across. Preliminary findings from a principal component analysis using the R software statistical package showed three cluster of Cocoa farmers in the impact zones. The results also show varying intensity of implementation of CSC practices which determines the efficiency of the clusters. The first cluster of cocoa farmers is characterised as the least efficient in production in terms of Cocoa productivity (248.3 kg ha^{-1}) and Cocoa income (USD 981.32 per annuum) while the second cluster of farmers are the most efficient with the highest cocoa income (USD 3000.31 per annum) and Cocoa productivity (583.7 kg ha^{-1}). The third Cluster represent farmers with the most resources in terms of land under cocoa (3.7 ha) and hired out labour ($\simeq$4 people from the household). In all clusters, access to hybrid seedlings, financial challenges and extension service delivery were identified as challenges hindering adoption of CSC recommendation. It is recommended that farmer typologies aligned with CSC recommendations in the climate impact zones should be taken into consideration for effective adoption.

Keywords: Climate change, climate impact zones, climate smart cocoa, cocoa productivity, farmer typologies

---

Contact Address: Mustapha Alasan Dalaa, International Institute of Tropical Agriculture (IITA), Climate Change, Agriculture and Food Security Program, CSIR Main Campus- Augustinho Neto Road Airport Residential Area, Accra, Ghana, e-mail: m.dalaa@cgiar.org

Filling Gap and Removing Trap: The Interface between Expert and Farmer Knowledge on Cocoa Pruning

FAUSTINA OBENG ADOMAA[1], SIETZE VELLEMA[1], MAJA A. SLINGERLAND[2], RICHARD ASARE[3]

[1] *Wageningen University and Research, Knowledge, Technology and Innovation Group (KTI), The Netherlands*

[2] *Wageningen University and Research, Plant Production Systems, The Netherlands*

[3] *International Institute of Tropical Agriculture (IITA), Ghana*

Cocoa trees are generally pruned to increase yields through maximising light capture, reducing unproductive branches and facilitating harvesting. Pruning can reduce the increasing incidence of pests and diseases and an over reliance on pesticides and fungicides use. Additionally, pruning can increase yields by 30 percent positively affecting incomes of cocoa farmers. Hence, governments of producing countries, scientists and lead firms among others, have committed resources to develop pruning recommendations and intensified farmer trainings on such recommendations in the bid to fill the gap in pruning knowledge among cocoa farmers. Yet, the practice of recommended pruning largely remains low among cocoa farmers in Ghana. However, the years of technical trainings on pruning received, coupled with farmers' situated knowledge of farms and fields, have resulted in various adaptations of pruning practices based on other physical conditions that influence tree response to pruning. These adaptations recognise pruning as a practice influenced by slope of land, soil type, presence or absence of shade trees on farm, age of farm, local weather conditions and accompaniment of other farm practices such as fertiliser application and manual pollination. Thus, while expert knowledge on pruning seemingly remains a generic-unitary-static practice, farmer knowledge on pruning is largely a situated-composite-dynamic practice which mostly varies from recommended expert knowledge. In Ghana's cocoa sector, mass trainings, one-on-one and/or group coaching, demonstration farm visits and gang pruning activities by lead firms and government, have become important interfaces where expert knowledge meets the situated knowledge of farmers on pruning. However, these interfaces are generally designed and operated as channels of transferring technical knowledge on pruning to farmers. Rooted in a path of seeing extension as a one-way transfer of knowledge from experts to farmers, these interfaces offer little to no room for manoeuvre, flexibility, hybridisation and/or co-creation of knowledge and solutions between expert and situated knowledge on pruning. Designing an interface that is inclusive and gives recognition to the knowledge and priorities of all actors is thus necessary in removing the trap of lock-ins and path-dependency approaches to filling knowledge gaps on pruning in cocoa farming.

Keywords: Cocoa farming, cocoa yields, extension, farmer knowledge, knowledge co-creation

Contact Address: Faustina Obeng Adomaa, Wageningen University and Research, Knowledge, Technology and Innovation Group (KTI), Hollandseweg 1, 6706 KN Wageningen, The Netherlands, e-mail: fobenga@gmail.com

Barriers to Implementation of Climate Change Adaptation Measures - The Case of Madagascars' Agricultural Sector

MAIKE VOSS

Humboldt University of Berlin, Albrecht Daniel Thaer-institute of Agricultural and Horticultural Sciences, Germany

With climate change being one of the main challenges of the current century, strategies responding to climate change impacts, like adaptation or mitigation, are becoming increasingly important. Especially developing countries are in need to adapt to avoid increasing vulnerability and negative effects like decreases in agricultural production and the impacts of extreme weather events. Though many adaptation projects exist, few examples of successful implementation of adaptation measures can be found in literature. The present work analysed what barriers to implementation of climate change adaptation measures can arrive and offered recommendations on how to overcome them. By applying a framework for diagnosing barriers to climate change adaptation measures, six adaptation options were analyzed. This was done for the agricultural and livestock sector of two regions in Madagascar, one of the most vulnerable countries to the impacts of climate change, with more than three-quarter of the population working in the agricultural sector. Further, seven expert interviews with employees in international organisations and development agencies in Madagascar were conducted to identify implementation processes in climate change adaptation projects. The results show that the main barriers to implementation of climate change adaptation measures in the agricultural and livestock sector in Madagascar are resource, financial and crosscutting barriers. In addition, several externalities seem to hinder successful implementation. Moreover, a certain path dependence enforced by sociocultural factors might block the uptake of new practices. An existing dichotomy in the strategic approaches of the interviewed international organisations and development agencies seems to enforce existing barriers for adaptation. Several recommendations were proposed to resolve this operational dichotomy and path dependence. A transmission of responsibility to local stakeholders can act as prerequisite for successful long-term implementation. Participatory approaches, development-focused collaboration and a facilitating role of project organisations can support the creation of this responsibility for implementation. Lastly, the development of a common framework to support adaptation processes and reduce barriers to implementation is highly recommended.

Keywords: Adaptation, barriers to adaptation, climate change, Madagascar

Contact Address: Maike Voss, Humboldt University of Berlin, Albrecht Daniel Thaer-institute of Agricultural and Horticultural Sciences, Berlin, Germany, e-mail: maike.voss.1@hu-berlin.de

Digital Extension Service: Delivering Agro-Advisory by an Automated Hotline and Asynchronous Communication via Voice Messages

JONATHAN STEINKE[1], BERTA ORTIZ CRESPO[2], CARLOS QUIROS[1], JACOB MUSYOKA[3], JACOB MUTEMI[3], JOY PHILOMENA WANZA[2], HAPPY DAUDI[4], GASPAR MAJUTO MGIMILOKO[4], OMARI MPONDA[4], JACOB VAN ETTEN[1]

[1]*Bioversity International, Information Services and Seed Supplies, Costa Rica*

[2]*Bioversity International, Kenya*

[3]*Lutheran World Relief, Kenya*

[4]*Tanzania Agricultural Research Institute, Tanzania*

Agricultural advisory services in developing countries need to serve the diverse knowledge and information needs of the farming population. At the same time, policy-makers and donors increasingly expect extension services to produce evidence on their efficiency. The wide availability of mobile telephony services has great potential to support agricultural extension: Digital information services can provide timely agricultural advice to large numbers of geographically dispersed farmers, while generating information on the delivery of advisory messages. Together with extension agents and farmers in Kenya and Tanzania, we have designed a digital agro-advisory service, called "Ushauri". This service integrates two components: In an automated call-in hotline, farmers navigate an interactive voice response (IVR) menu to access a set of pre-recorded audio messages. Additionally, farmers may submit questions to a speech mailbox. These farmers recordings are sent to an online dashboard for extension agents. Here, registered agents listen to the questions and reply by sending ad-hoc voice messages via automated calls. Moreover, each farmer question is given thematic keywords online. We tested and evaluated "Ushauri" during a 6-month pilot with 260 farmers in Makueni County, Kenya, and a 1-month pilot with 97 farmers in Mtwara region, Tanzania. 37 % of registered farmers used the service at least once, listening to 146 advisory messages and leaving 176 questions, in total. We draw three main insights from the pilot implementations: First, providing agro-advisory through automated hotlines using IVR is feasible. But the menu must be simple, to limit the rate of farmers hanging up during IVR navigation. Second, asynchronous communication between farmers and advisors through voice messages can allow efficiency gains for extension services. Because multiple farmers asked the same questions, extension agents recorded certain answers only once, and then sent that message each time a farmer asked the same question. Third, the frequency data on farmers' access to the pre-recorded audios, as well as the keywords assigned to questions, exposed farmers' information needs. This knowledge can be used to prioritise further information interventions by the extension service. Our findings are relevant for the ongoing development of digital advisory services for smallholder farming in developing countries.

Keywords: Agricultural extension, digital media, ICT, mobile phones

Contact Address: Jonathan Steinke, Bioversity International, Information Services and Seed Supplies, Turrialba, Costa Rica, e-mail: j.steinke@cgiar.org

Filling Knowledge Gaps between Model Farmers and their Communities: Implementing the Concept of Social Debt

CHECK ABDEL KADER BABA

TMG Research, Soil protection and rehabiltation for food security, Germany

In Benin like in most sub-Saharan countries, agriculture is dominated by smallholder farmers whose capacities to sustain and intensify food production relies on healthy soils. Yet, soil degradation in this region is reaching alarming rates recalling the urgent need of innovative approaches in promoting sustainable land management practices at local scale.

While numerous challenges contribute to the low adoption of SLM good practices in sub-Saharan Africa, technology diffusion beyond the circle of direct project beneficiaries appears as a major challenge of SLM technology long-term adoption by small farmers.

This paper presents the results of a pilot-experimentation of a technology diffusion model implemented within the frame of an accompaniment research to a GIZ soil protection and rehabilitation project in northern Benin. The core of the model is the so called Tem Sesiabun Gorado, that is a community-based agent (CBA) selected by their respective communities, and the concept of "social debt". The concept of social debt and the decision to share gained knowledge to new farmers is not an injunction or a simple requirement of project implementers or technology promoters. Rather, it is the result of a careful facilitation of village meetings and a deep reflection process involving both project beneficiaries and non-beneficiaries on the implementation strategy, the risk of technology non-diffusion and mechanisms to best address soil degradation challenges that farmers are facing.

Findings of the model implementation show that giving clear responsibility to farmers' representatives in technology diffusion through a carefully designed and adapted accountability mechanism (social debt) can help to reach farmers beyond the circle of direct project beneficiaries and out-scale project activities to new intervention areas. The model proved also to be effective in reaching more women and other disadvantage groups.

Though the model proved to be effective in addressing challenges of technology diffusion at local scale, further research is needed to apprehend conditions under which it can be self-sustained and successfully replicated in different agro-ecological and socio-cultural contexts and conditions.

Keywords: Social debt, sustainable land management, technology diffusion, Tem Sesiabun Gorado

---

Contact Address: Check Abdel Kader Baba, TMG Research, Soil protection and rehabiltation for food security, EUREF-Campus 6-9, 4.OG 10829 Berlin, 10829 Berlin, Germany, e-mail: cheikad@gmail.com

Factors Influencing Youths' Participation in Agricultural Training Programs in Nigeria: The Case of Fadama GUYS Program

DOLAPO ADEYANJU[1], JOHN MBURU[1], DJANA MIGNOUNA[2]

[1] *University of Nairobi, Department of Agricultural Economics, Kenya*

[2] *International Institute for Tropical Agriculture (IITA), Project Monitoring, Evaluation & Learning, Benin*

Many scholars have come to believe that Africa may truly be in dilemma. It is not known whether youth bulge is a time-ticking bomb waiting to explode or it is a blessing in disguise. This study was necessitated by the high rate of youth unemployment and low youth engagement in Agribusiness in Nigeria despite the numerous interventions by both government and non-government organisations to promote youth entrepreneurship particularly, agripreneurship. The paper investigates the factors influencing youth participation in agricultural training programs using the case of Fadama GUYS program. The programme was introduced in 2016 across 23 states in Nigeria. Given the fact that most of the studies that have investigated similar topic using a case study were based on relatively small data sets and focused on a single state in the country. This study progressed to take cases from three different states, representing three of the regions covered in the program. A multi-stage sampling technique was used to select a total of 977 respondents comprising of 455 participants and 522 non-participants across the regions. Thus, providing a basis for policy recommendation on a broader scope. Data collected were analysed using descriptive and inferential analysis. The study showed that high asset index reduces youths' likelihood of participation. Also, participation in the programme was negatively but significantly influenced by sex and household size but positively by Age, Years of education, Employment status, farm ownership and youth intention to start a farm. The study also analysed youths' willingness to engage in agribusiness, the result showed that more than 70 % were willing to engage in agribusiness. Based on this, the study went further to identify factors impeding youths from engaging in agribusiness. The result showed that majority of the respondents (56 %) indicated lack of access to finance as their major barrier to participating in agribusiness and hence do not see the need to participate in agribusiness training. Other barriers include lack of mentor-ship and information. Thus, the study concluded by recommending relevant interventions to tackle and overcome these challenges.

Keywords: Agribusiness, agripreneurship, training, unemployement, youth

Contact Address: Dolapo Adeyanju, University of Nairobi, Department of Agricultural Economics, Kangemi, 29053-00625 Nairobi, Kenya, e-mail: dolapo.adeyanju@students.uonbi.ac.ke

Role of Extension in Enhancing Positive Perceptions of Innovations; A Case of Hexanal in Kenya

JANE KAHWAI, JOHN MBURU, MARTIN OULU
University of Nairobi, Dept. of Agricultural Economics, Kenya

About 40–50 percent of horticultural produce is lost along the value chain in sub-Saharan Africa. This is due to the short shelf-life of the fresh produce and lack of effective storage methods thereby leading to the high post-harvest losses. Promotion of technologies to reduce these losses is necessary in improving food security and economically empowering smallholder farmers. Hexanal is one of the technologies developed to prolong shelf-life of fresh produce such bananas. However, farmers' perceptions towards such technologies are not yet fully understood especially when they are new to them. The study assessed the role of extension in enhancing positive perceptions and their determinants in using Hexanal to prolong fruits' freshness among banana farmers in Meru County. Data was generated from cross-sectional survey of 130 households who were categorised into two sub-samples, 'Aware' and 'Not Aware' of Hexanal. Three components of perceptions namely 'effectiveness' 'acceptability' and 'environmental safety' were extracted for both categories of farmers using Principal Component Analysis. Perception scores were later regressed against explanatory variables using Ordinary Least Squares. Despite both groups of farmers having positive perceptions towards some attributes of Hexanal, farmers already aware of Hexanal strongly agreed the technology was socially acceptable. However, farmers not aware of Hexanal had negative perceptions on 'acceptability' of the technology as they cited the need for more education on the use and benefits of Hexanal. Perceptions were influenced by different sets of variables such as age, access to credit, distance to input shop, income among others. To enhance positive perceptions, it is necessary stakeholders invest in factors such as supplying information about Hexanal through increased contact between extension agents and farmers which will increase uptake of the technology.

Keywords: Banana, enhanced freshness formulation, Hexanal, Kenya, perceptions, principal components,

---

Contact Address: Jane Kahwai, University of Nairobi, Dept. of Agricultural Economics, Nairobi, Kenya, e-mail: janekahwai77@gmail.com

Seeing Is Believing

SELINA BRUNS, FREDERIK SAGEMÜLLER, MUSSHOFF OLIVER
University of Goettingen, Dept. of Agricultural Economics and Rural Development, Germany

It is well documented in the literature that agrarian systems in the region of South East Asia are undergoing transitions from subsistence agriculture towards market oriented agriculture. The adoption of innovative financial services and technologies by smallholder farmers has emerged as a core element in the quest to sustainable economic growth. The rise of Information and Communication Technologies (ICTs) gives promising opportunities to governments and businesses to design services that suit the need of smallholders. In remote areas of Cambodia however, this process appears to be particularly challenging. Some farmers appear to be resistant to innovative and highly benefitting technologies and services, while others are willing to accept them and even act as first movers. While explanations for this difference in behaviour are manifold in the literature, e.g. risk attitudes and loss aversion, we strive to examine a new possible reason: Eyesight. Due to cultural reasons, but also e.g. lack of financial services, awareness, and infrastructure, the rural population in the North-Eastern district of Cambodia, Ratanakiri, seldom wears glasses.
Using a standardised eyesight-test, we recorded the ability to see of 305 smallholder farmers in 2018. We examine the relationship between vision and attendance at agricultural trainings; vision and uptake of formal financial services as well as vision and crop/input portfolio. We expect farmers with low vision to be less likely to diverge from traditional patterns.
If eyesight is a determinant factor, then policies that do not account for the effects of poor vision might not support all farmers equally. In fact, ignoring the importance of eyesight when developing policy implications might cause leaving those most burdened behind, while merely supporting those with less limited resources.

Keywords: Agriculture, behaviour, development, eyesight, financial services

Contact Address: Frederik Sagemüller, University of Goettingen, Dept. of Agricultural Economics and Rural Development, Platz der Göttinger Sieben 5, 37081 Göttingen, Germany, e-mail: frederik.sagemueller@agr.uni-goettingen.de

Land Tenure Security, Soil Water Conservation Technology Adoption and Household Welfare Linkages in Nigeria

EWERE EVELYN ANYOKWU
University of Bonn, Center for Development Research (ZEF), Germany

The challenges of climate change will exacerbate the problems of water availability for agricultural production especially in Nigeria where the predominant farming system is rain-fed. There are therefore, calls to promote soil water conservation (SWC) practices under the concept of 'conservation agriculture'. Yet, lack of land tenure security may hinder its adoption and continuous usage. This study assessed the extent to which land tenure security affects the farming systems and welfare outcomes of rural smallholder farmers, through the adoption of SWC technologies in South Western region of Nigeria. In contrast to the existing literature on conservation adoption, the study differentiated between perceived and legal land tenure security and allows for joint use of different SWC technologies since these are not-mutually exclusive. 240 structured questionnaires were administered to head farmers in the study area. Using structural equation and double hurdle models, the study showed that land tenure security in terms of legal documentation induced the adoption of more expensive SWC practices such as controlled irrigation while perceived tenure security induced cheaper SWC technologies adoption. Based on combined adoption, tenant farmers tended to adopt several SWCT but the cheapest ones such as straw mulching on ridged land. The study further found that farmers who adopted more expensive SWCT, cropped their fields at least twice yearly. However, these were dependent on their crop choices. Given the same crop type, adopting SWCT increased cropping intensity, use of other inputs and gross margin of the farmers. On average, farmers with secure land tenure and SWCT adoption, had better welfare outcome in terms of per-capita expenditure. This study thus concluded that agricultural innovation adoption driven by land tenure security has the potential to improve the livelihood of smallholder farmers.

Keywords: Land tenure, Nigeria , soil water conservation, welfare

---

Contact Address: Ewere Evelyn Anyokwu, University of Bonn, Center for Development Research (ZEF), Genscherallee 3, 53113 Bonn, Germany, e-mail: evelyn.ewere@yahoo.com

Value chains and markets

Oral Presentations

Posters

The Role of the Integrated Maize-Soybean-Chicken Value Chains in Sustaining Diverse Diets in Tanzania

WILSON WILSON[1], MAJA A. SLINGERLAND[1], HANNAH VAN ZANTEN[2], FREDERICK BAIJUKYA[3], SIMON OOSTING[2], KEN GILLER[1]

[1] *Wageningen University and Research, Plant Production Systems, The Netherlands*

[2] *Wageningen University and Research, Dept. of Animal Sciences, The Netherlands*

[3] *International Institute of Tropical Agriculture (IITA), Tanzania*

In Tanzania, there is a large gap between food production and consumption, which contributes to high rates of undernourishment and micronutrient deficiencies. The dietary problems are mainly due to limited dietary diversity among households. Furthermore, increased urbanisation leads to an increase in demand for poultry products which is difficult to satisfy by domestic production. The objective of the current study was to understand the current maize, soybean and chicken value chain(s) and identify important entry points for value chain integration in the Southern Highlands of Tanzania to support nutritious diets. The focus was on these value chain(s) as integrating soybeans in the maize-chicken value chains might increase the productivity of chickens by providing nutrient-dense feed. We carried out an explorative study followed by a multi-stakeholder workshop with 54 stakeholders and experts involved in development of the value chain(s). The current maize, soybean and chicken value chains interconnected particularly at the levels of the smallholder farming system and at processing facilities. The production of one or more of these products contributes to farmers' food security and income. Poultry feed is an important entry point for integrating the three value chains, whereby maize (grain/bran) and soybean meal are the main sources of energy and protein for chicken, respectively. A small proportion of maize produced is exported to neighbouring countries, while the current amount of soybean produced is mainly marketed in the domestic market. Three systems of poultry keeping were identified in the study area i.e. extensive, semi-intensive and intensive systems, with diverse feeding strategies ranging from scavenging, home-made rations, industrial feed rations and combination of home-made and industrial rations. Furthermore, the informal chicken market dominates in both urban and rural location. Currently, there is inefficient soybean marketing and processing in Tanzania, mainly due to disorganised producer groups and lack of adequate processing plants. As a result, soybean meal is mainly imported and sold at almost three times higher prices than the whole soybean grain produced in the country. Improving soybean production and marketing and investment in soybean processing infrastructures has a great potential to increase local availability of soybean products, reduce the cost of feed in chicken farming, and increase the availability of nutritious animal-based foods for human consumption.

Keywords: Chicken, feed, marketing, soybean, Tanzania, value chain

---

Contact Address: Wilson Wilson, Wageningen University and Research, Plant Production Systems, P.o. Box 430, 6700 AK Wageningen, The Netherlands, e-mail: wilson.wilson@wur.nl

Analysis of Business Model Innovation among Indigenous Vegetable Producers in Southwestern Nigeria

Abiodun Adelekun, Adeolu Ayanwale

Obafemi Awolowo University, Dept. of Agricultural Economics, Nigeria

The study characterised the business models, assessed the financial outcomes and examined the effect of the socioeconomic variables on the net returns of business models existing among indigenous vegetable producers in Southwestern Nigeria. This is with a view to enhancing the competitive advantage of the business model the producers. Primary data was elicited from 320 respondents with the aid of pre-tested structured questionnaire, focus group discussion and key informant interview. The data were analysed using discource analysis, descriptive statistics, budgetary analysis and multiple regression analysis.

The study identified that the multimarket, the cartel, buyer-driven and the auctioneer business models operated in the study area. The multimarket model was characterised by the servicing of four different market segments. The cartel model leveraged social capital and ethnic bond to establish a strong entry barrier. The auctioneer model operated by selling vegetables in-situ. The buyer-driven model faced an imposition of farm gate marketing. The summary statistics revealed that the producers executing these models were about 50 years of age, had up to 9 years of formal education, had at least 6 persons in their family, travelled up to 10 kilometres to the nearest market, and allocated 0.2 hectares of land to vegetable production out of about 0.8 hectares available for cultivation. The financial ratios showed that the cartel business model was the most successful having the highest rate of return (12.64), followed by the multimarket business model (6.97). Returns from the business models were influenced by total farm land available for cultivation, years of formal education, household size, distance to the nearest market, primary occupation, and indigeneity. The study concludes that the optimum utilisation of tangible and intangible resources such as land, market and indigeneity would enhance the performance of the business models.

Keywords: Business models, competitive advantage, indigenous vegetables, socioeconomics

Contact Address: Abiodun Adelekun, Obafemi Awolowo University, Dept. of Agricultural Economics, Fac. of Agriculture, 220005 Ile-Ife, Nigeria, e-mail: xty.san@gmail.com

Options to Address Food Security on Sustainably Certified Farms in Food-Insecure Countries

TINA BEUCHELT[1], RAFAËL SCHNEIDER[2], LILIANA GAMBA[3]

[1]*University of Bonn - Center for Development Research (ZEF), Ecology and Natural Resources Management, Germany*

[2]*Deutsche Welthungerhilfe e.V., Policy and External Relations, Germany*

[3]*WWF, Germany*

Concerns regarding the environmetal and social impacts of biomass production, including food security, are increasing. Companies in industrialised countries sourcing biomass in developing countries are increasingly interested, or pressured, to take responsibility and address sustainability issues in their international value chains. Voluntary sustainability certification schemes for agricultural commodities emerged as reaction to these concerns but food security is often not or not sufficiently addressed. Since many agricultural products imported by Europe are produced in food insecure countries, we aim to understand the relationship between sustainably certified production sites, the Human Right to adequate Food (RtF) and local food security and to identify pathways to address food security in sustainably certified agriculture.

The study is based on field research in Malaysia, Guatemala, Bolivia, Kenya and Zambia targeting sugar, palm oil, coffee and cotton produced by smallholders, medium-sized or large plantations. Five multi-stakeholder workshops and over 80 interviews with workers, farmers, plantation managers, community representatives, certification bodies, standard initiatives, NGOs, ministries, enterprises and scientists were conducted.

The certified plantations are engaged to improve the labour conditions of their workers and address their food security, but the large degree of seasonal work means also periods of food insecurity during phases of non-employment. The effects of plantations on food security of surrounding communities were mixed and depended on management, socio-economic context and historical development. The situation of certified smallholders ranges from being relatively wealthy, food secure to being desperately hungry and experiencing RtF violations despite selling to certified markets. Specific food security criteria for standards help to assess food security aspects for certified plantations and better-off smallholders, raise awareness and foster action to fully implement the RtF. In regions with very high food insecurity, the solution of many of the problems of extremely poor smallholders and their experienced RtF violations is often beyond the support that first purchasers can provide and outside the scope of most certification systems. Other parties along the value chain would need to pay premiums that enable living incomes for smallholders but additionally, governmental action and alternative development models outside agriculture for the extreme poor are needed.

Keywords: Biomass, certification, food security, plantation, right to food, smallholder, sustainability standards

---

Contact Address: Tina Beuchelt, University of Bonn - Center for Development Research (ZEF), Ecology and Natural Resources Management, Genscherallee 3, 53113 Bonn, Germany, e-mail: beuchelt@uni-bonn.de

Organisational Structures, Gender and Upgrading Strategies for Smallholders in Local Value Chains

Ayobami Adetoyinbo, Verena Otter

University of Goettingen, Dept. of Agricultural Economics and Rural Development, Germany

Numerous smallholders in local value chains (LVC) still face the challenge of effectively coordinating their activities and remaining competitive, especially in dualistic sectors. In order to enhance smallholders' gain, there is a need for further upgrading and development of LVCs. We contribute to this by distinctively depicting LVC and its characteristics, assessing the role of complete governance typologies and hidden gender-roles in shaping and driving it, and deriving concrete strategies for smallholder upgrading. The qualitative analysis builds on Global Value Chain framework and uses data obtained from three focus group discussions and 21 interviews with key informants to further the understanding of LVC organisational structures in the Nigerian shrimp and prawn sector. We further extend the framework to include gendered value-web which was developed to capture disaggregated gender-roles and power-relations. The general result highlights that competitive traders- the lead actors- drive and ensure the functioning of the LVC. However, further in-depth analysis shows that hidden and strategic mutual reliance between the production and processing/marketing segments which are coordinated through relational governance and the activities of female-processors are crucial for the proper functioning of the LVC at the supply base. The result indicates a paradigm shift in the inter-relationship between small-scale producers and processors in LVCs and confirms the eruption of a new segment at the supply base of this case. Although there are potentials for smallholders' upgrading and overall LVC development, the study warns against several constraints which inhibit smallholders' process and product upgrading which consequently result in inequitable benefit distribution to them. To overcome these constraints, managerial and policy implications need to simultaneously consider segments that are mutually dependent and based on differentiated gender-roles.

Keywords: Gender, global value chain, Nigerian shrimp and prawn, organisational structures, smallholders

Contact Address: Ayobami Adetoyinbo, University of Goettingen, Dept. of Agricultural Economics and Rural Development, Heinrich-Düker-Weg 12, 37073 Goettingen, Germany, e-mail: adetoyinbo@gwdg.de

The Heterogeneous Effects of Standards on Agrifood Trade Flows

DELA-DEM FIANKOR, OLIVER HAASE, BERNHARD BRÜMMER
University of Goettingen, Dept. of Agricultural Economics and Rural Development, Germany

With falling average tariffs, measures of food safety, e.g., standards and technical regulations, are gaining relevance in agricultural trade. As a result, these food safety measures are often seen as non-tariff barriers to trade, with different political economy implications for producers, retailers, consumers, and national governments. As a result, how standards affect trade flows is of particular interest in the agricultural trade literature. Despite the increasing number of empirical estimates, the standards-trade effect remains ambiguous, e.g., public standards are usually observed to have a trade-reducing effect whiles private standards enhance trade. Taking advantage of recent theoretical and empirical developments in the trade literature and using a large sample of data, we revisit the standards-as-barriers to trade debate. We focus particularly on maximum residue limits (MRLs), the highest level of a pesticide residue that is legally tolerated in or on food or feed when pesticides are applied correctly. We exploit the bilateral difference in MRLs over the period 2005–2014 for 145 agrifood products across 60 countries. Empirically, we deviate from the usual CES gravity model and instead estimate a theory-consistent translog gravity model. Consistent with the literature, we find that stricter importing country residue limits have a trade reducing effect, i.e., conditional on exporting they reduce the value of trade. However, our translog specification allows for variable trade effects from food standards. Hence, as a novel contribution to the literature, we are able to show that the trade reducing effect of food standards is heterogeneous across import shares. The smaller the exporting country's share in the importing country's total imports, the larger the trade reducing effect, and vice versa.

Keywords: Agricultural trade, food standards, maximum residue limits, translog

Contact Address: Dela-Dem Fiankor, University of Goettingen, Dept. of Agricultural Economics and Rural Development, Heindrich-Düker-Weg 12, 37075 Göttingen, Germany, e-mail: dfianko@agr.uni-goettingen.de

Agroeconomic Viability of Irrigated Common Bean Production by Small Farms in the Micro-Region of Ceres, Goias State, Brazil

OSMIRA FATIMA DA SILVA, GLAYS RODRIGUES MATOS, ALCIDO ELENOR WANDER

Brazilian Agricultural Research Corporation (EMBRAPA), National Rice and Beans Research Center (CNPAF), Brazil

The objective of this study was to determine the economic viability of common bean production by small farms in the micro-region of Ceres, Goias state, Brasil. Common beans are grown in three different seasons: 1st season (summer, during rainy season), 2nd season (late summer until autumn, with last rainfall) and the 3rd season (winter, with irrigation). The 3rd season represents the greatest ability to increase production, since area is available, and irrigation enables production in several areas of Cerrado region in Brazil. Ceres micro-region in Goias state represents one of those potential expansion areas to grow common beans. Therefore, within the Ceres micro-region, including the municipalities of Ipiranga de Goias, Rialma and Ceres, we monitored the cropping systems used by three farmers (one in each municipality). Those farmers actively interact with researchers doing on-farm research in their farms. The information about cropping systems allowed us to assess the economic viability of those cropping systems using indicators like production costs (total and average), profitability and break-even point. All farmers received US$ 60 to 65 per 60 kg bag of common beans. The results indicate that: (a) The drip irrigated common bean production system, interspersed with the production of green maize, developed by the farmer of the municipality of Ipiranga de Goias, led to an increase of family income. The common bean break-even price was US$ 51.39/60 kg bag. (b) For common bean producers in the municipality of Rialma who used the conventional irrigation system to distribute sprinklers in the production system, the break-even price for common beans was US$ 21.28/60 kg bag. (c) The irrigated common bean producer in the municipality of Ceres who used the no-tillage system to grow irrigated beans, obtained the break-even price of US$ 37.66/60 kg bag. Considering these break-even prices, all three farmers had a positive profit with common bean production under irrigation during the winter season.

Keywords: Break-even point, economic efficiency, production costs

Contact Address: Alcido Elenor Wander, Brazilian Agricultural Research Corporation (EMBRAPA), Rodovia GO-462, km 12, 75375-000 Santo Antonio de Goias, Brazil, e-mail: alcido.wander@embrapa.br

Biophysical and Socio-Economic Mapping and Prioritisation of Areas for Sustainable Dairy Development in Ethiopia

OGHAIKI ASAAH NDAMBI[1], JELLE ZIJLSTRA[1], MICHIEL VAN EUPEN[2], TINSAE BERHANU[3], TOMASO CECCARELLI[2], ADRIAAN VERNOOIJ[1]

[1] *Wageningen University and Research, Animal Science Group, The Netherlands*

[2] *Wageningen University and Research, Environmental Research, The Netherlands*

[3] *Wageningen University and Research, BRIDGE Project, Ethiopia*

Milk demand in Ethiopia is currently rising due to the growing population and high urbanisation rate (4.63 %), and people from urban Ethiopia consume two times more milk than the national average. Meanwhile, the Ethiopian dairy sector is dominated by low yielding local cows, with more than 98 % of its milk being informal. The sector highly depends on natural pastures and by-products from rain-fed agriculture, and due to current climate change effects, it has become more challenging to maintain a sustainable dairy chain.

The aim of this study is to contribute to bringing more milk to the Ethiopian formal market by identifying geographical areas (clusters) with high potential to sustainably increase milk production. It combines biophysical data, agricultural data and expert opinion, through: (i) Biophysical mapping of current and potential milk production areas by combining several layers of GIS images representing conditions for dairy development: vegetative cover, temperature, humidity, altitude, pest prevalence, current cattle population, etc. (ii) Identifying and weighting of indicators for dairy development, and ranking current and potential dairy clusters based on these criteria using panels of regional and national experts. The key indicator themes covered are: current production situation, environmental conditions for cows, feed and fodder production and availability, market access, expansion in milk volume and access to inputs and services. (iii) Elaborating on strengths and weaknesses of each identified cluster and identifying the most limiting factors for dairy development in each cluster.

The combination of these factors led to a selection of 14 dairy clusters with high potential to increase milk production located in Amhara, Oromia, Tigray and SNNP regions. The top three clusters with highest potential were North Shewa, Adama and South and West Shewa due to their proximity to Addis Ababa, with relatively attractive geographical conditions for cows, high demand for milk, well developed infrastructure and a dense input supply network. Two clusters (Weldiya and Hararghe) had low scores, but were seen to be promising for future milk expansion of dairying due to a high cattle density, high proportion of cropland and high biomass yield seen as potential forage for cows.

Keywords: Ethiopia, milk production, sustainable dairy development

Contact Address: Oghaiki Asaah Ndambi, Wageningen University and Research, Animal Science Group, Droevendaalsesteeg 1, 6780 PB Wageningen, The Netherlands, e-mail: asaah.ndambi@wur.nl

Farm Production and Market Access of Certified Coffee Farmers in Dak Lak, Vietnam

HUNG ANH NGUYEN, WOLFGANG BOKELMANN
Humboldt-Universität zu Berlin, Albrecht Daniel Thaer-Institute of Agricultural and Horticultural Sciences (ADTI), Germany

Under the trend of market-oriented reform and industrialisation process, sustainable coffee farming provides opportunities for small-scale farmers to integrate successfully into the high-value markets. Most farmers understand the benefits of sustainable coffee production in term of providing bio-diverse farming practices, coffee agro-ecosystem, production risk mitigation, better product quality, and price premium. However, like many other agricultural sectors in Vietnam, several factors such as output maximisation behaviour, profit incentives, shortage of capital, traditional farming habits, and non-transparent markets make coffee farmers extremely vulnerable. Efficient production management and improved market access are vital. This study, therefore, attempts to analyse production efficiency and factors that influence market access of sustainable certified coffee farmers using stochastic frontier approach and seemingly unrelated regression (SUR) model. The results show that sustainable coffee farmers in Dak Lak obtained the average technical efficiency level of 88.24 %. The technical efficiency ranged from 45.5 % to 98.0 %, hence sustainable coffee farmers could reach 9.97 % of cost saving if they achieve the technical efficiency level of their most efficient counterparts. Certified coffee farmers are marketing their coffee beans to different markets of exporter/processor, buying agent, and local trader. Significant SUR estimated variables that influence certified coffee farmers' market access are transaction cost attributes (price uncertainty, market competition, transportation cost, payment speed, and sale agreement) and socioeconomic characteristics of farmer (age, ethnic, farming experience, location, and certificate ownership). Social relationships embedded in economic activities explain the most preferred access to spot markets of buying agent. The study suggests that improving education, credit access, and collective actions are essential for sustainable coffee farmers to mitigate the effect of small-scale production. Given the need for increased coordination, farmers should be engaged in more direct market channels.

Keywords: Certified coffee, market access, seemingly unrelated regression, stochastic frontier, technical efficiency

---

Contact Address: Hung Anh Nguyen, Humboldt-Universität zu Berlin, Albrecht Daniel Thaer-Institute of Agricultural and Horticultural Sciences (ADTI), Invalidenstraße 42 (Hauptgebäude), Berlin, Germany, e-mail: nghunganh@gmail.com

Using Cassava (*Manihot esculenta*) Value Chain Products in Filling Gaps for Food Security in Nigeria

DAMIAN AGOM[1], EDDY ATTE ENYENIHI[2], DUMKA BENSON[3]

[1]*Akwa Ibom State University, Dept. of Agricultural Economics and Extension, Nigeria*

[2]*University of Uyo, Agricultural Economics and Extension, Nigeria*

[3]*Federal College of Education (Technical), Omoku, Dept. of Agricultural Education, Nigeria*

This paper analysed the use of cassava (*Manihot esculenta*) value chain products to fill the food security gaps in South-South Nigeria, a region with high prevalence of poverty, hunger, malnutrition and general food insecurity. Primary data were collected from 240 respondents using a set of well structured questionnaires which were complemented by personal interview and observation to ensure accuracy and consistency of data used. Data were analysed using descriptive statistics, rankings and regression analysis. Results of the analysis showed that cassava is very widely produced, accepted and consumed in various forms in the area. The common products from cassava value chain include garri, starch, tapioca, fufu, abacha and boiled cassava tubers. These engage a wide range of household members who earn income and feed from the products. The analysis further showed that garri was the highest income generating product in the value chain, followed by edible starch, tapioca, fufu and abacha, respectively. The regression result revealed that farming experience, farm size and fertiliser were significant in determining income ($p < 0.01$) in the study area. Transportation of the bulky product was a challenge affecting 62.9 % of respondents amongst other challenges. The study recommends farmers to process more of their cassava to higher income yielding products. The factors that are critical to output need to be enhanced for better outcomes, notably increasing production scale, making fertiliser available and improving road and transport infrastructure. Cassava value chain products have the potential to improve not only the household food security but making impact nationally and globally.

Keywords: Cassava, filling gaps, food security, income, value chain products

Contact Address: Damian Agom, Akwa Ibom State University, Dept. of Agricultural Economics and Extension, Obio Akpa, Nigeria, e-mail: agomd@yahoo.com

he New EU Regulation on Cadmium in Chocolate: Impacts on the Colombian Cocoa Value Chain

CRISTHIAN DAVID PUERTA RODRIGUEZ, ATHENA BIRKENBERG, REGINA BIRNER

University of Hohenheim, Inst. of Agric. Sci. in the Tropics (Hans-Ruthenberg-Institute), Germany

For the first time, the EU cadmium regulation determines cadmium thresholds for cocoa-based-products, effective from January 2019 onwards. Latin America is particularly affected, since cadmium levels in soils are high and fine flavored cocoa is derived from this region. In Colombia, cocoa production has been considered as an alternative to coca production and a critical element of rural development for peace achievement in the post-conflict era. However, this alternative is at threat since around 50 % of the total cocoa exports of Colombia are shipped to Europe. This case study aims at understanding the effects of the new EU regulation on Colombian smallholders. Particularly: 1) What strategies have been developed or taken by the different actors along the cocoa export value chain to tackle the cadmium challenge? 2) How do information channels work and how well are producers and cocoa associations informed? To answer these questions, affected Colombian cocoa producers and exporters, were identified. Stakeholder analysis and semi-structured interviews were conducted and an innovative network mapping tool, the Net-Maps, was applied in combination with qualitative interviews. The preliminary results identify potential risks for stakeholders along the Colombian cocoa value chain derived from the new regulation. Many smallholders are still in the process of identifying cadmium levels in their farms, information flows are mainly dominated by exporting companies, smallholder producers benefit from memberships in cocoa associations, and mitigation strategies such as blending cocoa with different cadmium concentrations might not result very effective in the long-term. Finally, the cadmium regulation on cocoa seems to challenge the maintenance of peace in rural areas.

Keywords: Cadmium, chocolate, cocoa, Colombia, regulation, smallholders

Contact Address: Cristhian David Puerta Rodriguez, University of Hohenheim, Inst. of Agric. Sci. in the Tropics (Hans-Ruthenberg-Institute), 70593 Stuttgart, Germany, e-mail: david.puerta@daad-alumni.de

Transaction Costs in Cassava Production Chain: The Case of Araguaia Valley Region (Goias State, Brazil)

OSMAR DE PAULA OLIVEIRA JUNIOR[1], CLEONICE BORGES DE SOUZA[1], JOSE ELENILSON CRUZ[1], ALCIDO ELENOR WANDER[2], CLEYZER ADRIAN CUNHA[3]

[1]*Federal University of Goias (UFG), Postgraduate Program in Agribusiness, Brazil*

[2]*Brazilian Agricultural Research Corporation (EMBRAPA), Brazil*

[3]*Federal University of Goias (UFG), School of Economics, Brazil*

The objective of this study was to research the cassava production chain in Araguaia Valley Region (Goias state, Brazil) from the perspective of the Transaction Costs Economics. The Araguaia Valley Region comprises the municipalities of Aruana, Britania, Buriti de Goias, Corrego do Ouro, Fazenda Nova, Itapirapua, Jussara, Matrincha, Montes Claros de Goias, Novo Brasil and Santa Fe de Goias. The region has an area of 18,451 square kilometers and a population of about 72,000 inhabitants (26 % living in rural areas). Primary data were collected through field research, using two separate semistructured questionnaires, with open and closed questions. The first questionnaire was used to interview family farmers, and the second one for other regional agents of the cassava production chain. In total 101 questionnaires were filled out. This sample included 73 small cassava producers, 14 local retailers who sell cassava and its derivate products, 13 small cassava processors and one wholesaler. The cassava production chain has strong social and economic importance in the Araguaia Valley Region. Almost all small farmers of the study region cultivate cassava. Cassava is produced for self-consumption and marketing of surplus production. Most of transactions are done on spot market. The main transaction costs are mainly related to high asset specificity, bringing up hierarchical or hybrid governance structures of transactions. 90 % of transactions are done in informal spot market transactions, which leads to inefficient management of transaction costs. Agent reputation arises as a mechanism to diminish transaction costs, since transactions occur with high frequency. Uncertainty is mainly related to climatic conditions as well as lack of information on new potential marketing partners.

Keywords: Family farming, governance structures, new institutional economics

Contact Address: Alcido Elenor Wander, Brazilian Agricultural Research Corporation (EMBRAPA), Rodovia GO-462, km 12, 75375-000 Santo Antonio de Goias, Brazil, e-mail: alcido.wander@embrapa.br

Marketing of Edible Species Collected in Ancient Walnut Forest in Southern Kyrgyzstan

YANA CHERNYKH[1], DIETRICH DARR[2], ZBYNEK POLESNY[1], VLADIMIR VERNER[1]

[1] *Czech University of Life Sciences Prague, Fac. of Tropical AgriSciences, Czech Republic*
[2] *Rhine-Waal University of Applied Sciences, Faculty of Life Sciences, Germany*

Non timber forest products (NTFPs) are important sources of livelihoods for the rural population worldwide. It is estimated that 1.5 billion people depend on forests as their primary source of food, nutrition, medicine or energy. However, after centuries when most of the forest products supplied particularly households' needs, nowadays there is an increasing role of income generation documented. As timber marketing is extensively discussed in forestry literature, studies on economic botany and value chains of NTFPs are rather scarce. This is true also for the area of walnut forest in southern Kyrgyzstan. Our research aims to analyse use patterns, collection management and commercialisation practices of five most commonly collected products by the local households and to estimate the potential contributions to cash income generation. Transect walks, key informant interviews, value chain actors (n=25, five per each node) and household surveys (n=60) were used for data collection. Respondents were identified via convenient, snowball and voluntary sampling method. Data analysis is presented by mapping existing value chains and understanding of potential effects between household characteristics and NTFPs collection and use. Preliminary results show the important role of the forest for local households, as only two of them were not involved in the collection of any forest product. Main products collected by local households were walnuts, apples, rosehips, barberries, and mushrooms, whereby the mushroom and walnuts were of highest financial value. Walnuts and apples were intended for export, the other products supplied local markets. Cash income from forest was more important for households without remittances than for those who were receiving cash transfers from abroad with 61 % and 35 % contribution to total household income, respectively. Household characteristics (farm size, labour force, gender, age, dependent members, off-farm activities) and geophysical data (distance to collection place, distance to markets) affect volumes and number of collected products. Outcomes of the study are useful for a better understanding of the role of the forest in the local livelihood under currently changing environment (climate change, monetarisation of rural economy, land use policy).

Keywords: Central Asia, household survey, markets, non-timber forest products, value chain

---

Contact Address: Vladimir Verner, Czech University of Life Sciences Prague, Faculty of Tropical AgriSciences, 129 Kamycka street, 16500 Prague 6 - Suchdol, Czech Republic, e-mail: vernerv@ftz.czu.cz

Cocoa, Forests and Peace – The Role of Cocoa in Conservation and Peace Building in Colombia

Andrés Charry, Fabio Castro, Augusto Castro-Nunez
International Center for Tropical Agriculture (CIAT), Colombia

Since the beginning of the peace negotiations between the Colombian government and FARC, the country has been experiencing changes in multiple dimensions, bringing along new environmental, social and political challenges. Deforestation patterns have shown different trends across the country, with some areas decreasing total forest loss while others showing rapid increases in deforestation. Similarly, conflict, with its many actors and expressions has changed both geographically and in intensity. Experts believe that both the causes and effects of conflict and deforestation are highly interrelated, with common factors such as land tenure and rights, economic opportunities, state presence, rule of law, among others. Aware of the commonalities, government officials and stakeholders have promoted sustainable agricultural value chains as approaches to tackle both phenomena simultaneously, assuming that increasing incomes, market access, productivity and welfare may help in reconstructing the social tissue and, with the right mechanisms, reducing pressure on forests. In that sense, the cocoa value chain has been proposed as one of the main alternatives by the international cooperation, private investors and the public sector.
In this research, we provide evidence on the potential of the cocoa value chain to contribute in both dimensions. For this, we conducted a spatial analysis to understand the current relationship of cocoa production, deforestation and conflict in Colombia. Additionally, we analysed the narratives and perspectives of 30 key stakeholders on what they consider the role of the cocoa value chain as a tool to for peacebuilding and forest conservation. We found that there is no significant spatial relationship between cocoa production and deforestation, nor between cocoa production and conflict. We also found that the majority of stakeholders interviewed believe that cocoa is not a driver of deforestation, but instead can become a reforestation tool through agroforestry systems, and a conservation tool as an economic safeguard in buffer zones for high conservation values. Similarly, they believe that it has a role in peacebuilding as a means for cooperativism, increasing rural incomes and providing licit economic opportunities, nevertheless, there is a broad agreement that this can only be achieved if the productive activity becomes a "profitable business".

Keywords: Deforestation-free value chains, environmental peacebuilding, spatial correlation, sustainable agriculture

Contact Address: Andrés Charry, International Center for Tropical Agriculture (CIAT), Km 17 recta Cali - Palmira, Cali, Colombia, e-mail: a.charry@cgiar.org

System Assessment on Exchange of Commodities among Rural Households by Using a Village Social Accounting Matrix

GABRIEL SPECHT, STEPHANIE DOMPTAIL, ERNST-AUGUST NUPPENAU
Justus-Liebig-University Giessen, Inst. of Agricultural Policy and Market Research, Germany

While the importance of agriculture to combat poverty has been widely acknowledged, most of the world's poor still live in rural areas. The exchange of food and non-food commodities through formal or informal market channels influence the livelihood of rural households. While part of household production is consumed by the households themselves, most of the households achieve surplus production, which in turn is exchanged with other households inside or outside a given village. Thus, understanding the exchange flows of food and non-food commodities within a rural economy allows us to grasp the contribution of commodity exchange in the income portfolio of rural households. Yet, the question remains to what extent the exchange of commodities within a given economy benefits the rural poor. We propose to assess the exchange linkages among households and to identify pathways which promote rural development through the promotion of commodity exchanges within a rural economy. For this purpose, we assess household exchanges of food and non-food commodities, using village Social Accounting Matrix (SAM) framework. The SAM method has originally been conceived to map interactions between different sectors within a national economy, and has been further developed to show interactions at village level between different household groups, their production and consumption pattern and exchange of commodities. We use the example of a village in Southern Brazil to build the SAM, in which household economic activities are explicitly exhibited and households are grouped according to their livelihood strategy. This will allow us to understand how exchanges of food and non-food commodities take place with the different types of households. The study show that a system level analysis enables to look at how the poor benefit from exchanges of food and non-food commodity exchanges.

Keywords: Commodity exchange, smallholder farming, social accounting matrix, system analysis

Contact Address: Gabriel Specht, Justus-Liebig-University Giessen, Inst. of Agricultural Policy and Market Research, Senckenbergstr. 3, 35390 Giessen, Germany, e-mail: gabriel.specht@agrar.uni-giessen.de

Price Linkages in Indonesian Palm Oil Industry

RAKHMA MELATI SUJARWO, THOMAS KOPP, BERNHARD BRÜMMER
University of Goettingen, Dept. of Agricultural Economics and Rural Development, Germany

Environmental issues related to the Indonesian oil palm plantation expansion remain crucial. Thus, the European Union (EU) as one of the influential Indonesian palm oil and biodiesel importers took several measures. Trade policies were made to tackle the land use conversion and to prevent unfair market competition. Contrarily, the palm oil industry highly contributes to the Indonesian national income and increases the local labour usage. Domestic policies were formed to motivate an increase in palm oil quality and production. While the environmental aspect remains important, the economic aspect needs to be considered. Those policies may lead to a shock in the industry described by a structural break occurring in the relationship between palm oil price series. Observing the measurable effect of a policy implementation is able to let the industry effectively anticipate disadvantage and defend its market power. Thus, the study of the price linkages in Indonesian Palm Oil Industry is essential to be performed. Also, a multivariate price transmission study including local price in the price transmission model is scare. The international and Indonesian (Crude Palm Oil) CPO prices as well as Indonesian local oil palm Fresh Fruit Bunch (FFB) price were the variables observed, whereby all prices were stationary in the first difference. We further tested the cointegration pair-wisely with Johansen cointegration test. Later, we employed Gregory-Hansen test allowing a structural break as comparison. We evidently found that models with structural break gave better result with lower value of Akaike, Bayesian and Hannan-Quinn information criterions. The most plausible related event to the break is the biodiesel EU anti-dumping duty implementation, which started approximately one month prior to the breakdate. This presumably affects the Indonesian CPO and local FFB prices. The generated import tariff reduces the EU biodiesel import demand which may cause a decrease in the national CPO demand. Consequently, it may lead to a decrease in the national CPO and local FFB prices. This insight is supported by the result, that the EU import tariff negatively affects not only the Indonesian CPO price but also the local FFB price.

Keywords: Gregory-Hansen test, Johansen cointegration test, palm oil, price transmission, structural break

Contact Address: Rakhma Melati Sujarwo, University of Goettingen, Dept. of Agricultural Economics and Rural Development, Platz der Goettinger Sieben 5, 37073 Goettingen, Germany, e-mail: rakhma.sujarwo@agr.uni-goettingen.de

Tariff Impact on Industrialisation Development in Tanzania: Evidence from Edible Oil Sub-Sector

Charles Peter Mgeni[1], Klaus Müller[2], Stefan Sieber[1]

[1]*Leibniz Centre for Agric. Landscape Res. (ZALF), Inst. of Socio-economics, Germany*

[2]*Humboldt-Universität zu Berlin, Albrecht Daniel Thaer-Institute of Agricultural and Horticultural Sciences (ADTI), Germany*

The imposition of high tariff rates is considered as strategies to protect domestic producers, stimulate production and accelerate industrialisation in developing countries. Tanzania, among many developing countries, has a strong agricultural base and is in a verge of choosing which trade strategies will enable the country to attain among her population a high economic growth and develop her industrial potential. Therefore, the government has opted to implement a protectionism policy on the edible oil sector by gradually imposing 10 % tariffs in 2016/17 to 35 % in 2018/19 on both crude and refined edible oil with the aim of promoting domestic production. It is expected that the intervention will create more employment not only to the edible oil sub-sector but also to other sectors in the economy. This study assesses the impact of the imposed tariffs on edible oil in Tanzania using a recursive dynamic computable general equilibrium model (CGE) at the national level, dis-aggregating the edible oil sub-sector from the agricultural sector and other sectors and evaluates its multiplier effects in Tanzania's economy. Findings from this study show that tariff imposition has twofold outcomes; first, it incentivizes domestic producers to supply more due to rising in demand and prices for the commodities and eventually increase domestic production that suffices the material and quantities demanded by the agro-processing and other industries; second, such intervention erodes citizen welfare by limiting availability and access options to varieties of commodities that could make them better off in terms of prices, qualities, and quantities. protectionism policy when solely used as a solution for increasing domestic production to a sector that is not efficient in terms of productivity creates commodities supply deficit in the market, consequently, reduce consumers welfare. Therefore, for sustainability, and increase industrial competitiveness, it imperative for the government of Tanzania to promote policies and interventions that target increasing productivity to small, medium and large industries. Interventions that increase smallholder farmers productivity like the use of improved seed and other modern technologies that reduce costs of production are critical as commodities will be sold at a slightly competitive premium or same prices as imported commodities.

Keywords: Edible oil, industrialisation, recursive dynamic CGE model, tariff

Contact Address: Charles Peter Mgeni, Leibniz Centre for Agric. Landscape Res. (ZALF), Inst. of Socio-economics, Eberswalder Straße 84, 15374 Müncheberg, Germany, e-mail: charles.mgeni@zalf.de

Women's agency in rural households

Gendered Land Rights in Ghana: A Threat to Sustainable Agricultural Production

ANGELA AKORSU

University of Cape Coast, Dept. of Labour and Human Resource Studies, School for Development Studies, Ghana

Women and men contribute to, and are affected by threats to food security, climate change, soil degradation and depletion of natural resources differently and unequally. Sustainable resource management, which is a response to these threats, can therefore not succeed without interrogating the enormous gender-based inequities in agricultural production, which is fundamentally land-based. Land remains the most crucial resource for securing livelihoods for both females and males in any agrarian economy like Ghana's. Yet, gender social relations and their embeddedness in the agricultural production systems have always structured land allocation in ways that have meant limited access to and control over land for women. The importance of addressing this knowledge gap cannot be over emphasised in view of the fact that a notable proportion of agricultural workers in Ghana are women. Women in agriculture constitute an estimated 52 percent of agricultural labour force, 70 percent of food producers and 95 percent of agro-processors. This presentation highlights gendered land rights as a major threat to sustainable agricultural production by seeking to answer pertinent questions such as; (1) Why is gender an important narrative in agricultural production? (2) How does gender determine access to and control over land, particularly in Ghana? (3) How do women's limited land rights threaten sustainability in agricultural production? (4) What interventions can potentially remove the gendered land right trap to ensure sustainability in agricultural production?

Keywords: Gender, land access

Contact Address: Angela Akorsu, University of Cape Coast, Dept. of Labour and Human Resource Studies, School for Development Studies, Cape Coast, Ghana, e-mail: aakorsu@ucc.edu.gh

Small-Scale Processing and Marketing of Underutilised Vegetables – What Role for Women's Livelihoods in Ghana?

Bastian Lange, Oliver Hensel, Barbara Sturm
University of Kassel, Agricultural and Biosystems Engineering, Germany

A growing body of literature today is focusing on the potential of underutilised fruits and vegetables to improve nutrition in Western Africa and prevent micronutrient deficiencies among vulnerable populations. In Northern Ghana these crops, also often referred to as orphan crops, are commonly perceived as "women's crops". As such, they play a considerable role in women's agricultural production, processing and marketing routines. Despite growing evidence for their importance in terms of both on- and off-farm activities, their role in the context of rural livelihoods has been neglected in research to date. Market data for these underutilised vegetables is sparse and such crops are frequently omitted from agricultural and livelihood surveys, both in Ghana and Western Africa more broadly.

The study at hand addresses key knowledge gaps on the marketing of underutilised vegetables and their role within livelihood strategies of women in Ghana's Northern Region. For the purpose of this study, market surveys were conducted in two urban and three rural markets at the height of the local dry season. Between February and March 2019, data was collected on availability, supply chains and pricing for 10 species of underutilised vegetables, in both fresh and processed form. The survey was complemented by qualitative data obtained from Focus Group Discussions with four local women's groups.

Results emphasise the overall importance of production and marketing of underutilised vegetables for livelihoods of women in Northern Ghana. Small-scale processing of indigenous vegetable crops, in particular, provides a key source of income during the dry season, when prices for staple crops are high and income generating opportunities are otherwise constrained. Nonetheless, the economic potential of these underutilised vegetables continues to be curtailed by challenges in women's access to resources and raw materials, fragmented and volatile markets, as well as suboptimal quality and storability of traditionally processed products. In light of these constraints, the introduction of improved post-harvest processing practices has a significant potential to strengthen the livelihoods of women in Northern Ghana, by reducing losses and providing access to higher value markets.

Keywords: Marketing, post-harvest processing, underutilised fruits and vegetables, Western Africa, women's livelihoods

---

Contact Address: Bastian Lange, University of Kassel, Agricultural and Biosystems Engineering, Nordbahnhofstr. 1a, 37213 Witzenhausen, Germany, e-mail: bastian.lange@uni-kassel.de

Gender- and Age-Biased Land Tenure Systems and their Impact on Sustainable Agricultural Intensification

GUNDULA FISCHER[1], AKOSUA KESEBOA DARKWAH[2], JUDITH KAMOTO[3], JESSICA MZAMU-KAMPANJE[4]

[1] *International Institute of Tropical Agriculture (IITA), Africa RISING, Tanzania*

[2] *University of Ghana, Dept. of Sociology, Ghana*

[3] *Lilongwe University of Agriculture and Natural Resources, Dept. of Forestry, Malawi*

[4] *Lilongwe University of Agriculture and Natural Resources, Dept. of Human Ecology, Malawi*

Equitable outcomes are a critical component of working towards sustainability in agriculture. How does sustainable agricultural intensification's tenet of increased productivity on the same area of land relate to prevailing gender- and age-biased land tenure systems? Does the emphasis on a non-expansion of agricultural land imply redistributing control over land and the benefits gained from it to facilitate equitable outcomes? These questions guided a comparative qualitative study in intensification contexts in matrilineal and patrilineal communities in Ghana and Malawi. Using Kabeer's framework of institutional analysis, we go beyond a focus on the household to include other domains (community, market and government processes) that shape smallholders' access to land. We conducted a total of 102 semi-structured interviews and focus group discussions with male and female community leaders, market and government actors as well as household members. They reveal where respondents observed recent changes in agricultural land allocation and if/how they would imagine a fair redistribution of land. Possible pathways to redistribution mentioned by respondents include more equitable land inheritance patterns which village heads would champion in their families and communities, government control of land prices to ease market access for less well-off social groups, and gender/youth-sensitive legal education in rural areas. In the household domain, some interviewees proposed earlier land transfers from one generation to the next. Questions on a potential redistribution were also met with resistance, especially among advantaged landholders who justified prevailing systems. In general, limiting agricultural land expansion was seen as fostering gender imbalances if it is not mediated. We conclude that investments in agricultural intensification should facilitate equitable outcomes by supporting consensus-based institutional changes and creating positive synergies between multiple domains.

Keywords: Age, gender, Ghana, land tenure, Malawi, sustainable intensification

Contact Address: Gundula Fischer, International Institute of Tropical Agriculture (IITA), Africa RISING, Box 10, 0000 Duluti, Tanzania, e-mail: g.fischer@cgiar.org

Intra-Household Land Tenure Insecurity – The Mute Risk to Sustainable Food Production in Western Kenya

SERAH KIRAGU-WISSLER[1], BONIFACE KITEME[2]

[1] *TMG Research gGmbH, Germany*

[2] *Centre for Training and Integrated Research In ASAL Development (CETRAD), Kenya*

Land tenure security is a key and decisive factor in shaping land productivity, household food security and poverty levels in rural areas. In three counties of western Kenya – Kakamega, Bungoma and Siaya, intra-household land tenure insecurity is common - particular members of a household, especially women and youth, are denied the rights to own, access, use or inherit ancestral land due to their gender, marital status, age or other conditionalities set by the land owner, often the elderly male household head. This happens under cultural construction of a complex household structure upon which land access and control rights are allocated. The bundle of rights to land is based on customary believes and traditions which determine allocation according to gender, position and status of birth in the household, and relation with the household head. Legal pluralism is notable - modern and customary regimes prevail in the governance of land rights, often existing in contradiction while at times supportive of each other. Overall, traditional practices dominate. Complex and complicated succession laws only exacerbate the intra-household land tenure insecurity challenge. The results is that women, especially widows, and youth are unable to optimally partake to household food production decision making. A transdisciplinary approach to understand the complexity and drivers of intra-household land tenure insecurity and practical solutions to addressing the challenge were adopted by the study. A number of social innovations have been tested and found to possess promising potential for long term solutions. These include land clinics, community land lease guidelines and integrating youth and women into joint land investments with clear benefit sharing mechanisms.

Keywords: Intra-household, land tenure insecurity, youth, Kenya

Contact Address: Serah Kiragu-Wissler, TMG Research gGmbH, Egerländer Str. 2, 95448 Bayreuth, Germany, e-mail: serah.kiragu-wissler@tmg-thinktank.com

Entrepreneurial Potentials of Rural Women's Groups to Process Underutilised Species in Northern Ghana

Lydia Madintin Konlan[1], Rashida Chantima Ziblila[2], Margareta Lelea[3], Brigitte Kaufmann[3]

[1] *University of Kassel / University of Goettingen, Dept. of Agricultural Economics and Rural Development, Germany*

[2] *University for Development Studies Tamale, Agricultural Extension, Rural Development and Gender Studies, Ghana*

[3] *German Institute for Tropical and Subtropical Agriculture (DITSL), Germany*

According to the Global Entrepreneurship Monitor survey, women's engagement in entrepreneurship is highest in sub-Saharan Africa. Meanwhile, women are held responsible for the nutritional well-being of their families and many processing activities are gendered female. For these reasons, the transdisciplinary research project 'UPGRADE Plus' partners with women's groups to process underutilised species. These plant species that are available, but not utilised to their full potential for economic benefit are targeted with the broader goals of increasing agro-biodiversity, enhancing food access over the seasons, and improving women's livelihoods.

The aim of the study is to assess the entrepreneurial potential of the women's groups to process underutilised species by focusing specifically on context-specific factors related to their processing-related goals, needs and constraints. This research focuses on six women's groups in Tamale, northern Ghana where food security challenges are more acute than in the southern part of the country. Qualitative data was gathered from 37 group meetings in which multiple participatory rural appraisal tools were used in a sequence through which the women and the researchers could co-investigate business interest and options. This was complemented with 18 individual interviews.

Results showed variations between the different groups with regard to motivation to engage in a processing business and a learning process together with researchers. Differences were also found in groups' prioritisation of opportunities and their ability and capacity to navigate constraints. Different crops were evaluated in relation to their local supply and the availability of other necessary resources for production and development of their business.

The process of engaging the women in a co-inquiry of their business related goals, needs and constraints creates momentum for them to develop viable business plans and put them into action.

Keywords: Agri-food entrepreneurship, Ghana, processing, transdisciplinary

Contact Address: Lydia Madintin Konlan, University of Kassel / University of Goettingen, Dept. of Agricultural Economics and Rural Development, Steinstr. 19, 37213 Witzenhausen, Germany, e-mail: lydiam1988@yahoo.com

Rural Women's Access to Land and Household Food Security: Implications of Agricultural Intensification in Ghana

Rashida Chantima Ziblila[1], Margareta Lelea[2], Jasper Abembia Ayelazuno[1], Eliasu Mumuni[1], Brigitte Kaufmann[2]

[1]*University for Development Studies, Dept. of Agricultural Extension, Rural Development and Gender Studies, Ghana*

[2]*German Institute for Tropical and Subtropical Agriculture (DITSL), Germany*

From growing food to processing food, women's paid and unpaid labour in the food system is a critical part of food security. However, women's access to land is often mediated by men as is common in patrilineal societies such as the Dagbon in northern Ghana. With increasing pressure on land from new agricultural investment projects, men can shift priorities for land use. One such new investment is a European-Ghanaian joint-venture company, which cultivates and processes organic fruit for both domestic and export markets using both a plantation nucleus and out-grower scheme. Using this European-Ghanaian company as a land acquisition case study, this research empirically investigates the implications on women's access to land and food security. Fieldwork was conducted in six rural communities in the Savelugu Municipality and Tolon District of the Northern region. Drawing from the USDA's six-point scale measurement of food security, 185 questionnaires and 6 focus groups were conducted in 2018–2019.

The results revealed that because women rely mainly on male relatives for access to land and other common resources, they are vulnerable to food insecurities when there is a breakdown of relations or when the priorities of male landowners change. The land acquired by this company for its plantation led to a direct reduction in women's access to trees such as *Vitellaria paradoxa* (shea nut), *Parkia biglobosa* (locust bean known as 'dawa dawa' in the region) and various fruit trees. Because products harvested from these trees are important both for household consumption and for women's cash income, household food security diminished. Women who lost access to such trees faced longer travel distances as well as the need to negotiate access to new sources, further compounding scarcity and competition in the region.

Although the out-grower scheme of this company has been praised for keeping land and labour in the control of smallholders, the growing disadvantage of women has been overlooked. The removal and constriction of women's access to land and other resources led to diminished access to harvests needed for home consumption, processing and sale, which diminished food security.

Keywords: Agri-food processing, food security, Ghana, land access, locust bean, shea, women

Contact Address: Rashida Chantima Ziblila, University for Development Studies, Dept. of Agricultural Extension, Rural Development and Gender Studies, Tamale, Ghana, e-mail: razhida@gmail.com

Contributions of Home Gardens to Vegetable Biodiversity: A Case Study of Eco-Sustainable Gardens Empowering Minority Women in Cameroon

PRIDE EBILE, NANCY NAHSANG TAGU, JENS NORBERT WÜNSCHE
University of Hohenheim, Inst. of Crop Science, Germany

Biodiversity and natural resource management are critical to a sustainable livelihood of many rural communities because they depend highly on their ecosystems. However, the biodiversity of many communities is being threatened by many factors; one such factor is demand driven agricultural systems with an emphasis on high-value monocultures. This agricultural system has negatively affected the biodiversity of many communities thus fragilizing their ecosystems.

With the loss of plant biodiversity in many communities, especially of some indigenous vegetables, this study examined the potential of a garden project to reintroduce some indigenous vegetable crops back into the communities of rural regions in North-west Cameroon.

Eco-sustainable gardens that empower Mbororo minority women is a home garden project designed to help reduce the dependency of the Mbororo women on their husbands for food and income. The project also has an educational component that is mainly to provide the women with good horticultural management practices and the importance of dietary diversity and nutrition. Three groups of vegetables were planted in the gardens for the following purpose, i) marketable vegetables (fluted pumpkin, waterleaf, chili pepper), ii) nutrient-rich vegetables (aubergine, amaranthus, okra, Chinese cabbage, sweet bitter leaves), and iii) indigenous vegetables (Lalo, Folere, Caricachee).

Many families can depend on a diversity of vitamin rich vegetables for income generation and nutritional food. The project reintroduced and increased the variety of vegetables available within the project communities, adding to the plant biodiversity. The project helped to produce indigenous vegetable seeds that are scarce, and distributed them to many beneficiaries thus promoting the distribution of these vegetables.

Keywords: Biodiversity, home gardens, minority groups, vegetable, women empowerment

Contact Address: Pride Ebile, University of Hohenheim, Dept. of Crop Science Crop Physiology of Specialty Crops, Emil-Wolff-Str. 25, 70599 Stuttgart, Germany, e-mail: pride_ebile@uni-hohenheim.de

Impact of a Combined Nutrition and Agricultural Intervention on Time-Use and Dietary Diversity of Women Smallholder Farmers in Teso Sub-County, Kenya

M. Gracia Glas[1], Bárbara Fróes[1], Samwel Mbugua[2], Sahrah Fischer[3], Thomas Hilger[3], Ernst-August Nuppenau[4], Irmgard Jordan[1]

[1]*Justus-Liebig University Giessen, Center for International Development and Environmental Research, Germany*

[2]*Egerton University, Department of Human Nutrition, Kenya*

[3]*University of Hohenheim, Inst. of Agric. Sci. in the Tropics (Hans-Ruthenberg-Institute), Germany*

[4]*Justus-Liebig University Giessen, Inst. of Agric. Policy and Market Res., Germany*

Women constitute an essential agricultural workforce; they play a crucial role in the food provision of the entire household. Research has shown that interventions aiming to improve nutrition security without addressing gender may have adverse side effects and intensify women's fieldwork. Increased time invested in agricultural activities can unintendedly result in inadequate care and nutrition practices.

In June 2016 and 2017 cross-sectional agriculture-nutrition surveys were conducted in Teso sub-County, Kenya, targeting 421 farm households with children aged below five years. Minimum dietary diversity score for Women (max 10 groups) and child dietary diversity score (CDDS, max 7 groups) were calculated based on 24h-recalls. Time allocation was assessed based on 24h-physical-activity-recalls of the respondents.

Results were compared across three different groups: nutrition education only (NE), combined agricultural and nutrition education (AGNE) and control. Women of all groups had a mean cooking time of around 3 hours. The mean time for feeding children/breastfeeding was lower than 4 minutes. Mean time spent on children's care increased significantly in the combined intervention group (mean = 13.7 min). NE group and control group also increased the time spend in child care (6.4 and 8.3 minutes, respectively). None of the interventions had an influence on the Δcooking time of mothers. There was a significant difference in Δfarming time between the intervention groups. Changes in Δfarming time were significantly higher in the combined intervention compared to the NE Group. CDDS was not affected by an interaction between participation in the intervention and Δcooking time of mothers. There is also a statistically significant influence of Δeating time of mothers on ΔCDDSs in Kenya.

The analysis revealed that the project had a slight influence on the change of time-use suggesting that gender time-use data is decisive when promoting innovative agricultural approaches.

Keywords: Agriculture, dietary diversity, gender, nutrition, time-use

Contact Address: M. Gracia Glas, Justus-Liebig University Giessen, Center for International Development and Environmental Research, Giessen, Germany, e-mail: maria.g.glas@zeu.uni-giessen.de

Farm Household Decision Making on Crop Diversity in the Mt. Elgon Region, Uganda: A Participatory System Analysis Approach

CHRISTINE ARWATA ALUM, STEPHANIE DOMPTAIL,
ERNST-AUGUST NUPPENAU

Justus-Liebig University Giessen, Inst. of Agricultural Policy and Market Research, Germany

Members of farming households in Uganda face the need to decide collectively on the crops they want to produce yearly in order to meet their needs in terms of cash and food. This decision making process is characterised by perceived trade-offs among the existing production possibilities. Decisions are further constrained by land, farm labour and capital. We consider the farm and its household as a socio-ecological system and adopt a constructivist approach. In order to gain insights from households about the decision making process, we use an innovative method of participatory system analysis. With a view to empower farmers with limited writing and reading skills to participate, we used collages, instead of words. A collage elicits data using visual representations and offers vast ways of interpretation. In addition, it has rarely been used on farm household related studies. This study was carried out in two subcounties which represent the low and mid altitudes of Kapchorwa district in the Mt. Elgon region, Eastern Uganda. The study therefore examines how decisions about crop diversity and the area cultivated for each crop are made at the level of the farming household. The collected data consisted of the source (who makes the decisions) and the reasons (what drives the decisions) for the cropping diversity, the constraints, the trade-offs between cultivation options and cropping activities. The connections between the different elements of decision making were assessed as well. First, we collected narratives from individual interviews which we documented with our own pictures and observations. The subsequent system analysis took place in focus group discussions and made use of the pictures taken during the interviews. This research shows how important it is to involve farmers in participatory research if one aims to target the relevant constraints when formulating policies which influence cropping decisions.

Keywords: Collage, crop diversity, Farm household, Participatory system analysis, Socio-ecological system

Contact Address: Christine Arwata Alum, Justus-Liebig University Giessen, Inst. of Agricultural Policy and Market Research, Senkenbergstrasse 3, 35390 Giessen, Germany, e-mail: tinaalum@yahoo.com

Gender Specific Utilisation of Forest Products among Rural Dwellers in Osun State, Nigeria

WAHEED SULAIMON[1], OLANIKE DEJI[2], OLAMIDE OLOWOYO[1]

[1]*Obafemi Awolowo University, Dept. of Agricultural Extension and Rural Development, Nigeria*

[2]*Humboldt-Universität zu Berlin, Dept. of Gender Research in Rural Areas, Germany*

This study specifically determined male and female's level of utilisation of forest products among rural dwellers in Osun State, Nigeria, through the instrumentality of gender lens. Multi-stage sampling procedure was used in selecting a total of 240 respondents. Structured interview schedule and Focus Group Discussion guide were used to collect quantitative and qualitative data. Data collected were analysed using descriptive and inferential tools. The results show that the mean level of utilisation of forest products for male respondents was 267 ± 70.3, while that of female respondents was 248 ± 72.0 (with a maximum obtainable score of 747). The results of t-test show that calculated t-value of 2.042 was higher than the critical t-value of 0.042 at $p \leq 0.05$ with 238 degrees of freedom. This result implies that there was a significant difference in the utilisation of forest products by male and female respondents. The Focus Group Discussion results show that male respondents have a favourable disposition in their utilisation of economically viable forest products than females. This in turn has created a source of financial advantage through a robust economic engagement of males in the study area. However, females have always been at the receiving end of natural resource policies which tend to favour males at their detriment. This in turn has hindered sustainable utilisation of forest products in the study area. In conclusion, there was a significant difference in the level of utilisation of forest products by male and female respondents. Hence, there is need to address this inequality by engendering policy frameworks that focus on gender equality in order to ensure sustainable forest resource utilisation as a panacea for economic development.

Keywords: Forest products, gender, gender disaggregation, rural dwellers, utilisation

Contact Address: Waheed Sulaimon, Obafemi Awolowo University, Dept. of Agricultural Extension and Rural Development, House 2 Phase 2 Orisunmbare Peace Estate Ipaja-Ayobo, 100278 Ayobo, Nigeria, e-mail: sulaimon_suberu@yahoo.com

Gender and Bovine Livestock Production in Latin America: The Status Quo

NATALIA TRIANA-ANGEL, STEFAN BURKART

International Center for Tropical Agriculture (CIAT), Trop. Forages Program, Colombia

Bovine livestock production is a vital cultural and economic activity across Latin America and fundamental for the survival of rural households in developing regions. While both men and women perform crucial activities, women's contributions are still profoundly undervalued. Such overshadowing results from deeply rooted gender stereotypes and constricting family roles that determine both the division of labour and the productivity of dual-purpose producing farms. Gender inequity in the region, scholars state, prevents the sustained economic growth of livestock units and the possible impacts of new technologies and public policies, furthering conditions of poverty and disparity. This literature review summarises the contributions on gender and the bovine livestock sector in Latin America. We emphasise three salient perspectives and their respective inputs. First, we review value-chain studies and their role in tracking action nodes of both men and women. We focus on how such approaches underscore female participation in stages of manufacturing and commercialisation of milk byproducts in Central America. Second, we delve into familiarist approaches of household duties and responsibilities along gender lines. This perspective has served the vital purpose of understanding heritage, generational transference and continuity of bovine livestock activities in the midst of growing foreign-markets within the Pampean region. Thirdly, we look into more recent research on breeding and forage technologies and how they grasp the importance gender relations have in the appropriate implementation of silvopastoral systems, animal genetic improvement and mitigation strategies, especially in Colombia and Costa Rica. Such studies bring to light the different areas where women and men exert differential influence, underline the gendered impact of extreme weather events and reflect on women's potential as agents of change in the advent of climate change. Within the richness of the existing literature we were also able to identify the urgent need of new lines of research, as the impact armed conflicts, displacement and dispossession have had in cattle and dairy production in Latin American countries at war over the course of the twentieth century. Above all we strive to find connecting points across regions, research bottlenecks, and possible subject matters for further inquiry.

Keywords: Climate change, dual purpose value-chain, gender, livestock, mitigation

---

Contact Address: Stefan Burkart, International Center for Tropical Agriculture (CIAT), Trop. Forages Program, Km 17 Recta Cali-Palmira, Cali, Colombia, e-mail: s.burkart@cgiar.org

Cyclic Agroecosystems under Threat by Neo-Extractivism: The Answers from Peasant Women

Ana Alvarenga de Castro
Humboldt-Universität zu Berlin, Germany

The current *feminisation* of agriculture, poverty and hunger relies as a contradictory problematic of the contemporary global food and economic systems announced by international agencies such as FAO and the World Bank. Mining mega-projects represent one of the activities playing a core role in the dynamics of the International Division of Labor. They cause escalating conflicts over land and natural resources in the named *developing countries*, many times impacting peasant livelihoods. In Latin America, this process has been called Neo-Extractivism, especially to discuss the boom
of mining and monocultures for exportation in the last decade. This paper analyses constrasting metabolic dynamics within environmental conflicts: the one from mining megaprojects and the one from peasant agriculture, discussing then the concepts of Food Sovereignty in the Neo-Extractivist context.
The case study focusses on the two locations most affected by the Minas-Rio mining megaproject in the Southeast Brazil. The territories are inhabited by communities holding traditional farming rationalities and sociabilities which are under threat and give important answers for sustainable food systems. Through participative observation during six months and interviews with 26 peasant women, the author analysed three pillars of Agroecology: ecological rationality, social-economic efficiency and agrobiodiversity, specifically by looking at elements of reproductive work, gender relations and sexual division of labour inside farming units. The results demonstrate that social reproduction of peasant agriculture under threat by extractive megaprojects is dependent on cyclic agroecosystems caracterised by short circuits, cooperation and diversity. Those practices over food show specific gender relations which have been impacted by the linear metabolic dynamic of the mining megaproject. The second result is that mining megaprojects and agribusiness practices on land follow the same linear dynamics, what puts them both in the conceptualization of Neo-Extractivism. Additionally, the surveys indicate that peasant women in these affected territories represent potential responses to the Neo-Extractivist approach by re-creating farming and collective practices that question the global proposals to *end hunger* and suggest alternatives to peoples' food sovereignty through women's autonomy.

Keywords: Agroecosystems, food sovereignty, gender, neo-extractivism, peasant women

Contact Address: Ana Alvarenga de Castro, Humboldt-Universität zu Berlin, Philippstraße 13 Haus 12, D-10115 Berlin, Germany, e-mail: alvacasa@hu-berlin.de

Can Climate-Smart Aquaculture Enable Women's Empowerment in Rural Bangladesh?

JODIE COLGAN[1], CYNTHIA MCDOUGALL[2], UNA MURRAY[1], CHARLES SPILLANE[1], PETER MCKEOWN[1], MD. EMDAD HOSSAIN[2]

[1] *Ryan Institute, MScCCAFS Program, Plant & AgriBiosciences Research Centre (PABC), Ireland*

[2] *Worldfish, Malaysia*

Climate-smart aquaculture provides a means to ensure sustainable fish supply to those who experience negative impacts of climate change. However, there has been little research on possible benefits of climate-smart aquaculture for enabling the empowerment of women farmers. This WorldFish-NUI Galway MScCCAFS study investigated women farmers empowerment under two domains, decision-making power and time burden, with the aim to record farmers perceptions of any change in theses domains as a direct result from the WorldFish climate-smart aquaculture intervention in rural Bangladesh. The 'ladder of power and freedom' tool was used to determine whether women and men perceived an increase in household decision-making power since the implementation of the homestead pond intervention, and to understand the underlying reasons for any such change. Different qualitative tools were used to generate primary data, consisting of 16 focus-group discussions, 21 mini-focus group discussions, 32 semistructured interviews and 16 key informant interviews, with a sample size of 170 farmers across 9 villages in the Barisal and Khulna divisions of rural Bangladesh. The farmers involved were purposively selected among households implementing the WorldFish intervention programme. Two neighbouring villages that had no WorldFish climate-smart intervention were included for comparative purposes. The results indicate that while the climate-smart intervention enabled the empowerment process for women farmers, this was uneven across households. The main enabling factors included no perceived imposed time-burden and an increase in decision-making within the household since the intervention was introduced in 2016. However, barriers were identified for those women farmers who did not feel a positive change in decision-making power after adoption, based on household characteristics and dynamics, intra-household power relations and household income. This study highlights the need for integrating gender when introducing climate-smart interventions to address any gender disparities that could contribute to negative outcomes.

Keywords: Climate change, climate-smart aquaculture, empowerment, gender

Contact Address: Jodie Colgan, Ryan Institute, MScCCAFS Program, Plant & AgriBiosciences Research Centre (PABC)
current address: University College Cork, Dept. of Food Business & Development, Ireland, e-mail: 113741531@umail.ucc.ie

Food Species Richness and Commercialisation of Homegardens in Indonesia: Example of Central Sulawesi and West Sumatra

TEREZA VEJVODOVA[1], LUKAS PAWERA[1], AIN SYAHRIL[2], ZBYNEK POLESNY[1], VLADIMIR VERNER[1]

[1] *Czech University of Life Sciences Prague, Fac. of Tropical AgriSciences, Czech Republic*
[2] *Tadulako University, Fac. of Agriculture, Indonesia*

Homegardens represent old and rather subsistence-based farming systems, which play an important role in livelihood especially in rural as well as in peri-urban areas. Food security is often being cited as a one of the most crucial roles of homegardens. Globally, they are famous for their rich diversity of annual and perennial plant species. Nevertheless, the structure, species composition and their level of commercialisation differ according to different physical, ecological, cultural and socioeconomic factors. The aim of our study was to document food species richness and their level of commercialisation with respect to homegarden and household characteristics. Data were collected among 115 households, 71 around city of Palu in Central Sulawesi, 44 around city of Padang in West Sumatra. Data were collected via direct observation and interviews with focused households. Convenient purposive sampling method was used to identify suitable respondents. Shannon-Wiener, Margalef and Simpson diversity indexes were used to calculate agrobiodiversity. Multiple linear regression was applied to identify potential effects of household geophysical and socioeconomic characteristics on homegarden structure, species richness and level of commercialisation. A total number of 71 species was distributed accross 39 families. A typical homegarden in our study was 1,834 m^2 large and 28 years old, these from Central Sulawesi were smaller and younger compare to those in West Sumatra. Differences in valuation of agrobiodiversity, aesthetic role and ecological benefits were also observed among focused households in our two study sites. Preliminary results shows effects of gender on species composition and commercialisation. Dependent members, age, distance to city centre, terrain, size and age of homegardens influenced species structure as well.

Keywords: Agro-biodiversity, attitudes, geophysical characteristics, household characteristics, multiple linear regression, perceptions

Contact Address: Vladimir Verner, Czech University of Life Sciences Prague, Faculty of Tropical AgriSciences, 129 Kamycka street, 16500 Prague 6 - Suchdol, Czech Republic, e-mail: vernerv@ftz.czu.cz

Governance and sustainable resources management

Science-Policy Transfers and the Perpetual Tension between the Spheres of Activism and Academic Knowledge Production and Diffusion

DANIEL HAWKINS
National Union School (Escuela Nacional Sindical), Colombia

Gradually the sanctified neutrality of scientific knowledge production, wrapped in the notions of progress, modernity and rationality, has come to a crux as the growing reach of business interests, fake news and algorithmic calculations bend the idea of "objective" science to fit their ends. This growing tension is stronger still in the murky waters that often join the spheres of activism and academic knowledge production and diffusion. On one hand, social activism still often upholds the ideals of solid, theoretically and methodologically sound scientific research which is used as a means of strengthening the discursive and political resonance of proposals or demands made by diverse social groups. On the other hand, in the ever-more interconnected world of social webs and digital platforms that feed off exposure, science-policy transfer often occurs due more to the degree of diffusion rather than the quality of the research. How to navigate the often conflicting demands of producing sound science with the need to influence public policy debates with engaging ideas is an increasingly challenging task, especially for young budding scientists who are intrigued with the idea of working as both academics and activists. In the field of modern agriculture, these issues are also prevalent, especially if we examine the scientific and discursive development of popular concepts such as the governance of agricultural value (or supply) chains and how one should understand and measure the concept of sustainable certification systems in agriculture. Daniel Hawkins will offer some reflections into how these debates and tensions interfold from within the themes of social-labour standards in the Colombian coffee and palm oil supply chains.

Keywords: Social labour, value chain

Contact Address: Daniel Hawkins, National Union School (Escuela Nacional Sindical), Medellín, Colombia, e-mail: dirinvestigacion@ens.org.co

From Conflict to Engagement in Agricultural Water Governance: The Case of Zayandeh-Rud Basin, Iran

MOHAMMAD NASER REYHANI[1,2], PHILIPP GRUNDMANN[2,1]
[1]*Leibniz Institute for Agricultural Engineering and Bioeconomy, Germany*
[2]*Humboldt Universität zur Berlin, Division of Resource Economics, Germany*

With rapid population growth, expansion of agricultural areas, industrialisation and urbanisation, effective engagement of multiple stakeholders is crucial in ecological and political dimensions of spatial planning during the times of changing climate, particularly in countries facing water scarcity. The main objective of this paper is to enhance the potential contribution of stakeholder analysis to an effective stakeholder engagement by considering novel factors like stakeholders capacity for change, distinct formal and informal relationships, and perception of stakeholders. Our findings are based on a case study of the Zayandeh-Rud river basin in central Iran that suffers from growing socio-economic and environmental consequences of water scarcity. In this study we used a qualitative research design by conducting series of interviews, focus group discussions, and workshops with a total number of 74 people from water, agriculture and environment sectors, farmers and civil society. Additionally, three main groups (affecting, affecting and affected, and affected) of the identified key-stakeholders (n=156) were asked in form of a self-administrated questionnaire. Relying on common understanding of stakeholder engagement in water governance, the research design has been formed jointly, stakeholders have been identified and categorised based on their level of power, interest and capacity for change, and finally mapped by considering the type and degree of inter-relations using Social Network Analysis tool. Notably, we obtained these results after analysing the perceptions of the three groups of key stakeholders across the entire basin. An elaborated stakeholder map and social network proved to provide valuable findings for decision makers to understand: how the roles and responsibilities are shared, how they interact, and how they can get engaged effectively? As an outcome, an adaptive process of Participatory Stakeholder Analysis with respective tasks has been introduced for formulation of engagement strategies. Our analysis extends the literature on institutional change, and informs the debate over effective engagement process toward adaptive governance of resources.

Keywords: Engagement process, institutional change, participatory stakeholder analysis, water governance

Contact Address: Mohammad Naser Reyhani, Leibniz Institute for Agricultural Engineering and Bioeconomy, Max-Eyth-Allee 100, 14469 Potsdam, Germany, e-mail: nreyhani@atb-potsdam.de

Community Based Governance and Sustainablity in the Paraguayan Pantanal

Luca Eufemia[1], Izabela Schlindwein[1], Michelle Bonatti[1], Sabeth Tara Bayer[1], Stefan Sieber[1], Marcos Lana[2]

[1]*Leibniz Centre for Agric. Landscape Res. (ZALF), Germany*

[2]*Swedish University of Agricultural Sciences, Crop Production Ecology, Sweden*

The megadiverse biome of the Paraguayan Pantanal is in danger due to the expansion of cattle ranching and agricultural frontiers that threaten not only the fragile equilibrium of natural resources, but also the one of local governance and cultural identities. As a consequence, weak governance stresses the relations between natural resource dependent communities, generating socio-environmental conflicts. This perception study seeks to find community-based governance model standing for sustainability in the context of Paraguayan wetlands. Under the organisational principles of community based natural resource management (CBNRM) we applied qualitative approaches with the use of the Governance Analytical Framework (GAF) to identify problems and social norms. Our findings suggest that community-based governance is constructed by the Yshiro indigenous community relation to land use/management (e.g. Traditional Ecological Knowledge, TEK) and their self-organised group (Unión de Comunidades Indígenas de la Nación Yshiro, UCINY), as well as highly threatened by the impact of the national neo-extractivist economy. In addition to the reaffirmation of identities, resistance and knoweledge sharing, many structural changes need to happen (e.g social/economic standing of indigenous communities to contribute to TEK and cultural survival), since indigenous communities are clearly embedded in post-colonial settler relations in multiple ways. As for the findings of this work, the role of the developmental strategy of extractivism is having negative impacts on the equilibrium of local governance and cultural identities. Likewise, the historical role of identity of the Yshiro community should be included in dialogues of sustainability for its intimate bond to land and its entrenched connotations.

Keywords: Community-based natural resources management, community-based governance, identity, indigenous rights, Paraguayan Pantanal

Contact Address: Luca Eufemia, Leibniz Centre for Agric. Landscape Res. (ZALF), Inst. of Socio-Economics, Eberswalder Str. 84, 15374 Muencheberg, Germany, e-mail: luca.eufemia@zalf.de

Public Programs and their Influence on Agricultural Production and Practices: Evidences from Rural Brazil

Gabriel Specht, Stephanie Domptail, Ernst-August Nuppenau
Justus-Liebig-University Giessen, Inst. of Agricultural Policy and Market Research, Germany

Agriculture is one of the core activities characterising livelihood strategies of smallholder farmers in the developing world. Agricultural practices and production are partly facilitated and mediated by public policies implemented in rural areas, in general by loosening investment constraints and by granting market access to agricultural producers. Program outcomes on rural households may vary, as public policies vary in scope and objectives. As an example, the outcomes produced by rural credit and cash transfer programs, concomitantly implemented in Brazilian rural areas, are not very clear. On the one hand, PRONAF, a credit programme giving annual crop and investment loans for smallholder farmers, has participation biased towards asset-rich wealthier households involved in the production of cash crops, especially soybeans and maize. On the other hand, Bolsa Familia, a conditional cash transfer programme for poor and extreme poor households, especially supports asset-poor rural households involved in subsistence agriculture. Yet, the question remains to what extent the concomitant implementation of public programs in a rural village shape overall agricultural production and agricultural practices, and how it influences household's choice of livelihood strategy. We propose to assess household livelihood strategies based on their agricultural production activities and to explore their determinants with a focus on the role of public programs. As such, we aim at unveiling the role of concomitant public programme implementation in shaping agricultural production in rural areas, thus influencing household economic benefits from programme participation. An activity-based two-step cluster analysis will be conducted to identify different livelihood clusters, and regression models are performed to determine the major factor affecting the choice of livelihood strategies. The analysis uses survey data from 2017 of 101 households of a rural village in Southern Brazil. The results are expected to show the importance of rural credit and conditional cash transfer for smallholder farmers in influencing their livelihood strategy, and the factors affecting their livelihood choice. The findings of this study shed light on the role public programs play in determining agricultural production and their role in households' livelihood strategy, thus informing pathways out of poverty for smallholder food producers in southern Brazil.

Keywords: Livelihood strategies, public programs, smallholder farmers

Contact Address: Gabriel Specht, Justus-Liebig-University Giessen, Inst. of Agricultural Policy and Market Research, Senckenbergstr. 3, 35390 Giessen, Germany, e-mail: gabriel.specht@agrar.uni-giessen.de

A Multi Stakeholder Action Framework for Inclusive Action on Climate Resilient Agriculture with a Special Focus on the Coastal Areas of South Asia

RAJESH S. KUMAR[1], SHILPI KUNDU[2]

[1]*National Institution for Transforming India (NITI Ayog), Health Division/State Coordination Division, Tamil Nadu, India*

[2]*Griffith University, School of Environment & Science, Australia*

South Asia is projected by the Intergovernmental Panel on Climate Change (IPCC) to be home to the largest number of food insecure people on the globe due to the adverse impacts of climate change. With a population of more than 1.75 billion and with merely 4 % of the world's GDP coupled with high income inequality, the landscapes in the region are dotted with several traps such as poor inclusive policy development and governance, and fragmented resource management arrangements particularly in the context of agricultural production systems. Besides, the socio economic and political landscape is hit by the exacerbating risks and challenges posed by the climate change. In order to address the fast emerging food security challenges in South Asia on account of growing population, declining production and productivity, and urbanisation, it is imperative that productivity augmentation is key to the challenge rather than expansion of cultivable land. Although, the prospects for increasing farm production could be potentially constrained by the impacts of climate change; identification of scalable evidence based multi sectoral action strategies could provide impetus for policy development and promotion of climate resilient agriculture, especially in the highly vulnerable coastal areas of the region. However, only limited region specific studies have focused so far on the potential to harness multi–actor action synergy and to rationalize the investment of resource envelopes under a regime of a multi stakeholder framework for active promotion of climate resilient agriculture. In this paper, we have attempted to address these policy gaps through a documentation and analysis of prominent political discourses on Climate Resilient Agriculture (CRA) in the region and propose a multi stakeholder action framework for inclusive action on CRA. The outputs of this study are expected to contribute to the region specific evidence based efforts to remove the traps in the climate response policy landscape and foster dynamic operation of actors and stakeholders in development of climate resilient agricultural systems for robust food security.

Keywords: Climate resilient agriculture, coastal area, multi stakeholder framework, political discourses, South Asia

Contact Address: Rajesh S. Kumar, National Institution for Transforming India (NITI Ayog), Health Division/State Coordination Division, Tamil Nadu, NITI Bhavan, Parliament Street, 110001 New Delhi, India, e-mail: rskumarifs@gmail.com

Opportunities and Challenges of the Grain Sector in Kosovo - Managing Domestic Resources

ANIKA TOTOJANI[1], STÉPHANE FOURNIER[2]

[1]*University of Copenhagen, Dept. of Food and Resource Economics, Denmark*

[2]*Innovation, Montpellier SupAgro, INRA, CIRAD, Univ Montpellier, France*

This study sets out the opportunities and stakes of grain sector in Kosovo - the recent country in the map of Balkans. Kosovo has a transition economy and the government is tempting to restructure the economy according to the European Union policy. The research analyses the role of actors involved in the food and feed agri-sector involving supplying, production, processing, marketing and distribution of food and feed grain. The research employs qualitative methods. We conducted 68 semi-structured interviews, and respondents were selected randomly. Grounded theory and content analysis guided data analysis.

Research data refers to the chain governance, collaboration, integration and strategies between actors. Study highlights the implication into the regional market, and the mediums of actors being better off.

The study reveals that Kosovo depends on grain imports, and lacks an organised domestic grain market. Wheat is the major crop planted followed by maize and other small grain quantity. Actors are partly integrated and not very efficient in the grain production. They partially fulfil the domestic grain need.

The quality of the imported grain depends on the choice of individual traders. The price is the major factor influencing the quality of the imported grain. Furthermore, the existence of informal market influences the decision making of actors. Actors involved in the grain chain have mix governance types. Domestic actor's relations are based on trust mechanism. The research has drawn the importance of revised agriculture public policies to sustain grain domestic production and a better management of the domestic resources. It is encouraged the creation of public-private partnerships to manage domestic resources, restore the grain market and is recommended the revision of trading policies. In doing so, fair competition between domestic and foreign traders will be enhanced.

Keywords: Chain governance, domestic resources, grain sector, Kosovo, public policies

Contact Address: Anika Totojani, University of Copenhagen, Dept. of Food and Resource Economics, Rolighedsvej 25, 1958 Frederiksberg C, Denmark, e-mail: at@ifro.ku.dk

Influence of the International Environmental Agenda on Amazonian Cooperation Treaty Organisation (ACTO) Regional Governance

FREDY DAVID POLO VILLANUEVA, PRADIP KUMAR SARKER, LUKAS GIESSEN

University of Goettingen, Section Forest and Nature Conservation Policy, Germany

International regimes, both global and regional, can play a key role on the international environmental governance. However, due to the increasing frustration with these regimes, the topic is gaining also attention in political practice and academia. On this path, based on qualitative data from key regime policy documents developed between 1978 to 2018 by the regime, the aim of this study is to analyse the institutional design (ID) and forest policy developed within the Amazonian Cooperation Treaty Organisation (ACTO) and address the following empirical questions: 1) What is the ID of ACTO; and how is the environmental agenda influenced on its evolution? 2) Which policy goals, policy instruments and the precise settings of these instruments can be identified within the regime; and how were these influenced by the international environmental agenda since the last forest policy was adopted? The results show that along the lifetime of the regime the membership features did not change; meanwhile the "scope" and "flexibility" were slightly changed; and the "centralisation" and "control" features were strongly modified. Also, we identified that three main impulses, boost by the international environmental agenda, drove these important changes in ACTO ID. In addition, the forest policy developed by ACTO is characterised by a large number of issues covered, coherent and non-conflicting policy goals, short term and very limited quantity of donor projects as the only instruments; and an extreme lack of regulatory, (des)incentive, and informational instruments developed within the regime. In conclusion, the ID evolution helped to increase the formalisation of the regime and present a solid image of ACTO in front of the international community. Moreover, the policy development show that one of the main regime's informal goals is to raise funds from the international community through legitimacy provided by member countries. Therefore, we conclude that ACTO is a weak international regime that depends on donor funded projects, driven by the international agenda, to achieve its policy goals.

Keywords: ACTO, forest policy, governance, institutional design

---

Contact Address: Fredy David Polo Villanueva, University of Goettingen, Section Forest and Nature Conservation Policy, Büsgenweg, 37077 Göttingen, Germany, e-mail: fredypolo2018@gmail.com

Applying the MESMIS Methodology to the Nexus Approach: The Nexus Pampa Project

Vicente C Pires Silveira[1], João Garibaldi Almeida Viana[2], Cláudia Garrastazu Ribeiro[3], Claudio Marques Ribeiro[4], Fernando Luiz Ferreira de Quadros[5], Jean Paolo Gomes Minella[6], Jean Francois Tourrand[7,5]

[1]*Federal University of Santa Maria, Agricultural Education and Rural Extension, Brazil*

[2]*Federal University of Pampa, Agricultural Economic, Brazil*

[3]*Federal Inst. of Education, Science and Technology Sul-Rio-Grandense, Brazil*

[4]*Federal University of Pampa, Extension, Brazil*

[5]*Federal University of Santa Maria, Zootecnia, Brazil*

[6]*Federal University of Santa Maria, Soil, Brazil*

[7]*CIRAD, France*

The Brazilian Pampa is an old group of ecosystems that has its own flora and fauna, and is home to sizable biodiversity. It is a natural, genetic and cultural heritage of great national and global importance. Continuous introduction and expansion of monoculture, especially soybean, and of pastures with exotic species has led to rapid degradation and to the loss of distinctive characteristics of the Pampa. This issue has prompted the creation of the Nexus Pampa Project (Projeto Nexus Pampa), financed by the Brazilian Ministry of Science and Technology, with the goal of verifying and creating scenarios that are subsequent to the farming production systems used in the Ibirapuitã river basin, in the Brazilian Pampa biome. Accordingly, this paper describes the way in which the MESMIS methodology was applied to the Water-Energy-Food nexus in the Nexus Pampa Project. The research method used is the MESMIS (Framework for Assessing the Sustainability of Natural Resource Management Systems). The MESMIS operational structure used consisted of six stages developed through collaborative methods: i) definition and description of the systems that would be evaluated; ii) identification of the critical points of the systems; iii) selection of diagnostic criteria and indicators; iv) measurement and monitoring of the indicators; vi) conclusions and recommendations regarding the systems. The first three stages of the MESMIS method took place in group meetings between researchers and other project participants. Agricultural and livestock farming systems from the Ibirapuitã river basin in the Brazilian Pampa biome were evaluated. The identification of weaknesses and strengths of the production systems studied were evaluated, as well as their opportunities and threats, through a SWOT analysis. Subsequently, indicators were formulated in light of the Nexus approach, inside the water, energy and food axis, in an interdisciplinary approach. Presently, indicators in four systems have been assessed, validating the methodological construction of the Nexus Pampa Project. Initial results indicate a positive capacity of the methodology used in this project for assessing the sustainability of production systems, contributing to human development and to reaching the objectives of sustainable development projected by the UN.

Keywords: Animal production, Ibirapuitã river basin, MESMIS, nexus, Pampa Biome

Contact Address: Vicente C Pires Silveira, Federal University of Santa Maria, Agricultural Education and Rural Extension, Sala 5110 Predio 44, 97105-900 Santa Maria, Brazil, e-mail: vicentesilveira@smail.ufsm.br

Investigating Institutional Arrangements in Groundwater Resources Management in Iran

FATEMEH RAHIMI FEYZABAD[1], MASOUD YAZDANPANAH[1], SAEED GHOLAMREZAI[2]

[1]*Agricultural Sciences and Natural Resources University of Khuzestan, Agric. Extension and Education, Iran*
[2]*Lorestan University, Agricultural Extension and Education, Iran*

Globally, there has been a dramatic increase in groundwater use in the past half-century, especially in arid and semi–arid countries. As one of the most arid countries in the world, Iran currently ranks among the top groundwater users globally – in fact, it is estimated that Iranians have already used most of their groundwater reserves. The continued unchecked use of ground water resources may lead to serious problems, and if immediate action is not taken to address the issue, the situation could become disastrous in the near future. Sustainable groundwater resources management is therefore urgent and necessary in Iran.
While there are several reasons for the underlying groundwater crisis in the country, a number of studies show that the groundwater crisis in Iran is an institutional crisis. In this regard, evidence revealed that countries around the world have different institutions at the forefront of groundwater resources management. These institutions supply guidelines for human conduct, while also providing distinctive opportunities for monitoring social behaviour and controlling resources through changes in the institutional environment, in order to shape the way individuals behave within social entities.
Due to the importance of institutional arrangements in groundwater resources management, it is important to examine them in detail. The purpose of this research was therefore to identify the institutions involved in groundwater resources management and to examine the power and interest of each institution based on the Mendelow power and interest matrix. The study was conducted in the western part of Iran, specifically Lorestan Province. The data were collected through observation and interviews with the stakeholders involved (institutions) in groundwater management, and analysed using a stakeholder analysis based on the interests and power held by the actors. The results indicate that there are many stakeholders (institutions) involved in decision-making processes related to groundwater resources management that all play a key role in groundwater management, and their decisions largely determine the success or failure of any groundwater-related policy. Among these institutions, water authorities are the most powerful institution sin groundwater resources management, followed by agricultural and environmental authorities.The findings yield public policy for sustainable groundwater management in Iran.

Keywords: Groundwater resources management, institutions, Iran, Mendelow power and interest matrix, stakeholder analysis

Contact Address: Masoud Yazdanpanah, Agricultural Sciences and Natural Resources University of Khuzestan, Agricultural Extension and Education, Mollasani, 744581 Ahvaz, Iran, e-mail: masoudyazdan@gmail.com

Challenges of Sustainable Wastewater Management in Pakistan: A Case Study of Faisalabad

SAIMA JABEEN, REGINA BIRNER

University of Hohenheim, Inst. of Agric. Sci. in the Tropics (Hans-Ruthenberg-Institute), Germany

With the industrial revolution and having big cities for a good quality of life, we are living in a situation, which is deprived of environmental and ethical standards. The present scenario of industrial development and rapid urbanisation in most of developing countries cause wastewater production. Wastewater management remains an important challenge for the state, communities and other stakeholders within such economies. The wastewater generations, mixing with freshwater resources, and its usage in irrigation are critical issues and all-pervasive. The policies mostly target engineering-oriented solutions (treatment plants). There is a dearth of managerial and administrative solutions to handle the problems locally. The research, thus, sets out to explore the gaps in the institutional frameworks for wastewater management within the context of Pakistan. The study applies a combination of methods to assess and explore the gaps and traps within the wastewater management sector. Policymakers mainly face the problem of unreliable data and absence of specified information required for strategic planning in most of developing countries like Pakistan. This study provides a framework to analyse the current status of wastewater management and reasons of unsustainability. Notably, 'sustainable wastewater management' is a multifaceted problem having particular situation due to the nature of product or service. It advocates that sustainable wastewater management is a social/governance dilemma. The historical institutionalism of administrative organisations helped to find out the inherent challenges of ministries, departments, and agencies. Furthermore, the institutional analysis identified the constraints and challenges in the legal and administrative arrangements. Consequently, these gaps and traps identification direct towards future policy formulation in a sustainable manner.

Keywords: Governance challenges, IAD framework, institutional analysis, linear to circular water economy, public administration in wastewater management, qualitative analysis, sustainable wastewater management

Contact Address: Saima Jabeen, University of Hohenheim, Inst. of Agric. Sci. in the Tropics (Hans-Ruthenberg-Institute), Balingerstr.1, 70567 Stuttgart, Germany, e-mail: saima_sam79@yahoo.com

Implementation of a Multi-Criteria Assessment of Sustainability of Smallholder Organic and Conventional Farms in Kenya

John Ndungu[1], Anne Muriuki[2], Johan Blockeel[3], Christian Borgemeister[1], Lisa Biber-Freudenberger[1], Oliver Kirui[1], Justa Gitonga[2], Charity Gathambiri[2], Paul Kiuru[2], Marian Kamau[2], Irene Kadzere[3], Christian Schader[3]

[1] *University of Bonn, Centre for Development Research (ZEF), Ecology and Natural Resource Management, Germany*

[2] *Kenya Agricultural and Livestock Research Organization, Kenya*

[3] *Research Institute of Organic Agriculture (FIBL), Switzerland*

Organic agriculture, which is based on four principles: health, ecology, fairness, and care, is gaining importance in Kenya. However, objective and comprehensive assessments on the comparative sustainability of organic and conventional farming systems are still limited. Through the "Organic Food Systems Africa" (OFSA) project we aimed to evaluate the social, economic, environmental, and governance sustainability dimensions of 864 organic and conventional smallholder farms in Kirinyaga, Machakos and Muranga, in Kenya, and identify hotspots for sustainability improvement. We assessed each farm using the SMART-Farm Tool, which was developed to operationalize the FAO Guidelines for Sustainability Assessment in Food and Agriculture (SAFA). In this paper we present the implementation of the SMART approach and the lessons learned. Fifteen enumerators and 10 other staff participated in this implementation. We selected, oriented and trained the data collection team intensively for two weeks, and sensitized the farmers and their communities about the study. We integrated both classroom-based and field/farm practical testing of the tool and methodological approaches to ensure objectivity, and a good understanding by enumerators of the process and the tool's technical content, i.e. its sustainability dimensions, themes and subthemes, as well as indicators for rating different aspects of the farm performance. For hands-on experience, each enumerator assessed three farms during the practice sessions and exchanged their completed tools for peer review, followed by presentations and discussions on the experiences and outcomes of the field testing. We incorporated the feedback and practical lessons to improve the actual assessments, subsequently. The comprehensiveness of the tool, and efficiency considerations required each enumerator to use a laptop for direct data logging during the interviews. For an affordable workload, each enumerator visited, toured and completed data for one farm, on average, per day followed by data review before submission to a central database. The methodology, although comprehensive, requires good interpersonal skills and judgement, clear understanding of the technical terms, and objectivity by enumerators. The SMART-Farm Tool has been useful for the Kenyan context, but adequate time, human and financial resources as well as good technical capacity and peer learning and exchanges are requisites for its successful implementation.

Keywords: Kenya, organic farming, smallholder, SMART-Farm Tool, sustainability assessment, training

Contact Address: John Ndungu, University of Bonn, Centre for Development Research (ZEF), Ecology and Natural Resource Management, Genscherallee 3, 53113 Bonn, Germany, e-mail: s7jondun@uni-bonn.de

Impact of USAID/Markets II Intervention on Productivity of Rice Farming Households in Eastern Nigeria

ADUNNI SANNI, OGECHI NWAHIA, ZAKARI ABDULSALAM

Ahmadu Bello University, Agricultural Economics, Nigeria

Among the efforts made by Nigerian government to promote sustainable self-sufficiency in rice production is the USAID-Maximizing Agricultural Revenue and Key Enterprises in Targeted Sites (MARKETS), initiated in 2005. This study analysed the impact of USAID-MARKETS II project on the productivity of rice faming households in Ebonyi State, Nigeria. Multi-stage sampling technique was employed in the selection of the respondents for the study. A total of 491 rice farming households, comprising of 239 participants and 252 non-participants of the project, were included in the study. Structured questionnaire, focus group discussions and field observation were used in data collection. The data were collected on production and socio-economic variables relating to 2017/2018 production. Data collected were analysed using descriptive, Total Factor Productivity Model, Propensity Score Matching and Local Average Treatment Effect (LATE). The USAID-MARKETS II project employed several empowerment strategies towards improving the productivity of farming households in the project sites. These include, supply of improved rice seeds, fertilisers, technology development, training and extension support. The result of Total Factor Analysis shows a progress in total factor productivity growth of 69.7 % for farmers with the USAID-MARKETS II compared to 41.9 % by the non-participant farmers. Using PSM, the average impact estimation shows that USAID-MARKETS II have a positive and significant impact on productivity of the participants by 1.075. The Average Treatment Effect (ATE) on the treated for rice farming households from the overall population is larger with a value of 1.178 compared to the untreated category. The LATE estimate revealed a significant mean difference of 0.406 in rice productivity between participants and non-participants. The analysis also divulges that the average increase in total productivity brought about by participation in USAID-MARKETS II is 0.396. Major constraints to increased productivity among farmers are; inadequate capital/poor access to credit, inadequate land, high cost of inputs, poor access to technologies, poor market access and low producers price. Despite the significant impact of the project on productivity, there is the need for government to further intensify agricultural empowerment programmes by strengthening the public private partnership linkages that will address sustainable development along the rice value chain in Nigeria.

Keywords: Farming-households, impact, intervention, markets, productivity

Contact Address: Adunni Sanni, Ahmadu Bello University, Agricultural Economics, PMB 1044, Zaria, Nigeria, e-mail: adsanni@yahoo.com

Policy Issues, Actors' Public Claims and Informal Interests: Insights from the Sundarbans Mangrove Forests Management

MD FAISAL ABEDIN KHAN[1], MD SAIFUR RAHMAN[2], LUKAS GIESSEN[3]

[1] *Georg-August-University Goettingen, Dept. Forest and Nature Conservation Policy, Germany*

[2] *Ministry of Environment and Forests, Planning Division, Bangladesh*

[3] *European Forest Institute (EFI), International Forest Governance, Germany*

Comprising diverse ecosystems, the Sundarbans in Bangladesh are characterised as an unique mangrove forest, which becomes an area of national as well as international importance in policy and management perspectives. Considering the ecological as well as economic importance of Sundarbans Mangrove Forests, a number of issues have been originated now-a-days which are of policy concern. The aim of this study is to identify the most recent conversant issues with the Sundarbans and analyse the actors' claims/arguments on identified issues which were publicly stated only as an expression of formal interests or also part of their informal interests therein. Content analysis of Bangladeshi national newspapers, experts' deliberations and national policy documents were exerted as an empirical method for relevant issues, actors' arguments and interest's questions or perspectives while a theory based typology was applied to distinguish the present actors. The issues came out from this analysis are: Rampal Power Plant Project, Protection of Sundarbans Forest Area, Environmental Pollutions, Biodiversity Conservation, Reputation as World Natural Heritage Site, Climate Change Adaptation and Local People's Livelihoods. The analysis detected the Rampal Power Plant Project as being the most conflictive issue among all actors. Government Bureaucracies are found as the most active actor with all perceived issues and surprisingly Prime Minister, as a single entity of that actor type exercises intrinsic power in terms of displaying some informal interests with formal ones in her public statements. However, results recommend to conduct future empirical study on mapping about high degrees of bureaucratic rivalry for Sundarbans with formal and informal interests between the responsible ministry, Ministry of Environment and Forests' and other ministries' in relevant doings.

Keywords: Bangladesh, policy actors and interests, policy Issues, Sundarbans mangrove forest

Contact Address: Md Faisal Abedin Khan, Georg-August-University Goettingen, Dept. Forest and Nature Conservation Policy, Göttingen, Germany, e-mail: fkhan@uni-goettingen.de

Using Treated Waste Water for Agriculture in Tunisia: Farmers' and Consumers' Perspectives

MAJDI GOUJA[1], CHRISTIAN SCHLEYER[2], ANDREAS THIEL[1]

[1]*University of Kassel, International Agricultural Policy and Environmental Governance, Germany*

[2]*University of Innsbruck, Institute of Geography, Austria*

As climate variability escalates, depleting high-quality groundwater, the exploration of non-traditional water sources for irrigation agriculture has become a top priority in many drought-prone countries in the Middle East and in Northern Africa. Before the background of improvements in and better availability of wastewater treatment technologies, this contribution investigates institutional, social, economic, and governance drivers pertinent to understanding farmer´s acceptance and adoption of treated wastewater (TWW) in agriculture in Tunisia. Regionally, the research focused on the island of Djerba and on the region of Medenine. Here, the selected case study sites are suffering from an increasingly severe water shortage coupled with declining groundwater supplies; therefore, fostering the use of treated waste water for agricultural purposes has become an important policy objective. Empirically, it draws on an extensive review of primary written sources, on more than 12 semi-structured interviews with relevant stakeholders and other experts at national, regional and local level, and on a focus group with more than 20 participants carried out in spring 2018. The institutional analysis of the collected data reveals a set of common factors affecting the reuse of TWW at farm level. The poor quality and irregular availability of reclaimed wastewater, the lack of coordination between involved actors (e.g., farmers, managers of waste water treatment Plants, regional municipal and state authorities), the low economic incentives, and the low level of knowledge and social awareness regarding the benefits of TWW are the major reasons behind the difficulties of TWW reuse projects on the island of Djerba. The results also highlight differences between farmer´s and consumer´s acceptance of TWW reuse. A consumer survey conducted at weekly fruit and vegetable markets in two different regions on the island of Djerba shows a high reluctance to buy agricultural products irrigated with TWW. While farmer´s attitude is driven by freshwater scarcity and needs for alternative water sources, consumers show high levels of awareness and perception of risks, and insecurity about product quality.

Keywords: Agriculture, treated waste water, Tunisia

Contact Address: Christian Schleyer, University of Innsbruck, Institute of Geography, Innrain 52d, 6020 Innsbruck, Austria, e-mail: christian_schley@hotmail.com

Waterbody Management in Peri-Urban and Rural Areas of Metropolitan Bengaluru: Example of a Polycentric System

ARVIND LAKSHMISHA, ANDREAS THIEL

University of Kassel, International Agricultural Policy and Environmental Governance, Germany

Urban and rural areas were regarded as distinct areas in the past, which in the recent years has increasingly blurred. This is mainly associated with interdependence between the two areas leading to a functional relation due to complementarity, which comprise of economic, political, physical and social issues including resource use and sharing. The nature of interactions between rural and urban areas can be viewed as partnerships with two-way flows and establishment of variety of governance systems to manage these relationships leading to emergence of polycentricity. Our focus is on the Metropolitan region of Bengaluru, India. It covers an area of 8005 km ^{2}, across three administrative districts, comprising of numerous small and medium towns and villages in addition to the metropolis. The region is struggling to prevent the destruction of its waterbodies that are converted into wastewater dumps. The wastewater from the urban area has transformed a seasonal river into a perineal river leading to an increase in the agricultural production and dairy farming in the periphery and the rural areas. This increased availability of water [wastewater] to farmers in addition to improved access to services and market has changed the patterns of interactions among actors in relation to management of waterbodies in the region.

This paper tries to understand these interactions between peri-urban and rural areas through the lens of polycentric governance framework developed in Thiel & Moser (2019) based on Ostrom & Ostrom (1999) and Ostrom *et al.* (1961) to understand foundational aspects shaping the emergence of planning and governance of waterbodies in the region. Based on the preliminary analysis of the secondary and field interview data we identify three main forms of interactions, which affect the structure of polycentric governance in the region namely: competition, cooperation and coercion, between actors both in urban and rural areas which lead to coordinated outcomes. The paper further argues that polycentric governance is established first, due to overarching rules, which enables self-organisation by providers and consumers involved in governance. Second, the heterogeneity of the community referring to the capabilities, values and interests, income and information available which dictate how actors introduce into negotiations over governance.

Keywords: Polycentric governance, urban-rural interactions, waterbody management

---

Contact Address: Arvind Lakshmisha, University of Kassel, International Agricultural Policy and Environmental Governance, Kloester Gebaude Steinstr 19, 37213 Witzenhausen, Germany, e-mail: arvind@uni-kassel.de

Agricultural Sector Model: A Tool to Evaluate Agricultural Production and Food Security in Syria

KINDAH IBRAHIM[1], YIGEZU ATNAFE YIGEZU[2], MAHMOUD BABILI[3], RAID HAMZA[3]

[1]*Czech Univeristy of Life Sciences Prague, Faculty of Tropical AgriSciences, Department of Economics and Development, Czech Republic*

[2]*International Center for Agricultural Research in the Dry Areas (ICARDA), RALS Program, Egypt*

[3]*National Agricultural Policy Center, Rural Development Department, Syria*

Syria's conflict has entered its ninth year. The severe war has brought the country's economy to its knees. Only agriculture has been saving the country. No doubt, once the war ends the country will be exposed to recovery challenges. A big issue will be how to enhance food security during the post-war transition. The objective of this study is to explore various options for improving food security in the country shattered by this protracted crisis. The research methodology employs the Agricultural Sector Model (ASM) to depict the possible scenarios for the rehabilitation of Syrian agriculture and thus improving the state of food security. The framework of this partial equilibrium model consists of a quantitative analysis of the structure of agricultural production; the marketing system; government policies and programs; projections of demand for and supply of goods and considerations of alternative policies to improve the performance of the sector. The work in ASM started in 2009, as a part of the AgroSyr project, a research project jointly executed by ICARDA-CGIAR and the National Agricultural Policy Centre (NAPC) in Syria but stopped when the conflict erupted. Recently, the work in AgroSyr project has been resumed, and new data was collected in 2018. The survey covered 13 out of the 14 Syrian Governorates, where more than 1380 households were interviewed and a total of 2144 gross margins with detailed information about main crops production was gathered. The analysis was performed using GAMS software. The results helped to evaluate two dimensions of food security: availability and access; and simulate the future policy options for the stability dimension. The model also served to determine the optimal use of different resources to enhance agricultural production in Syria through policy scenarios.

Keywords: Agricultural policy, food security, sector model, Syria

---

Contact Address: Kindah Ibrahim, Czech Univeristy of Life Sciences Prague, Faculty of Tropical AgriSciences, Department of Economics and Development, Kamýcká 129, 165 00 Praha - Suchdol, Czech Republic, e-mail: kindah.ibrahim@gmail.com

Institutions

GIZ/BMZ: Opportunities and challenges in scaling agricultural innovations

Oral Presentations

Does Sustainable Land Management Lead to an Inclusive Rural Transformation? - Experience from Ethiopia

BORIS BÜCHLER

GIZ Gesellschaft für Internationale Zusammenarbeit GmbH, Germany

Rural Areas in Africa are transforming. Change processes are driven by population growth, urbanisation, domestic and regional migration and impacts of climate change. Consequences for the structure of the local economy, political governance and social fabric are arising and changing these areas.

Occurring transformation of rural areas needs shaping to avoid negative impacts on ecology and society in order to leave no one behind. Natural resources and the people managing them play a key role in sustainable development with regard to integrated ecological, social, economic and political/institutional dimensions. Experiences from the long-standing German support to Ethiopias' Sustainable Land Management in the Ethiopian Highland Regions underpin the relevance of natural resource management. This presentation identifies the main drivers, chances and challenges shaping inclusive rural transformation through sustainable land management.

Keywords: Ethiopia, GIZ, sustainable land management

Contact Address: Boris Büchler, GIZ Gesellschaft für Internationale Zusammenarbeit GmbH, Friedrich-Ebert-Allee 36, 53113 Bonn, Germany, e-mail: boris.buechler@giz.de

More Impact through Collaboration – How International Agricultural Research and GIZ Collaborate to Achieve Impact on Scale – A Case Study from Tunisia

UDO RUEDIGER

International Center for Agricultural Research in the Dry Areas (ICARDA), Tunisia

ICARDA is an International Agricultural Research Institute specialised in Dry Areas. Its key areas are conservation of agrobiodiversity, crop – livestock farming systems, value chain development and water and soil health management. GIZ in Tunisia collaborates with ICARDA through the support of a CIM – integrated expert and financing large and small grant projects. The CIM expert is responsible for the scaling of agricultural innovations as well as business development around these technologies. This is achieved through strong collaboration with national research and extension services (NARES), private sector, farmer groups and the national GIZ agricultural programs (Green Innovation Center and PAD). Scaling methods, like trainings, field visits and sending of SMS messages to farmers are tested and scientifically analysed regarding cost effectiveness and impact. Results are shared and recommendations are given to extension services and GIZ programs to help improving extension systems and adoption of innovations.

Keywords: Collaboration, GIZ, ICARDA

Contact Address: Udo Ruediger, International Center for Agricultural Research in the Dry Areas (ICARDA), Tunis, Tunisia, e-mail: u.rudiger@cgiar.org

Scaling of Improved Forages in Eastern Africa: An Activity of the GIZ/BEAF Task Force on Scaling

STEFFEN ENTENMANN[1], UWE OHMSTEDT[2], EDUARDO STERN[3]

[1]*Deutsche Gesellschaft für Internationale Zusammenarbeit (GIZ) GmbH, Advisory Service on Agricultural Research for Development (BEAF), Germany*

[2]*International Center for Tropical Agriculture (CIAT), Tropical Forages, Kenya*

[3]*Semillas Papalotla, Mexico*

In 2017 the Task Force on Scaling was launched by the Advisory Service on Agricultural Research for Development (BEAF) of GIZ to support the CGIAR in delivering their ambitious strategic goals of improving the livelihoods of smallholder households and using natural resources more sustainably. The Task Force on Scaling is a group of Integrated CIM Experts that are seconded to Agricultural Research Centers around the world. The Task Force comprises a broad range of different professional experiences and skills. It understands ‘scaling’ as pathways for widespread adoption of beneficial and customer-oriented innovations.

One activity of the Task Force aims at improving the productivity of dairy farms in East Africa by promoting large-scale adoption of improved forages, i.a. *Brachiaria* and Panicum Cultivars. Dairy cows fed on *Brachiaria* increase milk production up to 30 %. Scaling approaches focus on spreading adequate information about the forages to farmers and improving access to forage plants. This includes the establishment of demonstration plots, field trainings and radio transmissions. Partnerships with commercial seed distributors are part of the strategy to improve availability and distribution of seeds. Private sector actors like Semillas Papalotla have a high operational flexibility and good access to infrastructure, allowing them to respond fast to customers’ needs. On the other hand, the private firms may benefit from an extended product range. Further implications of engaging private sector actors in scaling activities are discussed, together with other lessons learned from the Task Force activities.

Keywords: Fodder plants, GIZ, Public Private Partnerships, scaling, Kenya

Contact Address: Steffen Entenmann, Deutsche Gesellschaft für Internationale Zusammenarbeit (GIZ) GmbH, Advisory Service on Agricultural Research for Development (BEAF), Bonn, Germany, e-mail: steffen.entenmann@giz.de

ICARDA: Cereals and pulses for climate-smart agri food systems

Cereals and Pulses for Sustainable Agri-Food Systems under Climate Change

JACQUES WERY[1], YIGEZU ATNAFE YIGEZU[1], SHIV KUMAR AGRAWAL[2], ANDREA VISIONI[1], MICHAEL BAUM[2], BARBARA ANN RISCHKOWSKY[3]

[1]*International Center for Agricultural Research in the Dry Areas (ICARDA), Egypt*

[2]*International Center for Agricultural Research in the Dry Areas (ICARDA), Morocco*

[3]*International Center for Agricultural Research in the Dry Areas (ICARDA), Ethiopia*

In dry regions food systems are largely driven by cereals, especially bread wheat. *Durum* wheat is also a major food crop in many parts of the Central and West Asia and North Africa (CWANA) region. Barley has a specific place as it both contributes to human food (grain and malt) and livestock feed (grain, forage and straw), especially under conditions of water scarcity and low input systems. In response to increasing imports in most of the countries of the region, food policies have been focusing on bread wheat, leading to both increasing areas and yield. A large part of this wheat expansion and intensification is occurring under irrigated agriculture while many countries in the MENA region are already above the threshold level for the use of renewable resources. At the same time the production of other crops such as pulses and forages are stagnating or declining, while they are increasingly recognised as smart foods for human and livestock nutrition. These crops are also essential for adaptation to climate change, improvement of soil fertility, management of weeds, and breaking the pest and diseases cycle of cereal-based cropping systems.

Presentations in this session will show how current research conducted by ICARDA and its partners in the CWANA region can support a transition towards nutrition-sensitive and climate smart cereal-based agri-food systems under irrigated and rain-fed conditions. We will show in particular how plant breeding, agronomy, livestock feeding and systems analysis can be combined to support this diversification and sustainable intensification of cereal-based agri-food systems.

Keywords: Climate smart agriculture, dry areas, transition

Contact Address: Jacques Wery, International Center for Agricultural Research in the Dry Areas (ICARDA), Cairo, Egypt, e-mail: j.wery@cgiar.org

BMEL: Scaling up, out and deep - lessons learned

Oral Presentations

Panel Discussion: Scaling Up, Out and Deep – Lessons Learned. Sharing Experiences from Research Projects Funded by the Federal Ministry of Food and Agriculture, BMEL

MAJA CLAUSEN[1], ANDREA ROTHE[2]

[1]*Federal Ministry of Food and Agriculture (BMEL), Division 121, Research and Innovation, Coordination of Research Area, Germany*

[2]*Federal Ministry of Food and Agriculture (BMEL), Division 622, World Food Affairs, International Food and Agric. Organizations, Causes of forced migration, Germany*

The Federal Ministry of Food and Agriculture (BMEL) established the funding instrument "Research Cooperation for Global Food Security and Nutrition" to strengthen the German contribution to agricultural and nutritional research. Within this framework, research projects aim to improve food systems in partner countries, while building long-term partnerships between agricultural and nutritional research institutions in Germany, Africa, South and Southeast Asia.

Three calls for research proposals have been published in 2013, 2016 and 2019. The first call focused on diversified agriculture for improved and nutrition-sensitive diets. The second call related to innovative processing technologies of local food to reduce seasonal food insecurity as well as food and nutrient losses. A total of 17 project consortia have been funded. The most recent third research call (launched in July 2019, application open until October 2019) focuses on food environments for improved nutrition.

Project executing agency is the Federal Office for Agriculture and Food (BLE).

The BMEL-Session at the Tropentag 2019 presents a panel discussion on the topic: "Scaling Up, Out and Deep – Lessons Learned". The aim is to share experiences and to develop recommendations on how research results and innovations can be spread more effectively for lasting and sustainable impacts. As the funded projects are achieving promising results and working in diverse contexts, an enriching exchange can be expected. The panel consists of members from different project consortia funded by BMEL:

- Dr. Etti Winter, Gottfried Wilhelm Leibniz University, Hannover – Project coordinator: "Food Security in rural Zambia – Integrating traditional Fruit and Vegetable Crops in Smallholder Agroforestry-systems (FoSeZa)"
- Dr. Klaas Dietze, Friedrich-Loeffler Institute – Coordinator for international relations, respresenting: "The importance of farm, domestic and wild animals as a source of Ebola virus infection (Ebola Foresight)"
- Dr. Stefan Sieber, Leibniz Centre for Agricultural Landscape Research (ZALF)

Contact Address: Ariana Bystry, Federal Office for Agriculture and Food (BLE), International Cooperation and Global Food Security 323, Deichmannsaue 29, Bonn, Germany, e-mail: Ariana.Bystry@ble.de

– Project coordinator: "Scaling-up Nutrition – Implementing potentials of nutrition-sensitive and diversified agriculture to increase food security (Scale-N)"
- Dr. Barbara Sturm, Department of Agricultural and Biosystems Engineering, University of Kassel – Project coordinator: "Decentralised postharvest processing of underutilised species into innovative value added products for improved food and nutrition security in West Africa (UPGRADE Plus)".

Keywords: Food and nutrition security, Global South, research cooperation, scaling

Out – Up – and Deep Scaling in a Reverse Auction: Experience from the Foseza Project

Etti Winter

Leibniz Universität Hannover, Institute for Environmental Economics and World Trade, Germany

Our objective was to identify community members in a remote rural area of northern Zambia willing to implement an agricultural innovation on their personal plots. We designed a reverse auction of banana seedlings as an intervention. The existing agricultural system is mainly based on Cassava and diets consists mainly of carbohydrates. Fruits and vegetables, including banana, are rarely cultivated with hidden hunger as a result of this deficiency situation. A reverse auction is a suitable instrument for revealing the risk attitude of individuals. We assume that low bids indicate an increased willingness to take risks so that winners of the auction may become early adopters supporting change. A network analysis will accompany the socio-economic experiment and different scaling strategies can be simulated with the help of an agent-based model. In the experiment, all community members could make an offer. The winners received 5 banana seedlings, training how to plant and propagate and all get the amount of the first rejected offer, the first half at the beginning of the term, the second half after successful performance of the contract. The experiment shows certain scaling outcomes: up-scaling options of the applied tool were identified at district and provincial level; one out-scaling activity was the dissemination of suckers to non-winners; deep-scaling could be observed by the demand of advanced training, demand of other species and the revealed willingness to integrate trees in the farming system. However, problems have also been identified which need to be addressed at several levels of the agricultural system, including conflicts with the traditional authorities that exploit high rents and thus weaken the potential of innovation, but also practical problems with irrigation, access to inputs, improved varieties and marketing options.

Keywords: Agent-based modelling, participatory action research, reverse auction, social network analysis

Contact Address: Etti Winter, Leibniz Universität Hannover, Institute for Environmental Economics and World Trade, Königsworther Platz 1, Hannover, Germany, e-mail: winter@iuw.uni-hannover.de

Ebola and Trust...
(How Do Those Two Get into the Same Headline)?

KLAAS DIETZE[1], KERSTIN FISCHER[1], JULIET JABATY[2], ALIMOU CAMARA[3], SANDRA DIEDERICH[1], THOMAS HOENEN[1], ANDREAS MÜLLER[1], ROLAND SULUKU[4], CÉCILE TROUPIN[3], MATHEW L. GBOKU[2], BASHIRU KOROMA[4], NOEL TORDO[3], THOMAS C. METTENLEITER[1], MARTIN H. GROSCHUP[1]

[1]*Friedrich-Loeffler-Institut, Office for International Relations, Germany*

[2]*Sierra Leone Agricultural Research Institute, Sierra Leona*

[3]*Institut Pasteur de Guinée, Guinea*

[4]*Njala University, Animal Department, Sierra Leona*

The Ebola virus disease (EVD) epidemic in West Africa (2013–2016) has been the largest known outbreak in history that severely affected Sierra Leone, Guinea and Liberia. The unprecedented size of the epidemic with more than 28,000 confirmed cases of EVD and more than 11,000 deaths caused disastrous economic and humanitarian consequences for the local population.

Interventions during and after the epidemic, in particular by international partners, trying to assist affected communities in the control as well as to increase the knowledge on potential virus reservoirs afterwards have encountered a major challenge beyond science – the lack of trust.

The "Ebola Foresight" project as a collaboration project between the Friedrich-Loeffler-Institut (FLI), the Institut Pasteur in Guinea, the Sierra Leone Agricultural Research Institute, and the Njala University in Sierra Leone aimed to study the role of livestock, domestic animals and wildlife in ebolavirus infection. Funded by the German Federal Ministry of Food and Agriculture, the aim was also to build capacities by training of local PhD students and laboratory staff in laboratory methods and blood collection from livestock and domestic animals. In the initial phase, the reservation towards the scientific activities in the target communities was underestimated and required a change in tactics. To overcome this obstacle, long-standing contacts of local partners to the villagers and the establishment of animal health clubs with ongoing community engagement prior to sampling activities were put in the centre of activities to implement a scientific project in a sustainable manner. In-country workshops teaching hands-on experiences in different laboratory methods in order to allow testing of the collected serum samples in the African laboratories could subsequently assure that local people performed the diagnostics locally.

The involvement of local communities is key for success in field-based projects, and sustainable support can only be achieved if trust is there first and action comes second. That is not a new finding in the field of international development yet scientists in the field of infectious diseases tend to miss out the fact that they might be dealing with communities that have a very different understanding of diseases.

Keywords: Animal sampling, community engagement, health interventions, local partnerships

Contact Address: Martin H. Groschup, Friedrich-Loeffler-Institut, Institute of Novel and Emerging Infectious Diseases, Greifswald - Insel Riems, Germany, e-mail: Martin.groschup@fli.de

Principles and Implementation Models for Out- and Upscaling of Innovations

STEFAN SIEBER[1,2], CONSTANCE RYBAK[2], HARRY HOFFMANN[2], MICHELLE BONATTI[2], HADIJAH MBWANA[3], NYAMIZI BUNDARA[3], VICTORIA GOWLE[3]

[1]*Humboldt-Universität zu Berlin, Dept. of Agricultural Economics, Germany*
[2]*Leibniz Centre for Agric. Landscape Res. (ZALF), Inst. of Socio-Economics, Germany*
[3]*Sokoine University of Agriculture, Dept. of Food Technology Nutrition and Consumer Sciences, Tanzania*

The objective of Scale-N project is to safeguard food and nutrition security for the local population in Tanzania by supporting the development of diversified and sustainable agriculture. This conference contribution describes relevant principles and an applied implementations model for effective out- and up-scaling, which were applied and tested within the frame of Scale-N and beyond. General principles for research models are:

1. Use of existing local knowledge & low-cost measures tailored to enhance resilience, increase adaptive capacity and integrate into regional agricultural systems.
2. Intrinsic motivation of all involved stakeholders through high level of participation to assure the link between local ownership and adoption.
3. Mobilizing adequate incentive structures to set up business models in a sustainable way (through micro-credit or non-monetary incentives such as market access or organisational memberships)
4. Efficiency of up- & out-scaling through site-specific methods tailored to the type of agricultural innovations, which ideally lead to self-outscaling through farmers only.
5. Ensuring a good collaboration within research consortia using Conflict Prevention and Management Systems (CPM-Systems, intercultural-sensitive collaboration, minimise the costs of inter-personal and inter–institutional conflicts).
6. Action research in a partnership approach, scientists accompany implementation processes using piloting. They advise and communicate risks for joint decisions.

Based on these principles we can propose a generalisable and adaptable implementation model for implementing organisation that allow sustainable adoptions of innovations:

Phase 1 - framing: To avoid efficiency losses we identify innovative farmers (or else public authorities) at the beginning to collaborate with. This cost-minimisation method "positive deviance" filters those farmers with higher performance. They can pose as out-scaling centres at community level (demonstration).

Contact Address: Stefan Sieber, Leibniz Centre for Agric. Landscape Res. (ZALF), PB 2, AG SUSLAND, Eberswalder Str. 84, 15374 Müncheberg, Germany, e-mail: sieber@zalf.de

Phase II - conceptualising: Stakeholder mapping and analysis is a prerequisite to properly setting up stakeholder involvement. Therefore social-relation mapping and actor analyses of the situation are performed first. In parallel local inventories on potential innovations are generated, which pose as base for participative stakeholder process to choose most adequate innovations at farm, policy and societal levels.
Phase III - operationalising: Researchers accompany stakeholders to test innovations using piloting programs (and up-scaling programs). Self-monitoring & external monitoring, tool-based impact assessment in ex-ante as well as ex-post identify impacts ideally in a joint settings (stakeholders, researchers).

Keywords: Food security, implementation model, outscaling, scale-N, Tanzania, upscaling

Participatory and Collaborative Strategies for Out-, Up- and Deep Scaling of Post-Harvest Innovations in West Africa

Barbara Sturm, Bastian Lange, Sharvari Raut, Oliver Hensel
University of Kassel, Agricultural and Biosystems Engineering, Germany

Participatory methods, sustainable use and replicability of developed innovations, and capacity development are core concepts within UPGRADE Plus .
The project's scaling out strategy builds on participatory approaches to involve key stakeholders throughout the process of co-development and -assessment of innovations, and the development of outreach materials. The continuous engagement of key stakeholders strongly depends on a mutually perceived value of activities and effective trust-building. Due to associated resource and time constraints, the necessary depth of engagement requires a careful stakeholder selection process, while simultaneously ensuring that applicability and relevance for a broader set of stakeholders are guaranteed.
Participatory methods and the resulting mutual learning serve as a key pillar for capacity development among target groups and building knowledge and insight among local researchers and institutions. Knowledge on nutrition and local plant species was identified as one of the most critical barriers towards a sustainable shift in perceptions and behaviour. In response, UPGRADE Plus has devised local partnerships with NGOs and foundations with a focus on health and education, to more effectively address existing knowledge gaps and raise awareness, capitalize on multiplier effects and support deep scaling of innovations.
Collaborations are also a key element of the project's scaling-up strategy. Influencing political and regulatory framework conditions is a challenge for research projects, especially considering the competition for policymakers' attention that exists among agricultural development initiatives and projects in the target countries. Thus, the project has taken a strongly collaborative stance, focused on issue-based networking. Visibility of project content and results, as well as awareness raising are achieved largely through a conscious effort on knowledge management knowledge sharing with interested parties. This has enabled the project to successfully initiate formal collaborations with NGOs and the FruVaSe project, in addition to the non-formalized knowledge exchange with many other organisations and projects active in Sub-Sahara Africa.

Keywords: Issue-based networks, knowledge management, knowledge sharing, participatory methods

Contact Address: Barbara Sturm, University of Kassel, Agricultural and Biosystems Engineering, Nordbahnhofstr. 1a, 37213 Witzenhausen, Germany, e-mail: barbara.sturm@uni-kassel.de

BMBF - CLIENT II: Partnerships for sustainable innovations

Oral Presentations

Implementing Climate-Sensitive Adaptation Strategies to Reduce Flood Risk in the Transboundary Lower Mono River Catchment in Togo and Benin – The Project CLIMAFRI

YVONNE WALZ[1], KOUAMI KOKOU[2,4], JULIEN ADOUNKPE[3,4], MOUMINI SAVADOGO[4], AYMAR BOSSA[4], AKPAMOU KOKOUVI GBÉTEY[5], MARTIN PÉPIN AINA[6], MARIELA EVERS[7], ADRIAN ALMORADIE[7], RHOLAN HOUNGUE[7], SOPHIE THIAM[8], SARAH VERLEYSDONK[8], GESINE SCHIEWER[9], VICTOR KPOKPOIA[9], KAJ LIPPERT[10], GERNOT BELGER[10], MICHAEL HAGENLOCHER[1], YANNICK SCHILLINGER[1], SIMON WAGNER[1], LORINA SCHUDEL[1], ZITA SEBESVARI[1]

[1] *United Nations University, Inst. for Environment and Human Security (UNU-EHS), Germany*

[2] *University of Lomé, Togo*

[3] *University of Abomey-Calavi, Benin*

[4] *West African Sci. Service Centre on Climate Change and Adapted Land Use (WASCAL), Burkina Faso*

[5] *Ministry of Environment, Sustainable Development and Nature Protection, Togo*

[6] *Ministry of Living Environment and Sustainable Development, Benin*

[7] *University of Bonn, Dept. of Hydrology and Water Resource Management, Germany*

[8] *University of Bonn, Centre for Development Research (ZEF), Germany*

[9] *University of Bayreuth, Chair of Intercultural German Studies, Germany*

[10] *Björnsen Consulting Engineers, Germany*

Floods have affected more people than any other type of disaster in the 21st century (CRED. Natural Disasters 2018. https://emdat.be/sites/default/files/adsr_2018.pdf). Many countries in West Africa suffer severely from the destructive violence of floods caused by heavy rainfall events, which are expected to increase in frequency and severity with projections of climate change. In the transboundary Mono River Basin of Togo and Benin, floods occur with very high frequency causing deaths and losses of existential livelihoods of the local people. The combination of frequent flood events together with the overuse of natural resources and a lack of information and knowledge about the interrelationships, demonstrate the great need for intervention in this area, which was communicated by African partners from science and national government and led to the joint development of the CLIMAFRI project presented here.
The German-African inter- and transdisciplinary CLIMAFRI project is funded by the German Federal Ministry for Education and Research (BMBF) under the CLIENT II program. The project will co-develop adaptation strategies for sustainable management of water and natural resources to reduce current and future flood risk considering scenarios of climate change in the transboundary Lower Mono river catchment

Contact Address: Yvonne Walz, United Nations University, Inst. for Environment and Human Security (UNU-EHS), Platz der Vereinten Nationen 1, 53113 Bonn, Germany, e-mail: walz@ehs.unu.edu

of Togo and Benin. Another scientific and technical objective is to collaboratively establish a River Basin Information System through the integration of science-based data with information and knowledge from local stakeholders and communities. To achieve the sustainable implementation of the River Basin Information System, it is a key objective of CLIMAFRI to train professional staff on multiple scientific and technical aspects during the process of establishing the information system and to embed the information system within the responsible authority(ies) in the transboundary region.

Keywords: Adaptation, climate change, flood risk, information system, West Africa

Increasing Climate Resilience via Agricultural Insurance – The Project KlimALEZ

IHTIYOR BOBOJONOV, LENA KUHN, THOMAS GLAUBEN

Leibniz Institute of Agricultural Development in Transition Economies (IAMO), Germany

The significance of the agricultural sector in Central Asia reflects in the very large contribution of agricultural production to the GDP and the large share of employment in agriculture. Rising frequency of dry spells associated with climate change are causing serious damage to the livelihoods of the rural population in semiarid and arid regions of Central Asia. The adaptation capacity of agricultural producers in the CA is very limited.

Agricultural insurance may contribute to increasing resilience of agricultural producers. However, insurance markets have several development challenges in the region. In particular, new forms of index products could solve existing problems of coping with intransparent and inefficient settlement processes, better meet the needs of producers, and solve problems of systemic risks.

In the scope of the project KlimALEZ, IAMO researchers are cooperating with science partners from both Germany and the target countries as well as German and local insurance companies to develop and implement an agricultural index-based insurance program. In a transdisciplinary approach, the project has two closely related objectives. First, the project aims at increasing the resilience of the Central Asian agricultural sector to climate risks by introducing innovations to the agricultural insurance markets, taking into account local requirements and capabilities. The second objective is to analyse and explore the influence of index insurances on the production and efficiency of resource use on the level of agricultural producers.

Several project activities are implemented in the first one and half years of the project. First, suitable index products are developed based on the basis of satellite data. Second, these products are adjusted to the demand of local farmers based on their opinions on this products obtained from surveys and interactive extension seminars conducted in the regions. Third, validated products are sold to selected number of farms in the pilot regions in cooperation with local insurance company. Further improvement of offered insurance products and planning of new piloting phase are in process. Furthermore, scientific transfer of project results is implemented.

Keywords: Climate change, drought resilience, scientific transfer, technology adoption

---

Contact Address: Ihtiyor Bobojonov, Leibniz Institute of Agricultural Development in Transition Economies (IAMO), Theodor-Lieser-Straße 2, 06120 Halle (Saale), Germany, e-mail: Bobojonov@iamo.de

FarmImpact - Development of Sustainable Water and Energy Solutions for Farms in South Africa

Maik Veste[1], Thomas Berger[2], Ben du Toit[3], Gerhard Kast[4], Thomas Littmann[5], Elmi Loetze[3], Eckart Priesack[6], Thomas Seifert[7], and the FarmImpact Team

[1] *Centrum für Energietechnologie Brandenburg e.V. (CEBra), Germany*

[2] *University of Hohenheim, Inst. of Agric. Sci. in the Tropics (Hans-Ruthenberg-Institute), Germany*

[3] *Stellenbosch University, Forestry and Wood Sciences, South Africa*

[4] *Umweltanalytische Produkte GmbH, Germany*

[5] *Dr. Littmann Consulting, Germany*

[6] *Helmholtz Zentrum München, Inst. of Biochemical Plant Pathology, Germany*

[7] *Scientes Mondium UG, Germany*

According to current climate scenarios, the Western Cape is the region in South Africa that will be particularly affected by declining precipitation and increased droughts in the future. In addition, it has alarming soil erosion rates and is therefore a region of the highest priority for measures to adapt to climate change. There is a need for measures to make the existing agricultural land more productive and sustainable, so that no further transformation of natural ecosystems is necessary to maintain agricultural production and food security. The rigorous implementation of water-efficient agriculture will be necessary to adapt to the predicted rainfall reductions. On the one hand, this includes the design of a water-efficient agricultural practice to improve the water consumption of crop plants through windbreak hedges, including improved tree selection and a technological part for predicting irrigation needs. The use of windbreaks can reduce the water consumption of vineyards by up to 20 %. A further increase in water use efficiency can be achieved through active technical measures in irrigation technology. Their optimisation potential depends both on the technology used and on the irrigated crop. FarmImpact will create an application-related basis that will provide agricultural enterprises with information on actual water consumption and future demand. For example, expert models are also used to calculate the current daily situation and the situation of the agricultural production system to be expected within a week on the basis of the weather forecast.

Keywords: Agriculture, CLIENT II, fruits, irrigation, wine

Contact Address: Maik Veste, Centrum für Energietechnologie Brandenburg e.V. (CEBra), Cottbus, Germany, e-mail: veste@cebra-cottbus.de

Innovations for Sustainable Agricultural Resource Utilisation and Climate Adaption in Dry Steppes of Kazakhstan and Southwest Siberia (ReKKS)

JANNIS FLORIAN CARSTENS, OLGA SHIBISTOVA, GEORG GUGGENBERGER
Leibniz University Hannover, Inst. of Soil Science, Germany

While the intensive agricultural use of temperate grasslands plays a pivotal role for the global food supply, the overexploitation of these steppe soils causes severe economic and ecological consequences. The project ReKKS is focused on the dry steppes of Kazakhstan and southwestern Siberia, which have been put under widespread agricultural use in the mid-20th century. The applied land use techniques were not adapted to the specific requirements of these marginal lands, which resulted in significant soil degradation. The soil organic matter content decreased considerably, so that the steppe soils became a source instead of a sink of atmospheric carbon, with repercussions for the global climate. Furthermore, the loss of macro nutrients, which are contained in soil organic matter, leads to a significant reduction in soil fertility and thus the agricultural yield. In order to address these negative developments, the aim of the ReKKS project is develop innovative, sustainable, and climate-adapted agricultural concepts in close cooperation between German, Kazakh, and Russian companies and scientists, together with local partners. The project is focused on the reduction of erosion, the improvement of water balance, the increase in carbon sequestration as well as the nutrient and herbicide use efficiency. Moreover, on severely degraded soils, steppe restoration treatments are investigated. In particular, ReKKS quantifies soil carbon and nutrient stocks as affected by land use techniques and develops agricultural machinery for ultra-shallow tilling as well as the targeted application of liquid fertilisers and herbicides. In field experiments, specifically designed instruments such as weighing lysimeters are deployed. Hence, ReKKS will deliver land use systems specifically adapted to the dry steppes of Kazakhstan and southwestern Siberia, which will simultaneously limit climate change and contribute to food safety on a global scale.

Keywords: Carbon sequestration, herbicides, land use techniques, liquid fertilisers, soil degradation, steppe restoration, steppe soils

Contact Address: Jannis Florian Carstens, Leibniz University Hannover, Inst. of Soil Science, Herrenhäuser Str. 2, 30419 Hannover, Germany, e-mail: carstens@ifbk.uni-hannover.de

Arbuscular Mycorrhizal Fungi in Phytoremediation of Mercury Polluted Soils in Ghana and Burkina Faso

SERGEY BLAGODATSKY[1], MIRIAM EHRET[2], FRANK RASCHE[1], IMKE HUTTER[3], REGINA BIRNER[1], BELOVED M. DZOMEKU[4], OBLÉ NEYA[5], GEORG CADISCH[1], JENS NORBERT WÜNSCHE[6]

[1] *University of Hohenheim, Inst. of Agric. Sci. in the Tropics (Hans-Ruthenberg-Institute), Germany*

[2] *University of Hohenheim, Research Center for Global Food Security and Ecosystems, Germany*

[3] *INOQ GmbH, Germany*

[4] *CSIR - Crops Research Institute, Ghana*

[5] *West African Science Service Center on Climate Change and Adapted Land Use (WASCAL), Competence Center, Burkina Faso*

[6] *University of Hohenheim, Inst. of Crop Science, Germany*

Small-scale gold mining in Ghana and Burkina Faso (West Africa) contributes significantly to national economies and people livelihood. This activity causes, however, high environmental risks. Unregulated surface gold mining contributes to deforestation and land degradation in these countries. In addition, illegal mining operators use a technology for gold amalgamation that pollutes the environment with mercury (Hg) and adversely affects human health. The overall objective of the BMBF-funded Mercury-AMF-project is to reduce the environmental damage caused by mercury used in gold mining in Ghana and Burkina Faso. This will be achieved by developing and implementing novel arbuscular mycorrhizal fungi (AMF) - plant systems as a strategy to reclaim mercury-contaminated sites. The cultivation of pioneer plants on contaminated soils can reduce the mercury pollution. Symbiotic mycorrhizal associations of those plants may strengthen the potential to remediate Hg-contaminated soils.

The implementation of the project is based on the following specific activities:

1. Characterisation of the arbuscular mycorrhizal fungus (AMF) candidates in the soils of Ghana and Burkina Faso;
2. Development of prototype AMF plant systems as an innovative strategy for the remediation of Hg-contaminated sites;
3. Testing of mycophytoextraction methods to reduce the Hg soil concentration below threshold values;
4. Examination of the return of Hg-contaminated sites to agricultural use and the promotion of sustainable land management in gold mining regions;
5. Set-up of modelling approaches for the efficiency of mycophytoextraction methods and Hg plant uptake;

Contact Address: Jens Norbert Wünsche, University of Hohenheim, Inst. of Crop Science, Stuttgart, Germany, e-mail: jnwuensche@uni-hohenheim.de

6. Exploration and communication of institutional and socio-economic framework conditions for the introduction of AMF plant systems.

During the first 6 month of the project soil and plant sampling campaigns were organised for screening the AMF-candidates capable for symbiosis with local plant species and tolerant to the mercury pollution. Clarification of possible mechanisms of phytoremediation is the next essential component of the research: several pathways of decontamination are possible including phytostabilisation, phytovolatilisation and phytoextraction. Based on the first results, field experimental trials with new AMF-plant systems will be established.

Keywords: Arbuscular mycorrhizal fungi-plant systems, gold mining, mercury-contamination, phytoremediation

Special projects

DAAD alumni seminar

Posters

Impact of Climate Change on Livestock Production in Pakistan Using Statistical Copulas

Shahzad Kouser[1], Faisal Nawaz[2], Abedullah Abedullah[3]
[1] *Comsats University Islamabad, Dept. of Agricultural Economics, Pakistan*
[2] *Comsats University Islamabad, Dept. of Finance, Pakistan*
[3] *Pakistan Institute of Development Economics (PIDE) University, Environmental Economics, Pakistan*

Livestock is produced on the world's largest land resource, covering 45 % area, which is mostly in harsh and vulnerable environment unsuitable for other purposes. Climate change can adversely impact livestock's natural resource base, quantity as well as quality. In Pakistan, livestock is the first largest sub-sector of Agriculture. It contributes 58.55 % to agriculture value addition and 11.60 % to the national GDP. Further, the country is the fourth largest milk producer in the world. Around 90 % of small ruminants are being reared by resource poor farmers. This sector is not only main source of livelihood but also provides financial security to small farmers. However, the country is ranked seventh among the top ten climate affected countries, therefore, climate change is considered a significant threat for sustainable livestock production of the country. Higher temperatures may decline livestock production by reducing animal weight gain, reproduction, and feed intake. Though the relationship between temperature and livestock production is assumed to exist at their extreme tails of distributions. Therefore, this study employs copula technique to model tail risk between climate change vulnerabilities and livestock production. FAO data of temperature (°C), livestock production index (US$) and milk yield (tonnes/milking animals) from 1972 to 2014 were used. Static and time-varying copulas are used to jointly model temperature with livestock production and with milk production. The best-fitted copula was time-varying t copula. The results revealed that increased temperature volatility due to change in climate is likely to influence the livestock production and milk yield negatively. As livestock is vital to eliminate food insecurity and rural poverty, it is necessary to identify and implement climate change mitigation strategies and solutions.

Keywords: Climate change, copula, livestock production, milk, Pakistan, temperature

---

Contact Address: Shahzad Kouser, Comsats University Islamabad, Dept. of Agricultural Economics, 45500 Islamabad, Pakistan, e-mail: drskouser@comsats.edu.pk

Towards a Sustainable Management of Olive Trees' Orchards

Ines Ksentini, Boutheina Lagneb, Ines Abida, Mohieddine Ksantini
University of Sfax, Tunisian Olive Institute, Dept. of Plant Protection, Tunisia

Advances in Science and especially molecular genetics were able to reconstruct the history and to determine the origin of many cultivated trees, such as the olive trees. In fact, it is from the wild olive that the cultivated olive trees Olea Europaea derive. However, while wild olive is found in forests, surrounded by many other trees and plants, the domesticated olive trees are mostly mono-cultivated in orchards around the Mediterranean basin. The olive trees monoculture led to the fields' natural vegetation destruction. In fact, entire hectares are completely devoted to the cultivation of olive trees in a monoculture way in Tunisia. In such system, no spontaneous plant could be tolerated because of water competition, except those growing on field borders. However, despite all efforts of trimming and plowing, olive orchards are often attacked by several specific pests, which are accented with the misuse of chemical insecticides. Because of the latter harmful effects on consumers and on the auxiliary fauna, several farmers have converted to organic methods. They therefore have resorted to the use of biological formulations and releasing parasitoids and predators in order to control pests. However, in the absence of plants diversification of the olive tree ecosystem, all these efforts may result in a failure, because as olive pests presence in the field is not that continuous, auxiliary fauna could not be easily maintained in the field. Thus, the need in the field, of associated plants acting as shelter for natural enemies, and increasing their environmental opportunities. In that purpose, introducing aromatic and medicinal plants in the olive trees orchards would be beneficial for both auxiliary arthropods and farmers which could gain additional income with the appropriate management of these plants inside their groves. In that context, experiments are being conducted in two organic olive orchards in which aromatic and medicinal plants, among other plants, were included. Farmers' testimonies* and the impact of these plants on soil arthropods and auxiliary fauna presence and abundance in the field are being documented.

Keywords: Associated plants, biological control, olive orchards, pests-beneficial insect interactions

Contact Address: Ines Ksentini, University of Sfax, Tunisian Olive Institute, Dept. of Plant Protection, Sfax, Tunisia, e-mail: ines.ksentini@gmail.com

Reprimarisation: Implications for Regional Agriculture Development in Argentina

ANA FLORENCIA STODDART
SME Secretariat, Argentina

In spite of the fact that Argentina ranks among the world top agriculture exporters, in reality, it is strongly reliant on a limited selection of products, mostly raw and low-processed commodities. In recent decades there has been a tendency in Argentina to return to primary commodities as the main source of export revenues, namely reprimarization. This has led to a great expansion of extensive crop production activities such as soybean and maize, most of which are genetically modified.
The need to generate greater export volumes has meant that in many cases priorities have emphasised short-term productivity expansion in natural resource intensive sectors, without due concern for environmental criteria. Over the years, there have been increasing problems related to environmental (land erosion, loss of biodiversity) and socio-economic issues (displacement of small producers, impact on rural development, malnutrition). Hence, local communities increasingly recognise that it is necessary to integrate environmental issues and start to set more focus on a sustainable pass.
This research aimed to identify the regionalism and agriculture development orientations which have occurred in the last 20 years. Presenting the agricultural trade patterns and analysing the evolution and structural relationships in the agribusiness sector for main five productive regions. The findings will contribute towards a better understanding of the need for a more diversified and sustainable agriculture.

Keywords: Agribusiness, organic agriculture, sustainable development

Contact Address: Ana Florencia Stoddart, SME Secretariat, Mendoza, Argentina, e-mail: anastoddart@outlook.com

Current Issues in Biodiversity Management in South and South East Asia: Sri Lanka

THAVANANTHAN SIVANANTHAWERL

University of Peradeniya, Sri Lanka

Sri Lanka is an island with 6.6 million ha of land area of which 21 % is covered with the natural forest. It's remarked as one of the biodiversity hotspots among the world, falling into "Western Ghats" which including western parts of India. Biodiversity of Sri Lanka could be explained by mainly four components; Ecosystem and habitat diversity, Species richness of plants and animals, Genetic diversity of plant and animals and Cultural diversity.

Ecosystem diversity mainly covers with the forest and related ecosystem ranging from lowland forest to thorn forest in the low rainfall area. Wetland ecosystem in the inland varies from swamp forest to riverrine forest where as coastal and marine ecosystem ranges from mangroves at the shore to coral reefs into the sea. Agricultural ecosystem covers a wide range of cultivation from paddy land to different perennial cropping system including the homegarden.

In account of species richness of flowering plants, around 3,154 species form 1369 genera present of which 28.3 % are endemic to Sri Lanka. Among the faunal diversity, 33 bird species, 21 mammal species, 125 reptile species, 95 amphibian species, 50 fresh water fish species, 205 land snails, 256 spiders, 47 dragonflies and 26 butterfly species are endemic to Sri Lanka. The genetic diversity is the other contributor to the biological diversity, as an example over 2500 land races and wild relatives of rice found of which most of them are resistant to disease, salinity and drought. Cultural diversity is the other important component, which leads to the biodiversity where variety of ritual ceremonies considered to be more diverse with different crops cultivated. Biological diversity is important for ecological and genetic conservation, speciation, food, medicine, fodder, raw material for industrial development and local livelihood and ecotourism. Unplanned agriculture and construction, pollution due to industrial wastes, fire, invasive species, overexploitation, pirating genetic materials are some of the reasons for loss of biodiversity in Sri Lanka.

Proper land use planning in certain areas, forest zonation from strict conservation to forest plantations and agroforestry, and establishment of conservation area PA network are the most successful remedies implemented to overcome the loss of biodiversity in Sri Lanka.

Keywords: Biodiversity, endemic, fauna, flora, Sri Lanka

Contact Address: Thavananthan Sivananthawerl, University of Peradeniya, Kandy, Sri Lanka, e-mail: sivawerl@yahoo.com

Genetic Fidelity Assessment and Phytochemical Bioprofiling of Field-Grown, *in vitro*-Grown and Acclimatised *Musa* spp. Accessions

IBUKUN AYOOLA[1], MUBO A. SONIBARE[2], BADARA GUEYE[1], RAJNEESH PALIWAL[1], MICHAEL ABBERTON[1], GERTRUD MORLOCK[3]

[1]*International Institute of Tropical Agriculture (IITA), Nigeria*

[2]*University of Ibadan, Dept. of Pharmacognosy, Nigeria*

[3]*Justus-Liebig-University Giessen, Inst. of Nutritional Science, Germany*

Musa spp. (bananas and plantains) are staple food crops for millions of people living in tropical countries. They are considered as important crop plant due to their high economic value and good dietary source. More so, they have significant medicinal importance and can serve as a source of phytoconstituents for the pharmaceutical industries. *In vitro* plant tissue culture can be a reliable alternative source of plant material for phytoconstituents production. However, true-to-type clonal fidelity is one of the most important pre-requisites for the use of micropropagated plant materials as source of phytoconstituents. This work aimed to assess the genetic fidelity of the *in vitro*-grown materials from the taxonomic reference collection; as well as the phytochemical characterisation of the accessions.

Diversity Array Technology (DArT) sequencing was used to generate single nucleotide polymorphisms (SNPs) markers that was used for the genetic fidelity study. In addition, the variation in the secondary metabolites profile of the leaves from field, *in vitro*-grown and acclimatised materials was compared. The effect-directed analysis using high-performance thin layer chromatography (HPTLC) coupled with high resolution mass spectrometry (HRMS) was used to identify compounds responsible for many biological activities.

The results of the pairwise genetic distance matrix among the *Musa* accessions were determined using 150.5K SNPs and found 23 out of the 26 accessions were true-to-type. Considerable variation of total phenolic content and antioxidant activity of the Musa spp. were revealed to be genotype-dependent. Accumulation of the useful secondary metabolites, in terms of antioxidant activity, was higher in *in vitro*-derived material. Four compounds with multiple biological activities were identified in the *Musa* spp. studied samples: Asparagine, pyrrolidine alkaloid, linolenic acid and a phenolic compound. *Musa* spp. leaves were also found to be rich in flavonoids (kaempferol-3-O-rutinoside and quercetin-3-O-rutinoside).

The results confirmed the genetic fidelity of *in vitro*-grown accessions that were true-to-type which validates the use of plant tissue culture as an alternative and reliable method for large scale production of plant material and supply of bioactive constituents.

Keywords: Antioxidants, bioactive constituents, *in vitro* culture, *Musa* spp., true-to-type

Contact Address: Ibukun Ayoola, International Institute of Tropical Agriculture (IITA), Ibadan, Nigeria, e-mail: ayoola.ibukun@gmail.com

Ecosystem-based Conservation of Myanmar's Southern Coastal Zone

LE LE WIN

Food and Agriculture Organization of the United Nations, Myanmar

Myanmar's recent Country Evaluation concluded that high economic growth has been achieved at the cost of increasing environmental degradation. Environmental impacts from climate change and pollution are additional and emerging threats, particularly to the health of the country's fragile marine and coastal ecosystems. The rate and intensity of coastal ecosystem degradation, biodiversity loss and fishery stock declines will continue to accelerate unless resource governance is greatly improved and species and habitat conservation regulations are actually enforced. If future coastal zone development and resource use are not aligned strategically to protect and enhance marine conservation more effectively, coastal ecosystems will become increasingly degraded, fish stocks will continue to decline and some may even collapse. Food and Agriculture Organisation of the United Nations in Myanmar is going to implement a project named "Ecosystem-Based Conservation of Myanmar's Southern Coastal Zone" responds to the urgent need to apply integrated coastal zone management (ICZM) as a governance and sustainable management approach to address these issues. This project aims to improve marine and coastal zone management to benefit marine biodiversity, climate-change mitigation, and food security, which will be delivered through the following two Project Components. Under Component One, national and region/state capacities will be developed to plan and implement strategic coastal conservation management based on integrated coastal zone management principles. Under Component Two, equivalent local capacities will be built and integrated coastal zone management (ICZM) will be demonstrated in practice in one or more selected geographic areas of southern Tanintharyi Region, with a focus on the Myeik Archipelago. The Project will provide capacity development for integrated coastal zone management policy, planning and implementation nationally to benefit the Union and State/Region levels, as well District to local levels in Tanintharyi. The project's strategic focus will be on the mangrove forests in Tanintharyi, due to their significance both within Myanmar and globally but will foster the use of harmonised multi-sector approaches to mainstreaming environmental and resilience considerations into fisheries, forestry and coastal land use and marine resources management and conservation.

Keywords: Integrated coastal zone management, mangrove forests

---

Contact Address: Le Le Win, Food and Agriculture Organization of the United Nations, 11061 Yangon, Myanmar, e-mail: queenlele@gmail.com

Using Genomic Prediction to Increase Genetic Gain in *Oryza sativa* for Increased Food Security in ESA

ROSEMARY MURORI
International Rice Research Institute (IRRI), Kenya

Rice (*Oryza sativa* L.) is the world's most important staple crop, feeding half of the global population. In Africa, its importance is evident from the continent's highest consumption growth rate of more than 34 % when compared to 8 % in Asia and 10 % as the world average. In eastern and southern Africa (ESA), rice is considered a cash crop owing to its importance. Despite the importance, the paddy productivity in ESA countries is below 2 t ha^{-1}. This is partly because farmers in ESA continue to use improved cultivars selected thirty or more years ago, or landraces selected generations ago, in a changing climate. In addition, breeders continue to use landraces in their breeding programs which, despite the fact that they contain enormous genetic variability, they are unsuitable for modern commercial agriculture because they lack the fertiliser responsiveness and yield potential farmers need for the changing climate. Productivity in ESA can at least be doubled by changing the current low yielding, old and obsolete varieties with modern resilient and high yielding varieties, and by modernizing the current traditional production systems. Although, exploring diversity in the ESA collection is imperative for identifying new genes, conservation and further improvement of the germplasm, utilisation of global breeding networks is critical to the developing world's capacity to adapt to climate change in crop production. Our objective was to use genomic prediction to improve genetic gain in rice in ESA. We tested 384 rice lines and used genomic prediction to envisage the performance of untested material in each selected ESA country and selected the subset to advance and also use as parents in our breeding program. We present preliminary results of the trials we are conducting in breeding zones in four ESA countries.

Keywords: Rice breeding program

Contact Address: Rosemary Murori, International Rice Research Institute (IRRI), Nairobi, Kenya, e-mail: rosemary938@yahoo.com

Genetic Diversity of Blackberry (*Rubus* subgenus *Rubus* Watson) in Kenya Using Simple Sequence Repeats (SSRs)

Jared Ochieng

MESPT/DANIDA, Kenya

Genetic diversity of blackberry (*Rubus* subgenus *Rubus* Watson) is essential for efficient breeding and improvement of its pomological traits and yield. In this research, simple sequence repeats (SSRs) were used to determine the genetic diversity of 90 blackberry accessions collected from six different counties in Kenya. From 11 SSR markers used to genotype the blackberry accessions, a total of 127 alleles were generated. The average number of alleles (A) per locus was 4.00 while the expected heterozygosity (HE) of the SSR loci varied between 0.34 and 0.50, with a mean of 0.467. Polymorphism information content (PIC) values ranged from 0.357 to 0.753 with a mean of 0.520. HE of the blackberry accessions were higher than the observed heterozygosities (HO), having 0.75 and 0.64, respectively. Analysis of molecular variance (AMOVA) revealed 95 % variability within accessions and 5 % ($p < 0.01$) among accessions. Cluster analysis using the Jaccard's similarity coefficient grouped the accessions into three classes: I, II and III, consisting of 31, 52 and 7 accessions, respectively. The clustering was random and did not group the accessions according to their geographical origin, indicating that accessions found in Kenya are closely related. This study detected considerable levels of genetic diversity within the analysed accessions, which could be exploited in a blackberry breeding program. Key words: Simple sequence repeats (SSRs), Rubus subgenus Rubus Watson, genetic diversity.

Keywords: Breeding, genetic diversity

Contact Address: Jared Ochieng, MESPT/DANIDA, Mombasa, Kenya, e-mail: jaredochieng58@gmail.com

Metabolic Profiling of Endophytic Fungi Acting as Antagonists of the Banana Pathogen *Colletotrichum musae*

CHRISTIAN JOSEPH CUMAGUN[1], KATJA LENZ[2], RUTH JAKOBS[3], CAROLINE MÜLLER[3], ANNA RATHGEB[2], PETR KARLOVSKY[2]

[1]*University of the Philippines Los Banos, Crop Protection Cluster, Philippines*

[2]*Georg-August-University Goettingen, Molecular Phytopathology and Mycotoxin Research, Germany*

[3]*Bielefeld University, Dept. of Chemical Ecology, Germany*

Three endophytic strains *Phomopsis* sp., *Fusarium proliferatum*, and *Tinctoporellus epimiltinus* isolated from various plants in the rainforest of the Philippines were investigated regarding their ability to repress growth of the pathogenic fungus *Colletotrichum musae* on banana. *In vitro* plate-to-plate assay and an *in vivo* sealed box assay were conducted. Volatile profiles as well as the composition of the secreted metabolome were studied in commercial versus natural potato dextrose medium. All tested endophytes were able to statistically significantly reduce *C. musae* growth compared to the control ($p \leq 0.05$). The type of medium could not be shown to have a statistically significant effect on lesion size of *C. musae* on banana ($p = 0.05$). No interaction between fungal strain and medium could be shown. On the commercial medium, no differences between the performance of the fungi and control treatments could be found ($p = 0.1405$), while there were statistically significant differences between the fungal strains on natural medium ($p = 0.04026$). Lesions on banana that were incubated with *Phomopsis* sp. on natural medium were statistically significantly but only slightly bigger than those incubated with *F. proliferatum* ($p = 0.0139$). The volatile profiles of these two strains and one pathogenic strain *F. graminearum* were assessed using polydimethylsiloxane tubes and analysed via gas chromatography mass spectrometry. Twelve different volatiles could be detected. To reveal if growth on the commercial PDA causes differences in the metabolism of the fungi in contrast to growth on natural PDA, metabolic footprints were assessed via high performance liquid chromatography. After excluding all signals from the media, a total of 388 fungi-associated mass to charge ratio features with chromatographic peak areas higher than 100,000 in the raw data in all replicates of at least one group were detected.

Keywords: Banana anthracnose, *Colletotrichum musae*, endophytic fungi, secreted metabolites, volatile organic compounds

Contact Address: Christian Joseph Cumagun, University of the Philippines Los Banos, Crop Protection Cluster, College, 4031 Los Banos, Philippines, e-mail: cumagun@daad-alumni.de

The Benefits of Mixed Farming Practices in Tanzania

JOHN MATHIAS MANGANA, AMINA HAMIS BAISI

Tanzania Development Organisation, Tanzania

Mixed farming is an agricultural system in which a farmer conducts different agricultural practice together, such as cash crops and livestock. The aim is to increase income through different sources and to complement land and labour demands across the year. Mixed farming systems can be classified in many ways. They can be based on land size, type of crops and animals, geographical distribution, market orientation, and so on. Three major categories are distinguished here. On-farm mixing enables the recycling of resources generated on a single farm. Between-farm mixing can be used to resolve waste disposal problems where by crop farmers use dung from animal farms for fertiliser.

Mixing within Crops and/or Animal Systems involve multiple cropping or keeping different types of animals together. For example, grain-legume association can provide grain with nitrogen. With plant inter-cropping farmers can make the most of the space available to them by selecting plants and cropping formations that maximise the advantage of light, moisture and soil nutrients. Examples of mixed animal systems include chicken-fish production where chicken dung serves as fish fodder.

In a diversified system some components exist as independent units. In an integrated system, maximum use is made of resources, making the system highly interdependent. Advantages of Mixed farming: it enhances the productivity of the farmer and reduces dependency on external inputs and costs. In the example of mixed cropping of animal husbandry and crop farming the crops and animals components can complement and support each other Mixed farming technology contributes to adaptation to climatic change because the diversification of crops and livestock allows farmers to have a greater number of options to face the uncertain weather conditions associated with the increased climate variability. Mixed farming can also give a more stable production because if one crop or variety fails, another may compensate. Livestock represents a means by which families can save and invest in the future. Livestock is a walking bank of assets that can be sold during periods of need such as if crops fail due to drought or flooding. This technology allows greater food security and improved household nutrition levels. In addition, farmers can generate a surplus of some products that can be sold at market. Among other benefits, this technology also allows farmers to grow fodder for livestock and poultry. An additional benefit of mixed rice—fish culture systems is that the fish may help reduce populations of existing and emerging disease vectors such as mosquitoes.

Keywords: Mixed farming, Tanzania

Contact Address: John Mathias Mangana, Tanzania Development Organisation, Dar es Salaam, Tanzania, e-mail: mangana1967@gmail.com

Towards a Sustainable Sugarcane Management in Thailand through Mitr Phol Model

SASITHORN LIMSUWAN
Mitr Phol Innovation and Research Center Co., Ltd., Thailand

Thailand is one of the largest exporters of sugar. Sugarcane growing period in Thailand varies from 10 to 13 months depending on the cultivar. The farmers put a lot of efforts towards making the plants through maturity. Farmers are paid on the basis of cane quality, as measured by CCS (commercial cane sugar) content. Mitr Phol is ranked as the world's third largest sugar producer company in 2018. The "Grow Together" philosophy that strives to uplift the quality of life for sugarcane farmers is derived from the "Mitr Phol Model", which promotes sustainable sugarcane productivity development in four key areas: 1) Having high productivity requires proper farming management: fertile soil, sufficient water, good quality sugarcane sprouts, timely planting and harvest, and effective pest controls. 2) Operations must inspire collaborations between farmers, Mitr Phol Group and government agencies, in line with the "Grow Together" philosophy. For example, educating farmers on cane farming, reducing taxes on loans, providing water sources in local communities and procuring sugarcane variety. 3) Efficient cost management, particularly planning and controlling every step of the process, collective bidding, and reducing operational costs from soil preparation, farming, maintenance, and harvesting. 4) Sugar cane farming management must take into consideration environmental and social responsibility, such as reducing the use of chemicals, green cane harvesting, and minimising field burning. The goal of this Mitr Phol Model is to build a sustainable resources management of sugarcane productivity in Thailand. Moreover, Mitr Phol has operated its cane and sugar business using the "From Waste to Value" philosophy. Recognizing the importance of utilising waste, we have expanded our core sugar business to encompass several other entities including wood substitute, biomass, ethanol, and fertilisers. At the heart of all of our business operations is our effort to be "friendly" to all of our stakeholders.

Keywords: Sugarcane, sustainable development

Contact Address: Sasithorn Limsuwan, Mitr Phol Innovation and Research Center Co., Ltd., Pathumthani, Thailand, e-mail: jjoy1117@yahoo.com

Future of Rural E-Commerce Development for Vietnamese Organic Agriculture

PHUONG NGUYEN

Ideal Farm Corp, Vietnam

Organic agriculture in Vietnam develops slowly, but spreads in some provinces. The slow growth of organic agriculture is partly a consequence of the orientation of agricultural development towards quantity and not quality, lack of a legal framework, and high investment costs for a system change along the value chain. On the consumers side, there is little willingness to pay more for organic products and little information provided in general. Farmers that are interested in changing to organic farming lack information and guidelines on how to produce, harvest, process, and organise logistics to ensure organic standard requirements. Moreover, those farmers who have changed from conventional to organic methods face challenges in the occurrence of diseases, yield loss and market access and given in an example on avocado. The Vietnamese Organic Farms project provides facilitation and consultancy for farmers, provides links to stakeholders along the values chain and develops communication platforms to raise awareness among stakeholders. The aim is to encourage the whole sector (like seeding and planting institutes, agriculture geography and techs, packaging, food processing, distributors, inspection, and logistics) to support a change to organic farming and sustainable land use. These platforms will gather and share information and guidelines, for a market place to connect rural e-commerce development, and for promoting organic agriculture to inbound and outbound. This will be a way to short cut those costs from farms to tables, and so help to provide consumers with a reasonable price and farmers with a sufficient income. Moreover, the additional values of organic agriculture related to ecosystem resilience under climate changes and sustainable development will be discussed.

Keywords: Organic agriculture, organic standards requirement, rural development, rural e-commerce, vietnamese organic farms

Contact Address: Phuong Nguyen, Ideal Farm Corp, Hanoi, Vietnam, e-mail: phuong.emily1@gmail.com

Sack Gardening as a Sustainable Agriculture Strategy for Food and Nutrition Security of Landless Rural Women in Saline Affected Coastal Region of Bangladesh

Shonia Sheheli

Bangladesh Agricultural University, Bangladesh

The effect of climate change particularly rising of sea level, cyclone and tidal floods are considered the main reasons for salinity intrusion in agricultural land in the coastal region of Bangladesh. Salinity and submergence in the coastal region makes the land unfavourable that restricts the growing of field crops and vegetables throughout the year. Consequently, crop yields, cropping intensity, production levels are much lower in this region and peoples have been suffering more for optimum food intake and faces challenges to maintain nutritional security compared to other parts of the country. In this study, a total of 100 landless rural women from the coastal region of Bangladesh were selected for growing vegetables in sacks in the winter and summer season. Structured interview schedule and group discussions were used to collect data on vegetable production, marketing system and income. Opinions of rural women were gathered regarding the benefits or contributions of sack gardening on improving their household food and nutrition security as well as livelihood status. Rural women produced tomato, brinjal, chili, indian spinach, okra, radish and beet in the sack (maximum of two plants/sack) in winter season. In summer time, cucurbit vegetables such as sweet gourd, bottle gourd, ash gourd, ridge gourd, bitter gourd, cucumber and snake gourd wre grown in the sack. Women can produce an average of 7 kilograms of vegetables in one sack/year. The produced vegetables were mainly used for consumption at family level. Rural women sold their excess vegetables in the local community (80 % vegetables) among friends, neighbours, and vendors, while 20 % vegetables were sold at the local market for additional income. Regression analysis indicated that (adjusted $R^2 = 0.684$) sack gardening significantly contributes to increase women's household income. Majority of rural women (73 %) observed that their household food and nutrition security was significant improved due to participation in sack gardening. Rural women (70 %) indicated that their livelihood status improved through participation in sack gardening. The position in the family, decision-making ability, participation in social activities and employment also increased remarkably. Therefore, sack gardening benefited landless coastal women through diversifying household diets and allowing them to make small savings due to reduce purchase of vegetable induced by self-production.

Keywords: Bangladesh, farming, food security, vertical agriculture, women

---

Contact Address: Shonia Sheheli, Bangladesh Agricultural University, Mymensingh, Bangladesh, e-mail: ssheheli@yahoo.com

The Scenario of Environment and Agriculture Activities in Santa Catarina State, Brazil: Development of Technologies, Programs and Lessons Learned

Luciana Camargo Castro

Ministry of Science, Technology, Information and Communication / UNESCO, Brazil

Santa Catarina State is known as the land of small farmers in the south region of Brazil. That is one of the reasons why sustainable production due to technical optimisation and resource conservation has been implemented and incentivized there. That kind of initiative has been promoted for such a long time since the first so called “Microbasins I Project” was implemented in the whole state in 1995. Afterwards, the “Microbasins II Project” took place and afterwards there was the third project, the “SC Rural Programme”, which was a huge programme dealing, mostly, in the rural areas supporting the best technologies applied to small farmers concerning a sustainable production way. Through the implementation of new methods to recovery soil nutrients, enhancing of best land practices, and transdisciplinary research in the agriculture field, that last programme started in 2011 and ended in 2016. Moreover, the Secretariat of Economic and Sustainable Development by the Department of Water Resources Development is continuously working with the river basin committees in order to establish and increase a better water management at local and regional levels. This means that promoting those types of actions, the state government intends to reinforce the importance to understand the idea behind the improvement of tropical agricultural technologies in order to contribute to the socio-economic and environmental development of countries like Brazil, and to make possible the participation of all stakeholders towards a sustainable development.

Keywords: Soil and water management, sustainable production

Contact Address: Luciana Camargo Castro, Ministry of Science, Technology, Information and Communication / UNESCO, Florianopolis - Santa Catarina, Brazil, e-mail: luca66@hotmail.com

Mixed Cropping System with Bamboo as a Strategy of Sustainable Development of the Tropical and Subtropical Regions

Adriana Arenas

INCIVA - Research Institute of Valle del Cauca, Colombia

Bamboo, like sugarcane, rice and wheat, are grasses that belong to the Poaceae family. They are extremely diverse, ecologically and economically important and they occur naturally in tropical and subtropical climates. *Guadua angustifolia* (Guadua) is a bamboo native to Colombia, Venezuela and Ecuador, recognised as one of the 20 best bamboo species worldwide due to its fiber quality and its size up to 30mt high and 22 cm diameter, offering a series of competitive advantages compared to timber trees, such as self-propagation, high and faster growth rates and productivity. Widespread Guadua-Bamboo cutting for agricultural and road expansion contributed to the loss of its biodiversity. This loss, the lack of nurseries for *Guadua* propagation with traceability and the weakening of the *Guadua* value chain have driven the creation of the first project of certified *Guadua* seedling production centre in the Department of Valle del Cauca, Colombia. This project has allowed us to collect and characterise different *Guadua* material in the field to establish a germplasm bank with 6 different biotypes. This material has been multiplied both in nursery and *in vitro*. The project already has 5 certified production protocols, it is engaged in bamboo plant massive multiplication in nursery, and it has a laboratory where we developed a protocol for bamboo *in vitro* multiplication. These propagated plants will be taken to highly degraded river banks for reforestation, and their social, ecological and economic impact will be determined through scientific study: a cooperation plan with the community and rural associations would be developed to recognise the importance of their local practices, needs and ancestral knowledge so that we can jointly design and execute a participatory reforestation plan. Reforestation will be done along bean, corn and pumpkin crops as mixed cropping, thus allowing families to improve their food supply while bamboo grows. Mixed cropping ecosystem services (ES) for carbon sequestration, soil recovery, water-quality improvement and biodiversity will be assessed while a payment model for the ES evaluated will be designed to offer the community additional income for bamboo conservation, thus contributing to the sustainable development of the region.

Keywords: Guadua-bamboo, *in vitro*, nursery, reforestation

Contact Address: Adriana Arenas, INCIVA - Research Institute of Valle del Cauca, Buga, Colombia, e-mail: arenasadriana@gmail.com

Bioinoculants Reduce the Need of Mineral Fertilisers While Maintaining the Yield of Finger Millet and Pigeon Pea in Intercropping

Jegan Sekar[1], Mathimaran Natarajan[2], Thimmegowda Matadadoddi Nanjundegowda[3], Prabavathy Vaiyapuri Ramalingam[1], Kathiravan Raju[1], Manjunatha Baiyapalli Narayanswamy[3], Bhavitha Nayakanahalli Chikkegowda[3], Bagyaraj Davis Joseph[4], Ansgar Kahmen[2], Thomas Boller[2], Paul Maeder[5]

[1]*M.S. Swaminathan Research Foundation, India*

[2]*University of Basel, Dept. of Environmental Sciences, Switzerland*

[3]*University of Agricultural Sciences (GKVK), India*

[4]*Centre for Natural Biological Resources and Community Development, India*

[5]*Research Institute of Organic Agriculture (FiBL), Soil Sciences Division, Switzerland*

Agriculture is the livelihood for two-thirds of the 1.2 billion Indian population, yet food security continues to be a major threat due to stagnant productivity. In our project "BIOFI" (Biofertilisation and Bioirrigation), funded by the Indo-Swiss Collaboration in Biotechnology (ISCB), we tested the combination of biofertilisation – by arbuscular mycorrhizal fungi (AMF) and plant growth promoting rhizobacteria (PGPR) – and bioirrigation in finger millet—pigeon pea intercropping to ensure stable and sustainable yield increases. Three consecutive field trials were conducted at two different field sites (Kolli hills and Bengaluru) to test the effects of biofertilisers (AMF and PGPR), combined with mineral fertilisers, on the yield of finger millet and pigeon pea in mono- and intercropping systems under rain-fed conditions. The trials were designed with 20 treatments, each with 4 replicates, in mono and inter-cropping, with 100 %, 50 % or 0 % of recommended dose of mineral fertilisers (RDF), with or without biofertilisers. The yield of both finger millet and pigeon pea was increased by biofertilisation, particularly in the intercropping system. In general, the yield with biofertilisers, combined with 50 % RDF treatment, was on par with 100 % RDF without bioinoculants in all the trials and both the sites. At 50 % RDF, combined application of AMF and PGPR showed a significantly higher yield than single inoculations. Our study indicates marginal farmers could save up to 50 % RDF without jeopardising grain yield.

Keywords: AMF, biofertilisers, bioirrigation, finger millet, intercropping, PGPR, pigeon pea

Contact Address: Jegan Sekar, M.S. Swaminathan Research Foundation, Chennai, India, e-mail: jegan.sekar@mssrf.res.in

Planting Time Options as Alternative to the Katam Terpadu at Muara Experimental Site in Bogor

Elsa Rakhmi Dewi

Indonesian Agroclimate and Hydrology Research Institute (IAHRI), Indonesia

The information system of KATAM Terpadu (Integrated Cropping Calendar) has been broadly applied in many fields in Indonesia to cope with climate change and maintain the stability of agricultural production. However, the information system allows limited options of planting time of one growing season. We used crop simulation model APSIM-*ORYZA* to evaluate scenarios of favourable planting time to improve productivity at Muara experimental site in Bogor. Options evaluated considered the early and late planting times in accordance with KATAM Terpadu. Yield was used as indicator of productivity. APSIM-*ORYZA* simulated yield with a significant accuracy (EF=0.74) across the different planting times following the growing season. For the irrigation system, the late planting resulted similar yield to the KATAM recommendation (5.1–5.8 ton ha^{-1}), whereas the early planting resulted the lowest yield (3–5 ton ha^{-1}). For the rainfed system, the early planting presented the same yield as the KATAM planting in the rainy season (2.9–3.2 ton ha^{-1}) and the dry season (4.0–4.3 ton ha^{-1}), respectively. This study confirmed that simulation model is a useful tool in determination of planting time and that planting time based on prevailing climate conditions will ensure the acceptable yield in rice cropping systems.

Keywords: Climate change, crop simulation model, planting time

Contact Address: Elsa Rakhmi Dewi, Indonesian Agroclimate and Hydrology Research Institute (IAHRI), Bogor, Indonesia, e-mail: elsarakhmi@gmail.com

Developmental Approaches of Commercial Potential of PGPR Based 'Organic Biofertiliser-Beyonic LIPI' Technology from the Test Tube to Farmer's Field in Indonesia and some Private Industries

Sarjiya Antonius, Tirta Kumala Dewi, Dwi Agustiyani, Nur Laili, Agung Adi Nugroho, Achirul Nditasari

Research Center for Biology, Indonesian Institute of Sciences, Indonesia

The beneficial bacteria are collectively known as plant growth promoting rhizobacteria (PGPR). They colonize in the rhizosphere of plants upon introduction and promote plant growth and control plant diseases. The dramatic negative effects of conventional farming are harms to human health and environmental pollution. Due to several health hazards, farmer preferences shift towards the use of the organic food grown without the use of any chemical fertilisers. They offer an economically attractive and ecologically sound alternative for providing benefit to the plants. Therefore semi or full organic farming system is the best choice for developing sustainable agriculture. This research aimed to support for the sustainability of organic biofertiliser technology programme in Indonesia, to establish mentoring and coaching programs and to promote introducing, training and application of organic biofertiliser (OB). The final purpose of this research was to synergize with farmer community and local government to adopt and implement the OB technology and commercialising OB product. In this report we have presented the development of OB-Beyonic LIPI formula and its production from the laboratory scale to community levels, its application and impact on crops yields in remote districts of Indonesia. In this process, several thousands of farmers were successfully trained to implement our strategies in application of OB in their fields. The results showed an increase in yields at about 20 to 30 % of many commodities of crops when tested and evaluated compared to conventional one, though with 50 % reduction of synthetic inorganic agrochemicals or even pure orgaic farming. Due to our successful adaptation strategies of implementing OB-Beyonic LIPI technology, we have achieved in transferring this technology to various farmers in several districts and to some private companies in Indonesia.

Keywords: Commercialisation, Indonesia, PGPR, sustainable agriculture, transfer technology

Contact Address: Sarjiya Antonius, Research Center for Biology, Indonesian Institute of Sciences, Cibinong, Indonesia, e-mail: sarj.antonius@gmail.com

Facing Desertification in Egypt

MAGDY ELGHANNAM

ESIIC, Egypt

Sugars of dates as another alternative source of sugar rather than sugars from sugar cane and sugar beet. Egypt suffers from a shortage in sugar production whereas there is a gap between the produced and consumed sugar measured by more than one million ton annually. This quantity imported by Egypt from outside and cost the national treasury huge amounts of money. Despite this there is a solution for this problem that save the money consumed in importing sugar. This solution is extracting sugar from dates which unfit for the human feeding. Egypt produces about 1.5 million tons of dates annually according to FAO in 2018. Although Egypt is considered a leader land in producing dates but it does not export any amount as most of the produced dates goes to animal feeding because it poor quality of dates. These low quality of dates is containing a sugar ratio ranging from 60%-80 %. If the sugar in the low quality dates were extracted, we can fill the gap that existed. The sugars found in dates are mostly monosaccharide sugars (Glucose+Fructose) which are very difficult to separate as crystals but they can be separated as concentrated sugar syrup. So it is possible to use these sugars in the confectionery industries which consume a part of sugars produced in Egypt approximately equal to the gap quantity.

Keywords: Date, sugar

Contact Address: Magdy Elghannam, ESIIC, Tanta, Egypt, e-mail: magdy_elghannam@yahoo.com

Scenarios of Shifting Cultivation and its Consequences on Biodiversity

JESHMA NTSOU BAKWOWI

National Centre For Education, Min. of Scientific Research and Innovations, Cameroon

In the Past, shifting cultivation was and still a traditional method of farming widely practice in Sub Sahara Africa. Farmers could cultivate a piece of land for subsistence and when the land loses its fertility, the farmer abandoned the land with no intention of coming back to it. However, this notion has changed; a piece land could be cultivated over and over in a very short period of time. On the premise that agriculture is profitable business, shifting cultivation is gradually given way to plantation agriculture. This shift in faming system is backed by government policy to promote agriculture in which wealthy civil servants destroy hectares of forest land to establish monoculture plants like Cocoa and plantains plantation. This evolution of farming system has weaknessed serious consequences on biodiversity. This works which is focused on empirical analysis is aimed at examine the evolution of shifting cultivation and the consequences of this evolution on biodiversity. It equally examine measures put in place by the government and NGOs to mitigate loss of biodiversity and to promote sustainable agricultural development.

Keywords: Loss of biodiversity, plantation agriculture, shifting cultivation

---

Contact Address: Jeshma Ntsou Bakwowi, National Centre For Education, Min. of Scientific Research and Innovations, Yaoundé, Cameroon, e-mail: jeshmab@yahoo.com

Major Pressures Driving Changes in Rangeland Biodiversity and Ecosystem Services

SAMUEL TUFFA KAWO

Oromia Agricultural Research Inst., Yabello Pastoral and Dryland Agriculture Research Center, Ethiopia

It is important to understand divers of rangeland biodiversity and ecosystem change in order to contribute to informed decision making about managing the causes of negative changes in rangeland biodiversity and ecosystem. Ethiopian rangelands are endowed with adapted flora and fauna that contribute to the good life of its inhabitants. Grazing, fire, invasive plants, weather and climate and human influences are major factors that cause rangeland biodiversity and ecosystem services to change over time. Identifying driving factors is important to understand the impact of climate variation and human activities on rangeland degradation. This paper assessed different direct and indirect pressures that are driving changes on the status of the biodiversity and ecosystem services in rangeland of Ethiopia. Furthermore, the positive and negative impacts of lifestyle changes and development activities on the biodiversity and ecosystem services were assessed in this ecosystem. Biodiversity and ecosystem services are sensitive to the way we utilise and manage their resources and landscapes. Identification of rangeland biodiversity and ecosystem services drivers is a significant first step in understanding more about the ongoing changes that are happening. Drivers represent the underlying causes of rangeland biodiversity and ecosystem services change, and dictate the ways in which people's lifestyle, including socially, economically, politically factors, affect the world around us. Existing knowledge about the direct and indirect pressures driving changes in the Ethiopian rangeland biodiversity and ecosystem services were reviewed. The review result indicated that rangelands have been suffering losses of soil organic carbon because of soil disturbance, vegetation degradation, fire, erosion, nutrient shortage, and water deficit. Bush encroachment, overgrazing, recurrent drought, land use change and ban of rangeland fire were among the major factors that caused rangeland biodiversity and ecosystem services change. Therefore, rangeland restorations, re-vitalisation of use of fire as a rangeland management tool, appropriate land use policy, improved grazing management and selective clearing of woody plants should be the directives to be followed.

Keywords: Bush encroachment, fire, grazing management, land use, policy, restorations

---

Contact Address: Samuel Tuffa Kawo, Oromia Agricultural Research Inst., Yabello Pastoral and Dryland Agriculture Research Center, Yabello, Ethiopia, e-mail: satukada@gmail.com

Promotion of Improved Teff Technologies for Smallholder Farmers in Amhara Region, Ethiopia: Evidence from Wag-Himra and North Wollo Zones

TSEGA DESALEGN WEDU[1], ASMIRO ABEJE FIKADU[2]

[1] *BENEFIT-REALISE Programme Bahir Dar University Cluster, Ethiopia*

[2] *University of Bonn, Agricultural Science and Natural Resource Management in Tropics & Subtropics (ARTS), Germany*

Teff is the main source of the Ethiopian economy in terms of human food and animal feed as well. Due to its high level of demand at the local and global market, teff considers as a cash crop. This study aimed to promote high yielding teff varieties to increase teff production and productivity. The study was carried out in Wag-Himra and North Wollo zones for the 2013 and 2014 main cropping season. We applied two improved teff varieties, which are Kuncho and Zoble in the study above area. A total of 254 participant farmers who had 0.25 ha average farm; 131 and 123 farmers from Wag-Himra and north Wollo zones were selected, respectively. Those varieties covered a sum of 69 ha land. The result showed that the performance of Kuncho and Zoble varieties had a significant difference over the local. The mean grain yield of Zoble and Kuncho in Wag-Himra and North Wollo zones were 1.21 ton ha $^{-1}$ and 1.45 ton ha^{-1}; Kuncho and Zoble had yield advantage of 0.599 ton ha $^{-1}$ over the local varieties. Also, there was a significant mean difference between female and male-headed farmers because female farmers had more workload both in the field and home and had a small labour force to accomplish farm activities than male-headed farmers. Thus, about 89.4 % of the sample farmers were highly enthusiastic to use improved teff technologies in their future teff farm production instead of local teff varieties and traditional cultivation system. The super interesting approach to popularize the target teff technology was organising farmers' field day and stakeholder workshops. Thus, due to the spillover effect, the aforementioned improved varieties disseminated for 670 farmers (both in-village and out-village) in the following years. Besides, we found each participant farmer had transferred the improved teff varieties averagely for two other fellow farmers in any exchanging system in terms of money or in-kind. Thus, we safely recommended for further scale-up/out of those varieties through establishing sustainable seed multiplying cooperatives in each respective zones to keep its sustainability.

Keywords: Cluster approach, farmers' field day, Kuncho, perception, spillover effect, Zoble

Contact Address: Tsega Desalegn Wedu, BENEFIT-REALISE Programme Bahir Dar University Cluster, Bahir Dar, Ethiopia, e-mail: tsegad7@gmail.com

GESI and Governance Mainstreaming in Community Forestry Regime

Poonam Pant Bhatta

Cofsun, Nepal

Community forestry in Nepal has been recognised as a successful conservation model that engages community and creates their ownership towards the conservation interventions. Department of Forests reports that nearly 35 percent of the population of Nepal is involved in community forestry management programme through 22,266 Community Forestry User Groups. Ensure the women leadership & Governance community forestry has the guideline. As a partner of Hariyo Ban Program, Federation of community forestry users Nepal (FECOFUN) has conducted base line survey to analyse the status of women leadership and governance in its 20 working district. And the result of base line survey had shown that only 14 % women are in decision making position while in matter of contribution more than 70 % women are involved. Similarly in governance and livelihood Majority of the CFUGs are observed not to have performed PG assessment; wellbeing ranking and PHPA.
Hence Hariyo Ban project support these initiative by organising women in informal forum named Community Learning and Action Canter (CLAC) a collaborative approach, try to seek solution together which mobilised to carry changes on forest women rights, CFUG governance, conservation of NRMs and CLAC proved effective to women leadership promotion which is noted from 10 field consultation and In 479 CLACs 50 sample CLAC had effectively managed local resources through Cooperatives they mobilised their fund by agro forestry and gaining income from non timber product. The project progress presents women leadership promotion which is noted from 10 field consultation sample in field survey increased by 42 % in vital position in various institutions. It is also proved crucial to raise the voice of gender based violence in Natural resource management .As 63 % women heard Gender based violence through CLAC. At conclusion the programme was evaluated to be able to bring positive changes in the cross cutting issues of the project and adopted its learning in second phase - livelihoods, gender and social inclusion.

Keywords: Collaborative approach, community forestry, gender, governance, social inclusion, women leadership empowerment

Contact Address: Poonam Pant Bhatta, Cofsun, Kathmandu, Nepal, e-mail: Poonambhatta18@gmail.com

Gender-Inclusive Agarwood Value Chain in the Community-Based Enterprise of Thailand

RATCHADA ARPORNSILP

Kasetsart University, The Center for People and Forests, Thailand

Agarwood or eaglewood is a keystone dense heartwood with special aromatic and medicinal qualities in which three-fourths of existing species can only be found in tropical evergreen forests of Southeast Asia. When infected with certain types of mold, the wood structure is damaged. Yet, in that process, the value of agarwood is exceptionally increased in the production for incense, perfumes, cosmetics and jewellery. High-quality agarwood is globally known for such uniqueness, rendering its expensive raw materials to serve both domestic and international market demands. Although efforts to remove agarwood grown in private forest plantation from a restrictive list under the Thai Forest Law is underway, it is still enlisted under the Convention on International Trade in Endangered Species of Wild Fauna and Flora which places approval requirements and determines some restrictions in pre- and post-harvesting activities. Due to these cumbersome processes, the agarwood farmers' group in eastern provinces of Thailand collectively formed into Agarwood community in 2006 to mobilise support for smallholders. Led and engaged with women in the community, the Agarwood community develops a variety of innovative products, including toiletries, personal care items, deodorant spray, herbal medicine, etc. to serve diverse and high-end markets. It explores a niche market accessibility and create additional value of agarwood throughout the chain from a sapling, pre- to post-harvest or extraction, manufacturing, processing and trade, including the establishment of community retail stores as ecotourism destination with homestay and learning centre for consumers to experience and expose to agroforestry farmers' lives and visions.

Employing a gender-inclusive approach, this presentation reveals how initiatives to commodify and market agarwood foster socio-technical innovations and knowledge along the value chain, contributing to environmentally-sound development and rural livelihood diversification. The innovations ground on strengthening smallholder farmers' collectivity, networking and institution-building for eco-agroforestry farming and rural poverty alleviation. The active participation of women in agarwood value chain development significantly boosts their leadership and entrepreneurial competence for economic empowerment and sustainable resource management.

Keywords: Agarwood, community-based enterprise, gender-inclusive value chain

Contact Address: Ratchada Arpornsilp, Kasetsart University, The Center for People and Forests, Bangkok, Thailand, e-mail: Ratchada.Arpornsilp@anu.edu.au

Sustainable Flour House: A New Concept on Cassava Processing

TAINA SORAIA MULLER, J.V. RIBEIRO DE CARVALHO, L.S. LIMA LEMOS, N.L. LOSS, I. LINHARES JUNIOR, P.F. ONOFRE, A.C.B. FERREIRA

Federal University of the South of Bahia, Brazil

Culturally, the cassava (*Manihot esculenta* Crantz.) is the most cultivated crop by local small farmers in the far South of the Bahia State, in Brazil. A mainly rural region, under Atlantic rainforest domain, where social and environmental conflicts are common. Most of the informal business trade is based on clandestine manioc flour processing units (known as Flour Houses) and responsible for the main source of income for smaller farmers. The process of cassava into flour generates Manipueira, a toxic liquid indiscriminatly discarded in the soil or watercourses. The lack of basic sanitation protocols, a very low quality undocumented flour, and enviromental poluting led the Department of Justice to order the closing and destruction of Flour Houses, which would cause a significant impact to poor population. These facts associated to technological barriers and defective value-chain motivated the creation of the Program for Sustainable Flour House. A sustainable processing system optimised to avoid waste and polution, to create jobs and increase income generation. To fulfil the sanitary requirements by law, the flour house was divided into dirty and clean areas, to recieve the roots from field for cleaning procedures, and the cleaned decorticated root for flour processing, respectively. The air circulation is kept by netting of the wide open windows. The rainwater is collected in a cistern and destinated to retro-fitting use. Solid residue and black waters are discarded into a evapotranspiration basin for fruit prodution. The Manipueira is collected in a system assembled by polyethylene water tanks for posterior multi-purpose use on the rural property. The wastewater is disposed on agroforest systems, which promote water percolation and food production. This adequacy aimed to optimise water cycle and residue management as employment and income increasement. So far, five families, have been benifited by adjusting the Flour Houses to current policies, it has saved their rural property from been notified by state prosecution and improved their main income activity.

Keywords: Employment, smallholder, toxic residue

Contact Address: Taina Soraia Muller, Federal University of the South of Bahia, Itabuna, Brazil, e-mail: tainamuller@ufsb.edu.br

Professional Grouping as a Tool for Bridging Gender Gaps in Pastoralist Communities in Northern Benin

GEORGES DJOHY[1,2], ANGE HONORAT EDJA[2], NIKOLAUS SCHAREIKA[3]

[1]*University of Parakou, School of Statistics, Planning and Demography, Benin*

[2]*University of Parakou, Dept. of Rural Economics and Sociology, Benin*

[3]*Georg-August-Universität Göttingen, Institute for Social and Cultural Anthropology, Germany*

Pastoralism faces major challenges that compromise its practice and drive actors into a vicious circle of vulnerability, marginality and poverty. Women in pastoralist communities in northern Benin have increasingly less access to cow's milk, their main socio-economic resource. Herd loss, large-scale migration and difficult livestock mobility make Fulani households more precarious and further weaken the socio-economic conditions of women.

This ethnographic study analysed how the advent of professional associations in this context helped Fulani women to influence gender power relations and (re)negotiate better access to resources. The study approach moved away from male-centred ideology, with the premise that a more processual analysis of the complexity of social relationships between men and women could enable better understanding of ideas and actions of Fulani women in various spheres of society, as they are embedded in greater diversity of socio-cultural and power relationships.

Fieldwork was carried out in three districts in northern Benin (Gogounou, Banikoara and Nikki) where Professional Groups of Female Herders of Ruminants (GPFERs) were set up and developed over the last two decades. Semi-structured and open-ended interviews as well as focus group discussions were conducted with Fulani women and men of various ages, educational level and membership status. Fulani women and their groupings were followed in daily activities over one year, as part of participant observation.

Results reveal that: (i) GPFERs are not homogeneous entities that embrace and reflect the development brokerage-based aspirations of the male Fulani leaders who put them in place. Rather, they are structures modified and renegotiated by women to serve various purposes of emancipation and empowerment vis-à-vis men; (ii) Despite the influence of men in the establishment and functioning of the groups, Fulani women take advantage of support and capacity building by external development actors to promote socio-economic activities (crop farming, agroprocessing, livestock fattening, petty trade), diversify their livelihoods and improve their incomes; (iii) GPFERs enable Fulani women to strengthen their contribution to household social needs (food, health and education) and gain political power at local and regional level.

These findings are useful for decision-making on food security, women's empowerment and sustainable development in pastoral areas.

Keywords: Food security, Fulani women, gender power relations, pastoralism, professional association, sustainable development

Contact Address: Georges Djohy, University of Parakou, School of Statistics, Planning and Demography, Parakou, Benin, e-mail: gdjohy@gmail.com

Status of Cassava Value Chains and Commercialisation in Africa: A Systematic Review of Nigeria and Kenya

FLORENCE ACHIENG' OPONDO[1,2], DANIEL JORDAAN[2], ANDRE LOUW[2], GEORGE OWUOR[3], PATIENCE MSHENGA[3]

[1]*Laikipia University, Kenya*
[2]*University of Pretoria, Dept. of Agricultural Economics, South Africa*
[3]*Egerton University, Agricultural Economics and Agribusiness Management, Kenya*

Cassava (*Manihot esculenta* Crantz) is one of the underutilised tuber crops being promoted because of its potential to address food insecurity and reduce poverty. In Africa, Nigeria is the leading producer of cassava followed by Ghana. In East Africa, cassava is mainly grown in Uganda, Tanzania and Kenya. In Kenya, production of cassava is concentrated in the arid and semi-arid regions. The prevalence of food insecurity and high poverty levels are high in the mentioned regions. The role of cassava as both a food crop and raw material for industrial uses is increasing in Africa. However, cassava value chains have not performed as expected by the international standards. Therefore, this systematic review aimed at analysing cassava value chains in stimulating commercialisation in Nigeria and Kenya. Articles were searched from six relevant data bases. Additional reports from FAO and World bank were also considered. The articles and reports were reviewed by three independent reviewers for objectivity and inclusiveness. In total, 25 articles were identified. They were further analysed qualitatively by use of Atlas ti. software. According to the findings, cassava value chain present numerous market opportunities, yet, the chains are still underdeveloped and inefficiently organised. In Nigeria, the demand for cassava products both locally and internationally has increased. Interventions such as import substitution policies have promoted the use of cassava flour and other derivatives. Though, the growth was experienced after the initiation of the initiative, this later on declined because of the slow growth in the market side. In Kenya, cassava sub-sector is dominated by a few players mainly farmers, traders and consumers. Value chain activities are poorly coordinated and with weak linkages of the actors. Production of cassava is mainly for household consumption with minimal commercialisation activities. In both cases, policies that can stimulate cassava market demand are very important. Involving other stakeholders is equally important as it gives a holistic approach to value chain development. The review revealed that research institutions have produced knowledge that is beneficial to the different actors along the cassava value chain. However, research output has not been practically and economically used.

Keywords: Cassava, commercialisation, systematic review, value chain

Contact Address: Florence Achieng' Opondo, Laikipia University, Laikipia, Kenya, e-mail: fopondo@laikipia.ac.ke

Water Productivity of Irrigated and Rainfed Rice in Southeast Asian Countries: Indonesia, Myanmar, The Philippines, Thailand, Vietnam

RAHMI AZZAHRA[1], JOHANNES KAHL[2], OLCAY ÜNVER[3]

[1] *Samarinda State Agricultural Polytechnique, Indonesia*

[2] *University of Kassel, Organic Food Quality and Food Culture, Germany*

[3] *FAO, Italy*

Rice is one of the staple foods consumed by the world population. The projection of a vastly growing population by 2050 will also increase the demand for food production. Furthermore, the food production balance from irrigation and rainfed farming systems are expected to meet food demand in the future. Since water is highly used as an agricultural input, primarily to produce rice both in irrigated and rainfed environments, its consumption needs to be sustainable. Therefore, the terminology of water productivity will be an indicator of how adequate water is used to produce high rice yield. This study aims to assess the highest water productivity values for irrigated and rainfed rice farming systems in five Southeast Asian countries, which expected to describe how appropriate the water management and policy established in the countries. The methodology to measure irrigated and rainfed rice water productivity is adapted from the water footprint assessment. The results show that Myanmar has the highest level of irrigated rice water productivity whereas Vietnam is the main contributor to the rainfed rice water productivity value. The support of the government in Myanmar for developing and maintaining proper irrigation scheme and water resource management helps Myanmar to use water effectively for producing irrigated rice. Proper control of rainwater used for rainfed rice production with unpredicted rainfall precipitation in Vietnam, help to produce rainfed rice with the most efficient use of water.

Keywords: Local crops, local market, modified cassava flour, nutritious food, quality education, rural development, tropical crops

Contact Address: Rahmi Azzahra, Samarinda State Agricultural Polytechnique, Samarinda, Indonesia, e-mail: rahmiazr@gmail.com

Does Mushroom Cultivation Sheds Light on Household Livelihood Security?

VIJAYALAXMI DUNDAPPA KHED, UMESH K B
University of Agricultural Sciences, Bengaluru, India

Mushroom has been praised as the food of God on an account of their nutritive value, special flavour and medicinal property. Promotion and development of agro-based enterprises in the rural sector can play an important role in diversification of economic activities and generation of employment opportunities for the rural youth. Mushroom cultivation is an emerging agri-business venture, which uses agriculture residue as raw material to yield protein rich food and creating employment and income generation opportunities. Cultivation is simple, involves low cost and labour intensive. Hence, it can provide employment in both rural and semi-urban areas. India in general and Karnataka state in particular, where vegetarians forms a predominant share, every attempt should be made to popularize a vegetable protein source like mushroom. Due to absence of proper handling techniques estimated loss during pre and post harvesting is up to 20 percent. Therefore, adoption of proper harvesting technique is essential. Farmers face the problem of getting a remunerative price due to their non-involvement in marketing and also lack of good marketing infrastructure, complex trade chains, timely availability of necessary inputs, technological knowhow and financial support hinders mushrooms cultivation. Hence, there emerges the need to study the existing value chain and find the hotspots where sustainability can be brought in. The study revealed that mushroom production enhanced the livelihood security of marginal farmers through generating regular income and employment generation. However, unorganised market structure, non availability of basic raw materials and lack of technical and financial support were major constraints in adoption of this technology. Hence, there is need to bridge mushroom producers to consumers through co-operatives, credit institutions, logistics, chilling and processing plants.

Keywords: Mushroom cultivation

Contact Address: Vijayalaxmi Dundappa Khed, University of Agricultural Sciences, Bengaluru, Bengaluru, India, e-mail: vkhed@gwdg.de

Ecosystem Based Climate Change Vulnerability Assessment for Adaptation and Mitigation: A Mobile App Based Approach for Hotspots of Kerala State, India

ARCHANA RAGHAVAN SATHYAN[1], CHRISTOPH FUNK[2], BHASKARAN CHENTHURAN[1], ASWATHY VIJAYAN[3], RADHAKRISHNAN THANU IYER[4]

[1]*Kerala Agricultural University, Dept. of Agricultural Extension, India*

[2]*Justus-Liebig University Giessen, Dept. of Statistics & Econometrics, Germany*

[3]*Kerala Agricultural University, Dept. of Agricultural Economics, India*

[4]*Indian Institute of Information Technology and Management-Kerala, Techno Park, India*

Climate change is the defining issue of our time and we are at a decisive moment. Kerala, the Southernmost state of India, faces a myriad of climate change impacts on agriculture, human health, biodiversity, coastal areas and water availability in the recent years. The state's unique geographical location with hilly Western Ghats bordering one side and coastal fragile ecosystems on the other side, erratic weather patterns, high population density, rapid urbanisation, environmental degradation along with climate change compound the vulnerability condition. However, the region lack ecosystem based climate vulnerability assessment and validation of adaptation strategies. Hence, the overall objective of this study is to assess and analyse the climate change vulnerability of each agro-ecological units of climate vulnerable hotspots of Kerala and to validate ecosystem based adaptation strategies. For this, the study aims to develop an open data mobile application tool kit for data collection. This tool will have the potential to manage the data efficiently and thus ensures seamless collection, storage and analysis of the data. This data will be used to map the climate vulnerability of various agro-ecological units in the state. The climate vulnerability assessment will then be carried out by developing a composite index with emphasis on three dimensions of vulnerability - adaptive capacity, sensitivity and exposure. Validation and documentation of adaptation strategies will be facilitated with the help of Farm Science Centres and line departments in the agro-ecological units. The outcome of the study will help to identify the thrust areas for developing sustainable policies for micro and macro level planning for climate change adaptation and mitigation programs. Furthermore, it leads to sensitisation of farming communities in the agro-ecological units towards climate change and adaptation strategies.

Keywords: Agro-ecological units, climate change adaptation, India, mobile application, validation

---

Contact Address: Archana Raghavan Sathyan, Kerala Agricultural University, Dept. of Agricultural Extension, Vellayani, 695522 Thiruvananthapuram, India, e-mail: archanasathyan@gmail.com

TOPAS: Theory and data fuelling practical agrarian education

From Theory to Practical Calculation of the Gross Margin for Agricultural Crops in Different Countries

VLAD NICOLAE ARSENOAIA[1], FLORIN DANIEL LIPSA[2]

[1] *University of Agricultural Sciences and Veterinary Medicine, Pedotechnics, Romania*

[2] *University of Agricultural Sciences and Veterinary Medicine, Plant Sciences, Romania*

Universities from Germany, the United Kingdom, Poland and Romania cooperate with partner universities in Armenia, Ukraine and Uzbekistan in adapting agrarian studies. The ERASMUS+ funded project TOPAS aims to fill up the gap between the theoretical, formal higher education in agricultural sciences and practical training. The project's concept focuses on concerted improvements in the interaction between teaching approaches, student placements and the collection of local empirical data for teaching and research. Teaching content in the partner countries relies predominantly on information and data from agricultural research stations and do not reflect the situation of existing farming enterprises sufficiently. Improved student placements offer the opportunity for the collection of local empirical data on agriculture and allow for the continuous improvement of teaching content. Preconditions are the definition of relevant data and the establishment of appropriate data management systems at the partner universities. This paper aims to determine the differences between gross margin calculations for wheat crop in two European countries. The output of a marketable crops production includes marketable output evaluated with farm gate prices and also, where it is available, direct payments. In the profitability calculations all output, costs and factor requirements are determined for one specific time or area unit. Usually the indicators are determined per 1 ha or per 1 year. The quality differences of some crops affect the use and the price of the product. A positive profit represents the amount of money remaining for the return of own production factors (family labour, own capital, own land) after the costs of the used supplies, machinery, buildings, equipment as well as wages, interest and rent payments are covered. A positive net profit shows that all production factors have a rate of return. The gross margin obtained for 1 ha wheat in one country is lower in comparison with the other country because both the total yield and the direct payments are lower even if the costs are almost the same.

Keywords: Agricultural crops management, ERASMUS+ TOPAS, gross margin

Contact Address: Vlad Nicolae Arsenoaia, University of Agricultural Sciences and Veterinary Medicine, Pedotechnics, Mihai Sadoveanu Alley 3, Iasi, Romania, e-mail: vlad.arsenoaia@yahoo.com

TOPAS – From Theoretical-Oriented to Practical Education in Agrarian Studies in Eastern Europe and Central Asia

Ralf Schlauderer, Henry Matthews

Weihenstephan-Triesdorf University of Applied Sciences, Fac. of Agriculture, Food, and Nutrition, Germany

Armenia, Ukraine and Uzbekistan share a common historical background and face similar challenges in the transformation of their agricultural sectors since the countries' independence in the 90s. The underlying social, economic and political changes had simultaneously impacts on the entire higher education system, which has to cope with significantly altered demands of employment markets on the competence profiles of their graduates.

Higher Education Institutes from Germany, the United Kingdom, Poland and Romania cooperate with partner universities in the mentioned non-EU countries in adapting agrarian studies. The ERASMUS+ funded project TOPAS aims at filling the gap between the theoretical, formal higher education in agricultural sciences and practical training. The objective is a better match between job market needs and the qualifications offered.

The project's concept focuses on concerted improvements in the interaction between teaching approaches, vocational student placements and the collection of local empirical data for teaching and research. Fostering partnerships between universities and local stakeholders, i.e. farming enterprises, industry and associations, by activities of TOPAS supports the suitability of the improvements for the demands of the local job markets.

Promoted teaching follows the learning outcomes and competencies approach, with flexible learning pathways and permeability among the different agrarian management programs. An additional element is the development of Massive Open Online Courses on core subject matters, which allow for distance learning by professionals and employees in agricultural enterprises.

The reorganisation of student placements includes the installation of networks with adequate enterprises, the definition of placement contents and reproducible assessments in curricula. The European Quality Charter on Internships and Apprenticeships (EQCIA) sets the framework, which consider also aspects of labour legislation and social care.

Teaching content in the partner countries relies predominantly on information and data from agricultural research stations and do reflect the situation of existing farming enterprises insufficiently. Improved student placements and related visits by supervisors offer the opportunity for the collection of local empirical data on agriculture and allow for the continuous improvement of teaching content. Preconditions are the definition of relevant data and the establishment of appropriate data management systems at the partner universities.

Keywords: Agricultural management, ERASMUS+ TOPAS, higher education

---

Contact Address: Henry Matthews, Writtle University College, Dept. of Agriculture, Lordship Road, Writtle, 91746 Chelmsford, United Kingdom, e-mail: Henry.Matthews@writtle.ac.uk

Farm Data for Sustainable Agricultural Development in Uzbekistan (TOPAS Project)

FARHOD AHROROV

Samarkand Agricultural Institute (Institute of Veterinary Medicine), Agricultural Economics and Accountancy, Uzbekistan

Official statistics on farm data, namely, agricultural input and output data is available from State committee of statistics of Uzbekistan and its regional departments and district branches. Commercial farms apply compulsory reports to local branches of committee, but data on family farms are rarely available. Commercial farms report on crop area, yield and value of major cash crops as cotton and wheat. Lack of farm data on production inputs and outputs limits research capacity on efficiency increase, optimisation of resource use and marketing of commodities. This paper examines the possibility of implementation of internship schemes of agricultural higher education in Uzbekistan to collect and use on farm data of agri-economic and statistic agri-structural data for educational and research purposes.
There is a strong hypothesis that improvement in higher education and research could help closing the gap between potential and actual production. There is worldwide evidence that agricultural research yields high returns as it helps to increase yields. Within this context, agricultural HEIs do not have information about the results of economic activity of enterprises. Statistics reflects only the aggregated data of sector development using the index of certain types of products produced on certain territory (district, region, country). Internships may be used to obtain real data of farm enterprises that are not available in the official statistical information.
By developing tools and paths for improvement of practical skills of graduates and fostering better links with agribusiness the project will enhance employability and thus contribute to the improvement of the livelihood of the local population through intensification of private farming based on ecologically, economically and socially sustainable land and water resources.
The project TOPAS will create of a databank for applied sciences on the field of agriculture management for common use of students, teachers and researchers. This databank will consist of data on production inputs, costs and outputs of major cash crops of Uzbekistan.

Keywords: Agriculture, farm data, higher education, TOPAS, Uzbekistan

Contact Address: Farhod Ahrorov, Samarkand Agricultural Institute (Institute of Veterinary Medicine), Agricultural Economics and Accounting, Mirzo Ulugbek 77, 104103 Samarkand, Uzbekistan, e-mail: fahrorov@yahoo.com

Adaptation of KTBL's Databases for Planning and Benchmarking the Sustainable Use of Farm Machinery in Developing Countries

INNA KOBLIANSKA, OLEH PASKO, LARYSA KALACHEVSKA
Sumy National Agrarian University, Ukraine

Machinery costs (fuels & oils, maintenance costs, depreciation and others) comprise a significant part of total farm costs. Dispersed among the different costs items, they reach on average more than 16 % of total production costs at Ukrainian enterprises. This justifies the need to control and manage them in an efficient way based on a strict theory and exact data.

The study examines the experience of Germany related to the development of databases that satisfy the farmer's needs for machinery use planning and looks at the ways those databases could be utilised by users from less developed countries of Eastern Europe, in particular Ukraine. Germany's KTBL, a government-funded industry association, provides all with free, comprehensive, practice-based information (prices, fixed and variable costs of machinery use, labour needs, etc.) needed for planning agricultural production operations, whereas in Ukraine, as of today, only either suppliers and manufacturer's data, or official retrospective statistical data can be used for these purposes. Our central premise is that for the purposes of agricultural activities planning KTBL's data can be used in Ukraine provided adjustment of certain variables is done. Our analysis is facilitated by classification of all KTBL indicators to three categories: 1) strict (unchangeable) / customized, 2) economic and physical in nature, and 3) those that need adaptation and those that need not.

Our results show that the development of appropriate correction coefficients for these indicators that need adaption will make it possible to use KTBL's for the needs of farmers in developing countries and in particular in Ukraine, in order to establish reserves for improving outputs and reducing costs, and increasing competitiveness. Our results, i.e. development of correction coefficients for those indicators that need adaption, have important implications, as they enable the full use of KTBL's databases for the needs of Ukraine and other developing countries.

Keywords: Costs management, database, decision-making, farm planning, informational support, KTBL, machinery cost

Contact Address: Oleh Pasko, Sumy National Agrarian University, Economic Control and Audit, 160 Gerasym Kondratyev St., 40021 Sumy, Ukraine, e-mail: pasko.ov@gmail.com

TOPAS Erasmus+ Project: Set-up and Requirements of Suitable Student Placement Schemes

STANISLAW MINTA, KRZYSZTOF RUTKIEWICZ, JULIAN KALINOWSKI
Wroclaw University of Environmental and Life Sciences, Inst. of Economic Sciences, Poland

Good education should provide both theoretical knowledge and practical skills tailored to the realities of a real economy. In many cases, the problem is that the theory is too strongly emphasised, while the importance of practice decreases. This applies to all levels of education, but is particularly visible at universities. The result is that students are well-prepared theoretically, but have problems after graduating with the practical use of their knowledge. Such a state is treated as a weakness of higher education and is reported as a problem by potential employers. The solution to this problem may be well-organised student internships, which should better prepare students for the real requirements of the labour market. As part of the TOPAS Erasmus + educational project, the existing ways of implementing student internships at agricultural faculties were compared and actions were taken to improve the organisation of such practices. These activities were carried out at 10 partner universities (4 from EU-countries - Germany, Poland, Romania, Great Britain and 6 from the post-soviet area - including Armenia, Ukraine and Uzbekistan). It diagnosed that the main problems hindering the organisation of effective student placements in agricultural fields of study are: need to simultaneously reconcile the interests of different stakeholder groups (agribusiness companies and farmers, universities, students, state institutions), selection for internships primarily in large farms and companies, too small contact of academic staff with agricultural practice, weak willingness of academic staff to change teaching methods and study programs because of market changes, problems with data collection during internships, systemic errors in the organisation and control of student placement, insufficient motivation of students to do internships without salary. Proposals for changes in student placement requirements are: creating special funds for reimbursement of travel and student expenses during the apprenticeships, changing the supervisors' system and emphasising the importance of internships in the higher education cycle, more discussions and study visits for academic staff and students with agribusiness practitioners, introducing additions motivating employees to continuous improvement, introducing the obligation to prepare a final report along with collected data by students and conducting an exam after completing the internship.

Keywords: Agricultural management, ERASMUS+ TOPAS, higher education

Contact Address: Julian Kalinowski, Wroclaw University of Environmental and Life Sciences, Institute of Economic Sciences, Plac Grunwaldzki 24a, 51-363 Wroclaw, Poland, e-mail: julian.kalinowski@upwr.edu.pl

Overcoming Gaps in Education for Sustainable Development of Agriculture: Synergy between University Education and Agribusiness in Ukraine

NATALIYA STROCHENKO, SVITLANA LUKASH, NATALIIA MASLAK

Sumy National Agrarian University, Economics, Ukraine

Nowadays agricultural sector is considered as one of the most perspective and most profitable sectors of the economy of Ukraine and the world. During the active automation and informatisation of agribusiness, adaptation of agricultural technology to new seasonal and weather conditions, the lack of professional staff becomes particularly dangerous. It causes not only additional losses in production chain, but also slows down its sustainable development and the introduction of new technologies. There are many reasons for this situation: poor quality of education, lack of teachers who understand the latest trends and developments in agriculture and can bring them to the students, and, moreover, problems in organising high-quality practices using the latest machines and technologies. The new challenges of the future require of the participants in the educational process to move on to active cooperation, implementation of joint projects with business partners in order to obtain the synergy effect of combining efforts. This is exactly the direction of the international project Erasmus + KA2 TOPAS: from theoretically-oriented to practical training in agriculture. The main task of this project is to facilitate the transition from theoretically oriented education in agricultural studies to a practically oriented education, with the corresponding content of curricula, and with an increased cooperation with agrarian business. A survey was conducted amongst agribusiness representatives to assess needed training knowledge and skills to increase students' employability. The raw data was analysed to revise and improve curricula with using formal or non-formal education methods. Benefits for agribusiness are - the creation of a single European agricultural system of business analysis, and access to an international analytical database for assessing the potential opportunities for sustainable development of production areas. The project also envisages the development of an interdisciplinary and open platform for all users with a view to promote cooperation between agribusiness and universities, expanding cooperation with agrarian groups, and introducing innovations in practical training and teaching. Within the framework of the project a special database will be created that will be useful both for the educational and research process, and for those who make decisions for the development of the agrarian sector.

Keywords: Agrarian management, agribusiness, ERASMUS+ TOPAS, higher education, sustainable development, synergy effect

Contact Address: Svitlana Lukash, Sumy National Agrarian University, Economics, Herasym Kondratiev Street 160, 40021 Sumy, Ukraine, e-mail: svitlana.lukash@gmail.com

TOPAS Erasmus+ Project: Sources, Bottlenecks and Outlook of Farm Data Availability for Education and in Armenia

Arman Tonikyan, Narine Gasparyan

Armenian National Agrarian University, Fac. of Agribusiness and Economics, Chair of Finances and Accountancy, Armenia

The effectiveness of agrarian management is also conditioned by the quantitative and qualitative characteristics of the data collected and used in the organisation of agricultural business, as well as in carrying out research activities within relevant educational programs.

The larger the scale of the agricultural business or the scope of the survey, the greater the role of quality information, based on which decisions are to be taken and the necessary measures to be taken.

The agricultural data system is a continuous process of collecting relevant information indicators when analysing, planning and implementing effective operational and administrative decisions for different aspects of the farming business.

The quality of data required in agrarian management is largely dependent on the criteria that should be maintained during the collection and processing of agricultural data.

These include data integrity, reliability, maturity, availability, compliance and compatibility.

The information is a complete set of data that reduces uncertainty in any area. In fact, uncertainty is a risk, so it can be stated that the risk of agricultural sector in Armenia is conditioned not only by seasonality, peculiarities of climatic conditions and other objective reasons, but also due to lack of data in accordance with the above-mentioned criteria.

The agrarian sector in Armenia is officially handled by the National Statistical Committee, which annually publishes data on the agricultural sector, but often these data are averaged. As for concrete farm data, unfortunately, there is no base in Armenia, which creates a number of organisational, administrative and supervisory issues in that area. As a result, incomplete farmer data hinder farmer business efficiency assessment and proper presentation of investment projects.

The scarcity of the farming data and its non-coordinated nature have its negative impact on education as well. The Armenian National Agrarian University, being the sole industry specialisation institution in the country, tries to maximise the academic / scientific-production practices of its students and scientific research of the employees properly, but the data-related problems hinder the effective organisation of targeted activities. Data collected in practice and research are usually not systematized and

Contact Address: Arman Tonikyan, Armenian National Agrarian University, Fac. of Agribusiness and Economics, Chair of Finances and Accountancy, Eriwan, Armenia, e-mail: arman.tonikyan@yahoo.com

do not summarise in the general information database; data collection is carried out with different methods and principles, which causes problems for reliable and well-grounded analyzes.

Nevertheless, it should be noted that Agrarian University, in collaboration with Weihenstephan-Triesdorf University of Applied Sciences, uses the advanced experience of agrarian data collection, processing and application within the Ägrarian ManagementÏnternational Master Program: the lecturer visits 5–8 German farms in the year with students, where personally meets with the farmer, collects reliable information on which they make the necessary calculations. Additionally, the TOPAS programme is a continuation of development, as this programme allows to develop current curricula based on practical studies and to define criteria for effective data collection in the agricultural sector.

This article deals with the possibility of raising the farm data collection process in the Republic of Armenia, the issues related to it, and the efficiency of the data collection process.

Keywords: Agrarian management, agricultural business, educational program, farming data

Sources, Bottlenecks and Outlook of Farm Data Availability for Education and in the Eastern Part of Uzbekistan

Abduaziz Abduvasikov, Bahodirjon Nosirov

Andijan Branch of Thashkent State Agrarian University (Uzbekistan), Uzbekistan

The Fergana valley is in the eastern part of Uzbekistan and an intermountain depression in Central Asia, between the mountain systems of the Tien-Shan in the north and the Gissar-Alai in the south. Its position makes it a separate geographic zone. The Fergana Valley is historically convenient for agriculture, but there are a number of problems for better organisation of production in the conditions of transition to a market economy.

One of serious problems in in the eastern part of Uzbekistan is the lack of reliable data available. There is a strong hypothesis that improvement in higher education and research could help closing the gap between potential and actual production. There is worldwide evidence that agricultural research yields high returns as it helps to increase yields. Within this context, agricultural universities do not have real information about the results of economic activity of enterprises. Statistics reflects only the aggregated data of sector development using the index of certain types of products produced on certain territory (district, region, country). Internships may be used to obtain real data of farm enterprises that are not available in the official statistical information.

By developing tools and paths for improvement of practical skills of graduates and fostering better links with agribusiness the project will enhance employability and thus contribute to the improvement of the livelihood of the local population through intensification of private farming based on (ecologically, economically and socially) sustainable land and water resources.

Andijan branch of Tashkent state agrarian university, located in Fergana Valley, good networking institution working with family farms, particularly including fruit production and labour intensive structures. It is experienced in national and intraregional network building. It has three faculties which are focused on education of broad spectrum of agricultural sciences and other complementary streams such as economics and teacher training.

In Uzbekistan in bachelor's degree normative length of study is 4 years and study process consists 204 weeks. Usually internships is planning in 4, 6, 8-semesters of study period and depending on course orientation. Students will choose objects of internships for themselves based on mutual contracts such as: outstanding nearly farms, orchards, experimental lands, livestock farms, provincial agricultural administrations, crop protection bio-laboratories, area irrigation associations, machine-tractor parks,

Contact Address: Abduaziz Abduvasikov, Andijan Branch of Thashkent State Agrarian University (Uzbekistan), Oliygoh Street 1 Kuyganyar, 170600 Andijan, Uzbekistan, e-mail: abduvasikov_aziz@mail.ru

agro-chemistry branches, cotton raw bases, corn bases, primary processing of fruits and vegetables, silk raw bases.
Every department prepares in the start of academic year education-methodical guides-programs for internships. Here also necessary to prepare daily notebooks for students, which they have to fill on base of work done daily. For every internship will appoint supervisor from corresponding department for 6–8 students. At the end of internship needs prepare detailed report, which will discussed in institute.
How we may solve above described problem of high education in Uzbekistan? First of all, it is urgently necessary to revise a charter of the country about high education and to shift main attention from rating students theoretical skills to getting professional experiences. A final examination form should be considered seriously by means of establishing criterions on the student's independent work. A subject teacher must rate student theoretical skills and his/her ability of critical thinking and how to find solutions for existing problems of the enterprises.
TOPAS project support management decisions through evidence based farm data analyses and evaluation, more precisely support practice oriented education and training.

Keywords: Agricultural management, ERASMUS+ TOPAS, higher education

The Influence of Credible Data on Research Outcomes (cereals case)

KLYMENKO MAKSYM, VITRIAK OKSANA, DIBROVA MAKSYM

National University of Life and Environmental Sciences of Ukraine, Fac. of Economics, Ukraine

This contribution outlines the importance of improvements in higher education of agricultural studies and makes the case for its impacts on the national agricultural policy. It focuses on the critical types of costs , and the significance of reliable data management for cereal production in Ukraine and advocates suitable teaching approaches. The latter include vocationally oriented practical programs and competencies approaches, targeting new and old graduates, with flexible learning pathways and permeability among different agrarian management programs. Both elements are part of the programme of the ERASMUS+ funded international TOPAS project.

Ukraine is one of the largest producers and exporters of grain and steadily increases grain production despite a bunch of domestic grain market problems (e.g. high infrastructure costs for grain exports, price volatility, large companies' dominance). Data to forecast gross harvest, harvested area and yields were collected by AGMEMOD programme for 2019 — 2030 period according to the State Statistics Service of Ukraine from 2004 to 2018. The results showed that the production potential for wheat would be about 25 million tons by 2030.

The manufacturer's nominal producer protection coefficient was defined as the ratio of the domestic purchase price to the world. The competitiveness of domestic grain on the world market ensured by lower grain prices (during 2015–2017 years 21 % lower than the world average). Agricultural producers compensate their price losses by lower cost of land lease and wages (take only 9 % of direct costs, which is 3–4 times lower compared to developed countries); wheat exporters (traders) compensate their price losses by lower grain quality and logistics costs through lower purchasing prices.

Furthermore, during the last two years, statistical reporting in Ukraine by agricultural enterprises in the form of 50-SG was cancelled, which complicates analytical and research activity, especially in training economic profile specialists. Implementation of the EU Erasmus+ KA2 TOPAS project aims to collect data on typical Ukrainian enterprises. The applied standard methodology to calculate profit margins for crops and data processing relies on the RegioMAX management decision-making programme and allows for solving this problem substantially.

Keywords: Agricultural management, ERASMUS+ TOPAS, farm data

Contact Address: Klymenko Maksym, National University of Life and Environmental Sciences of Ukraine, Dept. of Finance, Heroiv Oborony Str.15 Building 3 Of. 207, 03041 Kyiv, Ukraine, e-mail: maxklymenko@nubip.edu.ua

Promoting Modern Teaching in Agricultural Sciences at Universities in Countries of the Former Soviet Union

NIEKY VAN VEGGEL, HENRY MATTHEWS
Writtle University College, United Kingdom

The Higher Education institutions of countries of the former Soviet Union had a justified reputation of technical and scientific excellence. In the area of agriculture specialists graduated from universities well suited to a farming system in which decisions were taken centrally and their role was to focus on specific areas of production. Teaching was based on the delivery and acquisition of knowledge in a didactic system and assessment through the repetition of the information gathered in the lecture room. However post-Soviet agriculture requires more dynamic flexible employees capable of critical thinking and problem solving. The industry requires generalists with transferable skills rather than specialists with a narrow area of expertise. For the countries of Armenia, Ukraine and Uzbekistan the pre-Soviet systems of university education are no longer fit for purpose.
The TOPAS project has promoted and introduced teaching and learning methods in these countries' universities, which put the student at the centre of the process and where assessment is for learning, rather than of learning. This encourages the development of those skills which enhance employability and make graduates suitable for the workplace. Techniques such as the Flipped Classroom, Problem Based Learning and Case Studies have been at the centre of this approach. Use has also been made of digital technologies to widen the student experience. This has been achieved through workshops, observation, peer discussion and experiential learning. Academics in the partner universities have introduced many of these techniques into their teaching and learning with immediate positive demonstrable benefits to the students.
The challenge for the future is to disseminate these techniques across the institutions, encouraging staff ownership to include these within their teaching and to gain institutional support at all levels to ensure that graduates can contribute effectively to such a significant sector of the economies in each of these countries.

Keywords: Agricultural management, ERASMUS+ TOPAS, higher education, teaching approaches

Contact Address: Henry Matthews, Writtle University College, Chelmsford, United Kingdom, e-mail: Henry.Matthews@writtle.ac.uk

The Formation of Foreign Direct Investments on Agricultural Scale and its Economic Concequences

Aram Arakelyan, Nairuhi Djrbashyan, Aram Karapetyan

Yerevan State University, Dept. of Mathematical Modeling in Economics, Armenia

The paper was prepared in the framework of the Agreement number - 2017–3299 / 001–001.TOPAS - From Theoretical-Oriented to Practical education in Agrarian Studies Project and is devoted to the study of the foreign direct investments for agricultural sector in developing economies. It is addressed to those who are working in the area of agricultural production planning, activities adjustment and decision making in competitive market environment. The role of FDI for developing and emerging economy is significant because it provides investments flows from developed donor countries to the economy of developing country and allow advanced technology transfer, creation of new job places, establishment processing potentialities for agricultural products. FDI stimulates the advancement of agricultural business, growth of agricultural productivity and as a consequence reducing the level of the poverty. The analysis of the poverty of developing countries has been analyzed. The study based on the quantitative assessment of sources allowing to reduce the poverty level. Further, using data concerning to the agricultural value added index and the level of the agricultural employment the assessment of the productivity is implemented. The productivity of sectors considered for comparative analysis allowed ranking countries according to the level of the productivity. Further, the solution of the following tasks should became vital to provide accuracy of the assessment of the efficiency of the Agricultural foreign direct investments. In line of this approach the paper presents:

(i) a suitable analysing technique for the quantitative measurement and modelling of the employment in agriculture and Foreign direct investment in developing and emerging economies.

(ii) a suitable analysing technique for the quantitative measurement and modelling of the dependence of GDP, from Employment in industry, Employment in agriculture, Employment in services and Agriculture, forestry, and fishing, value added (% of GDP).

In conclusion we argue that approaches studied in present paper allow to consider the significance of the agricultural foreign direct investments for the GDP growth. We argue also that FDI provide increasing the employment in agriculture, poverty reducing in agriculture and provide the establishment of harmony in sectors as follows: agriculture, industry and services.

Keywords: Agriculture, developing, emerging, employment, FDI, GDP, poverty, value added

Contact Address: Aram Arakelyan, Yerevan State University, Dept. of Mathematical Modeling in Economics, 1 Alex Manougian, 0025 Yerevan, Armenia, e-mail: aram.arakelyan@ysu.am

Sampling of Data and Information on Agriculture on a Regional Scale in Armenia

Aram Arakelyan, Nairuhi Djrbashyan, Aram Karapetyan
Yerevan State University, Dept. of Mathematical Modeling in Economics, Armenia

The paper was prepared in the framework of the Agreement number - 2017–3299 / 001–001.TOPAS - From Theoretical-Oriented to Practical education in Agrarian Studies Project . The broad range of problems concerning to the sampling of data and information on agricultural scale brought into existence of different factors characterising economic sectors and especially agricultural sector. Wide variations of Armenian economy since 1920 until 2018 velvet revolution gave birth new approaches for the study of Armenian agricultural sector's data and information. Present paper is the result of the implementation of the research with the objective of data sampling and collecting strategy to provide identifying information required for the quantitative assessment of factors characterising Armenian agricultural sector. The study has been focused on the examination features of Armenian agricultural sector's post and current data and information collection strategies concerning to domestic production, state trade, marketing/ processing and consumption of agricultural products. We studied, analysed and interpreted issues taking into consideration features that Armenian economy could have to overcome though the implementation of current Government Program confirmed by the National Assemble of Armenia. Data and information concerning to the study were collected from the Government organisations, private enterprises/groups and international organisations. We argued that the study of comparative advantage of Armenian agricultural production allows to specify agricultural subsectors promoting Armenian export. We argued also that data and information on Armenian agricultural scale must be based on the current Armenian economy development targeted to the attraction of foreign direct investments and especially in agricultural sector. Therefore, the research given in this paper outlined data and information on the dynamics of foreign direct investments and simultaneously data related to Armenian Government. Armenian agricultural development plan covers issues of poverty alleviation. Following Government police we developed data and information to provide the assessment of the state of poverty and Government efforts to increase population income. Conclusion. We argue that the approach of the research is in line with data dynamics features of the Armenian Agricultural sector. Consequently data and information collection mechanism is important and significant for current stage of Armenian economic development.

Keywords: Agriculture, Armenian, data, developing, employment, GDP, information, poverty, production, sampling

---

Contact Address: Aram Arakelyan, Yerevan State University, Dept. of Mathematical Modeling in Economics, 1 Alex Manougian, 0025 Yerevan, Armenia, e-mail: aram.arakelyan@ysu.am

Ukrainian Farm Management Data Network - Practice Oriented Approach for an Innovative Practice Oriented Training of Students

KATERYNA TUZHYK[1], ALINA BRYCHKO[2], ALLA KUCHEROVA[3], OLGA KHRYSTENKO[4], RALF SCHLAUDERER[5]

[1]*National University of Life and Environmental Sciences of Ukraine, Cybernetics Economic, Ukraine*

[2]*Sumy National Agrarian Uiversity, Economics, Ukraine*

[3]*National University of Water and Environmental Engineering, Dept. of Agrochemistry, Soil Science and Agriculture, Ukraine*

[4]*Mykolayiv National Agrarian University, Economy of Enterprises, Ukraine*

[5]*Weihenstephan-Triesdorf University of Applied Sciences, Fac. of Agriculture, Food, and Nutrition, Germany*

Experiences of the last years showed, that students of agrarian universities need more practice oriented training to be well prepared for their future job demands. Four leading agrarian universities of Ukraine, NULES, SNAU, MNAU and NUWEE have therefore created the UFMD project (supported by the German Federal Ministry of Food and Agriculture) to collect farm management data with students for educational and research issues. The initiative is accompanied by the HSWT and an outcome of the Erasmus+ project TOPAS. For the assessment and planning of farms activities, a sound understanding of the economic estimation of production processes is essential. Although state statistic is often used, the high aggregation and the unsecure data quality of the data is often not sufficient for the educational purposes. As a result the lack of appropriate farm data or information at the level of production processes is not appropriate for the educational process and does not lead to a sound understanding of farm business and production reality. UFMD aims to create an e-information base of farm data and to establish a gross margin data collection. Universities are closely cooperating with all stakeholders, particularly with farms. The data collected are treated anonymous. The target is to obtain reliable information for farm business analysis as well as a closer cooperation between universities and the agro business. The collection of farm data is part of the study process and carried out with experienced lecturers who supervise students. The presentation of the gross margins will be similar to gross margin collections available from Germany (KTBL, LfL Bavaria etc.).
UFMF also aims to have a broad impact on all target groups. The approach and the development is presented on an own webpage (www.u-fmd.net). The results of the data collections are to be compared among the farms and continuously improved and expanded. It is expected that more ukrainian agrarian universities and colleges will supplement the network in the next years. The results are also expected to form reliable Bench marks for Ukrainian agriculture development and so support farms to improve their decision making and competitiveness.

Keywords: Database, farm data, gross margin, higher agrarian education

Contact Address: Kateryna Tuzhyk, National University of Life and Environmental Sciences of Ukraine, Cybernetics Economic, 15 Heroiv Oborony Str, 03041 Kyiv, Ukraine, e-mail: ufmdproject@gmail.com

Index of authors

B

C

D

E

F

G

H

I

J

K

L

M

N

O

P

Q

R

S

T

U

V

W

Y

Z

Index of keywords

D

E

F

G

H

I

N

O

P

T

U

V

W

Abstract IDs

Abstract ID **Page**